Feeding
FARM ANIMALS

Feeding
FARM ANIMALS

Dr. S.S. Kundu
P.S. and Head PAR Division
Indian Grassland and Fodder
Research Institute (ICAR), Jhansi (U.P.)

Dr. Sultan Singh
Scientist PAR Division
Indian Grassland and Fodder
Research Institute (ICAR), Jhansi (U.P.)

Dr. S.K. Mahanta
Scientist PAR Division
Indian Grassland and Fodder
Research Institute (ICAR), Jhansi (U.P.)

Dr. G.H. Pailan
Scientist PAR Division
Indian Grassland and Fodder
Research Institute (ICAR), Jhansi (U.P.)

SATISH SERIAL PUBLISHING HOUSE
403, Express Tower, Commercial Complex
Azadpur, Delhi-110033 (India)
Phone : 011-47073040, Fax : 91-11-27672046
e-mail : info@satishserial.com, hkjain1975@yahoo.com
Website : www.satishserial.com

Published by :

SATISH SERIAL PUBLISHING HOUSE
403, Express Tower, Commercial Complex, Azadpur, Delhi-110033 (INDIA)
Phone : 011-47073040 Fax : 91-11-27672046
E-mail : info@satishserial.com, hkjain1975@yahoo.com

Print Again, 2023

ISBN : 81-89304-05-4

Composed, Designed & Printed in India

Dr. P.S. Pathak
FNAAS,
Director
Indian Grassland and
Fodder Research Institute
(ICAR), Jhansi

FOREWORD

I am extremely happy in writing the foreword for the book entitled "Feeding Farm Animals" authored by Dr. S.S. Kundu, Dr. Sultan Singh, Dr. S.K. Mahanta and Dr. G.H. Pailan of IGFRI, Jhansi. It was a felt need to bring out a book having practical feeding aspects of livestock, focusing on forages, grasslands and roughages. In this volume twenty one chapters have been written and first chapter gives an insight of Animal Husbandry in different climatic conditions along with its perspectives. Since animal production is poised to play a very important, essential and urgent role during present quarter of the century for sustainable agriculture, and feeding farm animals will continue to constitute a major (60 – 70 per cent) share of inputs, has been dealt with reasonable details. The country has a rich heritage of having thirty cattle breeds including internationally acclaimed Tharparkar, Ongole, Kankrej and Sahiwal, ten breeds of buffaloes with best of world Murrah, Jaffarabadi, Bhadawari and Surti. Goat is another prolific animal with twenty breeds and to name few Barbari, Jamunapari, Black Bengal and Osmanabadi have been known for their potential. Similarly among forty two recognized breeds of sheep, Kashmir Merino, Malpura, Chokla, Muzaffarnagri and Nellore play a role in uplifting poor among the poorest of farmers. Having such a rich germplasm and by just supplementing their dietary regimen with few kilos of green fodder, preferably legume, these animals have potential to increase productivity tremendously. Authors have made considerable efforts to incorporate latest developments on the practical aspects of farm animal feeding as imbalance nutrition is the main cause for poor animal performance, high mortality and morbidity rates among our farm animals. This poor animal performance provides an opportunity as well as offers a challenge to animal nutritionists to formulate feeding strategies for long term sustainability of animal production systems based on forages and silvopastures on degraded lands.

The next six chapters describe the basic principles of Animal Nutrition which will help in understanding the various feeding practices. The eighth, ninth and tenth chapters include the basic principles of forage production and pasture development and utilization, since quality forage production is the basis of quality animal production. The eleventh chapter includes various agro-industrial byproducts which have significant potential in the feeding of farm animals. To bridge the gap between demand and availability of concentrate feeds, the use of agro-industrial byproduct is inevitable. This chapter describes their nutritive value and levels of inclusion in the rations of various farm animals. Great opportunity is ahead of us to improve the nutritive value of agro-industrial byproducts and to detoxify their inherent the anti-nutritional factors to make it a viable feed resource. The contents of the chapters twelveth to fifteen describes the practical feeding of various animal species viz; cattle, buffalo, sheep and goat. In these chapters, efforts have been made to incorporate the recent scientific approaches of farm animal feeding. Chapters on improvement of fibrous forages and complete feed and densification, and feeding standards describe what best is possible out of available feed resources and how to make suitable combinations to get desirable results. The next two chapters are based on fodder conservation i.e., on hay and silage making as fodder conservation is very important in traditional as well as scientific feeding of farm animals. Since fodder is not available throughout the year, therefore, fodder conservation plays an important role to reduce the variations in seasonal availability of fodder and to improve the production potential of our animals. The last chapter on codex alimentarius, mentions the challenges of feeding in an era of open economy and market.

Animal rearing is an endeavour of million of families in rural areas of our country who raise livestock for their livelihood. The selected topics of this book will be of immense help in scientifc animal feeding for improved animal production. The incorporated chapters have full relevance in the present day animal feeding and will definitely contribute in improving the production potential of our animals. Through this publication, authors have attempted to bridge the gap between researchers and the end users. I congratulate the authors for attempting this exercise and hope that this book will be highly useful to researchers, teachers, students and farmers alike.

(P.S. Pathak)

PREFACE

Farm animals are raised mainly for high quality food, but in India, use of cattle for draught, sheep for wool in addition to providing livelihood to the population living below poverty line (40 percent) also provide much needed resilience to agrarian economy. Animals are the primary source of organic fertilizers - farmyard manure for sustainable agriculture in our country. Although animal products are not essential as dietary components for humans other than very young, but milk, meat, curd, paneer and eggs etc. are very satisfying to the palate as well as providing satiety. Animal proteins are of higher nutritional quality than plant proteins since they contain large amount of essential amino acids. In fact, with the increase in population and rise in the living standard, there is going to be an increased demand for animal products.

In the economics of farm animal production, the cost of feedstuffs is an item of greatest consequence. Thus, the success of animal enterprises depends primarily on the efficient utilization of the feeds. In such a situation forage based animal farming seems to be one of the answers for the successful development of animal husbandry in many regions of the country. A considerable amount of work has been carried out on various aspects of forage/ herbage production and utilization at IGFRI/ NDRI/ IVRI and elsewhere. Through this book authors have made an attempt to conglomerate the pertinent information available from various sources as comprehensive document on feeding of farm animals.

The subject matter of the book has been presented in twenty-one chapters. In the first chapter the status of farm animals with perspective have been described. The next six chapters describe the basic principles of animal nutrition, to provide a scientific basis to understand the diverse aspects of feeding practices. These chapters contain adequate information on carbohydrates, protein, lipids, minerals and vitamins, feeds and fodder evaluation. The chapters 8 – 9 include the details of forage production and pasture/ silvopasture development and utilization. The eleventh chapter includes various agro-industrial by-products, which have significant potential in the feeding of farm animals, particularly to bridge the gap between the demand and supply of concentrate feeds. Chapters from twelfth to fifteenth describe the feeding of farm animals like cattle, buffalo, sheep and goat for optimum and economic production. Chapters on improvement of fibrous roughages, complete feeds and densification, and feeding standards and ration computation describe to exploit the available feed resources through treatments and combinations, for possible production from livestock. The nineteenth and twentieth chapters on hay and silage making have been included, since conservation of forage is essential for scarcity feeding. The last chapter describes codex alimentarius for quality feeding as well as organically produced products in an era of WTO.

Bibliography of relevant books, review articles and a limited number of significant research papers are given at the end of each chapters for further reading and information, and as an acknowledgement to the original contributors. It is hoped that the book will be equally and highly useful to the researchers, teachers, students, farmers and the organizations/ personnel's concerned with farm animal production.

The authors are sincerely grateful to Dr. P.S. Pathak, Director, IGFRI, Jhansi for constant encouragement, all out support, guidance and writing foreword of this book. It is our pleasure to extend sincere thanks to our colleagues who have helped in the preparation of this book. Drs. Anil Kumar and Ram Singh have contributed in diverse chapters. The following colleagues have assisted in correcting the manuscript of chapters of the book: Drs. A.B. Mojumdar, B.K. Bhadoria, L.K. Karnani, N.C. Verma, S.B. Maity, M.M. Das, S.K. Nag, A.K. Misra, B.P. Kushwaha, A.K. Samanta and K.K. Singh. Other divisional staffs/ students like Dr. Ranjan Kumar, Shri. Chandra Bhanu Singh, Sudhir Kumar, Mukesh Kumar, Ms. Shweta Singh Bhaduria, Shalini Trivedi and Suman are also thanked for preparing chemical structures, typing, reading and corrections. Above all we profusely thank our family members (Dr. Sushila Kundu, Dr. Meenakshi Singh, Mrs. Mitali Mahanta and Mrs. Phalguni Pailan and children) for their moral support and sufferings in many arduous hours of devotion to write this book.

S.S. Kundu
Sultan Singh
S.K. Mahanta
G.H. Pailan

ACKNOWLEDGMENTS

Authors are grateful to the following and others mentioned in bibliography of each chapter for consulting and using their materials in the present book. We are also grateful to Director, IGFRI, Jhansi and PCFC, for permission to use materials from various annual reports, technical bulletins and other publications of the institute.

AFRC (1993). Energy and protein requirements of ruminants. An adversary manual prepared by the AFRC technical committee on responses to nutrients. CAB international, walling ford, UK. AFRC (1998). The Nutrition of Goats. AFRC technical committee on responses to nutrients. Report No. 10. CAB international, walling ford, UK. Alagh,Y.K. 1989. Agro-climatic Regional Planning and Overview. National Planning Commission, Government of India, New Delhi, India. Ammerman, C.B; Henry, P.R. and Miles, R.D. (1998). Supplemented organically bound mineral compounds in livestock nutrition. In: Garnsworthy P C and Wiseman J (eds). Recent Advances in Animal Nutrition, Nottingham, Nottingham University Press, pp 67 – 91. AOAC (1995). Official Methods of Analysis (16th Ed). Association of Official Analytical Chemists, Arlington VA ARC (1980). The nutrient requirements of ruminant livestock. C.A.B. International. Augelitti, R. (1998). Proteins. Academic Press, London. Banerjee, G.C. (1988). Feeds and Principles of Animal Nutrition. Rev. edn. Oxford & IBH Publishing Co. Pvt. Ltd., New Delhi. Bhattacharya, N.K. (1989). Goat Rearing. CIRG, Mathura, U.P. India BIS. (1979). Indian Standard: Specification for Mineral Mixture for Cattle. Bureau of Indian Standards, New Delhi. Cheeke P.R. (1991). Applied Animal Nutrition; Feeds and feeding, Prentice- Hall of India Private Limited, New Delhi. Church, D.C. (1988). The Ruminant Animal Digestive Physiology and Nutrition. Prentice Hall, Englewood Cliffs, New Jersey. Close, W. and Menke, KH. (1986). Selected Topics in Animal Nutrition. The Institute of Animal Nutrition, University of Honehelm, Germany Codex Alimentarius (2005). Practices on Good Animal Feeding. www.codexalimentarius.net. Accessed January 2005. D'Mello, J.P.F. (1995). Amino Acids in Animal Nutrition. Oxford, CABI Publishing. Devendra, C. and Burns, M. (1983). Goat Production in the Tropics. CAB Slough, U. K. Eric, E. Conn and Stumpf, P.K. (1985). Outlines of Biochemistry. Willey Eastern Limited, New Delhi. Fahey jr., G. C; Collin, M; Mertens, D. R and Moser, L. E. (Eds.) 1994. .Forage quality evaluation and utilization. CABI Publishing. FAO /WHO (2001). Codex Alimentarius – Organically Produced Foods. Joint FAO/WHO Food standards Programme, Codex Alimentarius Commission, Rome. FAO/ WHO (2001). Codex Alimentarius –Food Hygiene: Basic Texts. Joint FAO/WHO Food standards Programme, Codex Alimentarius Commission, Rome. Farnum, N.R. (1975). Modern statistical quality control and improvement. Outbury Press, Belmount Calfornia. Gall, C. (1981). Goat Production. Academic Press, London U K. Georgievskii, V.I; Annenkov, B.N. and Samokhim, V.I. (1982). Mineral Nutrition of Animals, London, Butterworth. Ghosh, S.P. 1991. Agro climatic zone specific research – Indian perspective under NARP. ICAR. New Delhi. Givens, D.I., Owen, E., Axford, R.F.E. and Omed, H.D. (2000). Forage Evaluation in Ruminants Nutrition. CABI Publishing, Wallingford, U.K. Goering, H.K. and Van Soest, P.J. (1970). Forage Fiber Analyses (apparatus, reagents, procedures and some applications). Agriculture Handbook No

379. ARS, USDA Washington DC. Harry, B. Pfost (1976). Feed Manufacturing Technology Feed Production Council, AFMA, Inc. USA. ICAR (1997). Handbook of Animal Husbandry. Publications and Information Division, ICAR, New Delhi ICAR (1998). Nutrient Requirements of Livestock and Poultry. ICAR, New Delhi. ICAR, 2002. Hand Book of Animal Husbandry. ICAR Publication, New Delhi. ILSI (1997). A simple guide to understanding and applying the hazard analysis critical control point concept. ILSI Europe Sci. Com. and Microbiology. India. 2002. A reference Annual Publication Division. Ministry of Information and Broadcasting. Government of India. Kundu S.S., Mahanta S.K., Singh S., and Pathak P.S. (2005) Animal feed Technology. Satish Serial Publications, Delhi. Kearl, L.C. (1982). Nutrient Requirements Ruminants in Developing Countries. International Feedstuff Institute. Utah State University, Logan, Utah, U.S.A. Kundu, S.S; Misra, A.K. and Pathak, P.S. (2004). Buffalo Production under Different Climatic Conditions. International Book Distributing Co., Lucknow, India. L.L.Relwani (1979). Fodder Crops and Grasses ICAR, New Delhi. Lehninger, Albert L., (1988). Biochemistry (2nd Ed.). Kalyani Publishers, New Delhi. Leonard C. Kearl (1982). Nutrient Requirements of Ruminants in Developing Countries. International Feedstuffs Institute, Utah State University, Logan Utah, USA. Maynard, L.A; Loosli, J.K; Hintz, H.F. and Warner, R.G. (1979). Animal Nutrition. 7th edition. Tata McGraw-Hill Publishing Co. Ltd., New Delhi. McDonald, P; Edwards, R.A; Greenhalhg, J.F.D. and Morgan, C.A. (2002). Animal Nutrition. Pearson Education, Singapore, Pte, Ltd., Indian Branch, Delhi, India. McDowell, L.R. (1985). Nutrition of grazing ruminants in warm climates. Academic Press, INC, London. McDowell, L.R. (1989). Vitamins in Animal Nutrition, San Diego: Academic Press. McDowell, L.R. (1992). Minerals in Animal and Human Nutrition, New York, Academic Press. McDowell, L.R., (1985). Nutrition of grazing ruminants in warm climates. Orlando, F.L; Academy Press. McEllhinery Robert, R. (1994). Feed Manufacturing technology, AFIA, Arlington, VA. Mertz, W. (1998). Trace Elements in Human and Animal Nutrition, Vols. I & II. Orlando, FL: Academy Press. Mudgal, V.D., Singhal, K.K. and Sharma, D.D. (2003). Advances in Dairy Animal Production. International Book Distributing Co., Lucknow, India. Muthariah, S. 1988. A social and economical Atlas of India. Oxford Publishers, New Delhi, India. Nash, M. J. 1985. Crop conservation and Storage. Oxford, Pergmon Press. Nivsarkar, A.E., Vij, P.K. and Tantia, M.S. 2000. Animal Genetic resources of India – cattle and buffalo. ICAR Publication. New Delhi. NRC (1980). Mineral Tolerances of Domestic Animals, Washington, DC, National Academy of Sciences. NRC (1981). Nutrient requirements of goats. Angora, dairy and meat goat in temperate and tropical countries. In: Nutrient requirements of domestic animals, No. 15, Washington, D. C., USA, National Academic Press. NRC (1985). Nutrient requirements for sheep: 6th Revise Edition, Washington, D. C., USA, National Academic Press. NRC (2001). Nutrient Requirement of Dairy Cattle. National Research Council, Washington, D.C. O.I.E. (2002). Report of the Second OIE/ FAO-APHCA Workshop on WTO's Sanitory & Phyto-sanitary (SPS) Agreement. Chiang Mai, Thailand. Pathak, N.N. (1998). Feeding Standards for Buffaloes. Animal Nutrition Division, Indian Veterinary Research Institute, Izatnagar. Pathak, N.N. and Jakhmola, R.C. (1983). Forage and livestock production. Vikas Publishing House Pvt. Ltd., New Delhi. Pathak, P.S., Pateria, H.M., Solanki, K.R. (2000). Agroforestry systems in India: A diagnosis & design approach. ICAR, New Delhi. Pierson, M.D. and Corlett, D.A. (1996). HACCP Principles and Application. Chapman &

Hall, N.V. Pratap S. Birthal and P. Parathasarthy Rao, (2002). Technology Options for Sustainable Livestock Production in India. NACP, New Delhi, ICRISAT, (AP), Systemwide Livestock Program, Addis Adaba, Ethiopia. Pratap S. Birthal, Joshi, P.K. and Anjani Kumar (2002). Assessment of Research Priorities for Livestock Sector in India. NCAP, ICAR, New Delhi. Preston, T.R. and Leng, R.A. (1987). Matching Ruminant Production Systems with Available Resources in the Tropics and Sub-tropics. Armidale, N.S.N., Australia: Penambul Books. Punj M.L. (1988). Agriculture Byproducts and Industrial Wastes for Livestock and Poultry Feeding. All India Co-ordination Research Project, ICAR, New Delhi. Ranjhan, S. K. (1999). Animal Nutrition and Feeding Practises. Vikas Publishing House Pvt. Ltd. Ranjhan, S.K. (1990). Agroindustrial By-products and Non-conventional Feed for Livestock Feeding. ICAR, New Delhi. Robert, R. McEllhhiney (1985). Feed Manufacturing Technology – 3rd. American Feed Industry Association, Inc., Arlington, Virginia, Sastry, N.S.R. (1995). Livestock sector of India – regional aspects. International Book Distributing Co. Lucknow. Uttar Pradesh. Saxena, A.P. (1989). Strategies for Agricultural Research and Development- A Zonal Approach. ICAR, New Delhi, India. Sharma, N.K., Singh, R.P., Singh Manjit and Harsh, L.N. (2001). Silvopasture management in hot arid and semi-arid ecosystem. Agrotech Publishing Academy, Udaipur, India. Singh, R.P. (1995). Forage Production and Utilization. IGFRI, Jhansi. Steele, Mike. (1996). Goat. CTA Publication. Published by McMillan Education Ltd. London and Basingstoke. Stumpf, P.K. and Conn, E.E. (1980). The Biochemistry of Plants Vol 3. Carbohydrates: Structure and Functions, Academic Press, New York. Swenson, M.J. (1977). Dukes Physiology of Domestic Animals, 9th edn. Coruell Univ. Press, Ithaca, USA. Tangendjaja, B. (2003) The application of Hazard Analysis and Critical Control Point Principle in Feed Manufacturing. American Soybean Association, New Delhi. Underwood, E.J. and Suttle, N.F. (1999). The Mineral Nutrition of Livestock, 3rd edn., Wallingford, CABI Publishing. Vaarst, M., Roderick, S., Lund, V and Lockeretz, W. (2004). Animal Health and Welfare in Organic Agriculture. CABI Publishing, UK. Walsh, G. (2002). Proteins Biochemistry and Biotechnology. Jhon Wiley & Sons Ltd., West Sussex PO19 1UD, England.

CONTENTS

FARM ANIMALS

India is endowed with the largest bovine population of the world, with rich domestic animal diversity having 30 breeds of cattle, 10 breeds of buffalo, 42 breeds of sheep, 20 breeds of goat, 9 breeds of camel, 6 breeds of horse, 3 breeds of pig and 18 breeds of poultry. India ranks first in respect of cattle (198.8 million) and buffalo (89.9 m illion), second in goat (122.7 million), third in sheep (57.49 million) and seventh in poultry population (347.61 million). Total livestock population increased to the tune of 1.6 times from 1951 to 1997, the livestock population data indicate only 27 percent increase in cattle population against 104 percent of buffalo during same span. The population of sheep, goat and pig has also increased at the tune of 45, 155 and 209 percent respectively (Table 1). The milk production is highest in Uttar Pradesh including Uttaranchal and Punjab is second in milk production followed by Andhra Pradesh, Gujarat, Haryana, Karnataka, Madhya Pradesh including Chattisgarh, Maharastra and Tamil Nadu (Table 2). However, milk production, buffalo population and yield per animal were much lower in North Eastern states. Karnataka attained the maximum increase in milk production (20%) followed by Uttar Pradesh including Uttaranchal (17%), Punjab (9%), Bihar including Jharkhand (9%), Gujarat (6%) and Haryana (6%) during 1999-2000 to 2001-2002. However, total milk production in the country increase by 7.35 percent during the this period (Table 3). Punjab has the highest per capita milk availability and also achieved increasing trend from 1996 to 2001 whereas availability was lowest in North Eastern states of the country. Meat production figures from 1981 to 2000 indicated that the buffalo meat availability increased by 11 times whereas beef and veal, mutton and lamb, goat meat and pig meat availability increased by 18.5, 1.8, 1.7 and 7.5 times respectively. Total meat production in the country also increased by 5.5 times over the last two decades (Table 3). Total meat production is at the tune of 4.0 million tones per annum (Table 4). The slaughter rate per annum in relation to the population of animals is about 6, 10, 99, 31 and 39 percent for cattle, buffalo, pig, sheep and goat, respectively.

Trend in animal population

Trend in animal/ livestock numbers between 1961 and 2004 have been presented in Table 5 and their population projected up to 2025 taking into the growth rate of last ten years (1984-2004). A look at the growth rate (Table 6) of cattle reveals that it increased up to 1990 after which a declining trend has set and become more pronounced during last five years registering decrease of 1percent p.a. On the other hand buffaloes have shown a steady march since

1970 along with the Operation Flood Programme. But in fact, the combined population of cattle and buffalo have remained almost constant or exhibited only a marginal decline (during last five years). The attrition in cattle numbers have spared the valuable feed resources to be put to more efficient use through buffaloes and high yielding crossbred cows, providing income and employment to farmers. Therefore, whether the dung of cattle is superior to that of buffalo in supplying the organic matter to fields is not known.

The population of goat, over the years remained stagnant, but in recent decades the production system, which supported them, has come under threat. The public grazing land is getting reduced and eroded because of their over use, resulting a decline in the productivity and carrying capacity. Therefore, the demand for goat meat is not being met resulting in increase in its price. Hence, the traditional extensive goat rearing practice has to undergo a change and supplementation under stall-feeding condition may be promoted to get further increase in productivity. Pig in India is caught in its image as filthy animal surviving in drains and by-lanes, evoking strong aversion to the potential meat eaters. Its role in alleviating the food and nutritional security for the lower income groups and weaker sections of the society needs attention. Their excellent feed conversion efficiency need to be exploited as has been done in China and a facelift is required to make it more acceptable to consumers. The chickens have shown an impressive growth consistently over last three decades.

Trend in milk production

The Operation Flood programme revolutionized the dairy sector in India. The production now stands at 90 million tones (Table 7) registering an average growth rate of 3.95 percent p.a. during last ten years (Table 8). The share of buffalo milk is 55 percent, cattle 42 percent and goat 3 percent (Table 9). Based on the growth rate of last ten years, the total milk production is expected to reach 170 million tones by 2020.

Trend in meat production

The meat sector has shown impressive growth indicating the change in dietary habits of population. In 1987-88, only 43.7 and 31.5 percent of urban and rural households, respectively, consumed meat whereas more than half of both types of households consumed meat, in 1993-94. The total meat production (Table 10) stands at 6.03 million tones (FAO, 2004). Total production is projected to reach to 8 million tones in 2020 with an average growth rate of 2.8 percent p.a. during 1993 to 2020 (Table 11). A look at the different components of the total meat production reveals that poultry contributes 28 percent, Buffalo and

Beef and Veal each 24.6 percent and pig and goat each contributing 8 percent. While the broiler is growing at the rate of 12 percent during last ten years, the growth rate in buffalo and sheep meat is slightly above 1 percent. The growth rate in beef and goat meat has declined in recent years. It appears that the present production system of goat production has reached its zenith and therefore any increase in production may come through increased productivity in commercial manner for which supplementation through stall-feeding will be desirable.

Animal farming in different zones

An agro-climatic zone refers to a land unit in terms of its major climates and suitability for a certain range of flora and fauna. Agro-ecological region is defined as an area of the earth's surface characterized by distinct ecological responses to macroclimates as expressed by soil, flora, fauna and aquatic systems. The agro-ecological region or zone is the land unit carved out of the agro-climatic region when super imposed on land forms and on the kinds of soils and soil conditions that act as modifiers to climate and length of growing period.

The concept of dividing the world into various zones has a rich and varied history. Of particular interest is the FAO's Agro-Ecological Zones Project that set out criteria based on plant growth days, soil types, rainfall, and temperature regimes. Since the inception of formal planning in India in 1951, a number of attempts have been made to categorize the country into different agro-ecological regions and zones. Many different approaches were made to classify the land area into climatic regions. Carter (1954) divided India into six climatic regions ranging from arid to peri-humid based on the Thornthwaite criteria for climatic classification. Singh (1974) classified the country according to parameters like units of cattle, buffalo, draft animals, milch cows and milking buffaloes per 100 ha of cropped land. He also classified the country based on percentage of total crop area planted to fodder crops over the period 1961-1966 as well as cropping pattern zones. Krishan (1988) delineated 40 climatic zones based on major soil types and moisture index. Muthariah (1988) classified the country on the basis of livestock density using the 1977 livestock census as well as per capita milk production for 1985 - 1986.

The Planning Commission divided the country into 15 broad agro-climatic zones based on physiography and climate (Alagh et al., 1989). This classification is based on physical conditions such as topography, soil type, geographical formation, rainfall, cropping pattern and irrigation. Since these zones were too broad to serve the purpose of planning, project teams were set up for each agro-climatic zone to divide each zone into sub-zones under the National Agricultural Research Project (NARP) of Indian Council of Agricultural Research (ICAR) and suggested 120 micro agro-climatic sub-zones after taking into account the rainfall patterns, temperatures, soil types and existing cropping patterns of each state as a unit (Saxena, 1989; Ghosh, 1991).

The major limitation of these approaches has been non-uniform application of criteria used, and the use of states as a unit for sub-divisions. This resulted in the creation of many sub-zones having similar agro-climatic characteristics but occurring in different states. These approaches also did not give adequate consideration to soils. Hence, a need was felt to generate an agro-ecological region map of the country giving due importance to soil environments (Table 12). With this viewpoint, the National Bureau of Soil Survey and Land Use Planning of ICAR brought out an agro-ecological map of the country through several approximations, using physiography, soils, bio-climatic types and growing period. During 1989, the Bureau brought out a 54-zone map, which had too much delineation. The map and the write up were finally discussed by a committee and the agro-ecological region map with 21 delineations was approved (Sehgal et al 1990). It was hoped that the map would serve India's needs in generating agro-technologies and transferring these to other comparable areas. Further, it may help in substantial land use plan.

Sastry (1993) made a good attempt to collect secondary data and information on various aspects of animal husbandry and classified the data agro-climatic region-wise. The agro-climatic regions considered, were 15 as described by Planning Commission. These 15 regions were regrouped into five regions viz., Himalayan regions (1 & 2); Gangetic Plain regions (3 to 6); Plateau and Hill regions (7 to 10); Coastal regions (11 & 12); and Island region to integrate with livestock production. Some of the general conclusions drawn by Sastry (1993) were: -

- Cattle are important in all the agro-climatic regions, predominantly as work animals; cattle as dairy animals (crossbreds) are prevalent in the West Coast Plains and Ghats region and in some pockets in the Western Plateau and Hills, Southern Plateau, East coast Plains and Hills in the two Himalayan regions.
- Milk production and marketing is well developed in agriculturally better-developed parts of the North Western, Western and Southern parts of the country. The importance of buffalo as a dairy animal decreases as one-move eastwards and southwards, but their concentration in peri-urban areas is generally high.

- Excluding the two Himalayan regions, in all the other regions, appear to have feed constraints, especially grazing land.

India is very diverse with regard to land use patterns, cropping systems and livestock husbandry. Various workers have classified the country into different agro-climatic regions since the start of the planning process in the country. But the suitability of macro-regions from a farming system perspective, particularly livestock farming combined with crop husbandry, needs to be reconsidered. Parameters like livestock density; productivity of milking cattle and buffalo; sheep, goat and pig productivity; feeds and fodder availability, milk availability, resource infrastructure, marketing, etc. need to be considered for any perspective plan in context of livestock productivity. Besides the above parameters, socio-economic parameters like family size, social status, income, literacy rates, customs and traditions also need to be included while classifying the country into different macro and micro regions.

Since, soil and climatic conditions of a region largely determine the suitability of crops growth period and their yield potential; almost same is applicable to livestock husbandry. The degree of correlation / dependence of animal raising has decreased to some extent on soil and climate because of human interventions-

- Transportation of feeds and fodder towards its deficit region
- Animal housing systems have been developed to mitigate the vagaries of climate, to a certain extent.

Based on above aspects an effort has been made (Kundu et.al., 2004) to integrate and classify livestock production into different agro-climatic regions as identified by planning commission of India. In the proposed zoning (Table 13) i.e. animal agro-climatic regions, minor deviations /changes have been adopted.

1. Trans and upper gangetic plains have been merged because of similar animal husbandry practices of the farmers.
2. Lower gangetic plains region, a part of Eastern plateau and hill region specially the Jharkhand and Orrisa have been regrouped under the name of eastern plateau and plain region keeping in view the status of livestock sector in those particular areas.
3. East coast plains and hill region, west coast plains and ghat region, Island region are brought under the one region i.e. coastal and Island region because of similar status of livestock enterprise.

4. Dry areas of Gujrat plains and hills region have been merged with the Western dry zone while the rest portions have been placed under the western plateau and hills region.

Thus, the country is proposed to be classified into 10 Animal Agro Climatic regions on following grounds:

The regions are as follows: 1.Western Himalayan Region, 2. Eastern Himalayan Region, 3. Eastern Plateau and Plains, 4. Middle Gangetic Plains, 5. Trans and Upper Gangetic Plains, 6. Central Plateau and Hills, 7. Western Plateau and Hills, 8. Southern Plateau and Hills, 9. Western Dry zones and 10. Coastal and Islands Region.

1. Western Himalayan region

The region encompasses Jammu & Kashmir, hilly areas of Punjab, Himachal Pradesh and Uttaranchal. Indian hill and Tarai zones account for 26.8 million hectare. As a result of high rainfall, the hill soils are exposed to extensive erosion, at higher altitude causing the soil shallow in depth, leached in bases and making them moderate to highly acidic, light in texture with low water holding capacity, and thereby make the soil deficient in certain nutrients. In the cold and arid zones, the soils are neutral to alkaline (Katoch and Singh, 2000). The major crops include wheat, rice, maize, minor millets (ragi), pulses, oil seeds etc. Apple is the dominant fruit crop of Western Himalayas comprising of 38 % of cultivated areas of Himachal Pradesh.

Among all the livelihood activities operating in the western Himalayas, dairying is considered as creamy activity, which has lot of potential for household protein security as well as economic up-liftment of the inhabitants. In the low hills areas (Doon vally, Tarai, Bhabar and Shivalik areas), animal husbandry plays a crucial role for milk and meat. The region is rich in high yielding milch cattle and buffaloes. In the mid hills region, cattle are reared for milk as well as draught power, buffaloes and goats are maintained for milk production only. Higher altitude regions of Western Himalayas are predominated by sheep and goat due to abundant availability of alpine pasture. In Uttaranchal, among all the ruminants (43.22 million), cattle constitute 46 % whereas buffalo represented by 19 – 21 %, sheep and goat constitute 8.2 % and 25 % respectively (Vir Singh,2002). The livestock enterprises contribute 20 –40 % of the farm levels income; the herd size per family is 1 –2 buffaloes, 0.6 cows, 1.4 bullocks and 3.5 goats. Buffalo is the main milch animal in the Uttaranchal hills. Tarai breed of buffalo yielding 2 –3 liter of milk are mostly seen in Tarai areas. Presently, improved type of buffaloes (cross of Murrah and Bhadawari) is well received by the farmers in hills. The milk yield in buffalo is upto 6.0 liter in mid hills and 12.0 liter in Murrah buffalo in lower hills has been reported.

Cynodon dactylon and *Ischoemum angustifolium* are major green forage source for ruminants. In addition to that, farmer also own trees/ shrubs like *Grewia optiva, Celtis australis* or *Melia azadirachta* (Mitha neem) on the riser of the crop field to augment the green fodder supply during lean period (Mittal and Rai, 2002). The commonly cultivated fodder in Western Himalaya include Berseem (*Trifolium alexandrinum*), lucerne (*Medicago sativa*), oats (*Avena sativa*), Jowar (*Sorghum bicolor*), Maize (*Zea mays*) etc. and their nutritive value are comparable or marginally higher than plains.

2. Eastern Himalayan Region

It includes Assam, Darjeeling district of West Bengal, Sikkim and the states lying beyond Assam in the far east namely Arunachal Pradesh, Manipur, Mizoram, Tripura, Nagaland, Meghalaya. Soil erosion in the high altitude and flooding are the main problems of the region. Tea gardening and associated activities are important employment source in hilly parts of Assam and West Bengal. The Jhoom cultivation practiced in the nearly one third of the cultivated areas has caused denudation and degradation of soils, with the resultant heavy run off, massive soil erosion and floods in the lower reaches and basins. Rearing of livestock is important livelihood activity in hill farming system. Pig farming is well developed and goat rearing is the next for meat production. Co-operatives have been initiated among pig farmers in the states of Nagaland, Meghalaya and other far east states.

Livestock species include cattle, buffalo, goat, pig, mithun, and yak. In Brahamaputra valley of eastern Himalayas, swamp buffaloes (Kaziranga) are reared by the local inhabitants for both draught power as well as milk production. The milk production of these buffaloes is very low ranging from 2 –3 liter daily per head. These buffaloes are usually dark grey or black to albinoids in colour and can thrive well in high rainfall and high humid climate. Siri is the native cattle breed of Darjeeling district and Sikkim. Milk yield is around 2 – 6 liter per head per day. Ruminant livestock unit (RLU) is maximum in Assam (5.4 million) and least in Mizoram (0.02 million). The cattle population predominates in all the states followed by goat, buffalo and sheep. But in Manipur, buffalo population is higher than goat (Anandan et al, 2001).

Mithun and yak are predominantly reared through grazing on pasture. The dominant plant species in the grazing field are *Bromus trypogen, Potentilla fulgen, Fimbristylis diphyla, Fragaria vesca* (Basu and Chatterjee, 2001). Livestock are generally fed locally available raw feed materials like roots, stem, leaves collected from jungle, kitchen waste etc. Animal are left for grazing in the forest immediately after morning milking and allowed for grazing up to evening and returned back around 4 – 00 PM for the evening milking. At the time of milking the concentrate consisting of boiled maize, broken rice, radish with leaves and mustard cake is offered.

3. Eastern Plateau and Plains Region

It encircles the territorial boundaries of West Bengal, Jharkhand, Orissa. The food habit of the human beings tilted towards fish and poultry. Thus, the livestock enterprise received lesser attention for resource allocation by the farmers.

Rice is the staple food for this region. The other crop includes cultivation of wheat, maize, bajra, mustard, til, potato, jute, vegetables etc. The livestock are mostly dominated by non-descript or Zebu cattle. The important buffalo breeds are Kujang, Manda, Sambalpuri (Orissa). There is no known breed of cattle in this region however following strains of cattle are found; namely Ghumsuri, Bingharpuri (Lesser known strain of Orissa), Bengali (a lesser known strain of West Bengal). In some areas of West Bengal like Burdwan, Nadia, 24 parganas, Murshidabad district, crossbred cattle are also reared by the farmers. The goat occupies a significant proportion of livestock population because of their characteristics meat quality and ability to survive under poor management conditions. Black Bengal goat breed is of three types viz.; black, grey and white are famous for its meat quality. As a result of demand for milk, crossbred cows as well as Murrah buffaloes are reared in the peri urban areas. In certain districts, sheep (Garol) are reared both for coarse wool and meat. In rural areas the poorer community rear pigs as a secondary source of income and for meat.

The livestock are maintained mainly on grazing and agricultural crop residues like paddy straw with little access of agricultural byproducts and kitchen waste. The pasture consisted of mainly Dub grass (*Cyanodon dactylon*). In addition to grazing mostly on bunds and roadside grasses, the animal receives 4 - 6 kg paddy straw as a bulk roughage (Das et al, 2003). At the time of milking mustard cake or de-oiled rice bran either singly or along with maize grain or gram chuni are offered to lactating animals. The concentrate supplementation to the animal solely depends on the economic status as well as the productivity of animals. But in the peri-urban areas, lactating animal are maintained on paddy straw and concentrate based diet with little green forage like *hybrid Napier*, Dinanath grass, Bracharia etc.

4. Middle Gangetic Plain

This area encompasses plains of Bihar and eastern parts of Uttar Pradesh. The rainfall is high and 39 % of the gross cropped area is irrigated and the cropping intensity is also high. Rice in Kharif, maize and wheat in Rabi are the main crops. Pulses, oil seed, sugarcane, vegetables, fruit crops are also widely grown. The preference of the livestock is in the order of buffalo followed by cattle and goat. The people of this region rear Murrah buffaloes. The native cattle strain of Bihar is Purnea, Sitamarhi. Crossbred animals are widely distributed through out the region (Roychaudhary, 1997). The feeding of livestock is mainly from crop residues grown by the farmers. Wheat / paddy staws constitute the basal dry roughage of the animals. Pulse straw, is also used upto certain extent.

Keeping in view the relatively higher productivity of the animals, progressive farmers also cultivate green fodder like maize, sorghum, berseem, hybrid napier etc. Fodder production practices are popular by the farmers in Begusarai, Samastipur and Muzaffarpur because of provision of good quality seeds and extension services provided by milk cooperatives (Sohane et al, 2001). The farmers usually chaff the green fodder and offer to animal for better utilization. The use of concentrate pellets is common by the members of milk producer cooperative societies. The balanced concentrate mixture available in the market is mostly used by the livestock owner of peri-urban / urban areas. In rural areas home made concentrate mixture consisting of oil cakes along with wheat bran / wheat flour are usually offered to animals. The concentrate mixture is mixed with wheat straw and water to make "Sani" and offered at the time of milking. The green fodder, if available, is also added in the "Sani" to fortify the ration. In the areas of Uttar Pradesh, the farmer offer more concentrate mixture to milch buffalo during rainy season, followed by autumn, spring, summer and winter. The average milk production of the buffaloes varies from 3.6 to 5.5 liter per day (Verma et al, 2001).

5. Trans and Upper Gangetic Plain

This region is famous for high yielding animals in the country. It spreads over the areas of western Utter Pradesh, plains of Punjab, Haryana, Delhi, Sriganganagar district of Rajasthan and Chandigarh. Soils of Uttar Pradesh region are as fertile as those of Punjab and Haryana. Cropping intensity is 144%. Jamunapuri goat is famous in this region for its high milk yield. Livestock rearing is an important subsidiary occupation in the rural areas of Uttar Pradesh. Buffalo enterprise plays a crucial role in providing income to the marginal and small farmers. The 'Bhadawari' breed of buffalo is crossed with Murrah for enhancing the milk productivity. The milk and milk products are commonly used by the farmer's family and dung is applied in the crop field to increase the production. The buffaloes are maintained on green and dry fodder produced in own farms. The animals are fed with some concentrate like cottonseed cake, mustard cake, barley husk and wheat flour, rice bran, wheat bran, gram churi etc. depending upon their availability. In rest parts of the region, crop production is the important sector of economy. About 60 % of labour force generates 40 % of income from this sector. Cultivation of crops like wheat, rice, pearl millet, cotton, gram, rapeseed, mustard and sugarcane is the most important activity of this region. Buffalo population is higher than cattle by about 40 % (Sastry, 1995). Buffalo breeds are predominantly of Murrah type. Cattle are mostly crossbred. Besides, the list of native cattle breed includes Sahiwal, Haryana, Tharparker etc. High fat milk is generally preferred. 65 % of total milk production used for home consumption. The farmers use both dry and green fodder for feeding to their livestock. The concentrate is either home grown or procured from the market. Dry roughage includes wheat straw, sorghum kadbi, pearl millet kadbi etc. A number of livestock owner also allocate their substantial lands for cultivation of green fodder. The common fodder is berseem, oat, maize, bathua etc. Because of high genetic make up of both cattle and buffalo, the diet of animals shares a good quantity of concentrate to sustain the milk production. Homegrown concentrate mixture generally comprises of cottonseed cake, mustard cake, wheat flour, mineral mixture, common salt etc. (Singh and Kundu, 2003) Most of the farmers fed crop residues and concentrate mixture in the form of 'sani'. Livestock are fed on the basis of milk yield but dry stocks are fed mainly on wheat / paddy straw with little concentrate mixture. Compound feed pellets are also used in the region and recently the feed manufacturers also introduced the complete feed block for different milk production levels.

6. Central Plateau and Hills

This region lies in the center of Indian subcontinent and spreads over southern part of Uttar Pradesh, south-east Rajasthan, central and northern Madhya Pradesh and Chattisgarh. Hills and ravines are widely spread in this region and these areas are not available for cultivation. The tract is semi –arid and rainfall varies from 740 – 1200 mm. Wheat, barley, pearl millet, pulses like mung, urd, gram, oilseed like mustard, groundnut, soybean, fruits and vegetables are widely grown in this region. The livestock sector is mainly comprised of buffalo, cattle, sheep and goat. The ravines are predominated by the local buffalo breed (Bhadawari), local goat and Jalauni sheep. The local breeds of buffalo are being replaced by crossing with high yielding Murrah buffalo. Bhadawari buffalo is famous for its high fat (14 %) secretion in milk. Due to abundance of shrubs (Jharberi, Pilua and Deshi Karaunda) and trees (Ruani, Chhonkra, Vilayati Babul) in the huge tracts of this region goat and sheep are maintained and contribut significantly to the income of small and landless farmers.

A system of animal grazing known as "Chutta Pashu" grazing system or "Annapratha" or "Stray Animal Grazing" is common in Budelkhand region of this zone (Shukla et al, 2000). The animals are left free for grazing without any monitoring. The owner who have the high yielding cattle and buffalo, keep them under stall-feeding. The high yielding buffaloes as well as cattle are maintained on balanced ration comprising of roughage and branded concentrate mixture. The basal roughage like wheat straw, groundnut haulms, pulses straw are mixed with home made concentrate mixture and water to make the "sani". The home made concentrate mixture usually include oil cake (mustard / cottonseed cake), wheat flour, gram chuni / gram husk.

7. Western Plateau and Hills

This region is comprised of varying topography and covers most of the interior parts of Maharastra, south-western parts of Madhya Pradesh and Gujrat. The annual rainfall varies from 850 – 950 mm. The irrigated area is nearly 12.4 % canals being the main source. Wheat, bajra, sorghum, maize, guar, mung, sugarcane, cottonseed are widely cultivated.

The zone has made significant strides on co-operative milk production. It is the home tract of several important breeds of cattle and buffalo. This include Surti , Jafarabadi, Mehsana, Nagpuri, Pandharpuri breeds for buffalo and Kankrej, Gir, Khillari, Krishnavally breed for cattle (Nivsarkar et al., 2000).

Cattle and sheep are reared by both nomads and stationary farmers. Buffalo keeping is important activity among the local inhabitants. The productivity of cattle and buffalo is generally low. Large-scale slaughterhouse and meat processing center located in this region attract the farmer for rearing animals for meat.

The dry roughage resources are comprised of wheat straw, groundnut haulms, bajra straw, maize straw, sorghum stover etc (Garg et al., 2003). Green fodder commonly grown in this area is bajra, cowpea, guar. *Hybrid napier*, local grass, maize, lucerne, moth, sorghum. Creeper of sweet potato is also fed to animals as green supplement. The major concentrate commonly used for feeding includes oil cakes (cottonseed, coconut, sunflower, mustard), grains (bajra, guar, maize, barley), by-products (wheat bran, gram chuni). Molasses available from sugarcane industry is utilized by compounded feed manufacturer for pelleting and urea supplementation. Different unconventional feed resources like babul pods, corn steep liquor, casitora seed, mahua seed cake, subabool seed are also incorporated in the diet of animals whenever available (Pande and Gupta, 2001).

8. Southern Plateau and Hills Region

The region includes interior parts of Andhra Pradesh, Karnataka and Tamil Nadu with annual rainfall in the range of 670 – 1000 mm. Rainfed farming is mostly adopted in the area and cropping intensity remains around 111 %. Major crops are cereals like sorghum, bajra, maize, ragi, paddy, wheat, oilseeds like groundnut, sesame, rape, mustard, linseed, niger, safflower, sunflower etc. (Raju et al, 2002, Anandan et al, 2003). The other crops of this region are coconut, cotton and sugarcane. Animal husbandry activity is one of the major components in the overall agricultural production system. Livestock species include buffalo, cattle, sheep and goat. Zerangi, Peddakimedi, Paralakhemundi breed of cattle are commonly found in the territory of Andhra Pradesh. In West Godavari district, the dairying is highly developed and the major contribution of milk originates from buffalo. Godavari buffalo breed is developed by crossing of native buffalo with Murrah. The native cattle of Karnataka are Amrit Mahal, Hallikar, and Krishna Valley. In Tamil Nadu, native breed of cattle are Kangayam and Bargur. Toda breed of buffalo is found in Nilgiri hills of Tamil Nadu. In south Karnataka region, high degree of cross breeding of cattle is adopted by the farmers and with animal husbandry activities are practiced on commercial scale.

The main dry crop residues commonly fed to livestock are ragi straw, sorghum stover, rice straw, sugarcane tops, legume straw from pulses and groundnut haulms also fed to limited extent. The green forages are from the cultivated fodder mainly hybrid napier, lablab and rest from public primary grazing areas consisting of permanent pasture and tree crops like glyricidia, neem, drum stick, banana, mulberry, subabool etc. The oil cakes, brans and chunis constitute the major portion while grain constitutes a small fraction of concentrate mixture. In Nilgiri hills, the buffaloes are solely maintained on grazing. No supplemental feeding of concentrate is done.

9. Western Dry Zone

This region includes mostly desert parts of Rajasthan as well as northern parts of Gujrat. The zone is characterized by hot sandy desert, erratic rainfall and scanty vegetation. The average annual rainfall is only 390 mm with the fluctuation from year to year. Bajra, guar, moth are the main khariff crops while rabi crops include wheat and gram. Livestock assets are of immense importance in this region as they provide both inter-year and intra-year stability in income and employment. Due to the imbalance between animals and biotic energy, migration of animals is a common phenomenon. The livestock include camel, cattle, sheep and goat. The zone is home tract of cattle breed of Malvi, Mewati, Nagori, Tharperkar, Rathi (Nivsarkar et al, 2000). Among sheep important breed are Chokla and Nali. Sheep are mostly carpet wool producer. Camels are generally used for transport, ploughing of fields, tractive power and can survive under poor / fibrous feed resources. The lists of roughage resources include pala, moth chara, groundnut haulms, guar phalghati. During summer the leaves of phog, khess, bawali, khejri, neem are commonly fed to animals. In the rainy season, green grasses like bhusal, anjan, sevan, bokoria, gokhru are available

10. Coastal and Island Region

It runs along the peninsular India along its eastern and western fringes and is demarcated as coastal plains, hills and the interior ghats, plateau or plains. Rain fall in western region ranges from 2000 – 4000 mm and in the eastern it varies from 780 – 1300 mm. The island includes the territory of Andaman and Nicober islands, Lakhasadeep, which are typically equatorial. Coconut based cropping system are largely practiced. Fisheries are major livelihood activity among the inhabitants. Paddy, wheat, gram, banana, tubers, tapioca, rubber, sapota (chiku) are widely cultivated in this region. The livestock community is comprised of both large and small ruminants. The density of buffalo is higher in eastern region than western parts. The important breeds of cattle are Ongole, Vechur (small breed of cattle in the hills of Kerala). Vechur breed of cattle is famous for relatively higher milk production than other local breeds and its adaptation to the hot humid climate of the region. Another important cattle is Sunandini (crossbred of Brown Swiss bulls with non descript cows). Buffalo breed in costal Orissa is Paralakhemundi. Cross breeding of cattle has got monumental success in western region specially in Kerala while it has not made significant strides in eastern costal region. Sheep are comparatively lesser than goats. In islands, about half of the cropped area is under coconut plantation. Therefore, livestock sector is not much developed.

The availability of dry matter in terms of crop residues, concentrate and green are 273, 39.8 and 226 thousands tones in Andaman & Nicobar Islands (Ananthram et al, 2001). Greens accounted for 83 % of total dry matter and crop residues accounted for only 14 % of which paddy alone contributes more than 90 %. The higher availability of greens in Andaman & Nicobar Islands is due to greater forest cover. The concentrate ingredients in islands include grains, bran and chunnis as well as coconut cake.

Perspectives

Presently, India is richest in bovine population and attained highest milk production in the world. To achieve the targeted growth in animal agriculture sector increased production through improved feeding, breeding, management and health care, a regional focus appears to be imperative. For each region, separate species (Table 14) and thrust areas have been suggested for different categories of farmers (Table 15). Specific know-how, dovetailing, the peculiar needs of the region and their resources, are needed for equitable and sustainable growth of livestock industry in the country along with its length and breadth.

With the ever-increasing demand for food, clothing and shelter to support the burgeoning population in the country, there is a clear need to have a systematic appraisal of the soil, climate and livestock resources to develop an effective alternate land and livestock use plan. Since the soils and climatic conditions of a particular region largely determine the suitability of different crops and livestock for their yield potential. Intensive efforts have to be made to map agro-ecological regions (uniforms in their soil-site characteristics) in identifying suitable crops to develop suitable cropping pattern for increasing crop and livestock production.

Increase in human population rising per capita income, increasing urbanization and the middle income group are fueling rapid growth in demand of food of animal origin in developing countries. The trend has been almost similar in whole world (Table 16-21). In India despite being highest total milk producer per capita availability of milk and meat are still very low thereby pointing to large unexploited potential to grow in this area. Ever increasing population in past largely contributed to total increased production. Production potential of indigenous animals could not be realized mainly because of less availability of feeds, and animals are mostly raised of fibrous crop residues. Though there is increase in cross-bred cattle, but still these are only around 10% of the population. The level of adoption of improved breeding, feeding, management and health related technologies has also been poor. The importance of livestock in India goes beyond their main function as a food supplier. Livestock are important source of draught power, manure and fuel for domestic use. In land scare households livestock provide livelihood to millions. They also provide much needed resilience, to farmers where crops fail partly or completely because of weather or other situations.

To enhance animal productivity following issues in addition to feeding on balanced nutrition may be rewarding-

• Breeds of animals

There are a number of good indigenous breeds available, in addition to cross bred cattle. The latter are useful under resource rich situations and adequate veterinary care. Where resources are limited indigenous animals, which are very well adapted could be preferred (Refer chapters cattle, buffalo, sheep and goat).

• Feeding on balanced diets

Ruminant animals are herbage eaters. Supply of green forage, from sown crops, pastures, top feeds at least 20 to 30 percent of ration will be very rewarding. Legume sources in addition to being rich in protein are also a better source of minerals and vitamins. Berseem, lucerne, lobia, guar, stylo, leuceania, etc. should find a place in ruminant ration. In addition to ad lib crop residues at least 20 to 30 percent energy and protein should come from concentrates (chapter 12 to 15). A balance concentrate should consist of grains like, barley, maize, wheat, sorghum along with cakes such as mixture of two or three out of mustard cake, soybean, groundnut, sunflower cake, cotton seed cake, sesam, linseed, etc. Brans wheat/ paddy or chunnies of pulses can make upto one third of concentrate mixture. A good quality mineral mixture at the rate of 2 percent of concentrate mixture is generally recommended. Feeds shoulds never be changed abruptly, there should always be a gradual change. Providing rations in mixed form (total mixed ration), twice or thrice a day may be practiced. Soaking of concentrate overnight and adding it with moistened crop residues as straws many result in higher feed intake by animals. Pregnant animals must be fed well during last 3 months of gestation. The animal must gain in weight by two to three times of weight of calf/ kid/ lamb, during this period. The cattle should not be milked during this period, and at least two months of dry period should be allowed.

• Management

Prophylactic measures such vaccination against foot and mouth disease (FMD), haemorrhagic septiceamia (HS) or any prevalent diseases of the area may be practiced, generally before rainy season. Animals must be protected from cold winds, and adequate housing be provided during winter and summer. Sufficient clean water should be made available at least three times a day. As soon animals come in heat, it should be identified and artificial insemination or natural service be arranged. Buffaloes generally come into heat during nights. For crosses of Holstein Friesian cattle and other good dairy animals following standards/ measures may be kept in mind.

1. Calf Rearing

i. Zero or nominal female mortality

ii. Good birth-weight (40-45 kg)

iii Male calves to be disposed speedily

iv. Growth rate of 200 to 500 grams/ day

2. Heifers

i. Good skin luster

ii. Gaining body weight of 250 kg in 18-20 months

iii. First service at 18-20 months with 250 kg body weight

iv. Age at first calving of 28-30 months.

3. Reproduction

i. Inter-calving interval of 13-15 months

ii. Dry period of 60-90 days

iii. Calving 60-90 days

iv. Lactation length of 300 days

v. Conception per 1.2 to 1.5 insemination

4. Production

i. Herd lactation yield of 3000 and 5000 litres for buffaloes and cows, respectively

ii. Daily peak yield of 12-15 litres for buffaloes and 15-20 litres for cows

iii. Wet average of 7-8 litres for buffaloes and 10-12 litres for cows during flush season and 5-6 litres for buffaloes and 6-8 litres for cows during lean season.

iv. Annual milk production per animal (average for the herd) of 2500 and 4000 litres for buffaloes and cows, respectively.

5. Culling

i. Animal suffering from chronic reproductive disorder or incurable diseases

ii. No animal which is dry and non-pregnant must stay at the farm

iii. Animals need not become too old at the farm

6. Health cover

i. Emphasis on preventive rather than curative treatment
ii. Timely vaccination against against FMD, HS etc.
iii. Calves in particular and other animals in general to be routinely dewormed
iv. Effective control of ecto-parasites
v. Codex alimentarius standards of sanitation
vi. Periodic white-washing
vii. Constructive, effective and prompt veterinary aid at reasonable cost at the door step of producers

7. Housing and care

i. Good comfort of the animals at all time. (A Kachcha space for few hrs)
ii. Loose-housing with free-access to drinking water
iii. Protection from heat and cold
iv. Appropriate schedule of offering fodder, cleaning of shed and milking utensils

8. Marketing

i. Remunerative price F.O.R. farm
ii. Inputs like feeds, seeds and veterinary drugs at reasonable cost
iii. Bulk milk cooler at village collection center.

Bibliography:

Alagh, Y.K. 1989. Agro-climatic Regional Planning and Overview. National Planning Commission, Government of India, New Delhi, India.

Anandan, S., Ramchandra, K.S., Raju, S.S. and Angadi, U.B. 2001. Status of livestock and feed resources in the northern states of India. Proceedings of 10th Animal Nutrition Conference held at NDRI, Karnal from November 9 –11, 2001. Abstract no. 096.

Anandan, S., Raju, S.S. and Ramachandra, K.S. 2003. Status of livestock and feed resources in northern Karnataka region. Indian Journal of Dairy Science. 56: 230 – 234.

Anantahram, K., Raju, S.S., Anandan, S. and Ramchandra, K.S. 2001. Animal feed resources availability in the union territories of Pondicherry and Andaman & Nicobar Islands. Proceedings of 10th Animal Nutrition Conference held at NDRI, Karnal from November 9 –11, 2001. Abstract no. 083.

Basu, A. and Chatterjee, A. 2001. Nutritional evaluation of a pasture land for yaks in Arunachal Pradesh. Proceedings of 10th Animal Nutrition Conference held at NDRI, Karnal from November 9 –11, 2001. Abstract no. 058.

Carter, D.B. 1954. Climates of Africa and India according to Thornthwaite's (1948) classification. John Hopkins University publication in climatology 7 (4).

Das, A., Ghosh, T.K. and Haldar, S. 2003. Mineral distribution in soil, feeds and grazing cattle of different physiological stages in the red laterite and new alluvial agro climatic zones of West Bengal. Indian Journal of Animal Sciences. 73 (4): 448 – 454.

FAO (2004). fao.org.

Garge, M.R., Bhandari, B.M. and Sherasia, P.L. 2003. Macro-mineral status of feeds and fodder in Kutch district of Gujrat. Animal Nutrition and Feed Technology. 3: 179 –188.

Ghosh, S.P. 1991. Agro climatic zone specific research – Indian perspective under NARP. ICAR. New Delhi.

India. 2002. A reference Annual Publication Division. Ministry of Information and Broadcasting. Government of India.

Katoch, K.K. and Singh, C.M. 2000. Technologies for managing hill agriculture resources. Indian Farming. 2: 57 –61

Kundu, S.S., Das, M.M., Samanta, A.K. and Pathak, P.S. (2004). Animal agro-climatic zoning in India – A perspective. In: Buffalo Production Under Different Climatic Regions (Kundu, S.S., Misra, A.K. and Pathak, P.S. Eds.). International Book Distributing Co., Lucknow, India pp: 43 – 84.

Mittal. R.K. and Rai, S.D. 2002. Farming system approach for sustainable agricultural development in Uttaranchal. Indian Farming. 5: 9 – 10, 25

Muthariah, S. 1988. A social and economical Atlas of India. Oxford Publishers, New Delhi, India

Nivsarkar, A.E., Vij, P.K. and Tantia, M.S. 2000. Animal Genetic resources of India – cattle and buffalo. ICAR Publication. New Delhi.

Pande, M.B. and Gupta, R.S. 2001. Availability and utilization of livestock feed resources in different agro climatic zones of Gujrat. Proceedings of 10th animal Nutrition Conference held at NDRI, Karnal from 9 – 11, November 2001. PP 145 – 150.

Raju, S.S., Anandan, S., Angadi, U.B., Anantahram, K., Prasad, C.S. and Ramchandra, K.S. 2002. Assessment of animal feed resource availability in southern Karnataka region. Indian Journal of Animal Sciences. 72 (12): 1137 – 1140

Roychudhary, P.N. 1997. Milk wave in Bihar. Dairy India. 125 –130.

Sastry, N.S.R. 1993. Animal husbandry in different agro-climatic regions of India-report of a consultancy mission. Swiss development Corporation, New Delhi, India

Sastry, N.S.R. 1995. Livestock sector of India – regional aspects. International Book Distributing Co. Lucknow. Uttar Pradesh.

Saxena, A.P. 1989. Strategies for Agricultural Research and Development- A Zonal Approach. ICAR, New Delhi, India

Sehgal, J. L., Mandal, D.K., Mandal, C., Vadivelu, S. 1990. Agro-Ecological Regions of India. National Bureau of Soil Survey and Land Use Planning (ICAR), NBSS Publ. 24, Nagpur,

India

Sehgal, J.L. 1995. Cited in Vision 2020. IISS Perspective plan. ICAR

Sekhon . 1996. Cited in Vision 2020. IISS Perspective plan. ICAR

Shukla, S.K., Solanki, K.R. and Dwivedi, R.P. 2000. Plant multipurpose trees and shrubs for live fence in Bundelkhand region. Indian Farming. 1: 11 – 15

Singh, Jasbir. 1974. An Agricultural Atlas of India: A Geographical Analysis. Vishal Publications, University of Campus, Kurukshetra.

Singh, Ram and Kundu, S.S. 2003. Quality evaluation of some animal feed stuffs available in Haryana. Animal Nutrition and Feed Technology. 3: 143 - 150

Sohane, R.K., Jha, P.B., Choudhary, M.K. and Kumari, Asha. 2001. Data base information on feeds, fodders and livestock in north Bihar. Proceedings of 10th Animal Nutrition Conference held at NDRI, Karnal from November 9 –11, 2001. Abstract no. 097.

Tantia, M.S., Kumar, P., Joshi, B.K., Vij, P.K. and Nivsarkar, A.E. 1994. Population status of cattle and buffalo breeds of India. Journal of Dairying, Foods and Home Science. 13: 192 – 194

Verma, D. N., Singh, S.P., Srivastava, D.K. and Maurya, S.K. 2001. Seasonal variation in feed milk relationship in buffaloes in Gorakhpur region UP. Proceedings of 10th Animal Nutrition Conference held at NDRI, Karnal from November 9 –11, 2001. Abstract no. 088.

Vir Singh. 2002. Small dairy production systems in Uttaranchal hills. Indian Farming. 3:23-25

Table 1. Livestock Population trends in India

Species	1951	1956	1961	1966	1972	1977	1982	1987	1992	1997*
Cattle	155.3	158.7	175.6	176.2	178.3	180.0	192.5	199.7	204.6	197.7
Adult Female Cattle	54.4	47.3	51.0	51.8	53.4	54.6	59.2	62.1	64.6	NA
Buffalo	43.4	44.9	51.2	53.0	57.4	62.0	69.6	76.0	84.2	88.6
Adult Female buffalo	21.0	21.7	24.3	25.4	28.6	31.3	32.5	39.1	43.8	NA
Total Bovines	198.7	203.6	226.8	229.2	235.7	242.0	262.4	275.8	289.0	286.5
Sheep	39.1	39.3	40.2	42.4	40.0	41.0	48.8	45.7	50.8	56.8
Goats	47.2	55.4	60.9	64.6	67.5	75.6	95.3	110.2	115.3	120.8
Horses & Ponies	1.5	1.5	1.3	1.1	0.9	0.9	0.9	0.8	0.8	0.6
Camels	0.6	0.8	0.9	1.0	1.1	1.1	1.1	1.0	1.0	0.9
Pigs	4.4	4.9	5.2	5.0	6.9	7.6	10.1	10.6	12.8	13.6
Mules	0.1	0.0	0.1	0.1	0.1	0.1	0.1	0.2	0.2	0.1
Donkeys	1.3	1.1	1.1	1.1	1.0	1.0	1.0	1.0	1.0	0.6
Yaks	NC	NC	0.0	0.0	0.0	0.1	0.1	0.0	0.1	0.1
Total Livestock	292.8	306.6	335.4	344.1	353.6	369.0	419.6	445.3	470.9	479.9
Poultry	73.5	94.8	114.2	115.4	138.5	159.2	207.7	275.3	307.1	352.0

NC: Not Collected

*Based on provisional results from States/UTs

Source: Department Of Animal Husbandry & Dairying, New Delhi

Table 2. State-wise production of milk during 1999-2000 to 2001-02

State/ UTs	Milk ('000 tones)		
	1999 - 2000	2000 - 2001*	2001 - 2002**
Andhra Pradesh	5122	5521	5145
Arunachal Pradesh	45	46	55
Assam	733	738	894
Bihar including Jharkhand	3740	3878	4068
Goa	43	44	47
Gujarat	5255	5317	5573
Haryana	4679	4849	4976
Himachal Pradesh	741	760	810
Jammu-Kashmir	1286	1037	1088
Karnataka	4473	4598	5357
Kerala	2673	2771	2907
Madhya Pradesh including Chattisgarh	5600	5806	6091

State/ UTs	Milk ('000 tones)		
	1999 - 2000	2000 - 2001*	2001 - 2002**
Maharashtra	5706	5850	6024
Manipur	67	69	73
Meghalaya	62	64	71
Mizoram	18	14	11
Nagaland	50	50	54
Orissa	847	875	865
Punjab	7700	7984	8375
Rajasthan	7260	7455	6330
Sikkim	35	35	46
Tamil Nadu	4574	4899	4629
Tripura	49	51	53
Uttar Pradesh including Uttaranchal	16506	14153	14840
West Bengal	3465	3470	4079
A&N Islands	23	24	25
Chandigarh	42	44	46
D & N Haveli	10	10	11
Daman & Diu	1	1	1
Delhi	290	292	321
Lakshdeep	1	1	1
Pondicherry	36	37	38
All India	78779	81430	84570

*Provisional

**Anticipated achievement

Source: Department of Animal Husbandry and Drying, Ministry of Agriculture, New Delhi

Table 3. Meat production during 1981 to 2000 (x 1000 tones)

Year	Buffalo	Beef & Veal	Mutton & Lamb	Goat Meat	Pig Meat	Poultry Meat	Total Meat
1981	127	78	125	277	75	120	850
1982	130	80	132	298	80	130	865
1983	132	80	134	302	80	137	1010
1984	148	149	135	346	82	150	1047
1985	152	150	141	358	85	161	1106
1986	163	160	147	370	86	180	1261

Year	Buffalo	Beef & Veal	Mutton & Lamb	Goat Meat	Pig Meat	Poultry Meat	Total Meat
1987	207	239	162	380	80	193	1630
1988	290	232	148	378	357	225	2974
1989	936	845	160	385	359	289	3596
1990	1048	1271	173	410	360	334	3710
1991	1176	1185	168	455	364	362	3800
1992	1182	1216	167	456	397	382	3950
1993	1182	1276	169	466	403	454	4052
1994	1204	1292	171	470	408	507	4259
1995	1351	1365	194	450	420	479	4319
1996	1382	1370	218	454	514	479	4417
1997	1403	1378	222	458	533	527	4521
1998	1380	1401	226	462	543	540	4552
1999	1410	1421	228	466	560	559	4644
2000	1421	1442	229	467	560	575	4694

Table 4. Details of production of meat and meat products (MT) from 1994 to 1998

Source of meat	Year				
	1994	**1995**	**1996**	**1997**	**1998**
Buffalo	1200	1204	1204	1205	1010
Sheep & goat	637	647	669	670	675
Pig	365	420	420	400	420
Cattle	1290	1292	1292	1292	1295
Poultry	442	578	480	580	600
Total	**3569**	**4141**	**4065**	**4147**	**4000**

Table 5. Trend in livestock number 1961-2004, and their projection* 2005-2025 (number in million)

Item	1961	1971	1981	1991	2001	2004	2005	2010	2015	2020	2025
Cattle	176	178	189	204	191	186	184	176	168	161	154
Buffaloes	51	57	68	82	94	98	99	105	111	118	125
Sheep	40	40	46	50	60	63	64	69	75	81	88
Goats	61	67	91	114	121	120	120	122	123	124	126
Pigs	5	7	10	13	14	14	14	15	15	15	15
Chickens	108	132	197	300	377	425	441	529	635	762	915
Cattle+Buffalo	227	235	256	286	286	283	283	281	280	279	280

*projection based on growth rate of last ten years (1994-2004)

Table 6. Growth rate (%p.a.) in livestock no. during different periods

Item	1961-71	1971-81	1981-91	1991-2001	Last 20 Years 1984-2004	Last 10 Years 1994-2004	Last 5 Years 1999-2004
Cattle	0.13	0.60	0.76	-0.62	-0.25	-0.88	-1.00
Buffaloes	1.06	1.73	1.98	1.40	1.44	1.20	1.18
Sheep	-0.03	1.47	0.69	1.96	1.38	1.66	1.19
Goats	0.97	3.11	2.30	0.57	0.94	0.22	-0.30
Pigs	2.36	3.92	2.67	1.07	1.65	0.37	1.01
Chickens	2.06	4.09	4.31	2.30	3.06	3.72	4.26
Cattle+Buffalo	0.34	0.88	1.09	0.00	0.27	-0.22	-0.28

Table 7. Milk production trend 1961-2004, and their projection* 2005-2025 (million metric tonne)

Item	1961	1971	1981	1991	2001	2004	2005	Projected yield 2010	2015	2020	2025
Buffalo Milk	11	12	19	29	47	50	52	65	80	99	123
Cow Milk, Whole, Fresh	9	9	14	23	36	38	39	47	57	68	82
Goat Milk	0.5	0.7	1.0	2.3	2.6	2.6	2.6	2.5	2.5	2.4	2.4
Milk,Total	20	23	34	54	85	90	94	114	139	170	208

*projection based on growth rate of last ten years (1994-2004)

Table 8. Growth rate (%p.a.) in milk production during different periods

Item	1961-71	1971-81	1981-91	1991-2001	Last 20 Years 1984-2004	Last 10 Years 1994-2004	Last 5 Years 1999-2004
Buffalo Milk	1.11	4.31	4.28	4.98	3.80	4.39	3.06
Cow Milk	0.77	4.31	4.81	4.45	4.23	3.76	2.88
Goat Milk	2.35	4.31	8.60	1.02	3.79	-0.48	2.64
Total Milk	1.00	4.31	4.65	4.60	3.97	3.95	2.97

Table 9. Percent share of different component in milk production

Item	1961	1971	1981	1991	2001	2004
Buffalo Milk	54.4	55.0	55.0	53.0	55.0	55.3
Cow Milk	43.0	42.0	42.0	42.6	42.0	41.8
Goat Milk	2.6	3.0	3.0	4.3	3.1	2.9
Total Milk	100.0	100.0	100.0	100.0	100.0	100.0

Table 10. Meat production trend 1961-2004 (,000 metric tonne)

Item	1961	1971	1981	1991	2001	2004
Beef and Veal	506	640	863	1228	1452	1483
Buffalo Meat	551	683	847	1224	1433	1483
Goat Meat	235	244	313	434	469	475
Mutton and Lamb	122	138	159	185	230	239
Pig meat	154	194	279	434	483	497
Meat others	50	76	108	129	140	140
Poultry Meat	78	95	141	389	1307	1715
Meat, Total	1696	2070	2710	4023	5515	6032
Eggs Primary	170	308	598	1210	1870	1890

Table 11. Growth rate (%) in meat production during different periods

Item	1961-71	1971-81	1981-91	1991-2001	Last 20 Years 1984-2004	Last 10 Years 1994-2004	Last 5 Years 1999-2004
Beef and Veal	2.38	3.04	3.58	1.70	2.17	0.87	0.85
Buffalo Meat	2.17	2.18	3.75	1.58	2.27	1.16	1.18
Goat Meat	0.38	2.51	3.34	0.78	1.59	0.63	0.38
Mutton and Lamb	1.21	1.41	1.53	2.23	1.46	1.35	0.93
Pig meat	2.36	3.68	4.52	1.08	2.48	0.41	1.02
Meat, others	4.28	3.58	1.79	0.82	0.99	0.59	0.43
Poultry Meat	2.02	4.03	10.67	12.88	12.05	12.07	14.42
Meat, Total	**2.01**	**2.73**	**4.03**	**3.20**	**3.50**	**2.99**	**3.83**
Eggs Primary	6.12	6.86	7.30	4.45	4.48	2.84	2.35

Table 12. Estimated areas of broad soil groups in India

Soil type	Area (million hectare)
1. Alluvium derived soils	74.3
Inland	63.1
Deltaic alluvial	6.8
Coastal alluvial	4.4
2. Black soils	73.2
3. Red, yellow and laterite soils	87.6
4. Soils of desert region	30.0
5. Soils of Himalayan region	28.7

Table 13. Integration of animal husbandry with the agro climatic zones of India

Planning Commission agro climatic zones	IGFRI perspective
1. Western Himalayan region	1. Western Himalayan region
2. Eastern Himalayan region	2. Eastern Himalayan region
3. Lower gangetic plains region 7. Eastern plateau and hills region	3. Eastern plateau and plains region
4.Middle gangetic plains	4. Middle gangetic plains
5. Upper gangetic plains region 6. Trans gangetic plains region	5. Trans and upper gangetic plains
8. Central plateau and hills region	6. Central plateau and hills
9. Western plateau and hills region 13. Gujarat plains and hills region	7. Western plateau and hills
10. Southern plateau and hills region	8. Southern plateau and hills
14. Western dry region	9. Western dry zone
11. East coast plains and hills region 12. West coast plains and ghat region 15. Island region	10. Coastal and Island region

Table14. Species density and recommended, developmental priorities of various species in different regions

Agro clinatic region	Species density				Develop-ment Priority	Regions Product
	Highe	High	Medi-	Low		
1. Western Himalayan region	C	BG	S	P	S-G-B-C	Wool-Meat
2. Eastern Himalayan region	C	GP	—	BS	P-G-C	Meat-Draft
3. Eastern plateau & plains	CG	—	BS	P	G-C-P	Draft-Meat
4.Middle gangetic plains	CG	B	SP	—	C-B-G	Milk-Meat
5. Upper & trans gangetic plains	BC	G	S	P	B-C—G	Milk-Meat
6. Central plateau & hills	C	BG	—	SP	C-G-B	Milk-Meat
7. Western plateau & hills	CG	B	S	P	S-G-B-C	Milk – Meat
8. Southern plateau & hills	C	BSG	P	—	C-S-G	Meat-Milk
9. Western dry region	GS	CB	—	P	S-C	Wool –Meat
10. Coastal & island region	C	G	B	SP	C-G-P-S	Milk-Draft-Meat

C= cattle, B= buffalo, S= sheep, G= goat, P= pig

Table15. Vision for different categories of dairy farmers

Type of livestock rearing	Present status	Future demand / vision
Minimum input	• *Grazing land with less grass cover.	• Improved grazing land on forest and problem soils.
	• *Sheep are not prolific.	• Highly prolific breed.
	• High concentration of animal.	• Higher concentration of animals with high growth potential.
	• *very poor market.	• Better market (eliminating/ decreasing profits ofr middleman) organic food.
	• Organic animal farming	• Subsidized institutional support such as veterinary services (vaccination), training etc.
Low input	• Problem soils underutilized	• Improved grazing land required from forest or problem soils
	• *Shortage of quality fodder	• Availability of good quality feed and legume fodder.
	• Good local breeds available	• Improved breeds of livestock for milk and body weight gain.
	• *Poor market	• Better market (eliminating / decreasing profits of middleman) organic food.
	• *Organic animal farming	• Milk collection facility
		• Subsidized institutional support such as veterinary services, training etc.
		• Develop co-operative system.
Medium input	• Feed shortage and less green fodder availability	• Availability of good quality feed and legume fodder
	• *Chaffing practiced	• Balanced concentrate mixture
	• *Good breeds available	• Chaff cutter
	• *Costly veterinary help	• Improved breed of livestock for milk and body weight
	• *Poor market	• Good veterinary facilities
		• Better market (eliminating/ decreasing profits of middleman)
		• Milk collection facility
		• Paid institutional support such as veterinary services, training
		• Proper housing of animals

Type of livestock rearing	Present status	Future demand / vision
Commercial	• Fodder availability low	• Availability of good quality feed and legume fodder
	• Costly feed	• Improved breed with high productive (3000 liter per location) and reproductive efficiency (elite herd)
	• High rate of interest on loan	• Paid institutional suppor
	• Costly milking machine	• Intensive veterinary care
	• High rate of power	• Better and assured market
	• Shades available	• Balanced concentrate mixture/ mineral mixture
	• Costly veterinary aid	• Milking machine
	• *Rate of milk and milk products low	• Facilities for chilling of milk
		• Storage of fodder (hay / silage)
		• Location (near to market)
		• Grinder for concentrate mixture
		• Loaning facility from bank
		• Tractor / Trolley
		• Heavy duty chaff cutter
		• Cattle shed with proper scientific management system
		• Land with irrigation facilities
		• Generator for assured electric supply
		• Trained manpower availability

* Key response area, (Source: Kundu et al., 2004)

Table 16. Top twenty milk producer countries (MT)

SI.No.	Name	Milk	Production of Milk Protein
1	India	86.96	3.04
2	USA	77.25	2.70
3	Russian Fad.	33.10	1.15
4	Germany	28.38	0.99
5	Pakistan	27.81	0.97
6	France	25.42	0.88
7	China	17.96	0.62
8	UK	15.05	0.52
9	NewZealand	14.35	0.50
10	Ukrain	13.66	0.47
11	Itlay	12.04	0.42
12	Polland	11.80	0.41
13	Netherland	10.84	0.37
14	Australia	10.64	0.37
15	Mexico	9.99	0.34
16	Japan	8.36	0.29
17	Turkey	8.16	0.28
18	Argentina	7.90	0.27
19	Canada	7.88	0.27
20	Spain	6.91	0.24

Source: fao.org

NB: Milk contains average 3.5 per cent protein, MT = Million Ton

Table17. Top twenty meat producer countries (MT)

Name	Meat	Production of Meat Protein
China	69.20	14.53
USA	38.74	8.13
Brazil	18.68	3.92
Germany	6.59	1.38
France	6.51	1.36
India	6.03	1.26
Spain	5.44	1.14
Russian Fad.	4.93	1.03
Mexico	4.90	1.03
Canada	4.28	0.89
Italy	4.22	0.88
Argentina	4.16	0.87
Australia	3.85	0.80
UK	3.29	0.69
Polland	3.11	0.65
Japan	2.99	0.62
Netherland	2.42	0.50
Philippins	2.34	0.49
Thailand	2.20	0.46
Denmark	2.12	0.44

Source: fao.org
NB: Meat contains average 21 per cent protein

Table 18. Per capita availability of milk(Kg/yr)

Sl.No.	Name	Milk Milk	Availability of Protein
1	New Zealand	3,605.30	126.186
2	Ireland	1,372.50	48.038
3	Denmark	857.8	30.023
4	Montserrat	750.0	26.250
5	Netherland	674.8	23.618
6	Luxembourg	605.5	21.193
7	Australia	594.6	20.811
8	Switzerland	552.7	19.345
9	Lithuania	511.1	17.889
10	Malvinas	500.0	17.500
11	Finland	487.2	17.052
12	Belarus	480.1	16.804
13	Estonia	453.7	15.880
14	Uruguay	439.4	15.379
15	France	434.3	15.201
16	Austria	408.9	14.312
17	Iceland	385.9	13.507
18	Sweden	363.8	12.733
19	Slovenia	355.7	12.450
20	Narway	349.0	12.215

Source: fao.org

Table 19. Per capita availability of meat (kg/yr)

S.No.	Name	Meat	Availability of Meat Protein
1	Denmark	399.1	83.811
2	NewZealand	339.7	71.337
3	Malvinas	303.7	63.777
4	Montserrat	299.9	62.979
5	Ireland	248.5	52.185
6	Australia	195.3	41.013
7	Belgium	171.0	35.910
8	Netherland	166.6	34.986
9	Uruguay	159.7	33.537
10	Canada	137.2	28.812
11	USA	133.3	27.993
12	Cyprus	129	27.090
13	Spain	128.1	26.901
14	Hungary	119.9	25.179
15	Mongolia	117.8	24.738
16	France	109	22.890
17	Argentina	106.8	22.428
18	Austria	104.3	21.903
19	Luxembourg	102	21.420
20	Brazil	99.2	20.832

Source: FAO.org

Table 20. Production of milk and meat (,000 ton) in SAARC countries (2003)

Name	Milk	Milk Protein	Meat	Meat Protein
Bangladesh	2156.9	7.55	440.6	0.93
Bhutan	41.4	0.15	6.29	0.01
India	86960.0	304.36	6038.39	12.68
Maldives	-	-	1.00	0.002
Nepal	1246.5	4.36	250.87	0.53
Pakistan	27811.0	97.34	1891.58	3.97
Sri Lanka	298.4	1.04	114.43	0.24

Table 21. Per capita availability of milk, meat and total protein (both sources) in SAARC countries (Kg/person/year)

Name	Milk	Meat	Protein availability for milk and meat
Bangladesh	14.9	3.1	1.17
Bhutan	18.9	3.0	1.29
India	80.1	5.5	3.96
Maldives	-	3.2	0.67
Nepal	50.2	10.0	3.86
Pakistan	180.3	12.3	8.89
Sri Lanka	15.9	6.5	1.92

FEEDS AND FODDER EVALUATION

The notion that feeds and food could be evaluated by chemical analysis emerged quite early (about 1800) and has long preceded any such evaluations by balance trials or any detailed knowledge of the chemical composition of plants. Many modern texts erroneously credit the crude fibre method and the proximate system built around it to Henneberg. The most Henneberg could have contributed was standardized procedures that every one followed. Henneberg himself writes that he only adopted the methods of Einhof which was essentially maceration with water after solvent extraction, as used by person slightly earlier. This method gave much higher values than the standard crude fibre method and much closer to total plant cell wall. The person - Einhof method was used by Voeloker in 1860's but was generally superseded in this period by the method using dilute acid and alkali.

Scientists who developed fibre analysis methods:

Einhof	1806
Gorham	1820
Horsford	1840
Wolf	1850
Henneberg & Stohmann	1860, 1864
Morrison	1956
Van Soest	1967
Van Soest modified	1991
CNCP system	1992

Analytical Methods for Nutrient Composition:

For many nutrients required by animals there are direct analytical methods by which we can establish the potency of feedstuffs in these nutrients. There are three general types of analytical methods: (1) chemical procedures (gravimetric procedures, titration, colorimetry, chromatography, etc.); (2) biological procedures (employ animals such as chick or rat to give a much more accurate estimate of animal utilization but which makes procedures more tedious and expensive); and (3) microbiological procedures (similar to biological procedures but employ isolated bacteria).

Proximate analysis

A team of scientists under the leadership of Henneberg and Stohmann developed proximate analysis system in the sixties of last century at Weende Experimental Station. The different fractions that result from the proximate analysis include water, ash, crude protein, ether extract, crude fibre and nitrogen free extract (NFE). This procedure is probably the most generally used chemical scheme for describing feedstuffs in spite of the fact that the information it gives often is of uncertain nutritional significance, or may even be misleading. We should, therefore, consider in some detail the nature, peculiarities and limitations of proximate determination by proximate analysis: -

1. **Dry matter**

   Heat sample to a constant weight at a temperature above the boiling point of water (100 - 105°C). This removes the water, so loss in weight equals water. (100% - water = D.M.%)

   Sources of error: - Any materials that volatilise at the temperature are lost. (silage or other fermented products). Some substances in plants get oxidized when heated and therefore increase in weight. Other methods for determination of water in feeds include toluene distillation, drying under vacuum and freeze-drying.

2. **Ash (minerals)**

   Burn sample by placing a weighed amount in muffle furnace for 2 hours at 600°C. Ash is considered as the dry inorganic residue remaining. The process removes the water, fat, protein and carbohydrates. The high temperature may alter forms of some minerals and may even volatilise substances such as chloride, zinc, selenium and iodides.

3. **Crude Protein (Kjeldahl process)**

   Digest small dried sample in concentrated sulphuric acid until all organic matter is destroyed. Nitrogen from feed is now in the form of ammonium sulphate. The digest is neutralized with sodium hydroxide, distilled, driving the ammonia into standard acid and titrated. This determines the amount of nitrogen in the sample.

   Protein contains an average of 16% nitrogen; therefore, total nitrogen x 6.25 equals the crude protein in the sample. The analysis does not distinguish one form of N from another. Thus, we cannot tell if a feed mixture has true protein (amino acids) or other non-protein nitrogen NPN sources. In addition, nitrate N is not converted to ammonium salts by this procedure.

4. **Ether extract (Fat)**

   Extract dry sample with petroleum ether $(60 - 80)^0$ for a period of 4 hours or more. This removes the fat, so again the loss in weight after drying (evaporation of ether) equals the fat. This procedure includes "fat" as any ether-soluble compounds including some non-nutrient compound such as chlorophyll, volatile oils, resins, pigments and plant waxes, which are of little value to animals.

5. **Carbohydrates (CHO)**

   Not determined by analysis as such, CHO = CF+ NFE

i. Crude fibre (CF)

After removal of water and ether extract from a feed sample, the sample is boiled in 1.25% sulphuric acid (0.255 N H_2SO_4), then in 1.25 % sodium hydroxide (0.312 N NaOH) for 30 minutes each. This removes the proteins, sugars and starches, which are discarded. Cellulose, lignin and mineral matter are left in the feed residue. This material is dried and weighed, then burned in a muffle furnace at 600°C. Loss in weight is reported as crude fibre. This fraction consists primarily of hemicellulose, cellulose and some insoluble lignin.

ii. Nitrogen-free extract (NFE)

Nitrogen-free extract is found by difference - not by actual analysis. The percentages of water, ash, protein, fibre and fat are merely added together and subtracted from 100. NFE is made up primarily of readily available carbohydrates, such as the sugars and starches, but it may also contain some hemicellulose and lignin, particularly in such feedstuffs as forages. The calculation of nitrogen free extract (NFE) as the dry matter not accounted for by sum of ether extract, crude fibre, ash and crude protein (nitrogen x 6.25). In Weende's system of analysis "soluble carbohydrate" was called nitrogen free extract (NFE).

Main features of Weende's system of analysis

1. Lignin is indigestible and dissolved into the nitrogen-free-extract (NFE) by the use of sodium hydroxide.
2. Carbohydrate is divided into two fractions i.e. highly digestible (NFE) and less digestible (Crude fibre).
3. The highly digestible fraction calculated by difference is supposed to represent available carbohydrate.
4. The system is the basis on which total digestible nutrients TDN is calculated using the following assumptions:

a. Ether extract recovers lipids and fats, which contain 2.25 times the energy of carbohydrates.

b. All nitrogen is in protein, which contains 16% nitrogen.

c. Crude fibre recovers least digestible fibrous nitrogen.

d. The NFE represents highly digestible carbohydrates.

5. Ether extract includes waxes and pigments of little value and does not recover soaps in focus, which are the main form in which undigested fatty acids are extracted.

6. Forages contain no triglycerides and leaf galactolipids contain less energy than factor 2.25 would imply.

7. The error involved with ether extract is a relatively minor one unless a lipid constitute large component of the dry matter. Ether extract can be ignored in the analysis of most forage and many other ruminant feeds.

8. The error is most serious in faecal analysis where little true protein is ordinarily bound and the main nitrogenous constituents are microbial substance or Maillard products with only 7 - 11% nitrogen.

9. The NFE contains the cumulative errors of all of other determinations. The largest of these errors is due to the solubilization and loss of much lignin and hemicellulose in the preparation of crude fibre.

   Even cellulose is not wholly recovered and the behaviour of different plant materials is quite variable. Generally, the net solubility of lignin in grasses is greater than that in legumes. The error caused by the inclusion of cell wall fractions in the NFE is lowest in the case of concentrate feeds where about three quarters of NFE is starch and soluble carbohydrates. In alfalfa, available carbohydrate and organic acids comprise about 50% of the NFE whereas in mature grasses and straws very little of the NFE is available carbohydrate. The effect of this error is to cause the apparent digestibility of the NFE to be less than that of the crude fibre in a significant number of cases. The presence of a prominent metabolic fraction in faecal NFE contributes greatly to this effect. The proportion of cases where the digestibility of the NFE is about 30% for all feedstuffs but is notably greater in the grasses that contain more hemicellulose and soluble lignin. The error is largest in tropical grasses and straws.

10. The unequal recovery of lignin, cellulose and hemicellulose in the crude fibre leads to a variable relationship between crude fibre and plant cell wall. Grasses which have a high hemicellulose and moderate lignin content provide, on the average, the poorest recovery of cell wall components in

crude fibre, while legumes, lower in hemicellulose but higher in lignin, are intermediate. Non-legume dicots are chiefly unlignified vegetables and have a high rate of cell wall recovery because the principal component of their cell wall is cellulose.

11. The partition of carbohydrate in to fibre (CF) and nitrogen free extract (NFE) is presumed to represent a separation of less digestible starch and sugars. However, in 20 to 30 per cent feeds listed by Morrison, the NFE is less digestible than CF. This comes about for several reasons, the first of which is that CF method (successive boiling with dilute sulphuric acid and sodium hydroxide) does not recover all the fibre and large portions of fibrous constituents are extracted in to the NFE. The most important of these fractions are lignin and hemicellulose is dissolved by both acid and alkali. The basic error of the NFE concept is the assumption that if constituents are soluble they are digestible. Lignin, the rigid component of wood, not only is indigestible but lowers the digestibility of substances with which it is associated.

    Another portion of the indigestible part of NFE arises from an artefact in the calculation of faecal NFE, where it is presumed that faecal nitrogen is protein (N x 6.25). Actually, 80 - 90 per cent faecal nitrogen is NPN and is composed of bacterial residues in which the ratio of organic matter to nitrogen is about 1:4. Thus, part of the faecal NFE arises by an underestimation of the organic matter associated with faecal nitrogen.

The detergent system (Van Soest method)

The principal obstacle in preparing plant cell wall residues where indigestible components are recovered has been to remove contaminating protein. Detergents form strong protein complexes that are soluble under proper conditions. Anionic detergents form polyanionic complexes, the sodium salts of which are soluble above pH 6.0. Interferences by heavy metal and alkaline earth metal ions are prevented by ethylenediamine-tetraacetic acid (EDTA). Consequently, extraction of forage with a neutral (pH 7.0) solution of sodium lauryl sulphate and EDTA will allow the preparation of cell wall recovering the major cell wall components, lignin, cellulose and hemicellulose. The residue also contains minor cell wall components including some protein and bound nitrogen, minerals and cuticle. Pectin are removed, although they are a cell wall component. The neutral - detergent extraction is non-hydrolytic and recovers the insoluble matrix. The loss of pectin and its relatively complete nutritive availability is evidence of its lack of covalent bonding with the lignified matrix.

The proximate analysis system of feed analysis has served for many years and continues to serve a very useful purpose in predicting the nutritive value of feeds. However, there are some definite limitations of this system, especially with respect to both the crude fibre and NFE fractions. The material that was dissolved by the solvents (NFE) was assumed to be digestible and the residue (crude fibre) was assumed to be indigestible. Later studies showed that in some cases crude fibre was more digestible than was NFE. Low digestibility of the NFE fraction arises from the fact that nearly all of the fibrous hemicellulose and a variable amount of the lignin are dissolved in the weak acid and alkali. Both fractions have been shown to be relatively indigestible. Thus, predicting nutritive value from the proximate analysis figures became a major problem, especially with the more fibrous feeds. As a result, numerous workers over the past several years have tested various procedures that might provide a more definitive separation of feed carbohydrate than does not the proximate analysis.

To develop a chemical procedure that fractionates forages into relatively digestible and indigestible portions, Van Soest, P. J. (1967) proposed the extraction scheme mentioned in this chapter.

The portion of the plant material that is soluble in neutral detergent, cell contents, is virtually completely digested by the ruminant. The residue from this fraction, cell wall constituents, is of low but variable availability depending upon the species of plant and its stage of maturity. A similar statement can be made about acid detergent fibre and cellulose. Lignin is nearly indigestible. The neutral detergent fibre (NDF) is representative of the fibrous cell wall constituents and contains lignin, cellulose, hemicellulose and some fibre bound protein. That part of the sample not appearing as residue is termed as neutral detergent soluble, represent the cellular contents and contains lipids, sugars, organic acids, non-protein nitrogen, pectin, soluble protein and other water soluble matter. Crude fibre contains most of the cellulose and only part of the lignin so that ADF values are about 30 per cent higher than those for crude fibre the same feeds.

In modified detergent method Van Soest et al (1991) has recommended amylase treatment of the feeds for NDF estimation.

The neutral detergent fibre differs from the previously developed acid detergent fibre. It represents the cellulose and lignin portion of the cell wall plants. The ADF analysis will be required for evaluating the quality of fibre for ruminants and other fibre utilizing herbivores. For non-ruminants with little fibre utilizing capacity, the neutral detergent fibre will be only fibre value required. Partitioning of feed /plant material in to different chemicals by different schemes.

CLASSIFICATION - 1
Plant Material

Cell Wall

1. Hemicellulose
2. Cellulose
3. Heat damaged protein
4. Lignin

Cellular Contents

1. Sugars, Soluble carbohydrates, starch
2. Pectin
3. Non-protein nitrogen
4. Protein
5. Lipids
6. Other soluble

CLASSIFICATION - 2
Feed digest with neutral detergent

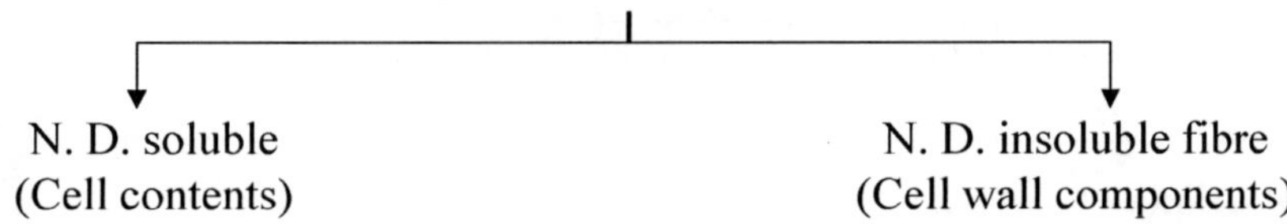

N. D. soluble
(Cell contents)

N. D. insoluble fibre
(Cell wall components)

Digest with acid detergent

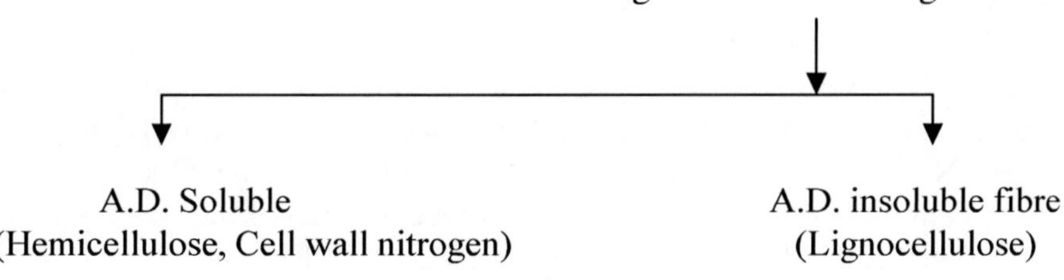

A.D. Soluble
(Hemicellulose, Cell wall nitrogen)

A.D. insoluble fibre
(Lignocellulose)

Digest with 72% H_2SO_4

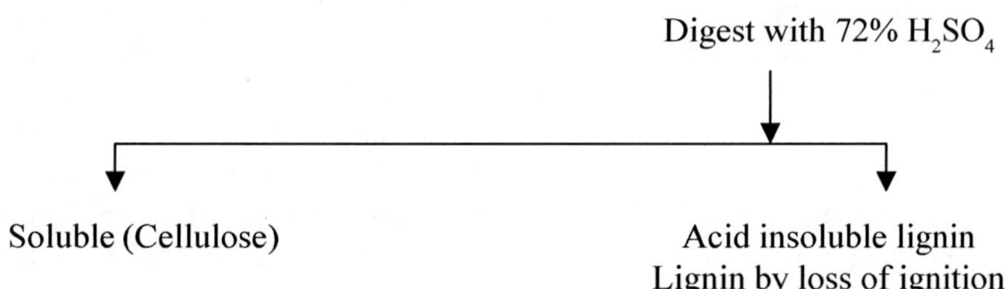

Soluble (Cellulose)

Acid insoluble lignin
Lignin by loss of ignition

CLASSIFICATION - 3

Basic scheme of Von Soest method (1967)

Fraction	Reagent	Treatment	Yield
Neutral dergent residue (NDR)	Na Lauryl suphate EDTA pH 7.0	Boil 1 hour	Plant cell wall less pectin
Acid detergent fibre (ADF)	Cetyl trimethyl ammonium bromide in I NH_2SO_4	Boil 1 hour	Lignocellulose + insoluble mineral
Unavailable nitrogen	Acid detergent	Kjeldahl N determination of ADF	Maillard products and lignified · N
Lignin	72% H_2SO_4 treatment on ADF	3h, 20°C	Crude lignin
Cellulose	None	Ash residue from lignin step	Loss in weight
Silica (SiO_2)	Conc. HBr (48%) treatment of ADF ash	Treat ash drop wise 1h, 25°C	Residue is SiO_2
Hemicellulose	None	Calculate as NDF - ADF	

Limitations

1. Although detergent fibre analysis is a valuable tool for evaluating the potential feed value of untreated agricultural by-products, it does not provide a good estimate of the effect of a treatment on improving the feed value of low quality roughages. Earlier studies illustrate that increasing levels of NaOH treatment had no effect on ADF or lignin content of straw, but did result in a constant decrease in neutral detergent fibre (NDF) content. The latter observation illustrates the effect of alkali treatment on solubilizing the hemicellulose fraction of the straw, which is defined as the difference between NDF and ADF.
2. Unfortunately, Van Soest's detergent techniques do not produce a short separation between cell content and cell walls nor between unlignified cell wall and lignified cell wall. Neutral detergent dissolves the major part of the proteins, which belong to the cell wall fraction. In the case of ruminant nutrition this is not serious because pectin can be fermented by the rumen microbes very easily, and further more the pectin content of grasses is low, but it can be high in legumes. More serious are filtration problems following treatment with neutral detergent, which may give technical difficulties, especially in the case of starchy foods.
3. The acid detergent treatment of neutral detergent fibre does not partition the unlignified and the lignified fractions, but rather produces fractions

containing the hemicellulose and cellulose and lignin as main constituents, respectively.

4. The complete Van Soest analysis, NDF, ADF and ADL, moreover is time consuming and cumbersome for routine purposes.
5. One of the difficulties of the NDF determination is the filtration after the detergent treatment, especially in case of starchy feeds. Several pre-treatment with enzymes (amylase, B-glucanase) had been tried, not always with complete success. Amylase resulted in a lower NDF recovery even in samples containing no starch. Good results were obtained with B-glucanase.

Cornell Net Carbohydrate Protein (CNCP) system

Ruminant utilize protein fractions of feed like, NPN, soluble peptide and soluble true protein as well as carbohydrate fractions like sugar, starch, non-starch polysaccharides and other structural carbohydrates at different rates and manner. Therefore, their estimation in animal feeds assumes significance. The development of the Cornell Net Carbohydrate and protein system (CNCPS) arose from the need to account for variation in forages, including environment and previous plans of nutrition effective dietary fibre, ruminally degradable carbohydrate and proteins, rate of digestion and passage and variable efficiency of use of absorbed energy and protein.

This is the recent method from Cornell University and seems to be most accurate method of evaluation that characterizes the carbohydrates and proteins of feeds and forage into different fractions as per their availability to the animals. This system characterizes the feeds into different protein (A, B_1, B_2, B_3 and C) and carbohydrate (Sugar, fructant; Fast degradation – starch pectin, oligosaccharides; slow degradable starch; Fast degradation fiber; and lignin) fractions. This method helps to rank the feeds based on carbohydrates and proteins degradability and thus guides in dietary formulation accordingly. This will assist in optimization and in efficient utilization of feeds/forages in nutrient feeding.

Conventional proximate analysis system of feeds and cell wall determination partially accounts for factors that influence animal requirements and feed utilization. The CNCP system meets these requirements as it characterizes the feeds and forages in to different protein and carbohydrate fractions. The protein fractions of feeds such as NPN content, soluble peptide, soluble protein as well as carbohydrate fractions like sugar content, starch, non-starch polysaccharides and other carbohydrates are utilized by different rumen microbes and animals differently, their estimation on feedstuffs is therefore important. The degradation rate of these stuffs and their passage rates are also one of the determinants of net energy and protein available from feeds. The CNCP system is an application model that uses a combination of mechanistic and empirical approaches and account to the effect of variation in animal factor as well as feed chemical entities.

The CNCPS (Chalupa *et al.* 1991 Sniffen *et al,* 1992) has developed the need for uniform procedures to partition feed nitrogen into A, B, C fraction (Pichard and Van Soest, 1977) while carbohydrate fractions are relatively standardized (based on NDF, ADF with corrections for ash, protein and lignin).

Protein and carbohydrate fraction of feed stuff.

Based on chemical constituent as well as potential degradation rate Pichard and Van Soest (1977) and Krishnamoorthy *et al,* (1982, 1983) fractionated feeds into various categories

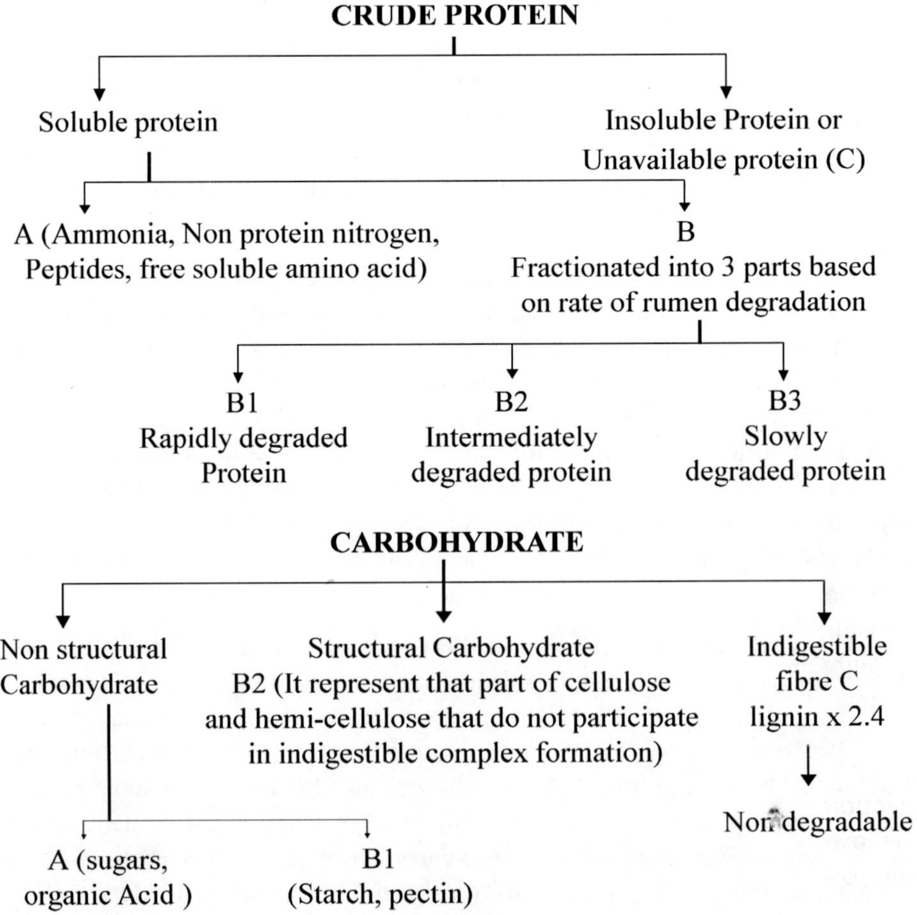

Fraction	Protein	Carbohydrate	Rumen Availability
A	Ammonia-N, Protein, soluble amino acids	Sugar, fructant, some starch	Soluble (generally highly available, 4% minimum to 2.0% maximum)
B1	Very rapid degrading protein, oligopeptides	Fast degrading starch pectin oligosaccharides	Insoluble, potential digestible 1-30 % hour
B2	Intermediate	Slow degrading starch	
B3 C	Slow degrading protein ADF bound protein	Fast degrading fibre Lignin	Non digestible (0 % hour)

Although this gives some useful indication of degradation rate, the various B fractions are based on arbitrary criteria and vary considerably.

Protein fractions

Crude protein (CP) contents of feeds/fodders can be fractionated into two categories namely, soluble and insoluble based on their solubility. The Cornell net carbohydrate and protein system partition CP of the forages into non-protein nitrogen (A), soluble CP (B1), insoluble CP that was acid detergent but soluble in neutral detergent (B2) and CP insoluble in acid detergent (C).

The soluble protein is further fractionated into fraction A and fraction B. Fraction A of protein is mostly composed of NPN, peptides and free amino acids. In forage most soluble CP is actually NPN but contain a considerable N which is true protein.

Fraction B of protein is further subdivided in to B1, B2 and B3 based on the degradation rate. Fraction B_1 is rapidly degraded in the rumen (Van Soest et al., 1981). Licetra et al. (1996) recommended use of borate-phosphate buffer for estimating soluble protein.

Fraction C of protein is the unavailable or bound protein fraction that is insoluble in the acid detergent (acid detergent insoluble protein, ADIP) fraction. This fraction (Fraction C) contains protein associated with lignin. Fraction B_3 is insoluble in neutral detergent but soluble in acid detergent [neutral detergent insoluble protein (NDIP) minus ADIP]. Fraction B_3 is slowly degraded in the rumen because it is associated with the cell wall. In the CNCPS, a high percentage of fractions B_3 escape degradation in the rumen. Buffer insoluble protein minus the protein insoluble in neutral detergent is used to estimate fraction B_2. Part of the B_2 is fermented in the rumen and remaining escapes to the lower gut. The fate of fraction B_2 depends on the relative rates of digestion and passage in GI tract.

Feed carbohydrate fractions

In CNCP system ccarbohydrates can be classified according to degradation rates, as with protein. Fraction A is fast and it is the sugar and organic acids, fraction B_1 is intermediate and is starch and soluble fibre, fraction B_2 is slow and is available cell wall (digestible fibre) and fraction C is unavailable cell wall (indigestible residue). These fractions are computed from feed content of non structural carbohydrate (NSC) and structural carbohydrate (SC) and indigestible fibre C. The fraction C is lignin x 2.4.

The NSC fraction represent the carbohydrates that are insoluble in neutral detergents and it can be estimated as 100 minus protein, NDF corrected for protein, lipids and ash in a feed stuff. The calculated NSC is usually in close agreement with direct measurements, but feed containing large amount of pectin will have a somewhat lower true NSC. Given proportion of starch and pectin in the NSC, the fraction A (Sugar and organic acids) can be determined by difference (Sniffen et al. 1992).

The CNCP system assumes that feedstuffs are composed of protein, carbohydrate, fat and ash. Protein and carbohydrate DM are further subdivided by chemical composition, physical characteristics, ruminal degradation and postiminal digestibility characteristics. NDF % of DM, lignin % of NSC, starch % of NSC are part of carbohydrate, while soluble % of CP, ADFIP% of CP, NPN% of SP, NDFIP% of CP are part of protein. Fractions of protein carbohydrate, ash fat in a feedstuff can be derived from the different method

1. NDF and lignin (Van Soest *et al.,* 1991)
2. Total nitrogen as assayed by macro and micro Kjeldahl (AOAC, 1980)
3. Soluble protein determined by the procedure of Krishnamoorthy *et al.,* 1983)
4. Nitrogen that is insoluble in neutral detergent and acid detergent (Van Soest *et al.,* 1991)
5. Ash (AOAC, 1980)
6. Solvent soluble fat (AOAC, 1980)
7. Determination of NSC from the determinations of NDF, protein, fat and ash (Van Soest *et al* (1991); 100 – [NDF- NDF protein + fat+ ash]
8. Starch by method of AOAC (1995)
9. NPN, soluble protein, NDFIN and ADFIN as per Licitra *et al.* (1996)

Estimation of Non Protein Nitrogen:

Non protein nitrogen (NPN), soluble protein, Neutral detergent insoluble nitrogen (NDFIN) and acid detergent insoluble nitrogen (ADFIN) were estimated as per Licitra *et al.* (1996).

Apparatus

Erlenmeyer flask (125ml), What man filter paper 54 or 541, analytical balance, pH meter, filter funnels, Kjeldahl apparatus.

Reagents:

1. Sodium tungastate ($Na_2 WO_4 2H_2O$) (100 g/ L) solution in water, 0.30 M
2. 0.5 M sulphuric acid (H_2SO_4).

Procedure:

Put about 0.5 g of dry ground sample into a 125 ml Erlenmeyer flask having 50 ml of cold distilled water. To it add 8 ml of 10 percent sodium tungstate solution. Keep flask at room temperature for 30 minutes. Adjust pH of the flask content to 2 with 0.5M sulphuric acid. Keep the flasks over night at room temperature. Filter the contents through Whatman filter paper 54 or 541. Wash the residue of filter paper twice with cold distilled water. Transfer this filter paper to kjeldahl flask and determine nitrogen Micro-kjeldahl procedure. Estimate NPN of the sample by subtracting residual nitrogen from total nitrogen.

SOLUBLE NITROGEN AND PROTEIN

Apparatus:

One Erlenmeyer flask (125 ml) Whatman filter paper 54 or 541 analytical balance, water bath, vacuum source, filter manifold fitted with conical funnels (50 ml), Kjeldahl apparatus.

Reagent:

1) Borate-phosphate buffer, pH 6.7-6.8 consisting mono sodium phosphate ($Na_2PO_4 . H_2O$) 12.20 g/l; sodium tetra borate ($Na_2B4O_7 .10H_2O$) 8.91 g/l and tertiary butyl alcohol 100 ml/l
2) Sodium azide to 10 percent solution freshly prepared.

Procedure:

Weigh approximately 0.5 gm ground sample and put into 125 ml Erlenmeyer flask. To the flask add 50ml borate-phosphate buffer. Then add one ml of sodium azide solution to the flask. Allow the flask to stand at room temperature for 3 hours. Filter the contents through Whatman filter paper 54 or 541 using mild vacuum. Wash the residue left after filtration with 250 ml cold distilled water. Estimate nitrogen in residue by Kjeldahl method. The N estimated in residue gives insoluble protein fraction. Calculate soluble protein by difference from total CP. The soluble true protein is obtained by subtracting the NPN estimated by tungstic acid procedure.

DETERMINATION OF ACID DETERGENT INSOLUBLE NITROGEN (ADIN)

Regents:

1) Acid-detergent solution.
2) Acetone.

Procedure:

For the estimation of ADIP, procedure of ADF estimation is used (Van Soest, 1973). For this 2 g of sample utilized for ADF refluxion is transferred to a Kjeldahl flask. Estimate the nitrogen in residue following the standard Kjeldahl procedure. Express ADIN of sample as percent of total nitrogen or as N x 6.25.

DETERMINATION OF NEUTRAL DETERGENT INSOLUBLE NITROGEN (NDIN)

The procedure is exactly same as that for acid detergent insoluble nitrogen except that neutral detergent solution was substituted for acid-detergent reagent (Van Soest *et al.*, 1991). Use of sodium sulphite is omitted. Pre-treatment with amylase is not done, although amylase may be used in the boiling step to aid filtration.

STARCH ESTIMATION

The starch is estimated as per the method of AOAC (1995)

Reagents:

$CuSO_4$ Solution: Dissolve 34.639 g of 34.639g of $CuSO_4.5H_2O$ in water and make volume up to 500 ml and then filter through filter paper.

Alkaline Tartrate Solution

Dissolve 173 g of potassium sodium tartrate, $4H_2O$ (Rochelle salt) and 50 g NaOH in the H_2O and dilute to 500 ml. Allow solution to stand for 2 days and then filter.

Potassium Permagnate Standard Solution:

Prepare approximately, 0.1573 N and containing 4.988 g/l and standardize with 0.35 g Na oxalate solution.

Ferric Sulphate Solution

Dissolve 135 g $FeNH_4 (SO_4)_2$ 12 H_2O in H_2O and dilute to 1 liter. Acidify fifty ml Fe $(SO_4)_3$ solution with 20 ml 4 N H_2SO_4. Titrate with $KMnO_4$ solution and use this titre as '0' point correction.

Ferrous Phenanthrelene Indicator:

Dissolve 0.7425 g O-phenanthrelene. H_2O in 25 ml, 0.025 M $FeSO_4$ solution (6.95 g $FeSO_4.7H_2O$)

Procedure:

Stir about 2.5-3 g dry material for 1 hr in 250 ml beaker with 50 ml cold water. T Filter the contents and wash with 250 ml cold water. Heat the soluble residue for 2.5 hour with 200 ml water and 20 ml HCl (Specific gravity 1.13) in a flask provided with reflux condenser. Cool the flask contents and nearly neutralized with NaOH. Transfer to 250 ml volumetric flask, dilute to mark and filter. Weight of dextrose obtained x 0.925 = weight of starch. Add 25 ml each of $CuSO_4$ and alkaline tatrate solution to 400 ml beaker (alkaline resistant glass) and then add 50 ml reducing sugar solution. Heat the beaker on a pre-adjusted electric heater so that boiling begins in 4 minutes and continue boiling for two minute. Cover the beaker with watch glass during heating. Filter hot solution through sintered glass crucible using suction. Wash the precipitates of Cu_2O thoroughly with water and determine the amount of reduced copper by following method.

Transfer the crucible in the sample with glass rod and add 50 ml $Fe_2(SO_4)_3$ solution. Stir vigorously until Cu_2O is completely dissolved. Remove the crucible after making clear solution. Add 20ml, 4N H_2SO_4 solution and titrate with standard KMnO4. As end point approaches, add one drop of ferrous phenanthelene indicator. At the end point, brownish solution changes to green. The weight is reducing sugar equivalent to weight of copper obtained from the standard table (AOAC, 1995).

The equations that are used to estimate the different fractions of protein and carbohydrate of feeds are as:

CALCULATION

NON-STRUCTURAL CARBOHYDRATE (NSC)

NSC = 100-[(NDF-NDIP) + protein + fat +ash)]

Protein Fraction

PA (%CP) = NPN (%SOLP). 0.01. SOLP (%CP)

PB1 (%CP) = SOLP (%CP) – A (%CP)

PC (%) = ADIP (%CP)

PB3 (%CP) = NDIP (%CP) –ADIP (%CP)

PB2 (%CP) = 100-A (%CP)-B1 (CP)-B3 (%CP) - C (%CP)

Where, CP (%DM) = Percentage of crude protein of the feedstuff;

NPN (%CP) = Percentage of crude protein of the feedstuff that is non–protein nitrogen x 6.25; SOLP (%CP) = Percentage of crude protein of the feedstuff that is soluble protein; NDIP (%DM) = Percentage of the feedstuff that is neutral detergent insoluble protein;

ADIP (%DM) = Percentage of the feedstuff that is acid detergent insoluble protein.

PA (%CP) = Percentage of crude protein in the feedstuff that is non-protein nitrogen;

PB1 (%CP) = Percentage of crude protein in the feedstuffs that is rapidly degraded protein.

PB2 (%CP) = Percentage of crude protein in the feedstuffs that is intermediately degraded protein.

PB3 (%CP) = Percentage of crude protein in the feedstuffs that is slowly degraded protein.

PC (%CP) = Percentage of crude protein in the feedstuffs that is bound protein.

Carbohydrate Fractions

CHO (%DM) = 100-CP (%CP)-FAT (%DM)- ASH (%DM)

CC (%CHO) = 100. (NDF (%DM). 0.01. LIGNIN (%NDF). 2.4)/ CHO (%DM)

CB2 = 100. {NDF (%DM) – NDIP (%CP). 0.01. CP (%DM) – NDF (%DM). 0.01. LIGNIN (%NDF). 2.4/ CHO (%DM)

CNSC (%CHO)	=	100 – B2 (%CHO) – C (%CHO)
CB1 (%CHO)	=	STARCH (%NSC). {100 – B2 (%CHO) – C (%CHO)}/100
CA (%CHO)	=	{100-STARCH (%NSC)}. (100 – B2 (% CHO) – C (%CHO)}/ 100

Where, CP (%DM) = Percentage of crude protein of feedstuff; CHO (%DM) = Percentage of carbohydrate of the feedstuff; FAT (%DM) = Percentage of fat of the feedstuff; ASH (%DM) = Percentage of ash of the feedstuff; NDF (%DM) = Percentage of ash of the feedstuff; NDF (%DM) = Percentage of the feedstuff that is neutral detergent fiber; NDIP (%DM) = Percentage of neutral detergent insoluble protein of the feedstuff; feedstuff; LIGNIN (%NDF) = Percentage of lignin of the feedstuff NDF; STARCH (%NSC) = Percentage of starch in the non-structural carbohydrate of the feedstuff; SUGAR (%NSC) = Percentage of sugar in the non-structural carbohydrate of the feedstuff; CA (%CHO) = Percentage of carbohydrate of the feedstuff, that is sugar; CB1 (%CHO) = Percentage of carbohydrate of the feedstuff that is starch + NSP;CB2 (%CHO) = Percentage of carbohydrate of the feedstuff that is available fibre; and CC (%CHO) = Percentage of carbohydrate in the feedstuff that is unavailable fibre.

In vitro Dry matter digestibility

Enzymatic methods

One-stage methods consist of incubating a feed sample (generally ground through a 1mm screen) for 48 to 72 h in a buffer solution containing cellulase. The solution is filtered and the residue is weighed. Disappearance of DM is generally substantially less than either in vitro or in vivo digestibility. The relatively low disappearance values obtained from one-stage methods led to the development of two-stage procedures. Treatment of samples with pepsin in HCl prior to cellulase treatment increased disappearance of DM as compared to cellulase treatment alone. Cellulase digestion followed by pepsin digestion did not increase DM disappearance as much as did pretreatment with pepsin. Generally, pretreatment with HCl-pepsin involves incubating the samples in a mixture of HCl (approximately 0.1 N) and pepsin (0.2% wt/ vol) for 24 to 48 h at approximately 40 to 50 °C. The pepsin solution is removed and cellulase added. Extracting samples with ND solution prior to cellulase treatment also increased DM disappearance as compared to no pretreatment. Pretreatment involves incubating the sample in hot, not boiling, ND solution for about 1 h at 110 °C. The ND solution is removed and cellulase solution is added. For both pepsin and ND pretreatment, cellulase incubation usually lasts 24 to 48 h.

Pretreatment of forages prior to cellulase treatment probably increases the access of the enzyme to cell wall components, perhaps by solubilizing some hemicellulose (pepsin-HCl treatment), removing interfering compounds such as proteins and starch, or by disrupting the structure of the cell wall. The cellulase enzyme technique for estimation of dry matter digestibility of grasses was given by Jones and Hay (1973).

Tilley and Terry method (Two stage technique)

This technique to estimate the dry matter digestibility of feeds/forages was proposed by Tlley and Terry in 1963. It involves two stages. In the first stage the ground sample of feed is incubated for 48 hours with buffered rumen liquor under anaerobic conditions, while in the secod stage the microbial fermentation is stopped by acidifying with hydrochloric acid to pH 2.0 and then digested by incubating with pepsin for another 24 hours. The incubated material is filtered and the loss in dry matter of the feed samples is taken as in vitro dry matter digestibility.

A weighed quantity of the forage sample is added to 100 ml centrifuge tube containing buffer and rumen liquor while CO_2 is being passed into the solution. The tubes are then placed in the incubator at 39°C for 48 hours. After completion of incubation the reaction is terminated by addition of saturated mercuric chloride solution. The supernatant is discarded after the centrifuging the tubes at 2000 rpm for 15 minutes and acid pepsin solution is added. There are again incubated for another 48 hours at 39°C. Then the content is filtered through a sintered crucible, residue is weighed to constant weight after drying the crucible at 100°C for 24 hours. A blank containing all the reagents but the sample, following the same procedure, is also run during each trail.

Gas Production Technique (Pressure Transducer)

This technique was proposed by Theodorou *et al* (1994) to estimate the digestion kinetics of feeds and forages. The principle of this technique is alike to the Menke gas production procedure where ground substrates (feed samples), anaerobic media and rumen fluid medium are source of inoculum. The difference is that here feed material is incubated in gas tight culture bottles, the gas produced during incubation is accumulated in the head space. A pressure transducer connected to a digital read out volumeter and a gas tight syringe assembly is used to measure release and accumulated gas pressure from the incubated culture bottles. By repeating the "gas-measurements + gas release" procedure at regular intervals, it is possible to construct gas accumulation profiles for feeds. The rate and extent of fermentation can also be estimated by this technique.

In this method about 1 gm of ground samples (1 mm sieve) is put in culture bottles containing an anaerobic medium (consist of micro minerals 0.1 ml, buffer 200 ml, macro minerals, 200 ml, resazurin 1 ml and distilled water 500 ml). The bottles are incubated with 10 ml of rumen inoculum and than the bottles are incubated at 39^{0}C for stipulated periods of incubation (2 h to 116 or hour/ less depending upon the feed characteristics). The gas is measured at different hours. At the end of incubation period, the substrate is filtered through the pre weighed Gooch crucible and less in DM is expressed a DM degradability of feed at that particular time.

Nylon bag technique

This in situ technique was given by Mehrez and Orskove (1977) to determine the rate and extent of feeds and forages degradability. This is the most appropriate method that provides information on the nutritive value of feed for ruminants and the efficiency of rumen ecosystem. This method also provides the information on both carbohydrate and protein status of feed and the degree to which it will be utilized/ digested in the rumen or escape to the intestine.

In this technique a weighed quantity (2-5 g) of substrate (feed sample) is put in a Dacron/nylon bag (length 12-14 cm, mesh size 20-40 micron), insuring that the bag is tightly closed and then put the bag in the rumen of a fistulated animal for a particular period. After a stipulated period of incubation bags are removed and loss in dry matter of substrate is calculated as the digestibility/ degradability of the feed. By this method digestion of feed along with rate and extent of digestion can be measured. The limitation of this technique is that the sample is not mixed with saliva, neither chewed nor ruminated and also there is no post ruminal digestion. Sample size, substrate particle size, bag length and its mesh size, inoculum donor, position of suspended bag in rumen, length of cord tied with etc are some of the factors that governs the digestibility/ degradability of feeds.

Near infra red reflectance spectroscopy

In spectrometer quantitative determinations, we run a set of standards containing various concentration of the analyte for example protein, then compare the results of unknown with the standard results using a standard curve. In such cases, the assumption is made that only the concentration of interested analyte is changing and the absorbance changes linearly with the changing concentration of the analyte according to Beer's law that states:

A = abc

Where:

A = absorbance or (log10 C 1/ transmittance)

b = the cell path length

c = concentration

a = molar extinction co-efficient for the analyte at the wavelength used

in such determinations, the wavelength used are generally in the ultraviolet (uv) or visible portion of the electromagnetic spectrum (EMS)and the wavelength needed for each determination is known.

The near infrared (NIR, 1100-2500 nm) portion of EMS occur at wavelength larger than visible (VIS, 400-700nm) and shorter than the mid-infrared (MIDIR, 2500-25000 nm) regions. It means the near infrared region of the EMS lies between the visible and infrared region and is normally defined by the wavelength range between 1100 and 2500 nm.

Near infrared spectroscopy (NIRS) is based on the absorption of C-H, N-H and O-H groups found in organic constituents. These absorptions are the overtones and combination bands of much stronger absorptions found in the MIDIR spherical region. Since NIR absorptions are weaker, longer path lengths can be used for liquids (particularly where water is involved) and for solids unlike the MIDIR, dilutions with KBr or prepration of KBr pellets are unnecessary.

NIRS involves the use of an optical device and computers to gather spectra data from a calibration set of samples and relate their reflectance spectra to known composition measured in traditional ways. These relationships are then used to relate the reflectance observations (readings) obtained for individual foods with their compositions. This technique is now in use in many laboratories for analysis of chemical constituents (nutrient concentration) of feeds and fodders.

Bibliography

AOAC (1980). Official Methods of Analysis 13th Edition Association of Official Analytical Chemists Washington, DC: 1094

AOAC (1995). Official Methods of Analysis (16th Ed). Association of Official Analytical Chemists, Arlington VA

Chalupa, W. and Sniffen, C.J. (1996). Protein and amino acid nutrition of lactating dairy cattle today and tomorrow. Animal Feed Science and Technology, 58: 65

D' Mello, J. P. F. (Ed) 2000. Farm animal Nutrition and Metabolism CABI Publishing.

Fahey jr., G. C; Collin, M; Mertens, D. R and Moser, L. E. (Eds.) 1994. .Forage quality evaluation and utilization. CABI Publishing.

Fox, D.G., Barry, M.C., Pitt, R.E., Roseler, D.K. and Stone, W.C. (1995). Application of the Cornell net carbohydrate and protein model for cattle consuming forages. Journal of Animal Science, 73: 267-277

Fox, D.G., Sniffen, C.J., O'Connor, J.D., Russel, J.B. and Van Soest, P.J. (1992). A net carbohydrate and protein system for evaluating cattle diets. III. Cattle requirements and diet adequacy. Journal of Animal Science, 70: 3578-3596

Givens, D.I. Owen, E; Axford, R.F.E. and Omed, H.M. (eds) (2002). Forage Evaluation in Ruminant Nutrition. CABI Publishing.

Goering, H.K. and Van Soest, P.J. (1970). Forage Fiber Analyses (apparatus, reagents, procedures and some applications). Agriculture Handbook No 379. ARS, USDA Washington DC

Jones, D. I. H. and Hayward, M. V. 1973. A cellulase digestion technique for predicting dry matter digestibility of grasses. J. Science Food and Agriculture 24: 1419-1426

Krishnamoorthy, U., Sniffen, C.J. and Van Soest, P.J. (1982). Nitrogen fractions in selected feedstuffs. Journal of Dairy Science, 65: 217

Krishnamoorthy, U., Sniffen, C.J., Stern, M.D. and Van Soest, P.J. (1983). Evaluation of a mathematical model of rumen digestion and in vitro simulation of rumen proteolysis to estimate rumen undegraded nitrogen content of feedstuffs. British Journal of Nutrition, 50: 555-568

Licitra. G., Harnandez, T.M. and Van Soest, P.J. (1996). Standardizations of procedures for nitrogen fractionation of ruminant feeds. Animal Feed Science and Technology, 57: 347-358

Mehrez , A. Z. and Orskove, E. R. 1977. A study of the artificial bag technique for determining the digestibility of feeds in rumen. J Agric. Science (Cambridge) 93: 217-222

Pichard, D.G.R. and Van Soest, P.J. (1977). Protein solubility of ruminal feeds. In Proc, Cornell Nutrition Conference: 91-98

Sniffen, C.J., O'Connor, J.D., Van Soest, P.J., Fox, D.G and Russel, J.B. (1992). A net carbohydrate and protein system for evaluating cattle diets II carbohydrate and protein availability. Journal of Animal Science, 70: 3562-3577

Theodorou, M. K., Williams, B. A., Dhanoa, M. S., Mc Allan, A. B. and France, J. 1994. A simple gas production method using pressure transducer to determine the fermentation kinetics of ruminant feeds. Animal Feed Science and Technology 48: 185-197.

Tilley, J. M. A. and Terry, R. A. 1963. A two-stage technique for the in vitro digestion of forage crops. J. British Grassl. Soc. 18: 104

Van Soest, P.J. (1963). The use of detergents in the analysis of fibrous feeds II. A rapid method for the determination of fiber and lignin. Journal of Association Official Analytical Chemistry, 46: 829-835

Van Soest, P.J., Robertson, J.B. and Lewis, B.A. (1991). Symposium: Carbohydrate methodology, metabolism and nutritional implications in dairy cattle. Methods for dietary fiber, neutral detergent fiber and non starch polysaccharides in relation to Animal Nutrition. Journal of Dairy Science, 74: 3583-3597

Van Soest, P.J., Robertson, J.B. and Lewis, B.A. (1991). Symposium: Carbohydrate methodology, metabolism and nutritional implications in dairy cattle. Methods for dietary fiber, neutral detergent fiber and non starch polysaccharides in relation to Animal Nutrition. Journal of Dairy Science, 74: 3583-3597

Van Soest, P.J., Sniffen, C.J., Mertens, D.R., Fox, D.G., Robinson, P.H. and Krishnamoorthy, U. (1981). A net protein system for cattle: The rumen sub model for nitrogen. (Eds Owens F.N.). Protein Requirements for Cattle. Proceedings of an International Symposium, Mp-109: 265. Division of Agriculture, Oklahoma State University, Stillwater

Van Soest, P.J., Sniffen, C.J., Mertens, D.R., Fox, D.G., Robinson, P.H. and Krishnamoorthy, U. (1981). A net protein system for cattle: The rumen sub model for nitrogen. (Eds Owens F.N.). Protein Requirements for Cattle. Proceedings of an International Symposium, Mp-109: 265. Division of Agriculture, Oklahoma State University, Stillwater

Vansoest, P.J. (1967). Development of a comprehensive system of feed evaluation and its application to forages. J Animal Science 26: 119-128

Vansoest, P.J. (1973). Collaborative study of acid-detergent fibre and lignin. Journal of Association of Official Analytical Chemist, 56: 781.

CARBOHYDRATES AND THEIR UTILIZATION

The name carbohydrates (carbo = carbon; hydrate = hydrogen and oxygen) is derived from the fact that these compounds contain carbon, hydrogen, and oxygen. Hydrogen and oxygen are nearly always present in the ratio in which they are found as in water (2:1, H_2O). Sugars are the first physiological product of photosynthesis and thus represent the basis for only renewable source of food energy. The photosynthesis reaction may be represented as follows:

$$6\ CO_2 + 6\ H_2O + 673\ \text{kcal} \xrightarrow[\text{Chlorophyll}]{\text{Solar energy}} C_6H_{12}O_6 + 6\ O_2$$

Carbohydrates may be defined as polyhydroxy aldehydes or ketones, alcohols or acids or as substances that yield one of their compounds on hydrolysis.

TYPES

The carbohydrates are usually classified as monosaccharides, oligosaccharides, polysaccharides and complex carbohydrates based on the multiples of five or six carbon sugars contained in the molecules.

A. MONOSACCHARIDES

The hexoses (six-carbon sugars) and pentoses (five carbon sugars) are the most important monosaccharides. The most important hexoses are glucose, fructose, galactose and mannose. Some important pentoses are ribose, xylose and arabinose.

1. HEXOSES

a. **Glucose:** Also called dextrose or grape sugar, is surpassed in sweetness only by fructose and sucrose. It is the sugar found normally in blood, in non-ruminant at a level of 70-100mg/dl, with higher level present in case of diabetes mellitus and lower level due to excessive insulin. In ruminant, the concentration of blood glucose ranges from 40 – 80 mg/dl. Glucose is found in nature, normally in combination with other compounds.

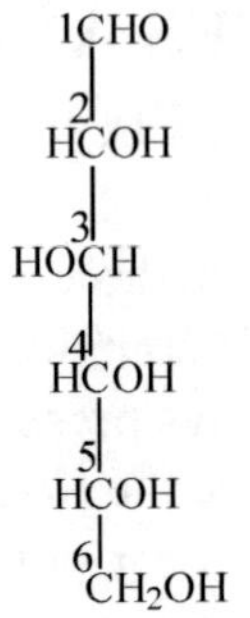

D-Glucose	L - Glucose
1CHO	1CHO
2 HCOH	2 HOCH
3 HOCH	3 HCOH
4 HCOH	4 HOCH
5 HCOH	5 HOCH
6 CH_2OH	6 CH_2OH

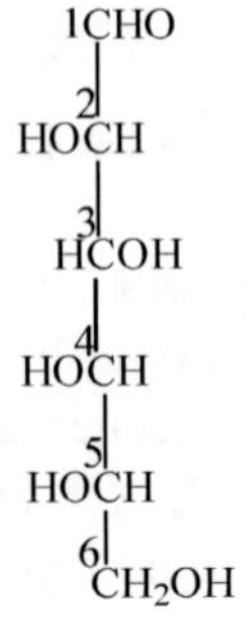

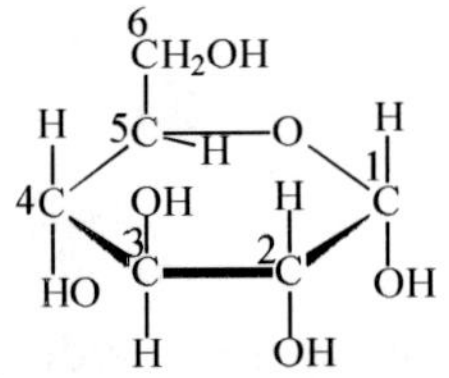

α - D- Glucose 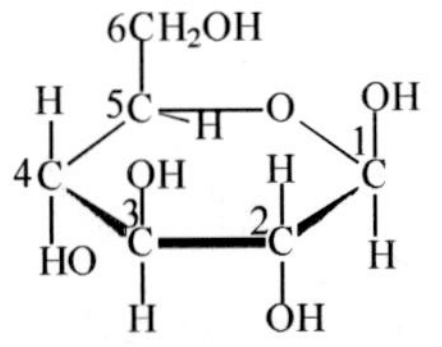 α-D- Glucose

b. **Fructose:** Also called levulose or fruit sugar, is sweetest sugar. In combination with glucose, fructose forms sucrose, which is found widely in nature. It is now possible to convert corn-derived glucose enzymatically to fructose. Many soft drinks use fructose instead of sucrose as a sweetener. The exceptionally sweet taste of honey is due to this sugar.

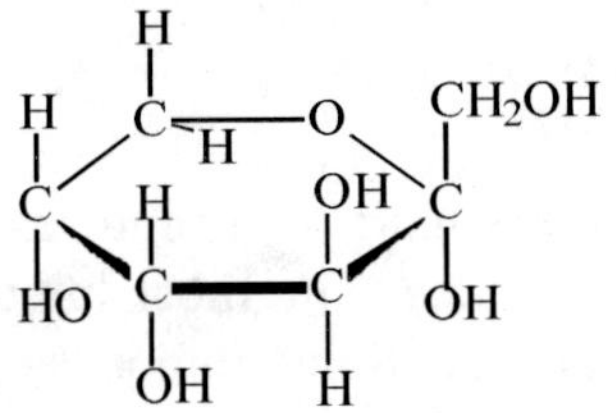

α – D- Fructose

c. **Galactose:** When lactose or milk sugar is hydrolyzed, galactose is formed. In nervous tissue it is found in combination with lipids. The only difference between glucose and galactose is the spatial relationship of the oxygen bridge to the rest of molecule. D-galactose is epimer of D-glucose, because these compounds differ from each other by their configuration on a single asymmetric carbon atom.

d. **Mannose:** This is an aldohexose occurring in a group of polysaccharides. It is apparently associated with some animal proteins. This is also epimer of glucose.

2. PENTOSES

The pentoses, or five carbon sugars, are distributed widely in nature, occurring in both plant and animal tissues, usually as components of a much larger molecule. Acid hydrolysis of complex pentosans yields primarily arabinose and xylose. Ribose and deoxyribose are present in nucleic acid; ribose is a part of riboflavin. Pentoses polysaccharides (hemicelluloses) are important constituent of the diet of herbivorous animals.

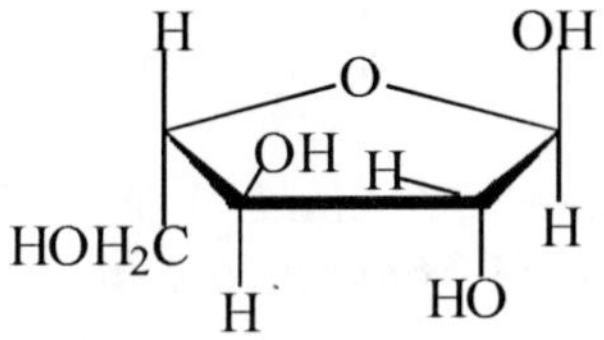

Arabinose 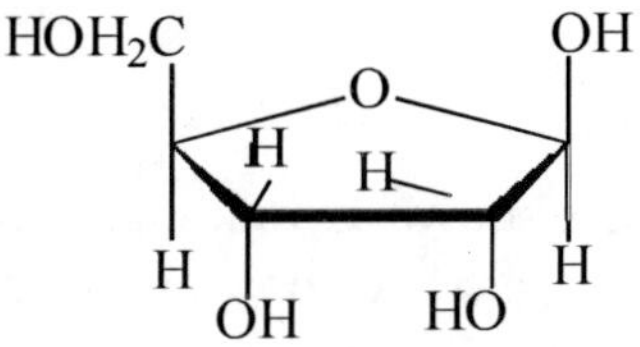 Ribose

B. OLIGOSACCHARIDES

They yield 2-10 monosaccharide molecules on hydrolysis

1. Disaccharides

Most disaccharides are sugars with general formula $C_{12}H_{22}O_{11}$, which upon hydrolysis yield 2 moles of hexoses.

a. Sucrose

Sucrose is distributed widely in nature (sugar cane, sugar beets, sugar millets and some maples). Neither salivary nor gastric enzymes will hydrolyze sucrose, but intestinal enzymes convert it in to glucose and fructose.

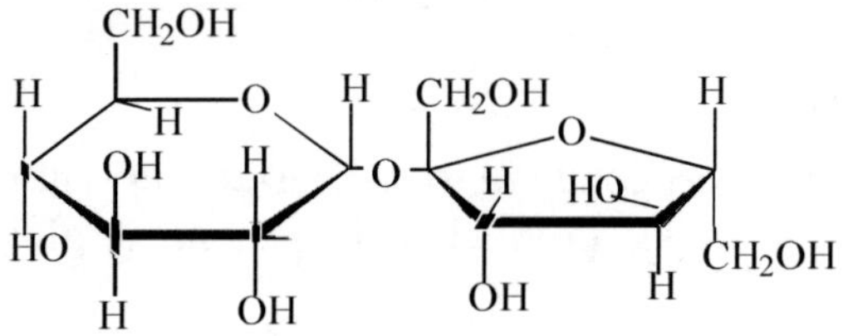

Sucrose

b. Maltose

Two glucose molecules are produced upon hydrolysis. Maltose is produced from starch by action of digestive enzymes. It doest not occur in free state but occurs in the animals as principle sugar formed by the enzymatic (ptyalin) hydrolysis of starch.

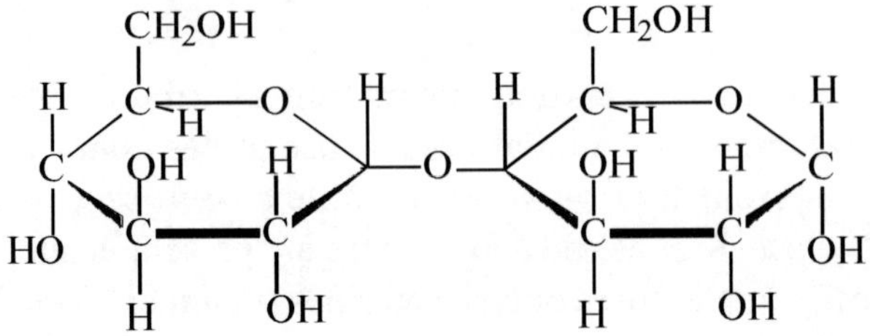

Maltose

c. Lactose

It is known as milk sugar because it constitutes approximately 6% of human milk and 4-5% of cow and goat milk. It is formed from glucose in mammary gland and does not occur in nature except as a product of this gland. The hydrolysis of lactose by lactase yields 1 mole each of glucose and galactose. Lactose promotes acidity in the small intestine by facilitating the growth of acidophilus organisms, and its presence enhances the intestinal absorption of calcium, magnesium, and the alkaline earth metals. Although older animals, whether mammalian or avian, cannot completely digest lactose, it is considered as the choice of carbohydrate for use in diets of young mammals.

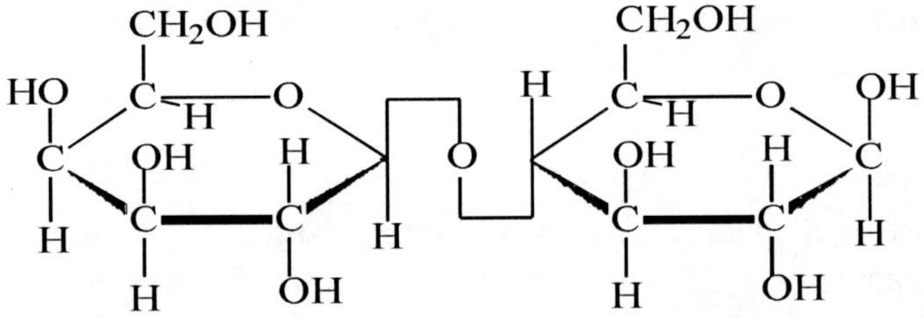

Lactose

2. Trisaccharides

Raffinose is the most common trisaccharides present in, small amount in sugar beet. The hydrolysis product of this sugar is glucose, fructose and galactose.

C. POLYSACCHARIDES

This group of the carbohydrate family represents higher molecular weight complex combination of primary carbohydrates, is the polymer of monosaccharides. These are either insoluble in water or at best, capable of forming colloidal suspensions.

1. STARCH

It is found in most plants, particularly in the seeds. Starch on hydrolysis yields glucose. Starch is distributed widely in nature and supply readily digestible energy in most livestock feeding systems, as exemplified by corn, wheat, oat, barley, and milo. Starch consists of two components, amylose (10 – 20%) and amylopectin (80 – 90%). Amylose consists of D-glucose units linked in a linear fraction by μ 1 – 4 linkages. On the other hand amylopectin is a branched polysaccharide having primarily μ 1-4 linkages and a small proportion may be linked by μ-1-6 linkages. Maltose and isomaltose respectively are produced by them during hydrolysis.

$$\text{Starch} \xrightarrow{\text{Amylase}} \text{Dextrins} \xrightarrow{\text{Maltase}} \text{Glucose}$$

2. GLYCOGEN

Glycogen is the storage form of carbohydrate in animal, hence it is called animal starch. It is stored primarily in liver but also in measurable quantities in muscle. Storage in animal body is facilitated by rest plus liberal feeding. However, stored glycogen is used rapidly when the body is exercised or at time of stress such as fright. It is similar to amylopectin in structure, but more branched.

3. CELLULOSE

Cellulose has been shown to consist of a large number of beta-glucose units joined together in a non-branched chain by 1, 4 – glycosidic bonds. Such bonds cannot be broken by any enzyme secreted by mammalian tissue. Cellulose is perhaps the most abundant organic substance, and is more efficiently digested by herbivores than carnivores. The cellulose form the fundamental structure of plant cell walls. It is composed of 4-linked glucopyranosyl units and has a wide molecular weight distribution. The building blocks of cellulose is D-glucose and the complete hydrolysis of cellulose yield D-glucose, however, cellobiose is the product when it is partially hydrolysed.

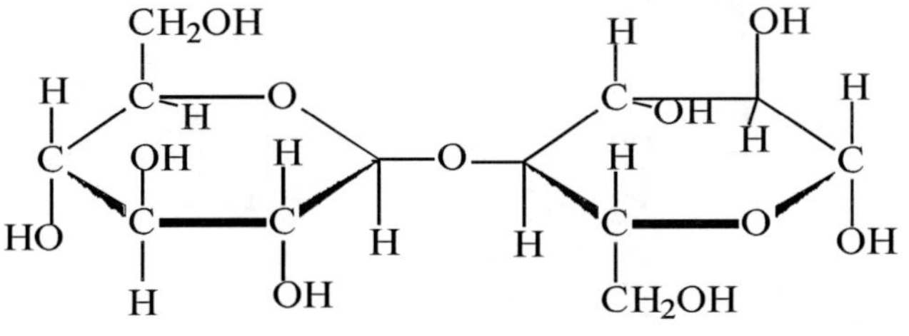

Cellobiose

4. LIGNIN

Lignin is not a carbohydrate. It is a collective term which embraces a whole series of closely related compounds. Generally, the lignin is said to lower the digestibility of cellulose and other products probably through complex linkages. Grains and oil meals contain very little lignin, where as straw may contain 8% or more. Lignin content of legumes is comparatively higher and the digestibility also remains higher. The percentage of lignin in a plant increases as the plant matures, probably to provide rigidity to plant stems for a reasonable period of time after seeds have matured. The rapid increase in lignin content with maturity points to the issue that harvesting of plant materials at a stage somewhat before full maturity so that its digestibility remains high. Lignin is a polymer that originates from three derivatives of phenyl propane; coumaryl, coniferyl and sinapyl alcohols.

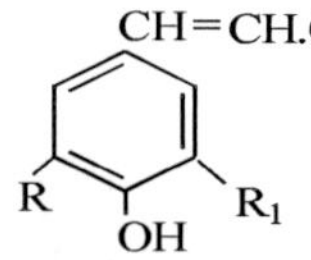

$CH{=}CH.CH_2OH$

R R_1

OH

Coumaryl alcohol, where R = R1 = H; Coniferyl alcohol, where R = H, R1 = OCH3 and Sinapyl alcohol, where R = R1 = OCH3

5. HEMICELLULOSE

These are defined as alkali-soluble cell wall polysaccharides and remain in close association with cellulose. Structurally, hemicelluloses from grasses are composed of a main chain of xylans ß-(1-4) linked D-xylose units with side chains of methylglucuronic acid and frequently glucose, galactose and arabinose.

D. COMPLEX CARBOHYDRATES

1. **Glycoproteins:** It is complex carbohydrate containing one or more heteroglycans. In general the heteroglycans are glucosamine or galactosamine or both, in addition galactose and mannose may be present.
2. **Glycolipids:** Glycolipids are the compound where two alcohol groups are esterified by fatty acids and the other is linked to a sugar residue. In general galactose is present as sugar and called galactolipids. In grasses and clovers about 60% of lipids present as galactolipids.

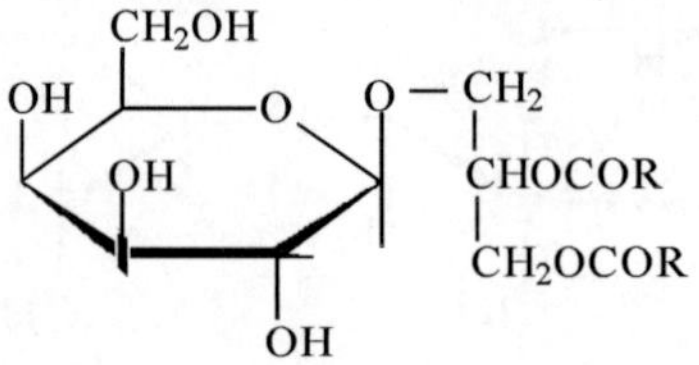

CH_2OH

OH O O — CH_2

OH CHOCOR

CH_2OCOR

OH

Galactolipid

DIGESTION

A. Nitrogen – Free Extract

Nitrogen free extract (NFE), or the more soluble fraction, is broken down into six carbon sugars primarily glucose, for absorption across the wall of small intestine into the blood stream. Some salivary breakdown of starch to maltose occurs in monogastric animals whose saliva contains amylase. This does not occur in ruminants, since their saliva contain no amylase.

The major portion of the breakdown of the more soluble carbohydrate or NFE occurs in the small intestine. Pancreatic juice contains a starch-splitting enzyme, similar to ptyalin found in human saliva, called amylopepsin or pancreatic amylase, which is responsible for the hydrolysis of starch to maltose. Sucrase, maltase, and lactase are found in the intestine and are capable of hydrolyzing the respective disaccharides i.e. sucrose, maltose and lactose to six-carbon sugars. In older animals, sucrase and maltase are present in much larger quantities than lactase, whereas lactase is quite evident in young milk consuming animals. As the animal matures, lactase activity declines.

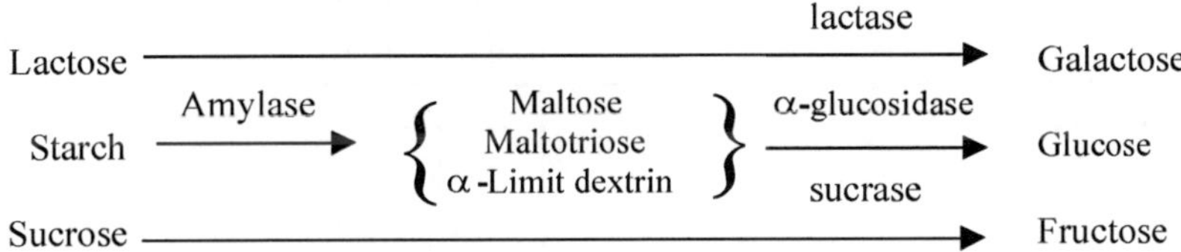

Fig.1 Digestion and absorption of carbohydrates

B. Cellulose and Crude Fiber

Cellulose and crude fiber digestion is confined largely to the ruminant animal. Essentially, the microscopic organisms (Table 1& 2) that inhabit in the rumen and reticulum of the ruminants and the cacum and colon of the horse, secrets enzymes capable of hydrolyzing the glycosidic bonds, holding cellulose molecules together, although none of these enzymes has a marked effect on lignin. Thus, cellulose, which is associated with higher level of lignin, as in wood, is not acted on to any marked degree. The large intestine of many animals contains bacteria capable of aiding in cellulose breakdown. However, since this area is beyond the area of nutrient absorption, such products are not utilized by the animal. The rabbit, however, consumes its portion of faeces and thus can benefit from these large intestine or caecal breakdown products.

Typically, ruminal microorganisms hydrolyze cellulose to mono-saccharides, which in turn form volatile fatty acids as end products via fermentation. With normal diets among volatile fatty acids, acetic acid is the principle fatty acid (65 – 70%) followed by propionic acid (18 – 20%) and butyric acid (15 – 17%). By changing the ratio i.e. high concentrate and low roughage diet will increase the proportion of propionic acid to acetic acid. This condition favours body fattening and lowers the milk fat percentage. However, the coccidiostat, monensin, when administered in the diet at a level of 30 ppm, has been shown to alter the pattern of fatty acid production, resulting in increased production of propionic acid.

Table 1. Rumen bacteria and their substrates

Kind of substrates utilized/ fermented	Bacteria
A. Cellulose degrading	*Bacteroides succinogenes* *Ruminococcus flavefaciens* *Ruminococcus albus* *Butyrivibrio fibrisolvens*
B. Hemi cellulose degrading	*Butyrivibrio fibrisolvens* *Bacteroides ruminicola* *Ruminococcus sp*
C. Pectin degrading	*Butyrivibrio fibrisolvens* *Bacteroides ruminicola* *Lachnospira multiparus* *Succinivibrio dextrinosolvens* *Treponema bryantii* *Streptococcus bovis*
D. Amylose (starch) degrading	*Bacteroides amylophilus* *Succinimonas amylolytica* *Bacteroides ruminicola* *Streptococcus bovis*
E. Urea degrading	*Succinivibrio dextrinosolvens* *Bacteroides ruminicola* *Selenomonas sp.* *Ruminicoccus bromii* *Butyrivibrio sp.* *Treponema sp.*
F. Methane producing	*Methanobrevibacter ruminantium* *Methanobacterium formicicum* *Methanomicrobium mobile*

Kind of substrates utilized/ fermented	Bacteria
G. Sugar utilizing	*Treponema bryantii* *Lactobacillus vitulinus* *Lactobacillus ruminus*
H. Acid utilizing	*Megasphaera elsdenii* *Selenomonas ruminantium*
I. Protein degrading	*Bacteroides amylophilus* *Bacteroides ruminicola* *Streptococcus bovis* *Butyrivibrio fibrisolvens*
J. Ammonia producing	*Bacteroides ruminicola* *Megasphaera elsdenii* *Selenomonas ruminantium*
K. Lipid utilizing	*Anaerovibrio lipolytica* *Butyrivibrio fibrisolvens* *Treponema bryantii* *Eubacterium sp.*

Source: Church (1988)

Table 2. Rumen ciliate protozoa/ fungi and their substrates

Genera	Kind of substrates utilized/ fermented
A. Protozoa	
Isotricha	
intestinalis	Starch, sucrose, glucose, pectin
prostoma	Starch, sucrose, glucose, pectin
Dasytricha	
ruminantium	Starch, maltose, cellobiose, glucose
Entodinium	
bursa	Starch, hemicellulose
caudatum	Starch, cellobiose, glucose, maltose, sucrose
furca bilobum	--- --- --- ---
simplex	Starch
Diplodinium	
Polyplastron	Cellulose, glucose, starch, sucrose
Diplodinium	Cellulose, hemicellulose, starch

Genera	Kind of substrates utilized/ fermented
Diploplastron	Cellulose, hemicellulose, starch
Eudiplodinium	Cellulose, hemicellulose, starch
Ostracodinium	Cellulose, hemicellulose, starch
Eremoplastron	Cellulose, hemicellulose, starch
Epidinium	
Ecaudatum caudatum	Cellulose, hemicellulose, starch, sucrose, maltose
Ophryoscolex	
caudatus	Cellulose, hemicellulose, starch
B. Fungi	
Neocallimastix frontalis	Degrading plant (fibrous) particle, cellulose, xylan
Sphaeromonas communis	Degrading plant (fibrous) particle, cellulose, xylan
Piromonas communis	Degrading plant (fibrous) particle, cellulose, xylan

Source: Church (1988)

Relatively large amount of methane gas, CO_2, and heat are produced during the digestion of cellulose. The heat produced aids the host animals in maintaining body temperature, which is especially important in colder environments. Neither the methane nor carbon dioxide produced is of any economic value to the animal. Production of methane represents a large loss of energy, perhaps as high as 10% or more of the total energy.

On a high roughage diet, the addition of sugar results in decreased digestion of roughage material, probably because the microorganism utilizes sugar before fiber. Thus it is obviously advantageous that ruminant animals have virtually no ptyalin in their saliva to breakdown starch into maltose; otherwise, much of the resulting maltose might be utilized in the rumen by the microorganisms.

In calf rumen and reticulum are relatively undeveloped and milk is channeled by a tube like fold, known as oesophageal or reticular groove to omasum and abomasum. As calf begins to eat solid feed these first two compartments enlarge greatly until in adult. They comprise 85% of the total capacity of stomach. The feed is diluted with copious amounts of saliva (150 liters/day), first during eating and again during rumination. Rumen contents contain 85–93% water, and the breakdown of feed is accomplished partly by physical and partly by chemical means. During rumination coarses solid material is drawn back into oesophagus and returned by a wave of contraction to mouth. The feed is again chewed thoroughly before being return to the rumen. Each bolus of feed regurgitated is chewed 40-50 times. Like any other continuous culture system, pH of rumen (5.5 – 6.5) is maintained mainly by inflow of saliva, which contained phosphate and bicarbonate salt acts as buffer. Rumen contents of adult animals under normal feeding conditions contain about 10^{11} bacteria per ml, and up to 10^{6} protozoa per ml. Most of bacteria are non-spore forming, anaerobes and protozoa are ciliates of family isotrichidae (holotrichs) and ophryorcalecidae (oligotrichs). Protozoa are smaller in number but being larger size, may equal to bacteria in total biomass.

ABSORPTION

The simple six-carbon sugars probably combine with phosphorus prior to absorption across the villi of small intestine into the blood stream. Absorption occurs by two methods: i) facilitated diffusion and ii) sodium-dependent active transport. Not all of the sugars are absorbed at the same rate. Galactose is absorbed most rapidly, followed by glucose and fructose. The product of crude fibre digestion –VFAs are absorbed or across the wall of rumen into the blood stream.

METABOLISM

Absorbed carbohydrates move rapidly to the liver, where they are stored, temporarily. During rapid absorption, the glucose level of the portal blood may reach 200 mg%, whereas the remaining blood contains not more than 100-mg%, indicating that the liver is removing glucose from the blood. The carbohydrate thus stored in the liver cells is deposited in the form of glycogen, commonly called animal starch. During such times, liver glycogen storage may equal 10% of the weight of the liver. However, this level may fall to practically zero when carbohydrates are not being absorbed. The muscle also store glycogen, and their content may reach 2% of their weight.

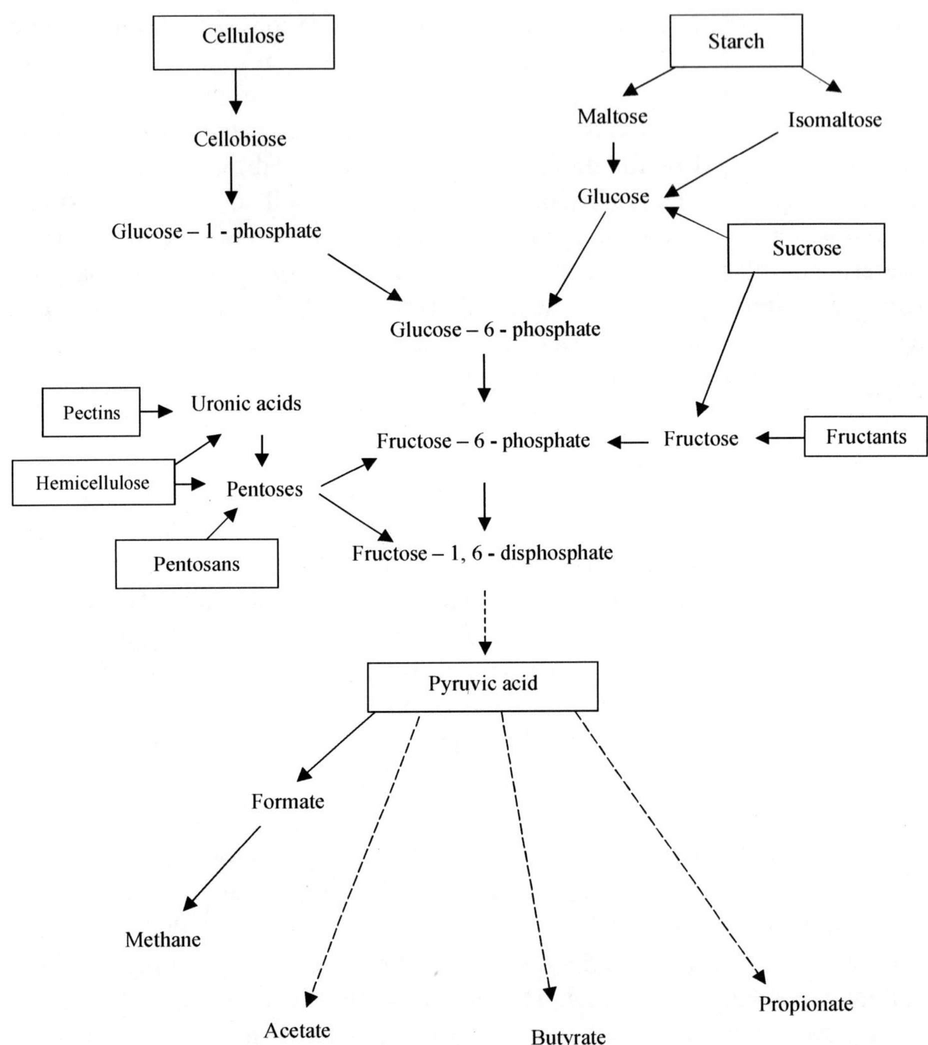

Fig 2. Digestion of carbohydrates in the rumen

The immediate fate of glucose is regulated by the level of at least two hormones, insulin and epinephrine, also called adrenalin. Most researchers believe that insulin accelerates both conversion of glucose to glycogen and carbohydrate oxidation in the liver and muscles, and that adrenalin increases the rate of hydrolysis of liver glycogen to glucose and also the conversion of muscle glycogen to hexoses phosphate. The most important step in carbohydrate metabolism is oxidation to CO_2 and H_2O, with a concomitant release of energy. The following equation is an oversimplified from the net reaction:

$$C_6H_{12}O_6 + 6O_2 \xrightarrow{\text{Enzymes}} 6CO_2 + 6H_2O + 673 \text{ kcal}$$

Glucose is catabolized to pyruvate through glycolytic pathway, which, in turn, enters the tricarboxylic acid cycle. The latter cycle occur in mitochondria of cells and yield 30 ATP per mole of glucose. While in glycolysis 8 mole of ATP are produced, thus total 38 mole of ATP are produced per mole of glucose. By a series of reactions in which as many as a dozen enzymes participate, a molecule of glucose yields two molecules of three-carbon pyruvate and eventually two molecules of lactic acids in the muscle under anaerobic conditions.

Of the three major volatile fatty acids formed in the rumen, glucose is formed primarily from propionic acid. This transformation occurs in the liver. This reaction is critical to ruminant animals, which may consume only limited quantities of starch as a source of blood glucose. Butyric acid is also metabolized primarily in the liver, whereas acetic acid may bypass hepatic action and enter the pheripheral circulation to be oxidized as an energy source. Acetic acid can be used by the mammary gland in producing milk fat. These three fatty acids may also be used as direct energy sources via the tricarboxylic acid cycle.

Propionate → Succinyl CoA → TCA cycle → 17 moles of ATP

Butyrate → Acetyl CoA → TCA cycle → 25 moles of ATP

Acetate → Acetyl CoA → TCA cycle → 10 moles of ATP

IV. CONVERSION TO FAT

The nutritional basis of fattening livestock for slaughter is that carbohydrate is readily converted to fat in the animal's body. This necessitates the formation of two types of compounds, glycerol and fatty acids. Glycerol may arise from glucose metabolism, possibly from phosphoglyceraldehyde, a product of carbohydrate metabolism. Acetyl-CoA formation from pyruvate probably provides the starting point for fatty acid synthesis. Carboxylation of acetyl-CoA takes place first, followed by a series of complicated condensations, by which two carbon fragments are attached to form the typically even numbered carbon chain fatty acids. In animals mechanism for synthesizing palmitic, stearic and oleic acids from acetate and glucose are present. However, synthesis of linoleic and arachidonic acid is not possible since mammals have no enzymes. Hence, it is suggested that the so-called essential fatty acids cannot be synthesized from carbohydrates.

LACTIC ACIDOSIS

Ruminant animals that are introduced to a highly soluble carbohydrate diet too rapidly may develop toxic symptoms, which are often fatal. This condition is known as lactic acidosis, the acute symptoms result when excessive consumption of fermentable carbohydrate is followed by a lowering of the pH (increased acidity). Roughage tends to have some buffering effect; however, acidosis appears to occur most readily when glucose accumulates in conjunction with a ruminal pH of 5 or less. Unfortunately, animals that recover from acidosis may be plagued by other ailments, such as laminates (and later, elongated hoofs), rumenitis, and liver abscess.

SOURCES OF CARBOHYDRATES

A. Cereal grains

'Cereal' is called to those members of the Gramineae which are cultivated for their seeds. Cereal grains are essentially carbohydrate concentrates, the main component of the dry matter being starch, which is concentrated in the endosperm. They are the primary energy source for human and nonruminants. The protein content in the grain is very variable; expressed as crude protein it normally ranges from 8 to 12%. Cereal proteins are deficient in certain indispensable amino acids, particularly lysine and methionine. The lipid content of cereal grains varies with species. Wheat, barley, rye and rice contain 1-3%, sorghum 3-4% and maize and oats 4-6%. Cereal oils are unsaturated, the main acids being linoleic and oleic, and because of this they tend to become rancid quickly, and also produce a soft body fat in pigs and poultry.

Among cereal grains, oats has the lowest metabolisable energy value and maize has the highest, the respective values (MJ/kg OM) for poultry being around 12 and 16 and for ruminants 12 and 14. Starch occurs in the endosperm of the grain in the form of granules, whose size and shape vary with different species. Cereal starches consist of about 25 per cent amylose and 75 per cent amylopectin. The main difference between a grass seed and cereal grain seed is that though plant breeding, seeds with a high starch (endosperm) have been developed. The cereals are deficient in calcium, containing less than 1 g/kg OM. The phosphorus content is higher, being 3-5 g/kg DM, but part of this is present as phytic acid, which is concentrated in the aleurone layer. Cereal phytates have the property of being able to bind dietary calcium and probably magnesium thus preventing their absorption from the gut; oat phytates are more effective in this respect than barley, rye or wheat phytates. The cereal grains are deficient in vitamin D and, with the exception of yellow maize, in provitamins A. They are good sources of vitamin E and thiamin, but have a low content of riboflavin. Most of the vitamins are concentrated in the aleurone layer and the germ of the grain. Cereals generally form a lower proportion of the total diet of ruminants, although they are the major component of the concentrate ration.

Barley ***(Hordeum sativum)***

Barley has always been a popular grain in the feeding of farm animals, the metabolisable energy value (MJ/kg DM) is about 13.3 for ruminants. The crude protein content of barley grain ranges from about 6 to 16 per cent with an average value of about 12 per cent. Like other cereal grains, the protein is of low quality, being particularly deficient in lysine. The lipid content of barley grain is low, usually less than 2.5 per cent. In many parts of the world, barley forms the main concentrate in the diets of pigs and ruminants. In the 'barley beef' system of cattle feeding, beef cattle are fattened on concentrate diets consisting of about 85 per cent barley without the use of roughages.

Maize ***(Zea mays)***

A number of different types of maize exist and the grain appears in a variety of colours, yellow, white or red. Yellow maize contains a pigment, cryptoxanthin, which is a precursor of vitamin A. The yellow varieties being preferred for animal feeding because they supply vitamin A. The yellow maize is particularly useful in diets for laying hens, where it contributes to the production of the orange coloration of egg yolk. Maize is an excellent source of digestible energy, it is low in protein and the protein is of poor quality. Maize contains about 73% starch, is very low in fibre and has a high metabolisable energy value. The starch in maize is more slowly digested in the rumen than that of other grains, and at high levels of feeding a proportion of the starch passes into the small intestine, where it is digested and absorbed as glucose. This may have advantages in conditions such as ketosis. When the starch is cooked during processing it is readily fermented in the rumen. The oil content of maize varies from 4 to 6% and is high in linoleic acid, which is an important factor in the diet controlling the egg size of hens. However, it tends to produce a soft body fat. The crude protein content of maize ranges from about 9 to 14%. The maize contains two main types of protein Zein and Glutelin, occurring in the endosperm.

Oats ***(Avena sativa)***

The oat has always been a popular cereal for ruminant animals and horses, but has been less popular in pig and poultry feeding because of its high fibre content and low energy value. The nutritive value of oats depends to a large extent on the proportion of kernel (groat) to hull. Oats of high hull content are richer in crude fibre and have a lower metabolisable energy value than low-hulled oats. The crude protein content ranges from 7 to 15%. Oat proteins are of poor quality and are deficient in the essential amino acids methionine, histidine and tryptophan, the amount of each of these acids in oat protein being generally below 2%. The lysine content is also low but is slightly higher than that of the other cereal proteins. Glutamic acid is the most abundant amino acid of oat protein, which may contain up to 20%. The oil content of oats is higher than that of most of the other cereal grains, and about 60 per cent of it is present in the endosperm. The oat oil is rich in unsaturated fatty acids and has a softening effect on the body fat.

Wheat *(Triticum aestivum)*

Wheat is the world's most important cereal crop. It is grown primarily for human consumption and is the preferred grain of most societies. Grain of wheat is variable in composition. The crude protein content may range from 60 to 220 g/kg OM, though it is normally between 80 and 140 g/kg OM. Climate and soil fertility as well as variety influence the protein content. The most important proteins present in the endosperm are a prolamin (gliadin) and a glutelin (glutenin). The mixture of proteins present in the endosperm is often referred to as 'gluten'. All glutens possess the property of elasticity. Strong glutens are preferred for bread making, and form dough which traps the gases produced during yeast fermentation. This property of gluten is considered to be the main reason why finely ground wheat is unpalatable when given in any quantity to animals. Wheat, especially if finely milled, forms a pasty mass in the mouth and rumen and this may lead to digestive upsets. Although, wheat is too expensive to be used as livestock feed. It is used as feed only when the price is low because of surpluses or when its quality is inadequate for milling purposes.

Other cereals

Rice *(Oryza sativa)*

Rice, the main cereal crop of Asia, requires a sub-tropical or warm temperate climate. Rough rice may be used as a food for ruminants and horses. However, most of rice is used for human consumption and little is available for farm animals. The two main by-products obtained from rice milling are the hulls and rice meal. The hulls are high in fibre content and can contain up to 210 g/kg OM of silica. They also have sharp edges which may irritate the intestine, and should never be given to animals. Rice meal or rice bran comprises the pericarp, the aleurone layer, the germ and some of the endosperm, and is a valuable product containing about 120-145 g crude protein/kg OM and 110-180 g oil/kg OM. The oil is particularly unsaturated and may become rancid very quickly. If it is removed a product, extracted rice bran, of better keeping quality is obtained. The crude protein content of about 280 g/kg OM and low in crude fibre and oil contents. It is suitable for ruminants and pigs.

Sorghum ***(Sorghum bicolor)***

Sorghum is the main cereal food grain in Africa and parts of India and China. It is more drought resistant than maize, for which it is used as a replacement. There are many different types of sorghum (e.g. dari, milo) and they vary in grain size, typical analyses can be misleading when applied to individual samples. The kernel of sorghum is very similar to that of maize, although smaller in size. It generally contains more protein but less oil than maize and has no pigmenting xanthophylls. Dark varieties contain tannins, which reduce protein digestibility. Whole sorghum grains can be given to sheep and poultry but are ground for other animals. Care is needed in the grinding process as this may produce a fine powder, which is pasty and unpalatable.

Bajra ***(Pennisetum typhoides)***

This is a millet, which is grown in warm season and relished by all categories of farm animals. As seeds are hard, they should be ground or crushed before being fed to cattle, while whole seeds can be fed to poultry. Bajra grain still has the hulls after threshing should be finely ground, otherwise the hard hulls splinter into sharp fibres which can lead to internal irritation. The crude protein content of bajra grain ranges from 8-12% and feeding value resembles to that of sorghum. It is also rich in tannins and requires detannification.

B. Roots, tubers and related by-products

The important root crops used in the feeding of farm animals are turnips, swedes, mangels and fodder beet. Sugar beet is another important root crop, but grown primarily for its sugar content and not given as such to animals. However, the two by-products of the sugar extraction industry, sugar beet pulp and molasses (from both sugar beet and sugarcane), are important and nutritionally valuable farm animal foods. The main tubers are potatoes, cassava and sweet potatoes, the last two being tropical crops.

Root

The main characteristics of roots are their high moisture content (75-94%) and low crude fibre content (4-13% on OM basis). The organic matter of roots consists mainly of sugars (50-75% on OM basis) and is of high digestibility (about 0.80-0.87). Roots are generally low in crude protein content although like most other crops this component can be influenced by the application of nitrogenous fertilizers. In the past, root crops have been considered as an alternative to silage in ruminant diets but their value as cereal replacements is now recognized.

Swedes and turnips

Swedes (Brassica napus), which were introduced into Britain from Sweden about 200 years ago, and turnips (Brassica campestris) are chemically very similar, although turnips contain less OM than swedes. Of the two types of turnip that are grown, the yellow-fleshed cultivars are of higher OM content than the white-fleshed cultivars. The ME value of swedes is higher than that of turnips, i.e. about 13 and 11 MJ/kg OM, respectively. Both swedes and turnips are liable to taint milk if fed to dairy animals at or just before milking time.

Mangels, fodder beet and sugar beet

These three crops are all members of the same species, Beta vulgaris, and classified according to their dry matter content. Mangels are the lowest in OM content, richest in crude protein and lowest in sugar content of the three types. Fodder beet can be regarded as lying between mangels and sugar beet in terms of OM and sugar content, while sugar beet is richest in OM and sugar content, though poorest in crude protein. On a DM basis the ME values range from about 12 to 14 MJ/kg, the higher values applying to sugar beet.

Mangels

'Low dry matter' mangels have a DM content of 9-12%. 'Medium dry matter' mangels have a DM content of 12-15%. This group is usually smaller in size than the low dry matter group, but usually develops fairly large tops. It is customary to store mangels for a few weeks after lifting, since freshly lifted mangels have a slightly purgative effect.

Fodder beet

'Medium dry matter' fodder beet contains 14-18% of dry matter, whereas the 'high dry matter' varieties may contain up to 22%. In addition to varietal type, the DM content is also influenced by the stage of growth at harvesting and environmental conditions. Fodder beet is a poor source of protein. Care is required in feeding animals on high dry matter fodder beet, since excessive intakes may cause digestive upsets, hypoca1caemia and even death. The digestive disturbances are probably associated with the high sugar content of the root.

Sugar beet

Most sugar beet is grown for commercial sugar production, though it is sometimes given to animals, especially cows and pigs. Because of its hardness the beet should be pulped or chopped before feeding. After extraction of the sugar at the sugar beet factory, two valuable by-products are obtained which are given to farm animals. These are sugar beet pulp and beet molasses.

Sugar beet pulp

The sugar beet on arrival at the factory is washed, sliced and soaked in water, which removes most of the sugars. After extraction of the sugar the residue is called sugar beet pulp. The water content of this product is 80-85% and the pulp may be sold in the fresh state for feeding farm animals. The crude fibre content is relatively high (about 20% on OM basis); the crude protein and phosphorus contents are low, the former being about 10% DM. Molasses sugar beet pulp is extensively used as a food for dairy aniamls and is also given to fattening cattle and sheep.

Beet molasses

After crystallization and separation of the sugar from the water extract, a thick black liquid termed beet molasses remains. This product contains 70-75% on OM basis, of which about 50% consists of sugars. The molasses dry matter has' a crude protein content of only 2- 4%, most of this being in the form of non-protein nitrogenous compounds, including the amine betaine, which is responsible for the 'fishy' aroma associated with the extraction process. Beet molasses is a laxative food and is normally given to animals in small quantities. Usually molasses is added to beet pulp or to other foods that include bran, grains, malt culms, spent hops and sphagnum moss. Molasses is also used, generally at levels of 5-10 per cent for preparation of cubes and pellets. The molasses not only improves the palatability of the product but also acts as a binding agent. An additional use of molasses is as a constituent (and binding agent) in the compressed feed blocks that are used as protein, mineral arid vitamin supplements for ruminants. Since beet molasses is a rich and cheap source of soluble sugars it is sometimes used as an additive in silage making.

Cane molasses

Although sugarcane is not a root or tuber crop, one of its major by-products, cane molasses, is a similar food to beet molasses. In the tropical and subtropical countries where sugarcane is grown, there is often a serious shortage of sugars or starch-containing animal foods, so cane molasses is a valuable resource. In addition to its uses as a carrier for supplements such as urea and as a conditioner of compound foods, cane molasses is used as a source of supplementary energy for forage diets, and even as the main component of diets. In Cuba, for example, diets for beef cattle have been based on cane molasses offered ad libitum and providing 50-80% of total dry matter intake, together with some forage and a supplement of protein (preferably protein of low rumen degradability, as in fishmeal). Such diets can give live weight gains of about 1 kg/day, but sometimes induce a condition in cattle known as molasses toxicity. This is characterized by incoordination and blindness caused by deterioration of the brain similar to that seen in cerebro-cortical necrosis. Its origins lie in the unusual rumen fermentation of molasses diets, which gives rise to volatile fatty acid mixtures rich in butyrate and poor in propionate. Like cerebro-cortical necrosis, the toxicity may be due to thiamin deficiency, but it may also be due to a shortage of glucose caused by the poor production of propionate in the rumen. The condition is best avoided by ensuring that animals have sufficient good-quality forage available.

Tubers

Tubers differ from the root crops in containing either starch or fructan instead of sucrose or glucose as the main storage carbohydrate. They have higher dry matter and lower fibre contents.

Potatoes *(Solanum tuberosum)*

In potatoes the main component is starch. The starch content of the dry matter is about 70%. The crude protein content of the dry matter ranges from about 9 to 12%. About half of this is in the form of non-protein nitrogenous compounds. One of these compounds is the alkaloid solanidine, which occurs free and also in combination as the glyco-alkaloids chaconine and solanine. Solanidine and its derivatives are toxic to animals, causing gastroenteritis. The alkaloid levels may be high in potatoes exposed to light. The risk of toxicity is reduced considerably if potatoes are steamed or otherwise cooked, the water in which the tubers have been boiled being discarded. Ensiling also destroys some of the toxin so that inclusion of slightly greened potatoes with grass should be acceptable. Ruminants are more resistant to toxicity than monogastric animals, presumably because of partial destruction of the toxin in the rumen. The crude fibre content of potatoes is very low, usually less than 40 g/kg OM. Potatoes contain a protease inhibitor that reduces the digestibility not only of potato protein but also of protein in other components of the diet. The inhibitor is destroyed by heating; it is normal practice to cook potatoes for monogastric, although cooking is unnecessary for ruminants, because the inhibitor is destroyed in the rumen. Potatoes are a poor source of minerals, except for potassium. The calcium content is particularly low. The phosphorus content is rather higher, since this element is an integral part of the potato starch molecule, but same 20 per cent of it is in the form of phytates.

Cassava *(Manihot esculenta)*

Cassava is a tropical shrubby perennial plant which produces tubers at the base of the stem. The chemical composition of these tubers varies with maturity, cultivar and growing conditions. Cassava tubers are used for the production of tapioca starch for human consumption although the tuber is also given to cattle, pigs and poultry. Its ME value is similar to that of potatoes but it has higher dry matter and lower crude protein contents. Cassava plants and tubers are to a certain degree poisonous since they contain varying proportions of two cyanogenetic glucosides (linamarin and lotaustralin), which readily break down to give hydrocyanic acid. In all cases care must be taken in the use of the tuber and wherever the plant is grown, indigenous methods of removing the glucosides have been devised. Such treatments include boiling, or grating and squeezing, or grinding to a powder and then pressing. Large amounts of dehydrated cassava meal have recently become available for feeding farm animals. Because of its high crude fibre content (about 270 g/kg OM) its use in the diets of non-ruminant animals should be restricted.

Sweet potatoes *(Ipomoea halalas)*

The sweet potato is a very important tropical plant whose tubers are widely grown for human consumption and as a commercial source of starch. The tubers are of similar nutritional value to ordinary potatoes although of much higher dry matter and lower crude protein contents. Fresh tubers, surplus to requirements, are often cut into small pieces, sun-dried and then ground to produce a sweet potato meal, a high-energy food of low protein content. Sun-drying does not destroy the trypsin inhibitors believed to be present in the tubers, and levels in the diets of farm animals are usually restricted.

Bibliography

Aman, P. (1993). Composition and structure of cell wall polysaccharides in forage cell wall structure and digestibility (eds. H.G. Jung, D.R. Buxton, R.D. Hatfield and J. Ralph), Madison, Wiscousin, USA. 183 – 196

Barber W P and Lonsdale C R 1980 By-products from cereal, sugarbeet and potato processing. In: 0rskov E R (ed.) By-products and Wastes in Animal Feeding, British Society of Animal Production Occasional Symposium Publication No.3. Gohl B 1981 Tropical Feeds, Rome, FAO.

Church, D.C. (1988). The Ruminant Animal Digestive Physiology and Nutrition. Prentice Hall, Englewood Cliffs, New Jersey.

Lehninger, A.I. (1978). Biochemistry. 2nd edition. Kalyani Publisher, New Delhi.

Maynard, L.A; Loosli, J.K; Hintz, H.F. and Warner, R.G. (1979). Animal Nutrition. 7th edition. Tata McGraw-Hill Publishing Co. Ltd., New Delhi.

McDonald, P; Edwards, R.A; Greenhalhg, J.F.D. and Morgan, C.A. (2002). Animal Nutrition. Pearson Education, Singapore, Pte, Ltd., Indian Branch, Delhi, India.

Preston T Rand Leng R A 1986 Matching Livestock Production Systems to Available Resources, International Livestock Centre for Africa, Addis Ababa, Ethiopia.

Stumpf, P.K. and Conn, E.E. (1980). The Biochemistry of Plants Vol 3. Carbohydrates: Structure and Functions, Academic Press, New York.

PROTEINS AND THEIR UTILIZATION

The term protein is taken from the Greek word, "proteus", which means first. Nearly half of the body weight of a typical animal cell is protein. Protein is necessary for life in more fundamental way than could be attributed to either carbohydrates or lipid. Without it no life is possible. Proteins are defined as complex nitrogen-containing organic compounds present in all animals and plant cells. Structural components of the cell, antibodies and many hormones are protein. Proteins are fairly constant in their elementary composition, particularly in nitrogen (16%), carbon (52.5%), hydrogen (7.0%) and oxygen (21%) and they contain traces of sulphur, phosphorous and iron. On hydrolysis proteins yield approximately 23 amino acids. Characteristically, proteins exist as large molecules or perhaps cluster of molecules. Corn protein (zein) has a molecular weight of over 53,000 compared to the carbohydrate glucose, which has a molecular weight of 180. Hemoglobin has a molecular weight greater than 70,000. Hydrolysis of proteins yields a number of intermediate products, namely, peptones and peptides.

Proteins differ in the number, kind, and arrangement of amino acids, which are present in the alpha form. The amino acids are chemical compounds, which contains both acidic carboxyl (-COOH) and a basic amino ($-NH_2$) group. The amino group (NH_2) is attached to the carbon atom adjacent to the carboxyl group (COOH), with two exceptions only proline and hydroxyproline. When proteins are hydrolyzed by enzymes, acids or alkalis, amino acids are produced. There are over 200 amino acids have been isolated from biological materials, however, only 20 of these are commonly found as components of proteins (Table 1). The structures of these amino acids are presented in figure 1. The amino acids may be classified into five groups on the basis of side chain structure.

Table 1: The commonly occurring amino acids

Rgroup Classification	Amino Acid	Abbreviated names	Molecular mass (Da)	% Occurrence 'average' protein
Non-polar, aliphatic	Glycine	Gly	75	7.2
	Alanine	Ala	89	8.3
	Valine	Val	117	6.6
	Leucine	Leu	131	9.0
	Isoleucine	Ile	131	5.2
	Proline	Pro	115	5.1
Aromatic	Tyrosine	Tyr	181	3.2
	Phenylalanine	Phe	165	3.9
	Tryptophan	Trp	204	1.3

Polar but uncharged	Cysteine	Cys	121	1.7
	Serine	Ser	105	6.0
	Methionine	Met	149	2.4
	Threonine	Thr	119	5.8
	Asparagine	Asn	132	4.4
	Glutamine	Glu	146	4.0
Positively charged	Arginine	Arg	174	5.7
	Lysine	Lys	146	5.7
	Histidine	His	155	2.2
Negatively charged	Asparatic acid	Asp	133	5.3
	Glutamic acid	Glu	147	6.2

Fig No.1: The chemical structure of the 20 amino acids commonly found in proteins:

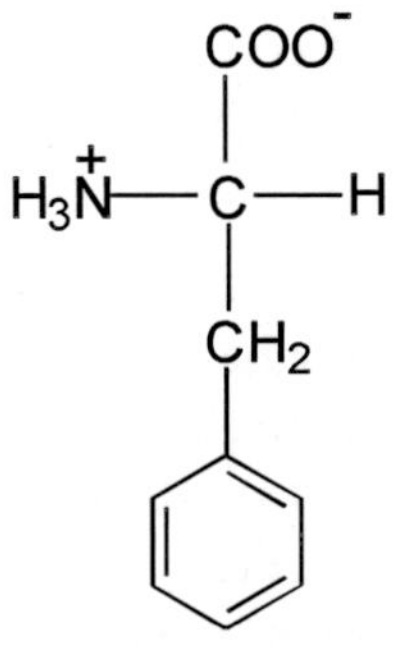

Phenylalnine 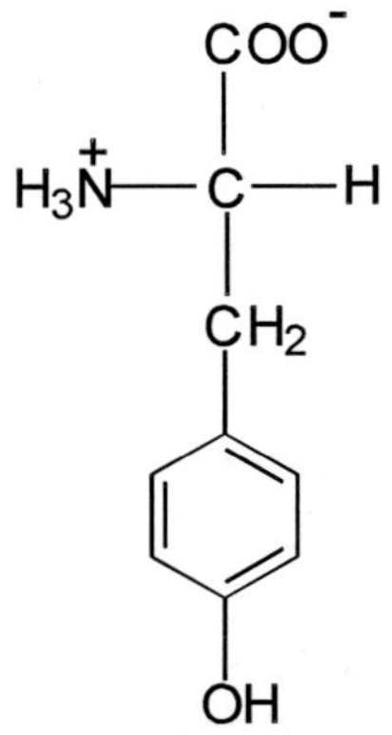 Tyrosine 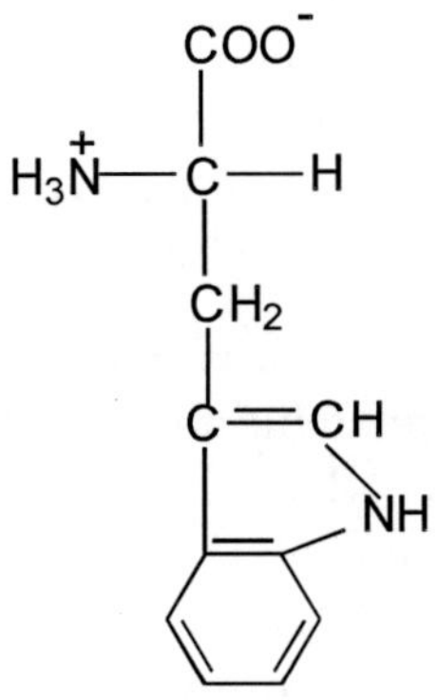 Tryptophan

Aromatic amino acids

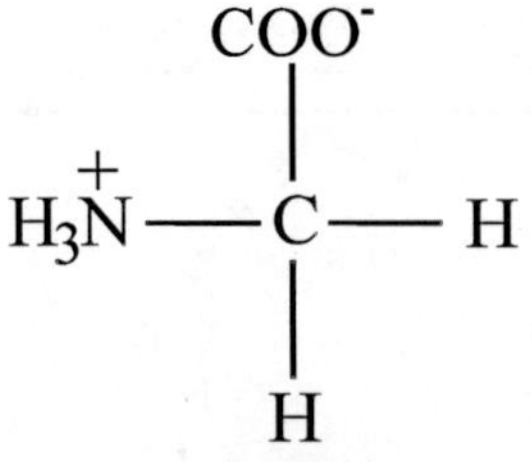

Glycine 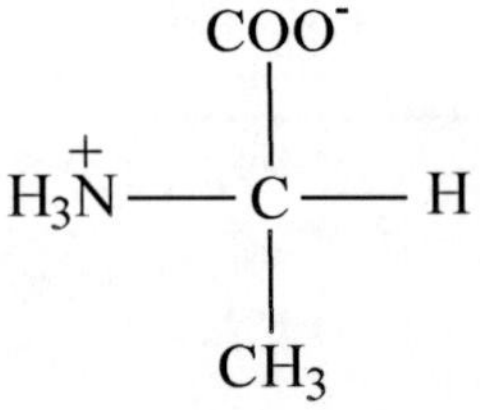 Alanine 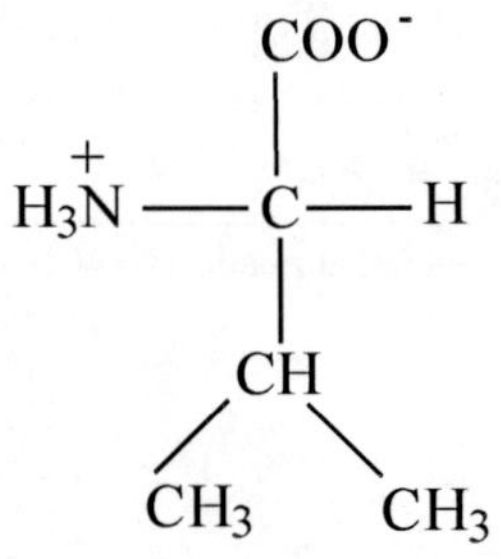 Valine

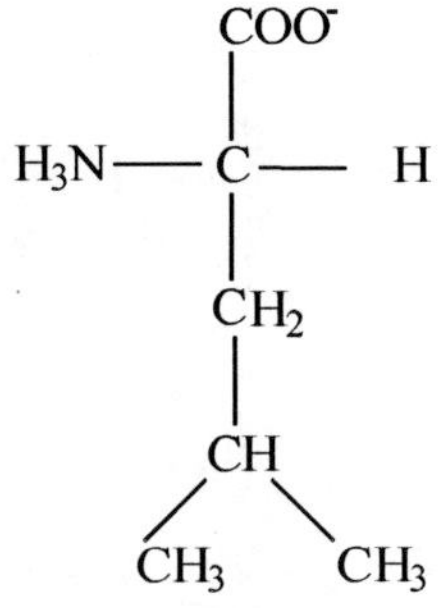

COO⁻
H3N—C—H
CH2
CH
CH3 CH3

Leucine

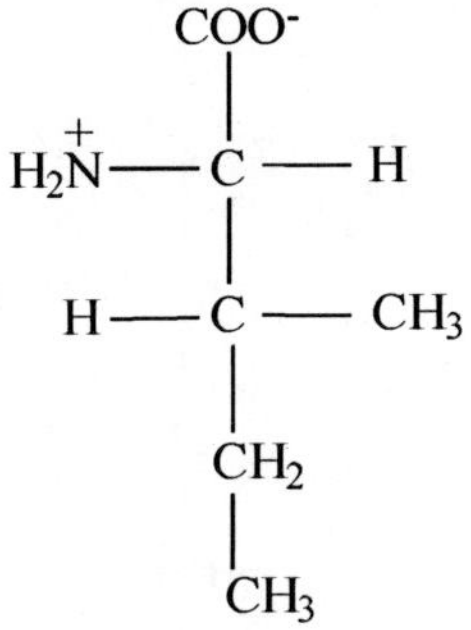

Isoleucine

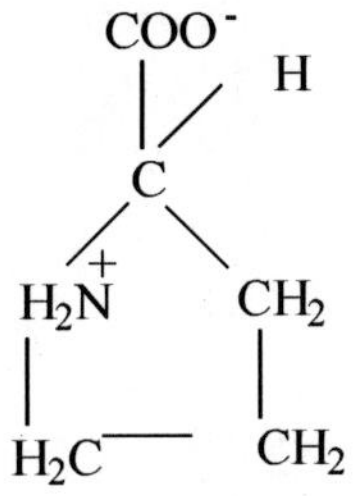

Proline

Non-polar, aliphatic amino acids

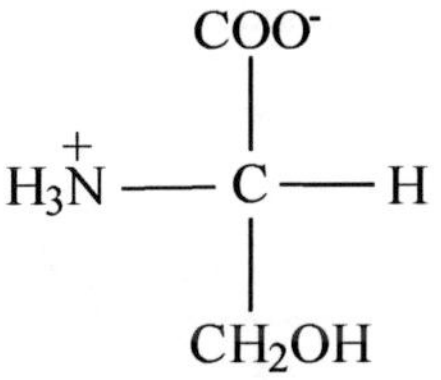

Serine

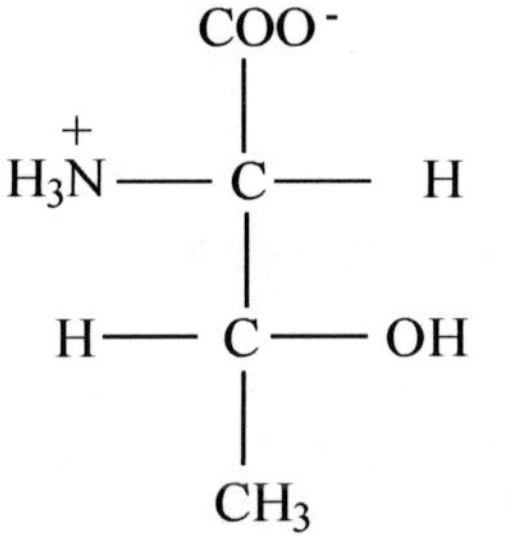

Threonine

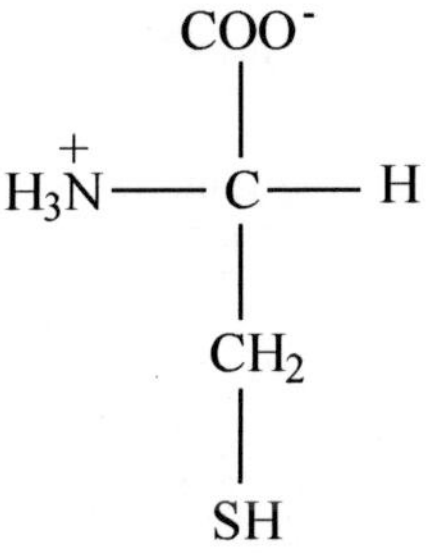

Cysteine

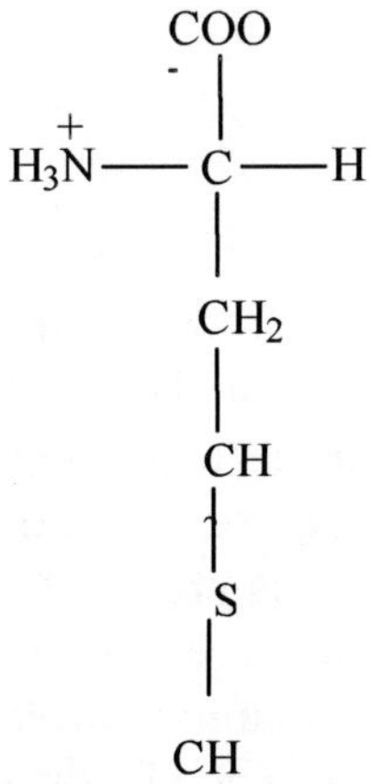

Methionine

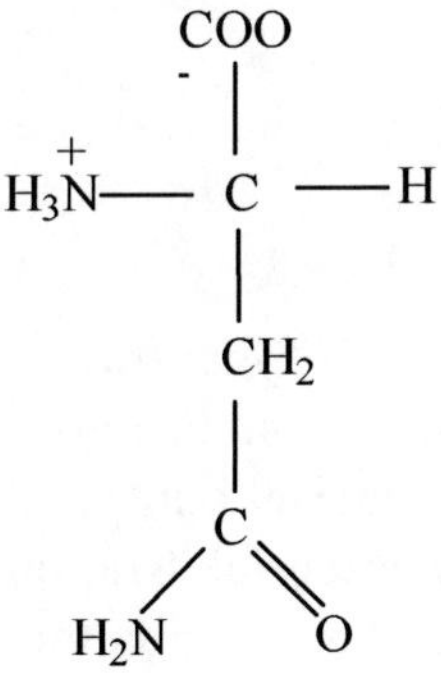

Asparagine

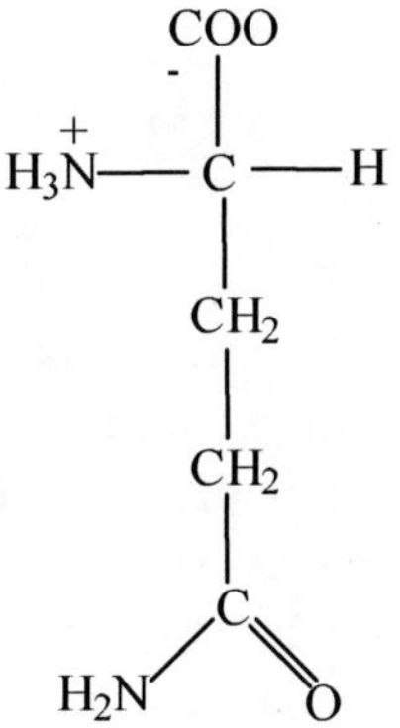

Glutamine

Polar but uncharged amino acids

COO^-
$H_3\overset{+}{N}$— C —H
CH_2
CH_2
CH_2
CH_2
NH_3

Lysine

COO^-
$H_3\overset{+}{N}$— C —H
CH_2
CH_2
CH_2
NH
C$=NH_2$
NH_2

Arginine

COO^-
$H_3\overset{+}{N}$— C —H
CH_2
C—N—CH
C=N
H H

Histidine

Positively charged amino acids

COO
$H_3\overset{+}{N}$— C —H
CH_2
COO

Aspartate

COO
$H_3\overset{+}{N}$— C —H
CH_2
CH_2
COO

Glutamate

Amino acids are required by animal body for performing various physiological functions. Essential or indispensable amino acids are those, which body cannot synthesize in adequate amounts and they are required to be supplied in the diets so they are dietary essential. Ruminants do not require the essential amino acids to be supplied in their diet as the microorganisms present in the rumen can synthesize all essential amino acids and when they are digested in the lower G.I. tract there amino acids are available to the host animals. Table 2 presents the various amino acid patterns contained in some common feedstuffs.

Structure of Protein

The structure of proteins can be considered under four basic headings:

Primary structure

The primary structure of a polypeptide refers to its exact amino acid sequence, along with the exact positioning of any disulphide bonds present Successive amino acids are joined together during protein synthesis via a peptide bond, with the condensation reaction by eliminating one molecule of water during each bond formation. A number of amino acids can be joined together by this means.

Secondary Structure

It refers to a regular, recurring arrangement in shape of the polypeptide chain along one dimension. The most commonly observed secondary structure elements are termed α – helix and β- strands. Fibrous proteins often display only one type of secondary structure, however globular proteins generally exhibit several structures of regular secondary structure (α – helix and / or β- strands).

Tertiary Structure

It refers to how the polypeptide chain is bent or folded in three dimensions, to form the compact, tightly folded structure of globular proteins.

Quaternary Structure

It refers to how individual polypeptide chain of protein having tow or more chains are arranged in each other. The forces stabilize these aggregates are hydrogen bonds and electrostatic or salt bonds formed between residues on the surfaces of the polypeptide chains.

I. DIGESTION

A. Monogastric Digestion (Calf)

The monogastric stomach contains pepsin, rennin, and hydrochloric acid. Rennin is more prevalent in young, or milk consuming animals, than in mature ones. It causes casein, the primary milk protein, to split into paracasein, which unites with calcium to form and insoluble curd, calcium paracaseinate (clotting). This slows down the passage of milk through the gastrointestinal tract, which permits more nearly complete digestion. Pepsin is primarily a proteinase that acts on the peptide linkages of protein, but only to a limited extent. In vitro pepsin breaks some of the peptide linkages to give amino acids. However, its primary action results in formation of peptides and peptones. The optimum pH for its action is acidic, 1.5 - 2.4.

The most complete digestion of proteins occurs in the small intestine. When peptides and other digestive products reach the duodenum, hormone cholecystokinin (Pancreozymin) is liberated from the mucosa. This hormone stimulates the secretion in to the pancreatic juice of live trypsinogen, chymotrypsinogen, carboxypeptidase A and B, Proelastase. The optimum pH for these enzymes is around 7 – 9. Trypsin is very specific and only acts upon peptide linkages involving the carboxyl groups of lysine and arginine. Trypsin also converts chymotrypsinogen into the active enzyme chymotrypsin which acts specifically on peptide bonds involving the carboxyl groups of tysosine, tryptophan, phenylalanine and leucine. Procarboxy peptidase after being converted into carboxy peptidase by trypsin attack the peptidase from the end of the chain, splitting off the terminal amino acid which has a free carboxyl group. The nucleic acids are hydrolysed by the polynucleotidase (DNase and RNase) into nucleotides. Nucleosidases attack the linkage between sugar and nitrogen bases, into purines and pyrimidine. Phosphatase complete the hydrolysis by separating the ortho phosphoric acid from the ribose or deoxyribose. The hydrolysis of small peptides to amino acids is brought about by enzymes associated with the intestinal villi. Amino peptidases act on the peptide bond adjacent to the free amino group of simple peptides whereas dipeptidases complete the breakdown of dipeptides to amino acids.

TABLE 2: Amino Acid Content of Selected Feedstuffs

Feedstuff	Arginine (%)	Lysine (%)	Methionine (%)	Cystine (%)	Tryptophan (%)	Threonine (%)
Alfalfa meal,	0.94	0.94	0.30	0.42	0.38	0.84
Barley	0.57	0.40	0.18	0.29	0.13	0.40
Blood meal	2.35	4.50	0.65	1.00	0.72	3.38
Corn, yellow	0.52	0.26	0.18	0.16	0.09	0.39
Cottonseed meal	4.49	1.70	0.52	0.71	0.46	1.12
Distillers dried solvent (corn)	1.15	0.95	0.50	0.40	0.30	1.00
Fish meal,	3.62	4.67	1.67	1.03	0.62	2.34
Canola seed meal	2.08	1.94	0.71	0.87	0.44	1.53
Oats	0.80	0.50	0.18	0.22	0.16	0.44
Rice bran	0.96	0.59	0.26	0.27	0.12	0.48
Rice Polish	0.74	0.43	0.22	0.21	0.10	0.36
Soybean meal	3.70	2.84	0.71	0.73	0.73	1.96
Wheat	0.60	0.40	0.17	0.22	0.16	0.43
Sorghum grain	0.35	0.21	0.16	0.17	0.08	0.29
Safflower seed meal	2.21	0.90	0.42	0.45	0.37	0.85

Proteins must be broken before they are capable of passing into the bloodstream. In general, the absorption of protein into the bloodstream causes an allergic reaction. One exception is lacto globulin of the colostrum, which is allowed to pass across the wall of the small intestine and into the bloodstream of the very young animal for not more then the first 3 days of its life. Lacto globulin provides the newborn animal with protection against disease.

B. Rumen digestion

Rumen digestion of protein differs entirely from gastric digestion of protein in the monogastric animal. The rumen contains microbes (refer carbohydrate digestion) that can modify or supplement the amino acid of the ingested protein and also alter the amount of protein that becomes available to the animal. Furthermore, the rumen microbes can utilize non-protein nitrogen and synthesize most of the amino acid in adequate quantities. Figure 2 present the possible fates of ingested protein. First, the intact dietary proteins may enter directly into the abomasum and intestine (UDP-undegradable proteins) or it may be degraded (RDP-rumen degradable proteins) to peptides, amino acids, and then to ammonia. Amino acids can be transformed, with the help of the rumen microflora, to microbial protein and then passed into the abomasum and intestine. Ammonia may take any one of three routes. First, it may enter rumen microbial protein. Second, it may go through the wall of the rumen to the liver, via the bloodstream, where it will be converted to urea. The urea then may excrete either the urine from the body or enter in the saliva to recycle through the rumen, along with dietary non-protein nitrogen. Last, the ammonia may enter the abomasums and the intestine.

Apparently, very little absorption of protein or amino acids occurs in the rumen. Once the protein and protein breakdown products of the rumen enter the small intestine, digestion and ultimate absorption of amino acid take place, comparable to the process, which was described for the monogastric animals.

TABLE 3: Estimates of Ruminal Degradation of Protein in Selected Feed Ingredients

SOURCE	RUMINAL DEGRADATION (%)
Casein	90.00
Cottonseed meal	60-80
Groundnut meal	78
Sunflower seed meal	75
Soybean meal	39-60
Fish meal	10-29
Zein	28-40
Corn	40

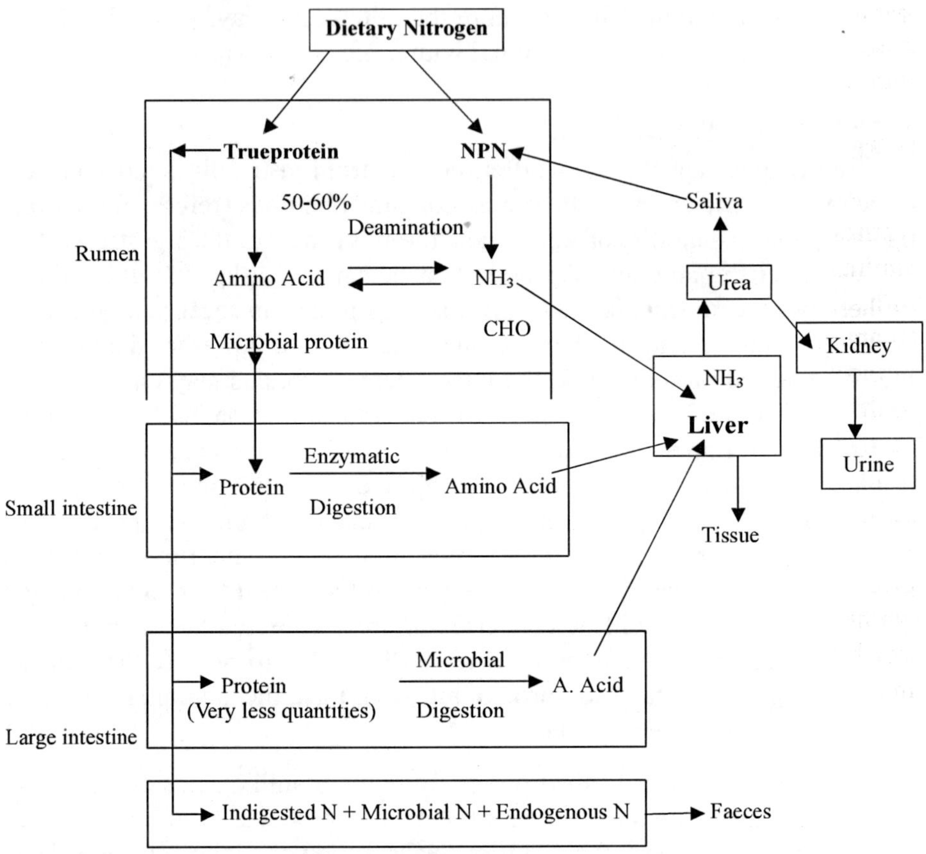

From Fig 2, one may deduce that a portion of the ingested protein may bypass the rumen, unchanged, passing into the intestine. Thus, all proteins are not degraded to the same degree in the rumen (Table 3). The proportion of dietary protein degradation/ hydrolysis in the rumen depends on a number of factors like solubility of proteins and quantity of proteins. On a normal diet about 50-60% of protein is attacked by the microorganisms. Protection from rumen degradation enables more intact amino acids, characteristic of the original dietary protein, to reach the intestine than would be the case without protection. Positive responses can be expected if the animal needs or can use more amino acids for maintenance plus production. However, if amino acid supplies are already adequate, additional "bypass" quantities may be of no benefit to the animal.

Several methods have been employed to decrease the degradability of protein in the rumen including: (1) Use of formaldehyde to react with free amino groups to form cross-linkages between protein chains (this reaction is reversed when enters in high acid medium in abomasum) (2) Use of heat treatment to create cross-linkages of free amino acid groups, (3) Treatments with tannins, (4) Encapsulation of amino acids, (5) Use of amino acid analogues.

II. ESSENTIALS AMINO ACIDS

An essential amino acid is defined as an amino acid that an animal cannot synthesize in sufficient quantities for normal functioning. Theoretically, the ruminating animal can synthesize all of its amino acid required if it is supplied with all of the necessary elements. On the other hand, there are approximately 10 amino acids that cannot be synthesized in sufficient quantities. These 10 amino acids are thus "essential" for the monogastric animal (11 in case of poultry): arginine, histidine, isolucine, leucine, lysine, Methionine (plus cystine), phenylalanine (plus tyrosine), threonine, Tryptophan, and valine. In the case of poultry, the amino acid glycine (plus serine) must be added to this list. (The amino acid listed in parenthesis can supply a portion of the amino acid with which they are associated in the list). In the ruminants, dietary needs of essential amino acids is less important since the rumen microorganisms are able to synthesize both essential and non-essential amino acids in adequate quantities which meet the amino acids requirement of the host animal when sufficient amount of nitrogen and carbohydrates are present in the diet.

Essential Amino Acids: - Essential amino acid can be remembered as a code, "AVHILLMPTT"

1. Argenine
2. Valine
3. Histidine
4. Isoleucine
5. Leucine
6. Lysine
7. Methionine
8. Phenylalanine
9. Tryptophane
10. Threonine
11. Glycine (additional required in poultry)

III. ABSORBED AMINO ACIDS

The amino acid absorbed into the bloodstream are primarily required for the manufacture of tissue protein and other nitrogen containing constituents, such as enzymes, milk, and eggs. Formation of tissue protein from amino acid is the reverse of the hydrolysis that is observed in the digestion process.

A second fate of absorbed amino acids is deamination. Excess of amino acids which are not required for synthesis of tissue protein, harmone, enzymes etc. are catabolized by the liver involving deamination process, whereby ammonia and a-keto acid are formed. The released ammonia is converted to urea or may be utilized by a-keto acid to form amino acid. The keto acid can be converted into (I.) fat, (ii.) carbohydrate, (iii.) resynthesized into an amino acid and (iv.) carbon dioxide and water.

A third fate of digested amino acids is storage. Although this capacity is quite limited compared to that of storage of fat, some capability for protein storage does exist. For example, heavy exercise can result in increased muscular bulk, which represents protein storage, Protein can also be stored briefly in blood.

IV. FATE OF AMINO NITROGEN - THE UREA CYCLE

Urea is the principal excretory product of proteins in most mammals, amphibia, and fish. Whereas, uric acid is the nitrogen excretory product produced in birds and reptiles. The liver is the main site of urea formation from ammonia through urea cycle. Ammonia per se is toxic, and 3 ATP are needed to excrete one molecule of it. The liver of animals that excrete urea contain arginase; those that do not excrete urea do not contain arginase. In urea cycle ornithine combines with carbamyl phosphate (ammonia + CO_2 + ATP) to form citrulline, that on energy addition is converted to arginosuccinate; this split into fumerate (which leaves the cycle) plus arginine. Arginine is hydrolyzed into ornithine and urea, with the latter being excreted in the urine. Ornithine is then recycled. In ruminants part of urea is secreted back in the rumen through saliva.

V. BIOLOGICAL VALUE

The efficiency with which a protein supplies the nitrogen requirements of an animal is defined as the biological value (BV) of the protein. The biological values are dependent on the amino acid composition. If all essential amino acids are present in right amount and proportion the biological value (BV) will be higher since the protein will be utilized for body tissues rather than being diverted for energy supply. In latter case the amino acids will be deaminated and there will be more excretion of urea. Animal proteins have higher BV since the essential amino acids present are very near to the proportion in which they are needed by the body. Deficiency or excess of any one of the amino acids lowers the biological value. Biological value is primarily applicable to monogastrics, since the rumen can rearrange the amino acid of ingested protein to meet its most if not all tissue and secretory demands.

In assessing the BV of a protein, the purpose of which the protein was intended must be taken into consideration since the array of amino acids required for work differs from the array used for growth or milk production.

The BV of a protein represents a measure of the digested and absorbed protein that is not excreted in the urine. It has been expressed historically by the Thomas Mitchell equation.

$$\%BV = \frac{100 \times \text{N intake} - (\text{fecal N-MFN}) + (\text{urine N-EUN})}{\text{N intake} - (\text{fecal N} - \text{MFN})}$$

Where,

MFN = Metabolic fecal nitrogen

EUN = Endogenous urinary nitrogen

VI. SPECIFIC DYNAMIC ACTION

Ingestion of all nutrients elevates the metabolic rate, while ingestion of protein has a more pronounced effect than ingestion of carbohydrate or fats. This tendency to increase the release of energy (increased metabolic rate) is called the specific dynamic action of nutrients. Because ingested amino acids are not stored to any great extent, ingestion of relatively large quantities of amino acid may temporarily overwhelm the need of the tissues for amino acids. Thus, excess amino acids are used for deamination and subsequent oxidation. This effect plus the formation of urea in the liver-and its excretion by the kidneys-probably accounts for at least one-half of the specific dynamic action effect and should be taken into account in calculating a calorically adequate diet. It may account for as much as 6-10% of the total calories. The specific dynamic action of nutrients varies according to the make up of the ration, the level at which it is fed, and the body function being supported-maintenance, growth, fattening or lactation.

Chemical Score

In chemical score the quality of protein is measured by the contents of the essential amino acids that is in greatest deficit when compared with standard. Egg protein is generally used as standard. The content of individual essential amino acids of a protein is expressed as a proportion of that is in the standard, the lowest proportion taken as the score. For example, in wheat lysine is the essential amino acid that is in greatest deficit. The content of lysine in wheat and egg are 2.7 and 7.2, respectively. Thus, the chemical score for wheat protein is 2.7/ 7.2 x 100 = 37.

The biological value and chemical score of some feed stuffs are given in Table 4.

Table 4: Biological value and chemical score of some protein feeds

Source	Biological value (rat)		Chemical score
	Rat	Growing Pig	
Oats	0.65	-	0.46
Wheat	0.67	-	0.37
Maize	0.55	0.49-0.61	0.28
Cottonseed meal	0.80	0.63	0.37
Groundnut meal	0.58	-	0.24
Soybean meal	0.75	0.63-0.76	0.49
White-fish meal	0.77	0.74-0.89	
Milk	0.85	0.95-0.97	0.69
Whole egg	0.95	-	1.00
Linseed meal	-	0.61	-
Barley	-	0.57-0.71	-
Peas	-	0.62-0.65	-

Non-protein nitrogen compounds as protein source for ruminants

Non-protein nitrogen compounds urea and biuret are recognized as useful sources, of nitrogen for ruminants. These compounds dissolve in rumen contents and with the help of microbial enzymes ammonia is produced that is used by microbes for microbial protein synthesis. Urea is a white, crystalline, deliquescent solid with the following formula and NH_3 is produced by urease activity.

$$\begin{matrix} NH_2 \\ | \\ C=O \\ | \\ NH_2 \end{matrix} + H_2O \xrightarrow{\text{Urease}} 2NH_3\uparrow + Co_2$$

Urea

Feedstuffs also contain naturally produced NPN and soluble nitrogen compounds that readily contribute to rumen ammonia. Furthermore, all natural feed proteins are partially degraded by rumen bacteria with production of ammonia. Ammonia production rate, when synchronized with rapid bacterial growth and degradability of fermentable carbohydrates in the rumen, will result in best conversion of NPN to protein. High-concentrate diets allow for better use of NPN than to high roughage diets; and certain steam-processed, starchy feeds improve ammonia utilization over natural grain source.

Because the percentage of ammonia in rumen fluid increases in proportion to the level of crude protein in the diet, at some level of protein in each dietary situation it will be found that more ammonia is accumulated than the rumen microbes can use. At this point it would be wastage to supplement the diet with NPN, because all that NPN can furnish is more ammonia accumulation, and absorbed into blood. Higher ammonia levels are toxic and even animals die. Therefore, if the dietary protein requirements exceed this critical level, supplementary protein from low solubility sources is necessary to ensure that the extra protein fed actually can be utilized by the animal. For cattle fed high roughage, low-energy diets the critical point of NPN utilization may be as low as 7 % dietary crude protein, whereas for those fed high-energy diets containing both roughage and concentrates the critical crude protein percentage could be about 13 percent. Optimum utilization of NPN will depend upon a relatively slow and constant intake of NPN, thoroughly mixed with the feed that will yield high energy from starchy sources and promote vigorous rumen bacterial growth.

Excessive levels of poorly mixed additions of urea to feed can cause toxicity due to rapid ammonia production such an effect may be avoided by limiting added NPN to 0.5 percent of the total dry matter fed (0.5 percent NPN is practically equivalent to 1 percent urea). Because rumen microorganisms require 2 or 3 weeks to adapt fully to diets containing added NPN compounds, abrupt changes to such mixtures should be avoided.

Protein fed above requirements

Experiments with cattle fed diets of mixed roughages and starch-rich concentrates indicate that most efficient digestion and fermentation in the rumen requires about 11 percent crude protein or more. When diets are too low in protein, digestibility of the total diet will be reduced, which will result in reduced feed intake and poorer efficiency of utilization of the feed energy. When milk cows were fed mainly corn silage and or/ high-concentrate diets, it was found that more than 12 percent dietary crude protein must be fed to maximize feed intake and production efficiency. Thus, although protein requirements per se may be met in some classes with a lower percentage of dietary protein, the aspect of feed efficiency it may be economical to feed at least 11-12 percent crude protein in the dry matter of mixed roughage-concentrate diets. Excess ammonia that results from high protein diets is converted to urea and excreted in urine; so high protein diets require increased water intake for waste removal and urine secretion.

Utilization of protein may be limited by inadequate energy in the diet, and the available energy for rumen microorganisms is an important factor in protein nutrition. Also, rumen microorganisms require dietary protein or non protein nitrogen (NPN) to grow and digest feed nutrients. Protein in excess of lactation requirements has been shown to have negative effects on reproduction. Several workers have reported that feeding diet containing 19 percent or more CP in diet DM, lowered conception rates. Plasma progesterone concentrations in early lactation cows were lower when diets contained 19-20% CP vs lower concentration of CP. Excessive rumen degradable protein (RDP) rather than rumen un-degradable protein was associated with decreased conception rates. It has been suggested that energy used associated with disposal of excess N, may be taken from body reserve and thereby result delayed ovulation and reduced fertility and plasma progesterone concentrations. Another theory is that excessive blood urea concentrations could have a toxic effect on sperm, ova or embryo. Blood urea; also have been shown to decrease uterine pH, and prostaglandin production. High blood urea may reduce the binding of leutinizing hormone to ovarian receptors leading to decrease in serum progesterone concentration and fertility. The function of hypophyseal pituitary-ovarian axis may be altered and thereby decreasing reproductive performance. Another theory states that high levels of circulatory ammonia may depress the immune system. Milk urea N above 19 mg/100ml and blood urea N above 20 mg/100ml have been linked with reduced conception rates in lactating cows.

Unavailable Feed Protein

Protein in feed that have been heated, charred, browned, or caramelized will be denatured to some extent, and may not be availability to animals. Low moisture silage or haylage, and hay put in storage with too much moisture, are frequently damaged as revealed by a brown color with a caramel odor. For calculating protein needs in diets containing moderately heat-damaged feeds, a reduction of 20 % in total quantity may be recommended. When the forage is dark brown or blackened, less than 50 percent of the crude protein may still be digestible. Diets containing highly heat damaged feed must be corrected for unavailable protein by nitrogen content of acid detergent fiber, which is an indication of unavailable protein.

Protein solubility

Solubility of feed crude protein also affects its nutritive value; highly soluble crude protein is quickly attacked in the rumen by bacterial enzymes and degraded to simpler compounds and to ammonia. When this released ammonia exceeds the capacity of rumen bacteria to use it to synthesize protein, the result is a loss of part of the dietary nitrogen via blood in the urine. Direct cut grass silage and high moisture corn or sorghum silage often contain more than 60 percent of water soluble crude protein, and this is not utilized very efficiently on all silage diets. Some concentrate feeds have a high percentage of soluble crude protein, which make them less suitable for feeding with silage. Rumen degradable and un-degradable proteins are thus considered separately, and necessary formulations are recommended based in metabolizable protein.

Protein sources

Oilseed Meals and Cakes

India is the third largest producer of oilseeds after the USA and China. It accounts for 8.0% of world oilseed production and 7% of world oil meal production (Tables 5-6). The major oilseeds cultivated in the country are groundnut, rapeseed-mustard, soybean, sunflower, sesame, linseed, castor, and niger. Oilseed meals are major source of protein and other feed nutrients for livestock (Table 7). Oil is extracted from oilseeds by using pressure to force out the oil, or by using an organic solvent, usually hexane, to dissolve oil from the seed. Seeds such as groundnut, cottonseed, and sunflower have a thick coat or husk. This husk may be completely or partially removed by cracking and riddling, the process known as decortication. Removal of the husk lowers the crude fiber content (Table 7). In general, oil seed proteins have low cystein, methionine and variable but usually low lysine content. These oil seed meals also are high in phosphorous, which tend to aggravate the ration in comparison to generally low calcium content. They may provide useful amounts of the B-vitamins but are poor sources of carotene and vitamin E.

Table 5: Oil meal production in India and the world (000' tonnes)

Oil meal	India 1998 – 99	World 1998 – 99
Soybean	4141	105421
Cotton	3213	15200
Groundnut	2281	6796
Sunflower	481	11003

Rapeseed	3015	20700
Sesame	240	912
Palm kernel	-	2810
Linseed	163	1442
Copra	256	1625

Table 6: Chemical composition of oilseed cakes

Cake	Crude protein	Oil (%)	Crude fiber	General remarks
Groundnut cake				
Expeller-pressed	48.0	8.0	7.0	Low in lysine, arginine, histidine and S-containing amino acids
Solvent-extracted	51.0	1.0	10.0	
Rapeseed-mustard				
Expeller – pressed	37.0	5.0	10.0	Palatability problems due to pungent odor
Sesame cake				
Expeller – pressed	40.0	8.0	7.0	-
Linseed cake				
Expeller-pressed	29.0	8.0	10.0	-
Solvent-extracted	33.0	1.0	9.0	-
Cottonseed cake (decorticated)				
Expeller – pressed	41.0	5.0	12.0	Low in Ca, Zn, Mn
Solvent – extracted	42.0	1.5	16.0	Effect of gossypol is greater in monogastric animals
Safflower cake (decorticated)				
Expeller – pressed	48.0	5.0	15.0	-
Coconut cake				
Expeller – pressed	22.0	6.5	12.0	-

Source: AICRP (1988) on agricultural by-products and industrial wastes for livestock and poultry feeding. ICAR, New Delhi

Generally, the residual oil content of the cakes is higher in screw-pressed cakes than in solvent-extracted meals. India also has many unconventional oilseeds of which only a small proportion is presently exploited (Table 7). They are also rich sources of protein and other nutrients, but their use in the livestock industry is limited by the presence of one or more undesirable nutritional factors (Table 8). These, however, can be used in concentrate mixtures in varying proportions depending on their chemical composition. The quality of a protein in a particular oilseed is relatively constant, but may vary in the cake or meal, depending on the technique employed for the extraction of the oil. High temperatures and pressures of expeller processing may result in the lowering of the nutritive value for monogastric animals. For ruminants, such denaturation may be beneficial, owing to an associated reduction in rumen degradability. High temperature and pressure also degrade some deleterious substances such as gossypol. Solvent extraction does not involve pressing, temperatures are comparatively low, and protein value of meals is almost the same as that of the original seed. The oilseed cakes may also make a significant contribution to the energy content of the diet, particularly when the oil contents are high.

Table 7: Unconventional oil meals/cakes available for animal feed

Source	**Availability of oilseeds**	**By-product availability**
	...('000 t)...	
Sal seed meal	5504	4806
Mahua seed	2176	1414
Neem seed	418	334
Rubber seed	336	117
Karanj	111	81
Kusum	105	70
Khankan, Undi, Nahar, Tobacco seed	102	59
Mango seeds	1000	500

Source: AICRP (1988) on agricultural by-products and industrial wastes for livestock and poultry feeding, ICAR, New Delhi.

Table 8: Antinutritional factors in oilseeds/plant protein meals

Meals	**Antinutritional factors**
Soybean	Protease inhibitors*, allergins*, oligosaccharides, phylin, lipoxygenase*, lectins*, saponin, hemagglutinins, citrinin
Rapeseed	Erucic acid, glucosinolates, sinapine, tannins, oligosaccharides

Meals	Antinutritional factors
Canola	Glucosinolates, sinapine, pectins, oligosaccharides
Cottonseed	Gossypol, tannins, cyclopropenoid fatty acids
Sunflower	Chlorogenic acid, high fiber, achrotoxin
Groundnut	Mycotoxins, tannins, oligosaccharides, protease inhibitors*
Copra	Fiber, mannans
Palm kernel	Fiber and sharp shells, galactomannans
Sesame	Phytase, oxalate
Sal seed	Tannins
Linseed	Prussic acid

* Destroyed by heating.

Critical Amino Acids in Oilseed Meals

Oilseed cakes and extractions are the most important sources of essential amino acids. In general, oilseed proteins have low cysteine and methionine content and a variable but low lysine content (Table 6). As a result, these cannot adequately supplement the cereal proteins for non-ruminant with which they are commonly used. The crude protein value of some oilseed such as cotton, groundnut, and soybean is high. Cottonseed meal also has the disadvantage of a low cysteine, methionine, and lysine contents, with lysine being the first limiting amino acid. Different oilseed meals are degraded at varying rates in the rumen of livestock. Since high-yielding dairy cattle and fast-growing goats and sheep need higher levels of undegradable dietary protein, it is possible to select desirable components of protein meals and formulate rations for differential levels of productivity.

Limitations to the Use of Oilseed Cakes

The problem of aflatoxins in animal feeds is of concern. The use of groundnut meal, which is very susceptible to fungal infection and aflatoxin contamination, has drastically decreased recently overseas despite being rich in proteins. Improper harvesting, transport, processing, and storage all contribute to aflatoxin contamination. The high humidity or high moisture content appears to be main factor for growth of Aspergillus. Preventive measures during crop production and postharvest handling of seeds would greatly reduce the level of contamination in feeds. Creating awareness among farmers about aflatoxins would help in utilization of groundnut meal in animal feeds. While soybean meal is not normally found to contain detectable amounts of aflatoxins, the incidence of these metabolites has been on the rise lately. Sunflower meal is very rich in many essential amino acids, but its use has been restricted in poultry owing to its high fiber content. Decortications of sunflower seeds prior to expelling/solvent extraction produces oil with less gum content, better clarity, and a meal with lower fiber. Similar problems with meals such as safflower meal, and niger seed meal in animal feeds.

Oil Seed Cakes

(1) Soy bean meal (Glycine max)

The meal is regarded as one of the best sources of protein to animals. It contains all the essential amino acids except that the concentration of cystine and methionine are suboptimal. Latter is the first limiting amino acid and may be particularly important in high-energy diets. Two of the six-protease inhibitors, Kunitz anti-trypsin factor and the Bowman-Birk chymotrypsin inhibitor are of practical significance. Pancreatic hyperactivity as a result of these inhibitors, results in increased production of trypsin and chymotrypsin. Another substance, contributing to growth retardation, along with protease inhibitors, is a heamagglutenin (lectins) but it may not agglutinate red blood cells in sheep and calves. Lectins, affect cell membranes, including those of epithelial cells lining of small intestine and reduce absorption of nutrients. Body's defence system is also impaired. Lectins may also encourage colonization of small intestine by coliforms. The toxicity of lectins is probably highest in castor bean but relatively mild in soybean. Heating inactivate inhibitors, but for ruminants these inhibitors are not important and toasting is not warranted. The toasting must be carefully controlled, as over heating will reduce the availability of lysine and arginine. The urease activity less than 0.4 mg N/g per minute may be desirable in meal. The meal being a poor source of B-vitamins must be supplemented for pigs and poultry. Fishmeal is a good supplement. It is a better source of calcium and phosphorous than cereal grains, but poorer than animal protein foods. Soybean meal contains about 1g/ kg of genistein, which has oestrogenic properties and a potency of 4.44×10^{-6} times that of diethyl-stilboestrol. Soya protein concentrates (70% crude protein), produced by solvent extraction and removal of insoluble carbohydrates, are suitable for inclusion in calf milk replacers and baby pig diets, since antigenic and antinutritional substances get removed during processing.

(2) Cotton Seed Meal (Gossypium hirsutum)

As for other oilseeds, cytine, methionine and lysine content are lower. Lysine is the first limiting amino acid. Calcium content is low and calcium to phosphorus ratio is about 1:6. It is good source of thiamine but poor in carotene. Pigs and poultry do not readily accept the meal, largely because of its dry and dusty nature. It has been a traditionally preferred source in most of the cotton belt of country, probably because milk fat tends to become hard and firm a preferred quality during traditional home churning. The cottonseed cake is costive by nature, and fit well when fed in rations having laxative feeds as berseem.

Gossypol, a polyphenolic aldehyde (alkanal) present in meals (0.3 to 20 g/kg) is an antioxidant and polymeristation inhibitor toxic to non ruminants. Gossypol is distinguished between free (soluble in 70-30 v/v aqueous acetone) and bound form. The former determine the toxic potency of the material. The shearing effect of the screw press in the expeller process is an efficient gossypol inactivator at temperatures which do not reduce protein quality. At high temperatures gossypol may bind with lysine. Pig and poultry should not contain more than 5 to 10 per cent of meal. Olive green discoloration of the yolk may be because of laying bird's diet having cotton meal.

Ferrous sulphate can ameliorate the effects to 1 part gossypol. Ruminants do not show ill effects even when they consume large quantities of cotton seed meal. Cotton seed meal upto 1200 mg of free gossypol / kg are acceptable in UK, bit its concentration should not increase 100, for poultry and calves and 60 for pigs, for ruminants it may be up to 500 mg/kg.

Like groundnut cake ghani pressed, expeller pressed and solvent extracted cotton seed cakes are available for feeding the livestock. In cotton growing belt it is the main source of protein to the livestock. It contains 38 – 40 per cent protein. It is a good source of protein for cattle, buffaloes, sheep and goats. It is deficient in lysine and is rich in gossypol which is injurious to swine, poultry and young calves. Horses and mules are also susceptible.

Cottonseeds are fed as such to the lactating animals. It is a common practice to feed the seeds to the lactating animals so that butter can be separated easily from the milk.

(3) Rape seed/ Mustard cake

Rapeseed meal is produced from oilseed varieties, Brassica napus, and B. compestris. This is the second largest sown oilseed crop of the country and grown in Rabi season in areas where water resources are limited. Much of the work on development of this crop has been conducted in Canada, and successfully modified in desirable fashion. The rapeseed oil contains erucic acid in high concentration, which is useful in a number of industrial applications.

Erucic acid is toxic, therefore low erucic acid rape has been developed for edible purpose.

Another group of toxic compounds called glucosinolates, inhibit the metabolism of thyroid gland and thereby induce goiter, i.e. enlargement of thyroid gland. Liver and kidney may also be affected. Glucosinolates (thioglucosides) have a characteristic pungent and hot taste. The "double zero" varieties having low concentration of both erucic acid and glucosinolates, have been developed and named canola. Sinapine, an astringent phenolic compound may have adverse effects on palatability. This gets converted to trimethylamine, which if not oxidized get transferred to the eggs. Rhode Island breed, develop fishy flavoured eggs because of sinapine.

Rape seed meal per se, has lower protein 40% and metabolisable energy than soybean meal, but it rich and methionine. The balance of calcium and phosphorous is satisfactory and contains higher phosphorous than other oilseed residues.

The presence of antinutritional/ toxic compounds is not of much consequence for ruminants, and these cakes are being traditionally used in rations of cattle and buffalo, in many parts of country. The same has also been demonstrated in a series of experiments at NDRI. Initial decrease in feed intake or minor liver damage may be observed. With pigs the reduced food intake, growth depression and carcass quality are affected. Rapeseed meals frequently contain tannins. These are polyphenolic compounds which complex with proteins and carbohydrates to form enzyme resistant substrates.

Mustard cake is also an important source of protein for livestock feeding. It contains 34 – 37 per cent protein. In the case of poultry and pigs it is not a good source of protein. About 10 – 15 per cent of it can be incorporated in poultry and swine rations.

(4) Groundnut (Peanut meals) (Arachis hypogoae)

The crop is grown for human consumption and extraction of oil. The area under crop is highest among oilseeds in India. It is grown in kharif in light soils, and nuts are produced in underground pods. The cake contains 42 to 48 per cent crude protein, which is quite deficient in lysine. Cystine and methionine are also in suboptimal amounts. It has been found markedly inferior to soybean meal for pigs even when supplemented with lysine. The undecorticated cake high in fibre and produced by screw pressing, which may be followed by solvent extraction.

Groundnut cake is important source of protein for livestock feeding. Three types of oil cakes are available in India, namely ghani pressed, expeller pressed and solvent extracted cake. The content of oil is variable according to the process of extraction of oil. It is 10 - 12 per cent in ghani pressed, 6 – 8 per cent in expeller pressed and 0.5 – 0.7 per cent fat in the solvent extracted cake. The protein content is variable from 40 – 50 per cent. The groundnut cake may be decorticated or with outer hulls. In the undercorticated groundnut cake fibre content is very high.

Groundnut cake is fed to the cattle, buffalo, sheep, goats, poultry and swine. It is deficient in lysine and methionine. The cake is very palatable to pigs, but its use more than 25 per cent of diet tends to produce a soft body fat and may have a troublesome laxative action. For cattle it forms an excellent protein source, being laxative in nature, can be beneficial in a high fibrous diet. Presence of growth factor as well as antitrypsin factor has been reported. Cake also has antiplasmin activity and so shortens bleeding time, but it is destroyed by heating.

Poisoning of turkey poults and ducklings by groundnut meal was reported. The toxic factor was shown to be a metabolite of the fungus Aspergillus flavus and was named aflatoxin. This is now known to be a mixture of four compounds designated aflatoxins B_1, G_2, B_2 and G_2, of which B_1 is the most toxic. There are considerable species differences in the susceptibility to these toxins; turkey poults and ducklings are highly susceptible, calves and pigs susceptible whereas sheep are classed as resistant. Young animals are more susceptible than adults of the same species. A common feature in affected animals is liver damage with marked bile duct proliferation, liver necrosis and, in many cases, hepatic tumours. In fact, aflatoxin has been shown to be a potent liver toxin and a very active carcinogen.

A loss of appetite and reduced milk yield in cows, have been reported. Deaths have occurred in six-month-old steers given 1 mg/kg of aflatoxin B_1 in their diet for a period of 133 days and live weight gains were generally significantly reduced. Administration of 0.2 mg/kg aflatoxin B_1 in the diets of Ayrshire calves significantly reduced liveweight gains. A metabolite of aflatoxin known as aflatoxin M_1, which causes liver damage in ducklings, has been shown to be present in the milk of cows fed on toxic meals. There have been no reports of clinical poisoning in sheep. Aflatoxins are relatively stable to heat, and methods of eliminating them from meals are elaborate. The best method of control is suitable storage to prevent mould growth, although aflatoxins may be produced in the growing crop.

(5) Sunflower seed meal

The annual plant sunflower (Helianthus annus) is an erect being 1.8 meter or more in height. The size of large yellow flowers varies greatly. The seeds are enclosed in a fibrous hull. The seed contain 40 to 42 percent oil. The meal is produced when the oil is removed from the seed by hydraulic pressure or solvent extraction. The hulls are usually partially, rather than completely, removed but the resulting high-fibre meals (up to 420 g/kg DM) are readily accepted by older animals provided they are finely ground. As a result of the variation in the extent of dehulling before extraction, the meals vary widely in composition and nutritive value.

Solvent-extracted meals contain about 15 to 20 percent crude fibre and 40 to 45 percent crude protein in of dry matter and have metabolisable energy contents of 8.1, 10.4 and 11.2 MJ/kg DM for poultry, cattle and sheep, respectively. Sunflower oil has a high content of polyunsaturated fatty acids and may cause soft body fat in pigs. The oil is very susceptible to oxidation and the meals have a short shelf life owing to the development of rancidity, which renders them unpalatable. The meals are useful sources of protein, which is low in lysine, but has about twice as much methionine as Soya protein.

(6) Sesame seed meal

Sesame (Sesamum indicium) is grown in many parts of the country. The seeds are produced in capsules or pods that split open at maturity. The expeller cake have lower protein content (about 400 g/kg DM compared with 500 g/kg DM for the solvent extracted material), but have oil contents of more than 100 g/kg DM compared with 20 g/kg DM for the extracted meal. The protein in rich in leucine, agrinine and methionine but is very low in lysine. The meal therefore needs to be combined with foods rich in lysine when given to pigs and poultry. For ruminants, the protein has a degradability of 0.65 – 0.75 depending upon the rate of passage through rumen.

(7) Linseed meal

Linseed or flax (Linum usitatissum) grows well in high rainfall areas in Rabi. A well drained loamy to clay soil is suitable for the crop. The plant bears blue flowers is also grown for stem that is used to make thread, for linen. The meal has laxative effect attributed to mucilaginous gums present in the meal. The flax seed has the highest omega-3 fatty acids are characteristic of fish oil and have been linked to beneficial effects on cardiovascular health of humans beings.

Linseed meal contains two types of toxic factors. One is the dipeptide called linaline (linamarine) consisting of glutamic acid and 1 –amino- d – proline, latter being antagonist to pyridoxine (B-6). Another toxin is cynogenic glycoside, similar to those present in cassava. The meal is also reputed to have favorable effects on the hair coat of the animals, producing a good "bloom".

It has been reported that linseed meal has a protective action against selenium poisoning. The cynogenic glycoside increases the excretion of selenium in the urine, thus decreasing its toxicity. Signs of selenium toxicity include hoofs deformities, hair loss and gastrointestinal tract irritation. Thiocynate during the detoxification may be goitergenic to pregnant ewes.

The mucilage containing (30 to 100 g/kg) property in the linseed meal is unique among the oilseed residue, which is a almost completely indigestible by nonruminant animals but can be broken down by the microorganisms of rumen. It is a readily dispersible in water, forming a viscous slime.

Linseed cake is fed mostly to horses and young calves. It is rich in protein and has a conditioning effect on the animals. It is good source of protein for cattle, buffalo, sheep and swine but is not a good source of protein in the poultry rations.

(8) Coconut meal

Coconut meal is a byproduct remaining after extraction of oil from the dried endosperm of the coconut (Cocos nucifera). The coconut trees are grown in southern, peninsular states. The crude protein content varies between 20 to 26 per cent. Coconut meal contains from 2.5 to 6.5 % oil, thereby being very useful in the preparation of high energy diets. The protein is low in lysine and histidine but methionine content is higher than soya bean. The low protein together with high fibre content (12 %) limits the use of the meal for the monogastric animals. But, neither protein quality nor fibre content is limiting for ruminant animals and coconut meal provides an acceptable and very useful protein supplement. In dairy diets, it is claimed to increased milk fat content. The milk fats produces on diet containing considerable amounts of coconut meal are firm and excellent for butter making. Coconut meal has valuable property of absorbing upto half its own weight of molasses and as a result is popular in compounding.

Bibliography

Augelitti, R. (1998). Proteins. Academic Press, London.

Cheeke, P.R. (1991). Applied Animal Nutrition: Feeds and Feeding. Prentice – Hall, Inc.

Dass, M.M. and Singhal, K.K. (2005). Effect of feeding chemically treated mustard cake on growth, thyroid and liver functions and carcass characteristics in kids. Small Ruminant Research, 56: 31-38.

Pailan, G.H. and Singhal, K.K. (2003). Effect of chemical treatment of mustard cake on its palatability and degradation of glucosinolates. Indian Journal of Animal Sciences, 73 (4): 428 – 431.

Tyagi, A.K. and Singhal, K.K. (2001). Effect of glucosionalate of mustard cake and rapeseed on performance of crossbred cattle. Indian Journal of Animal Science, 71: 789 – 792.

D'Mello, J.P.F. (1995). Amino Acids in Animal Nutrition. Oxford, CABI Publishing.

International Livestock Research Institute (ILRI). 1995. Global agenda for livestock research (Gardiner, P. and Devendra, C., eds.). Proceedings of a consultation, ILRI, Nairobi, Kenya.

Lehninger, A.I. (1978). Biochemistry. 2nd edition. Kalyani Publisher, New Delhi.

Maynard, L.A; Loosli, J.K; Hintz, H.F. and Warner, R.G. (1979). Animal Nutrition. 7th edition. Tata McGraw-Hill Publishing Co. Ltd., New Delhi.

McDonald, P; Edwards, R.A; Greenhalhg, J.F.D. and Morgan, C.A. (2002). Animal Nutrition. Pearson Education, Singapore, Pte, Ltd., Indian Branch, Delhi, India.

Sudhakara Rao, M., Krishna, N. and Prasad, A. 1992. Determination of effective protein degradability of some unconventional protein sources in cattle by nylon bag technique. Indian Journal of Agricultural Sciences 52:280-281.

Walsh, G. (2002). Proteins Biochemistry and Biotechnology. Jhon Wiley & Sons Ltd., West Sussex PO19 1UD, England.

LIPIDS AND THEIR UTILIZATION

The lipid group (fats, ether-extractable) contains naturally occurring substances characterized by their insolubility in water and solubility in organic solvents such as ether, chloroform, boiling alcohol and benzene. This group includes not only the true fats but also materials that are chemically related (lecithin) and materials that have comparable solubility properties (cholesterol and waxes). The true fats are of interest not only because these are concentrated source of energy (providing 2.25 times more energy than that of carbohydrates and proteins) but also because a number of vitamins are associated with fat. These are the fat-soluble vitamins (A, D, E and K). In addition, most nutritionists recognize the need for certain essential fatty acids (linoleic and linolenic).

I. Classification of Lipids:

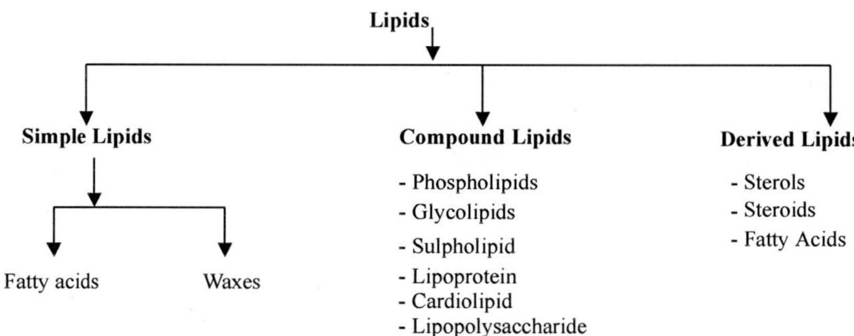

II. COMPOSITION

Fats and oils are esters of fatty acid with glycerol, waxes are esters of fatty acids with alcohol other than glycerol. Upon hydrolysis, typical fat yields three molecules of fatty acids and one molecules of glycerol. Hydrolysis of fat in the presence of alkali (sodium or potassium) results in the formation of alkali salts of the respective fatty acids or soaps (saponification). It should be noted that all the fatty acids listed in Table 1 have an even number of carbon atoms.

Glycerol	1-monoacylglycerol	1, 2, Diacylglycerol	Triacylglycerol
CH_2OH	$CH_2O-C(=O)-R_1$	$CH_2O-C(=O)-R_1$	$CH_2O-C(=O)-R_1$
$CHOH$	$CHOH$	$CHO-C(=O)-R_2$	$CHO-C(=O)-R_2$
CH_2OH	CH_2OH	CH_2OH	$CH_2O-C(=O)-R_3$

$$\begin{array}{l} \quad\quad\quad\; O \\ \quad\quad\quad\; \| \\ CH_2O-C-R_1 \\ | \\ CHO-C-R_2 \\ | \quad\quad\;\; \| \\ | \quad\quad\;\; O \\ CH_2O-C-R_3 \\ \quad\quad\quad\; \| \\ \quad\quad\quad\; O \end{array} \xrightarrow{3\ H_2O} \begin{array}{l} CH_2OH \\ | \\ CHOH \\ | \\ CH_2OH \end{array} + \begin{array}{c} R_1-COOH \\ + \\ R_2-COOH \\ + \\ R_3-COOH \end{array}$$

Triacylglycerol — Glycerol — Fatty acids

Hydrolysis of triacylglycerol

Table 1: The fatty acids

Fatty Acid	Formula	Occurrence
I. Saturated		
Acetic	CH_3 COOH	Vinegar
Butyric	C_3H_7 COOH	Butter
Caproic	C_5H_{11} COOH	Butter
Caprylic	C_7H_{15} COOH	Butter
Capric	C_9H_{19} COOH	Coconut oil
Lauric	$C_{11}H_{23}$ COOH	Coconut oil
Myristic	$C_{13}H_{27}$ COOH	Coconut oil
Palmitic	$C_{15}H_{31}$ COOH	Animal fat
Stearic	$C_{17}H_{35}$ COOH	Animal fat
Arachidic	$C_{19}H_{39}$ COOH	Peanut oil
Lignoceric	$C_{23}H_{47}$ COOH	Arachis oil
Cerotic	$C_{25}H_{51}$ COOH	Wool fat
Montanic	$C_{27}H_{55}$ COOH	Beeswax
Melissic	$C_{29}H_{59}$ COOH	Beeswax
II. Unsaturated		
1. One double bond		
Oleic	$C_{17}H_{33}$ COOH	Animal fat
Erucic	$C_{21}H_{41}$ COOH	Rapeseed oil
2. Two double bond		
Linoleic	$C_{17}H_{31}$ COOH	Linseed oil
3. Three double bond		
Linolenic	$C_{17}H_{29}$ COOH	Linseed oil
4. Four double bond		
Arachidonic	$C_{19}H_{31}$ COOH	Lecithin

III. PROPERTIES OF FATS AND FATTY ACIDS

The properties of fats and fatty acids depend largely on two things, the length of the carbon chain and the degree of saturation. The shorter chain fatty acids are volatile and water soluble whereas the longer chain fatty acids having low volatility and practically water insoluble. The hardness of the fatty acids at room temperature, which is correlated with melting point, also varies with the length of the carbon chain and the degree of saturation. Longer chain and saturated fatty acids tend to be solid at room temperature whereas shorter chain and unsaturated chain fatty acids tend to be liquid.

Unsaturated fatty acids can be saturated by the addition of hydrogen, utilizing nickel as a catalyst. Thus, oils or soft fats (due to unsaturation) can be made harder using this process. Many lard and butter substitutes made from unsaturated vegetable oils attain their firmness from hydrogenation of at least a portion of their unsaturated carbon chains.

The number of grams of iodine that combine with 100g of fat is known as the iodine number of that fat and indicates the degree of unsaturation of the fatty acid carbon chains. Unsaturated oils have a relatively high iodine number (butter, 22–38; lard, 54–70; mutton tallow, 32–50; coconut oil, 8–10; cottonseed oil, 104–114; olive oil, 74–95; peanut oil, 83–105; soybean oil, 115–145).

Oxidative rancidity of fats may occur in unsaturated fatty acids when oxygen attacks at double bonds. Such rancidity can be prevented by the incorporation of an antioxidant or by storage in an oxygen free environment. Rancidity is often associated with disagreeable odours, as in the case of rancid butter. The typically disagreeable odour of rancid butter is due to liberated butyric acid. Oxidative rancidity may destroy the potency of vitamins associated with such feedstuffs.

IV. FUNCTION

The obvious nutritional function of the lipids are three fold, namely, (i) for energy, (ii) for the formation of essential fatty acids, and (iii) as a carrier for the fat soluble vitamins. However, there are other uses, which will be discussed.

Energy supplied by the true fats is one of the more digestible sources. All animals, regardless of age or species, use fat readily. It is a very concentrated source of energy, containing 2.25 times the energy of either carbohydrates or proteins. Such a concentrated sources of energy permits storage of a given amount of energy with much less bulk and weight. However, since the total energy requirement of animals could be met by dietary carbohydrates and proteins, the energy required from the lipids is nonessential.

Essential fatty acids (linoleic and linolenic) must be supplied as such since there absence causes certain animals to have deficiency symptoms, such as hairlessness, scaly skin, deterioration of the tail, oedema, and in the case of chicks, poor feathering. The function of these two unsaturated fatty acids is not understood.

Another function attributed to fat is the carrying of fat-soluble vitamins. Fat is involved in the absorption of vitamins A, D, E, and K. Because the amount of fat required to meet this need is small, one would not anticipate such vitamin deficiencies merely because the level of fat in the diet was too low.

In addition to the three nutritional functions of fat, some other functions are: Heat insulation of the animal body against the environment by stored fat is important, as is cushioning of the vital organs against injury. A good example of this is the huge knob of fat that is used as a cushion to protect the kidney in the animal body. These functions of fat, along with its use as a concentrated source of energy, are vital to the hibernating animal, which consumes neither energy nor water for long periods of time. The fat is utilized for energy and for the production of metabolic water over the long hibernation period. Fat provides palatability to feed and food. It also decreases dust in final ground and dry feedstuffs.

V. ASSOCIATED COMPOUNDS

One group of compounds associated with the lipids is known as phospholipids. They contains phosphorous in addition to carbon, hydrogen and oxygen. The role of the phospholipids is primarily as constituents of the lipoprotein complexes of biological membranes. Phospholipids have the useful property of attracting both water soluble and fat soluble substances within the same molecule. They are therefore surface active (hence it is constituents of biological membrane) and may play an important role as emulsifying agents in biological system. Among phospholipids the most common is lecithin. Lecithin contains a variety of constituents, including several fatty acids (stearic, oleic, and palmitic), glycerol, phosphoric acid, and choline. Lecithin is a common cell constituent and is considered to play a role in the metabolism of fat. Choline

$$\begin{array}{l} CH_2.O.CO.C_{15}H_{31} \\ | \\ CH.O.CO.C_{17}H_{33} \\ | \\ CH_2.O.PO_3.CH_2.CH_2.N^+(CH_3)_3 \end{array}$$

Lecithin

$$\begin{array}{l} CH_2.O.CO.R_1 \\ | \\ CH.O.CO.R_2 \\ | \qquad\quad O \\ CH_2.O\overset{\|}{P}.OH \\ \qquad\quad | \\ \qquad\quad OH \end{array}$$

Phosphatidic acid

may be the key constituent in lecithin. In animals on a diet deficient in choline, neutral fats tend to accumulate in the liver leads to fatty liver disease. An excess of choline results in the opposite effect. One of the roles of lecithin seems to be that of emulsification of fat so that it can become water miscible.

In glycolipids the two alcohol groups of glycerol are esterified by fatty acids and other is linked to a sugar residue. The lipids of grasses and legumes which form the major part of dietary fat of ruminants is the galactolipids, where sugar residue is galactose.

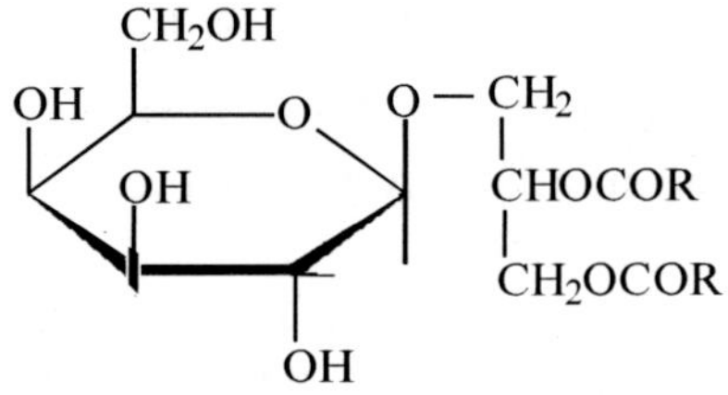

CH_2OH, OH, O, OH, OH, $O-CH_2$, CHOCOR, CH_2OCOR

Galactolipid

Glycolipids are also present in brain and nerve fibres called the cerebrosides, where the glycolipids have the amino group of the sphingosine linked to the carboxyl group of a long chain fatty acid, and the sugar residue, usually galactose linked to the terminal alcohol group.

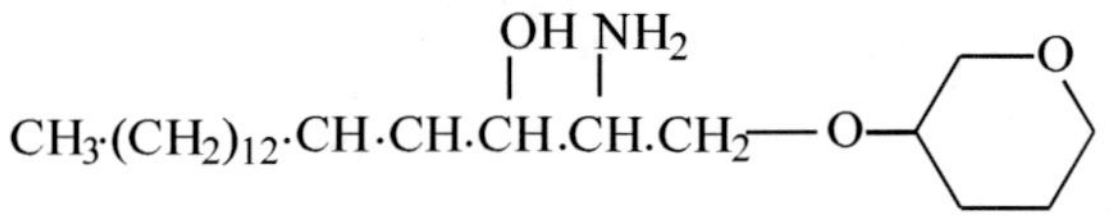

$CH_3{\cdot}(CH_2)_{12}{\cdot}CH{\cdot}CH{\cdot}CH(OH).CH(NH_2).CH_2-O-$ (Galactose ring, O)

Sphingosine Galactose

Cerebrosides

Another group of associated lipid compounds is the sterols, which are monohydroxy alcohol and are found in both animals and plants. The sterols and related substances are referred to as steroids (bile acid, sex hormone), which having a similar ring structure known as "Cyclopantane-per-hydrophenanthrene ring." The most important sterol in animal tissues is cholesterol particularly present in nervous tissue. It is found in the blood at the levels of 150 – 190 mg/ 100 ml. Cholesterol is the precursor of steroid hormones and bile acids. 7-dehydro cholesterol is an important precursor of vitamin D_3 which is produced in the body by the irradiation of U-V light. Vitamin D possesses only about one-seventh the antirachitic activity of irradiates 7-dehydro cholesterol for poultry. Irradiation of plant sterol (ergosterol) produces vitamin D_2 (ergocalciferol).

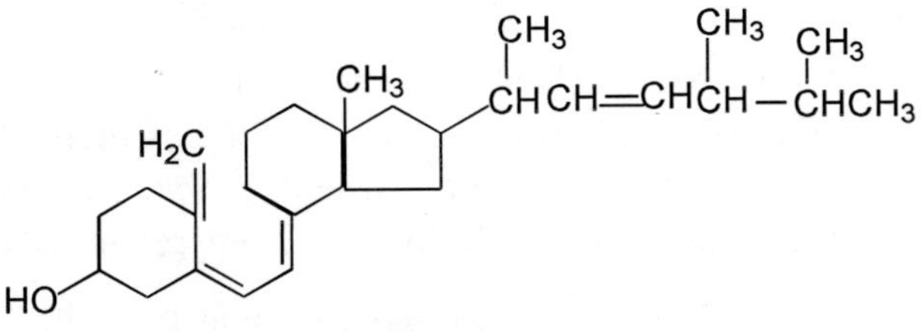

Vitamin D_2

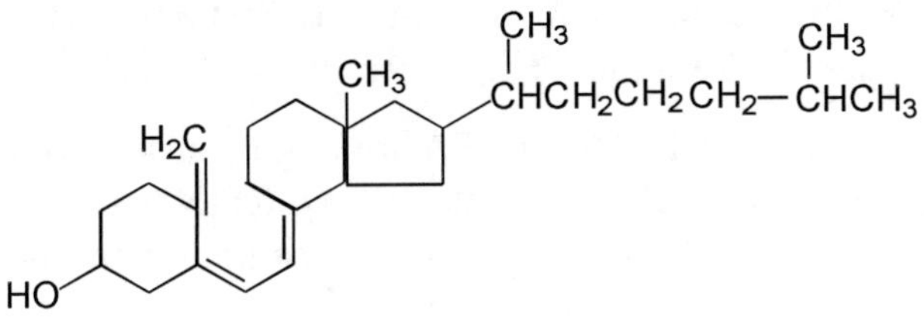

Vitamin D_3

VI. DIGESTION

The triacylglycerols present in the feed consumed by ruminants contain a high proportion of residue of the C_{18} polyunsaturated acids, linoleic, linolenic and oleic. These triacylglycerols are to a large extent hydrolyzed in the rumen by bacterial lipases, as are phospholipids. Once they are released from ester combination, the unsaturated fatty acids are hydrogenated by bacteria, yielding first a monoenoic acid and ultimately, stearic acid. Both linoleic and linolenic acid have all cis double bonds, but before they are hydrogenated one double bond in each is converted to the trans configuration; thus, trans acid can be detected in rumen contents. The rumen micro-organisms also synthesize considerable quantities of lipids, which contain some unusual fatty acids (such as those containing branched side chain) and these acids are eventually incorporated in the milk and the body fats of ruminants.

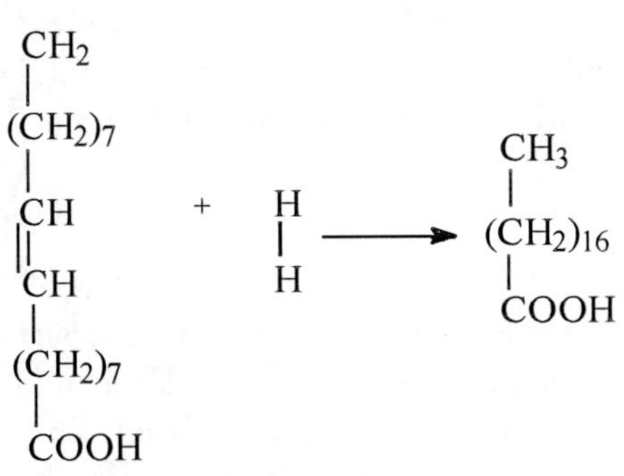

$$CH_2-(CH_2)_7-CH=CH-(CH_2)_7-COOH + H-H \longrightarrow CH_3-(CH_2)_{16}-COOH$$

Oleic acid Stearic acid

The capacity of rumen microorganisms to digest lipids is strictly limited (3 – 5%). The lipid content of ruminant diets is normally low (i.e. <50g/kg) and if it is increased above 100g/kg the activities of the rumen microbes are reduced because of coating of fatty acid upon fibre particles of feed leading to decreased crude fibre digestibility, particularly cellulose, resulting reduction of feed intake.

The principle site of lipid digestion is the small intestine, under the influence of the peristaltic action of the stomach, duodenum and presence of bile. In general, absorbability (digestion & absorption) of fat depends upon its ability to participate in micelle formation. Fat absorption is increased by phospholipid.

All short chain Fatty acids and VFA are largely absorbed through rumen wall, but long chain fatty acids are not absorbed directly from the rumen. When they reach the small intestine they are mainly saturated and unesterified but bacterial and protozoal lipids are esterified. The contents of the duodenum and upper jejunum are more acidic than that of non-ruminants, their conditions indicates there is little triglyceride available to be converted to monoglyceride (a potent emulsifying agent) and further pancreatic lipase will be less active in duodenum and upper jejunum. However, active micelle formation of fatty acids occur in the upper tract under the influence of bile salts, primarily taurocholate and biliary and ingesta phospholipid, largely lecithin and hence the absorption of long chain acids, is dependent on the phospholipids present in bile.

Depot fats in non-ruminant are formed from ingested fat and carbohydrate hence, it is possible to vary the fatty acid composition of body fats by altering the composition of dietary fats. In contrast to non-ruminants, fatty acids comparison of adipose tissue in ruminants is not markedly affected by the fatty acids composition of the diet, because of intense biodehydrogenation of dietary unsaturated fatty acids takes place in the rumen. The predominating fatty acid of ruminant depot fats is the stearic acid resulting from hydrogenation in the rumen. However, it is possible to treat dietary lipids in such a way that they are protected from attack in the rumen but remain susceptible to enzymic hydrolysis and absorption in the small intestine. If such lipids contain unsaturated acids, they will modify the composition of body and milk fats.

VII. STORAGE

Although fat may be deposited in various portions of the animal's body, it is stored primarily in (i) intramuscular connective tissue, (ii) the abdominal cavity, and (iii) subcutaneous connective tissue. Each animal tends to store fat that is characteristic of its species, regardless of diet. This is especially true of ruminants and monogastric, in which carbohydrate is the principle source of energy. However, swine depot fat can be altered by dietary fat. For example, if highly unsaturated fat such as soybean oil is fed to swine in excessive quantities, the resultant pork fat deposited will be intermediate in degree of unsaturation and softness between that typical of pork fat synthesized from a starch diet and that of the soybean oil consumed.

Adipose tissue always contains some water (4.5–14.4%). Thus, fat deposition in the body also involves deposition of water. At times when depot fat is called on to supply energy, there may be retention of water in place of the fat removed. This is the common occurrence observed by dieters. Dieters may often be discouraged initially because the scale does not show a loss in weight. Even though fat may be removed during the dieting process, the dieter may appear to be temporarily gaining weight because the same volume of water being stored despite the fact that depot fat was being used up.

Depot fat is in a constant state of turnover, fatty acids from the depots are mobilized, and newly formed fatty acids are constantly taking the place of the mobilized fatty acids.

VIII. OXIDATION

When fat is utilized as a source of energy, the first reaction is hydrolysis to glycerol and the three constituent fatty acids. Glycerol is then converted through L-glycerol- phosphate to dihydroxy acetone phosphate and then to acetyl-CoA, which then enter the tricarboxylic acid cycle (TCA cycle) and energy is released.

The major pathway for energy derived from fat is provided by ß-oxidation of fatty acids (Fig.1) which takes place in mitochondria. The theory of ß-oxidation of fatty acids, as formulated by Knoop in 1904, is still the most logical explanation of the method by which fatty acids are broken down eventually to CO_2 and H_2O. Actually fatty acid oxidation begins in the cytoplasm with formation of fatty acyl CoA (Fatty acid + coenzyme A = fatty acyl CoA). Fatty acyl CoA needs a special carrier mechanism in the form of carnitine to pass into the mitochondrion. In the mitochondria, fatty acyl CoA is successively dehydrogenated, hydrated, dehydrogenated again, and cleaved to acetyl CoA and a fatty acid shorter by two carbon atoms. This process is continued stepwise, each sequence producing a molecule of acetyl CoA, which further enters into TCA cycle and oxidized to $CO_2 + H_2O$. However, the presence of three or four carbon chain fragments, such as are found during a condition called ketosis is not clarified by the ß-oxidation theory. Some researchers suggest that three and four carbon chain fragments are broken off in the ß-oxidation process. Other suggest that the last four carbons remaining in the fatty acid chain are more difficult to dispose of and, therefore, tend to accumulate in various degrees of oxidation as ketone bodies (acetone, aceto acetic acid, ß-hydroxybutyric acid).

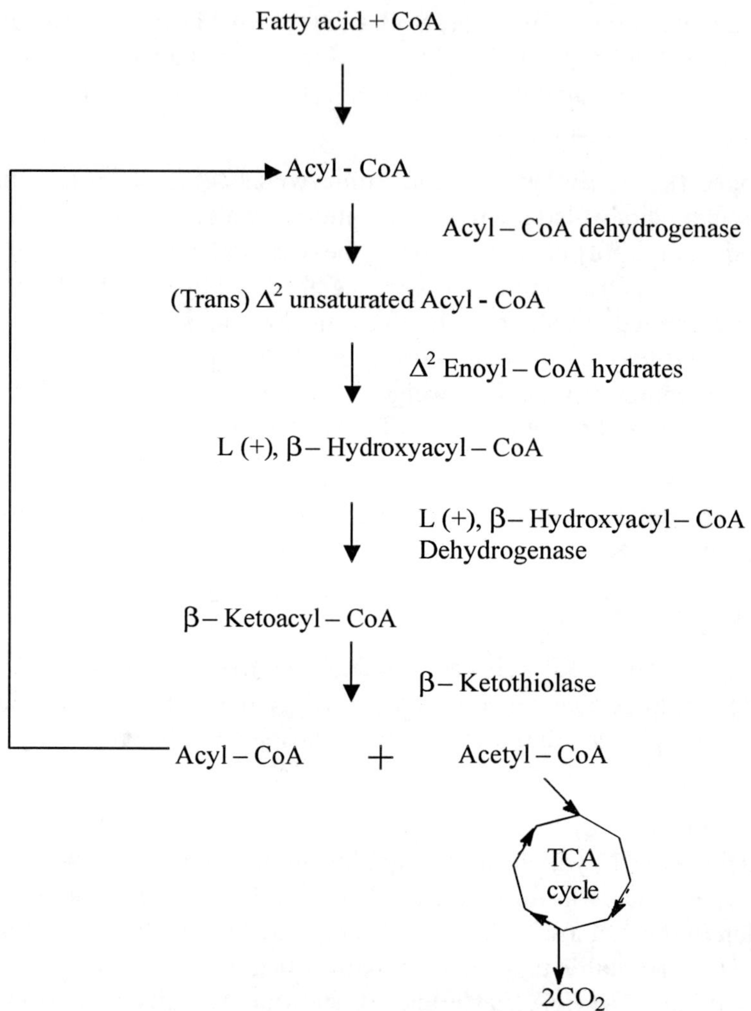

Fig-1: β – Oxidation of fatty acids

The energy yield in this sequence from palmitic acid as follows:

Moles	**ATP**	
1 moles palmitate to 8 moles acetyl CoA	+	-
	35	2
8 moles acetyl CoA to Co_2 and H_2O	96	
Total	131	2
Net gain	129	

IX. KETOSIS AND LIPID METABOLISM

Under certain nutritional or physiological stress, where energy demand exceeds the available tissue glucose, fat is mobilized to acetyl CoA and leading to produced Ketones bodies, condition called "Ketosis". This condition is quite common in pregnant ewes, one or two weeks prior to lambing, especially in those carrying two or more foetuses. It can also occur in heavy milking cows shortly after parturition. Ketosis can be explained by the fact that the animal's body does not have sufficient carbohydrates to meet its energy demand. Thus, it breaks down body fat to meet the increased call for energy. Hence, body fat is mobilized to acetyl CoA, it cannot be converted to energy via the TCA cycle and is thus converted to ketones. As fat is a readily available source of concentrated energy and can thus meet the demand.

The acetone bodies (acetone, ketobutyric acid, ß-hydroxybutyric acid, and acetoacetic acid) are normal products formed in the oxidation of fatty acid in the liver. However, the liver cannot utilize acetone bodies, and they are transported to the muscle for complete oxidation. Thus, if an overwhelming demand for energy requires excessive use of fat as the energy sources, then more acetone bodies are formed, which are absorbed into the blood stream. The bloodstream then contains an excessive amount of acetone bodies, which results into the condition of ketosis. Since, acetone is exhaled via the lungs and in severe cases, is readily detectable as sweet aroma of acetone on the breath of such animals.

X. FAT FROM CARBOHYDRATES AND PROTEIN

It is obvious that the animal body can convert carbohydrate to fat. As was discussed earlier, this is one of the bases of the livestock feeding industry by which finishing animals are fed predominantly carbohydrate diets from which they synthesize fat.

Protein can also be converted to fat. This pathway proceeds from protein to amino acid to carbohydrate to fat. However, there are some amino acids (leucine, tyrosine, or phenylalanine) that cannot form glucose, but that can be converted to fat without passing through the intermediate carbohydrate phase. Protein conversion to fat requires greater utilization of vitamin B_6 as well as other B vitamins.

Protected/ bypass fat

Energy density in the ration of lactating animals in developing countries is low and hence unable to meet its requirements. This lead to substantial loss in body weight which adversely affect production and reproduction in dairy animals. The protected/ bypass fats act as dense energy source and modify fatty acid profile of milk and body fat. The protected fats do not undergo hydrolysis and bio-hydrogenation in rumen, therefore, have very little effect on feed intake and digestibility when fed at higher levels. Feeding of bypass fat reduce weight loss of lactating animals at their first quarter of lactation and improve the body condition score. It also improves the conception rate of cows. Protected fat feeding to lactating cows and buffaloes resulted in sufficient increase in milk yield as well fat and protein content (Table 2). Feeding of bypass fat also increase the C_{18} unsaturated fatty acids in milk fat (Table 3). There are different types of protected fat i.e. crystalline fat, formaldehyde treated oilseeds, Ca salt of fatty acids, prilled fat.

Table 2: Effect of feeding bypass fat supplement on milk parameters

Meal	Yield (l/d)	Fat (%)	Fat yield (g/d)	Protein (%)	Protein yield (g/d)
Cow					
Control	8.2	3.95	324.3	2.90	238.09
Bypass fat	9.4	4.54	425.4	3.35	313.90
Buffalo					
Control	10.5	6.70	703.5	3.70	388.50
Bypass	12.3	7.80	959.4	4.30	528.90

Cows n = 12; Buffaloes n = 5 per group, were fed 1 kg per day of the bypass supplement

Source: Gulati et al.(2001)

Table 3: Fatty acids available for absorption at small intestine on feeding of bypass fat supplements

	Unprotected fat (g/kg)	Protected fat (g/kg)
Total fat	323.0	323.00
Rumen undegradable fat	16.0	255.17
Rumen degradable fat	307	67.83
Fatty acids available for absorption at the abomasums -		
Oleic (C18: 1 cis)	8.4	133.2
Linoleic (C18:2)	4.2	66.3
Linolenic (C18:3)	1.4	21.9
Bypass was made from canola/ soybean oilseeds (70/ 30)		

Source: Gulati et al.(2001)

Bibliography

Enser M. (1984). The chemistry, biochemistry and nutritional importance of animals fats, In: Fats in Animal Nutrition. J. Wiseman (ed.). Butterworths, London.

Gulati, S.K, Ashes, J.R, Kaur, R., Scott, T.W., Garg, M.R., Sherasia, P.L. and Singh, D.K. (2001). Role of protected nutrients in sustainable milk production. Proc. X Animal Nutrition Conference, Karnal, pp. 1-9.

Lehninger, A.L. (1982). Principles of Biochemistry, 1st edn. Worth Publ., New York.

Maynard, L.A; Loosli, J.K; Hintz, H.F. and Warner, R.G. (1979). Animal Nutrition. 7th edition. Tata McGraw-Hill Publishing Co. Ltd., New Delhi.

McDonald, P; Edwards, R.A; Greenhalhg, J.F.D. and Morgan, C.A. (2002). Animal Nutrition. Pearson Education, Singapore, Pte, Ltd., Indian Branch, Delhi, India.

Stryer, L. (1981). Biochemistry, 2nd edu. W.H. Freeman, San Francisco.

Sweuson, M.J. (ed.) (1977). Dukes Physiology of Domestic Animals, 9th edn. Coruell Univ. Press, Ithaca, USA.

MINERALS

Animal body contains a number of inorganic elements rather because present in the food. Nutritionally essential elements are only those, which are known to have metabolic role in the body and the deficiency symptoms develop in their absence. As many as forty or more elements may have metabolic roles in mammalian tissues. The essential minerals are classified into two groups, major and trace elements. Trace elements are present in animal body in a concentration not greater than 50 mg/ kg and are required at less than 100 mg/ kg diet. Major elements namely calcium, phosphorous, potassium, sodium, chlorine, sulphur, and magnesium are present in higher concentration than those of trace elements. The later group includes iron, zinc, copper, molybdenum, selenium, iodine, fluorine, manganese and cobalt. Elements chromium, arsenic, boron, lead, lithium, nickel, silicon, tin, vanadium, rubidium and aluminium have also been included in the list of essential, but these are required in so minute quantities that they can possibly be not required to be supplemented in natural diets.

The mineral group is defined as a portion of diet that will not be destroyed by ignition (620^{0}C) for 2h, and this technique is employed in assaying materials for ash content. This nonspecific assay has obvious built in errors. For example, extremely high temperatures will drive off much of the volatile fluorine. However, since fluorine seldom represents a significant quantity, this does not represent a very high percentage loss. The cooling of certain minerals, such as iron, increased slightly due to addition of the oxygen of oxidation.

I. CALCIUM

Calcium is the most abundant mineral element in the animal body and about 98% of the calcium is found in skeletal structure; thus, less than 2% is present in soft tissues. Small fraction of calcium found in soft tissues is also of tremendous importance. For example, transmission of nervous tissue impulse, excitability of the muscles, rhythm of the heart muscle, clotting of blood, maintenance of acid base equilibrium, regulation of the permeability of membranes, and activation of certain enzymes are calcium dependent.

Functions

Bone ash contains about 36% calcium. Living bone can contain as much as 40% calcium and as little as 15% water, depending upon its age and location in the body. The fat–free portion of bone contains about 40% organic matter. The remainder is ash or mineral. The ash or mineral portion of bone consists of about 85% (tricalcium phosphate), 10% (calcium carbonate), and 1.5% (trimagnesium phosphate). A part of calcium required for milk production or calcium and phosphorus may be needed by the developing faetus, can be drawn from bones.

With increasing age, a decreasing portion of the calcium enters the skeleton. Parathyroid hormone plays a major role in bone calcification. The level of dietary calcium will affect bone formation; the presence or absence of vitamin D, which is required for the absorption and utilization of calcium, will affect this process, as well. The calcium content of blood serum is between 9-12 mg/ 100 ml, but in infancy, it may be slightly higher; in extreme old age it may be somewhat lower. Long period of low calcium ingestion will result in lowered serum calcium. Calcium controls nerve and muscle excitability. Lowered levels of calcium result in increased excitability, where as higher level result in a pseudotranquilizing effect. Extremely low levels of serum calcium can result in tetany, this is followed by death unless the cause is determined and treatment given.

Calcium in the blood is essential in maintaining a pH of 7.4 – 7.6. Lack of basic elements (pH above 7.0) in the bloodstream causes acidosis. This is seen in lactating animals that are drawing blood calcium milk formation, thereby upsetting the acid-base balance. Animals that obtain too high a percentage of their energy needs from body depot fat produce higher levels of acidic intermediate breakdown products. High levels of acidic products in the blood stream can result in either acidosis or ketosis. Such animals may be supplemented additional calcium.

Absorption

Several factors affect the absorption of calcium, for example, the concentration of available calcium in the diet and the presence of vitamin D. The level of calcium available in the diet is inversely proportional to the absorbability of calcium across the wall of the small intestine.

When lactose, is present in the diet, animals will absorb a larger proportion of their dietary calcium. It has been demonstrated that the addition of lysine in particular, increases the level of calcium absorption.

Several factors interfere with calcium absorption. Sodium phytate, the hexaphosphate of the vitamin inositol, results in decreased absorption of calcium, probably by forming an insoluble complex with calcium, thus rendering it unavailable. Calcium, when combined with dietary oxalic acid, forms calcium oxalate, which is not soluble. High levels of phosphate form calcium phosphates of low absorbability. In ruminants oxalic acid is degraded in rumen, thereby ameliorating its effect. Excretion of calcium is primarily via faeces.

Rickets

Rickets, although associated with calcium and phosphorus malfunction, also generally associated with a deficiency of vitamin D. Osteomalacia or "adults rickets" is usually, but not necessarily associated with phosphorus deficiency. The condition is due to a negative calcium and phosphorus balance in the bone, which result in replacement of bone with soft tissue. In calves decreased growth stiffness of gait, enlarged and painful joints, arching of back appears due to rickets.

Milk Fever (Parturient Paresis)

Milk fever is misnomer, on the contrary body temperature may be depressed. It is a condition, which most commonly occurs in high yielding dairy cows and buffaloes shortly after calving. It is characterized by a lowering of the serum calcium level, muscular spasm, and in extreme case paralysis and unconsciousness. It is generally considered that, with the one set of location, the parathyroid gland is unable to respond rapidly enough to increase calcium absorption from the intestine to meet the extra demand. Injections of calcium gluconate help to improve blood calcium level. Adequate levels of phosphorus, and avoiding excessive Ca intake, reduce incidence of milk fever. Administration of very heavy doses of vitamin D_3 for 3 to 5 days prepartum and one day post partum have been claimed to produced some control of milk fiver. Short period prior to parturition has also proved beneficial.

Requirement and Sources

Calcium requirement are highly depended on the level of productivity. Increased growth rates and milk production will greatly increase mineral requirements. Farm animals require diets containing 0.24 – 0.77% calcium in the dry matter. Forages, particularly leguminous species are satisfactory source of calcium and contain on an average 1.01% calcium while grasses contain 0.38%. The leaf contains twice as much calcium as the stem. Again calcium concentrations of forage increased by the application of nitrogenous fertilizers and decreased with advancing maturity. Maize and sorghum silage commonly contain 0.10 – 0.40% calcium and are poor source than herbage, while cereal straws contain around 0.15 – 0.30% calcium. All the cereal grains and there by products, oil seeds and oil seed cake are low in calcium. Animal byproducts as fishmeal, meat meal are comparatively rich source of calcium, particularly products of low quality containing much bone. In practice rations may be deficient in calcium and should be supplemented with mineral source.

II. PHOSPHORUS

Phosphorus is mostly present with calcium.

Functions

About 80% of body phosphorus is found in skeleton, 10%in the muscles and 1% in the nervous system. The reminder is generally distributed throughout body cells. Within the cell, most of the phosphorus is found in the nucleus. Red blood cells and plasma are rich in phosphorus. Vitamins B_1 and B_2 are phosphoric acid compounds. It plays a key role in carbohydrate metabolism through the formation of hexose phosphates, adenylic acid, and creating phosphate. Phosphorus has more known functions in the body than any one other minerals.

Metabolism

Although absorption of phosphorus from the small intestine is quite rapid. The inorganic form occurs primarily in the plasma at a level of 3–4 mg/100 ml (pigs, 8mg%, sheep and cattle, 6mg%; horses, 2.3mg%).

Deficiency

Phosphorus deficiency is manifested by a decrease in the organic phosphorus content of the blood plasma and depletion of calcium and phosphorus stores in the skeleton. Severe, prolonged phosphorus deficiencies result in lowering of plasma phosphorus and an increase in plasma calcium levels.

Where as animals on calcium–deficient diet may gain weight quit well and appear to be thrifty, animals with a marked phosphorus deficiency will be emaciated, unthrifty and listless. This often leads to death of the animal. Livestock producers rarely observe such marked deficiency symptoms. A moderate phosphorus deficiency usually results in only depressed animal weight gain with no specific symptoms. The consumption of unusual materials such as dirty, wood and cloth by cattle is often characteristic of a phosphorus deficiency. This is referred to as depraved appetite, which is also known as 'pica'. Greatly reduced reproductive performance (upto 50%) often characterizes a phosphorus deficiency, especially in ruminant animals. A phosphorus deficiency often interferes with the estrous cycle. Lactating females will tend to produce milk as long as they can draw sufficient calcium and phosphorus from skeletal stores. Once such an available source is depleted, a sever decline in milk production can be expected.

Toxic Levels

Feeding with high level of phosphorus should also be avoided, as it is excreted and contribute to pollution by encouraging the growth of algae in water courses. Toxic levels of phosphorus, although possible theoretically, are not probable, since phosphorus continue to be the most expensive mineral element used to fortify diets. Excessive dietary phosphorus results in extremely high levels of bone resorption, which cause lameness and leg bone fracturing. Large intake of magnesium, iron and aluminium may interfere with phosphorus absorption.

Excretion

Secretion of endogenous phosphorus into the intestinal tract can occur, but this loss of phosphorus is fairly low. The kidneys are the principal regulators of blood plasma inorganic phosphorus levels. Increased amounts of phosphorus are excreted by the kidneys as the plasma inorganic levels become elevated.

Requirement and Sources animals require rations containing 0.18 – 0.50% phosphorus depending upon the level of productivity and physiological stage. The phosphorus content of forage very widely and are influenced by the phosphorus contents of the soil, climate and stage of maturity of a plant. Grazing animals or those fed high roughage diets are more likely to be deficient in phosphorus than in calcium because of the low phosphorus content of forage, but the cereal grains and there by products, oil seeds and oil seed cake are rich in phosphorus. Similarly, fishmeal and meat meal are comparatively richer source of phosphorus. Ration deficient in phosphorus should be supplemented with mineral source like dicalcium phosphate, sodium phosphate, phosphoric acid steamed bone meal etc, which are readily available for poultry and all other species of farm animals. Untreated rock phosphate and super phosphate may be harmful because of the formation of pyro and metaphosphate.

III. POTASSIUM

Potassium is present in adequate quantity, in herbage, thus it may not be a limited factor in usual dietary of the ruminants. Potassium content of the body is similar to that of sodium, but it is primarily present as a cellular constituent. Potassium is associated with nitrogen metabolism. When new protoplasmic tissue is formed, sufficient quantities of potassium should be available to meet the intracellular needs of the newly formed cells. The irritability of the nervous system is dependent upon a balance between calcium, potassium and sodium ions. Thus, a decrease in calcium ions increases irritability; an increase in potassium ions will cause the same effect. In contrast to sodium, potassium is present in animal tissue almost exclusively within the cells in ionic form. It is usually not found in extra cellular fluids, which is the case with sodium.

Functions

It appears to be the major mineral constituent within the cell concerned with osmotic pressure regulation and acid–base balance. The potassium content of the red blood cells is 20 times greater than that of the plasma, whereas sodium is found in the plasma outside the red cells. Potassium constitutes over one–half of the cations of saliva–over one–fourth of the total ions. In milk it constitutes 28% of the total cations and it is utilized in enzyme reactions involving phosphorylation of creatine and facilitates uptake of neutral amino acids by the cells.

Metabolism

Potassium is mostly absorbed through upper small intestine. The regulation of body levels of potassium has not been defined. However, excessive quantities that have been ingested are chiefly excreted via urine during profuse sweating a large portion may be diverted through this channel. Potassium deficiency also causes growth retardation and emaciation. Potassium is the third most abundant mineral in body and must be supplied in diet, as it cannot be stored and required in highest quantity of all the mineral elements. The tolerance level for excessive intake of potassium is quite broad, indicating that it would be difficult to create toxic levels, but it may cause problems with calcium and magnesium metabolism particularly for perparturient dairy cows and may cause under edema.

Requirement and source

The requirement for potassium is about 0.2 – 0.5% of DM in monogastric species and slightly higher, 0.6 – 0.8% in ruminants. Forage (.01-0.6%) is richer in potassium than concentrates. The common rations of all the farm animal species usually meet the requirements for potassium, but milch animals fed high concentrate diets may be marginally deficient in potassium. Energy source are low (0.3-.05%) in potassium and brewer's grain are exceptionally low (0.04%) in potassium. Protein source contain 1-2% potassium. Potassium concentrations are also influenced by the potassium status of the soil, the plant species and its state of maturity. The red soils may be poor in potassium.

IV. SODIUM

The metabolism of sodium, potassium and chlorine is closely inter-related. Symptoms of deficiency include a depraved appetite, depressed feed intake, emaciation and reduced production. Sodium and chlorine are commonly administered as the compound sodium chloride (NaCl). The adult bovine body sodium is distributed entirely in the extra cellular fluids.

Functions

Sodium functions with chloride and bicarbonate to control osmotic pressure and the electro neutrality of body fluids and tissues. Changes in osmotic pressure of plasma are due to and can be caused by varying the concentration of sodium. Under stress conditions of low sodium, the body can compensate with potassium, but only to a limited extent, and also decrease the excretion significantly. The level of sodium appears to be regulated by the adrenal hormones as with potassium heavy sweating may cause large losses of sodium.

Symptoms of Deficiency

Reduced growth rates and decreased efficiency of feed utilization are non-specific symptoms of sodium deficiency. Urinary excretion of sodium will drop to nearly zero under such conditions and affected animals develop an intense carving for salt, as manifested by an attempt to lap any urine that might be available. Assay of the blood indicates a greatly increased cell–packed volume. However, perhaps a more logical explanation is that total blood serum is reduced, providing the same net effect. The adrenal cells become greatly enlarged during sodium deficiency stress, perhaps indicating that the adrenal gland is attempting to compensate for the stress condition by increasing hormone secretion.

Requirement and source

The requirement for all farm animals for sodium is around 0.10-0.20% of the dry matter. The distribution of sodium in pastures is lower (< 0.15%). Low values are more common in tropical than temperate pastures, with legume particularly deficient (< 0.04%). There is also decrease in the sodium concentrations of grasses, but not in legumes as they mature. Cereal and vegetative foodstuffs (except sunflower meal) are poor in sodium. Root crops and animals products are generally much richer in sodium than cereals, while animal protein sources like fishmeal and meat meals are higher in sodium than most plant protein source. Thus sodium is not always present in common rations in adequate and it is a common practice to supplements ration with common salts.

V. CHLORINE

Chlorine is present at an estimated level of 0.12% of the adult body. It occurs especially in the extra cellular fluids and thus is found in association with tissues containing the greatest amount of interstitial water. The cerebrospinal fluid contains the greatest concentration of chloride, followed by interstitial fluids, lymph exudates and gastrointestinal juices, respectively.

Functions

The functions of chlorine are rather broad, considering the small quantity maintained by the body. One unique and fascinating function of chlorine in the animal body is what is known as the chloride shift. It has been previously pointed out that the acid – base balance (or pH) of the blood is maintained quite rigorously, within a pH range of 7.4 – 7.6, by a delicate balance of anions, cations, and buffers. The products of energy metabolism enter the blood stream at the cellular level. One such byproduct is CO_2, when dissolve in water gives a week carbonic acid effect. The chloride ion, too, is acidic and is a part of the balance of ions in the blood. However, only the chloride ion has the unique ability to cross the membrane of the red blood cell in either direction. Thus, when carbonic acid from the cell enters the bloodstream or plasma, the acidic chloride irons leaves the plasma by entering the red blood cells, eliminating the acidic effect of the chloride ion. The pH equilibrium of the plasma is thus maintained. When the venous blood, containing large quantities of carbonic acid, give up its CO_2 and some of its water in the lungs, the negative chloride ions leave the red blood cells, reenter the plasma, and exert their acidic effect.

Chloride is involved in osmotic pressure maintenance in the body. It constitutes approximately two–third of the total anions of blood plasma, as well as the other body fluids. Chloride is the part of hydrochloric acid (HCl), in the true stomach or abomasums. The acidity of gastric HCl may be as low as 0.17 N (equivalent to a pH of 0.9).

Metabolism

The metabolism of chlorine is related closely to that of sodium. Normally, chlorine is ingested in combination with sodium in the form of a salt. Absorption of the chloride ion takes place largely in the upper portion of the small intestine. There is far more resorption of secreted potassium, sodium and chlorine ions than initial absorption since all three are secreted directly into the digestive tract in fluids such as saliva, gastric, and intestinal juices. The chloride ion can be reabsorbed across the walls of the stomach, the small intestine, the even the large intestine.

Toxicity levels are difficult to attain if enough of water is available. Excess chloride is excreted in urine, usually accompanied by excess sodium or potassium. However, if either sodium or potassium is not present in excess, then the ammonium ion can be excreted with the chloride. Excretion may be greatly restricted in cases of chloride deficiencies.

Requirement and source

Like that of sodium, all farm animals require chlorine in the order of 0.10 – 0.20% of the dry matter, but pasture are appreciably rich in chlorine than they are in sodium with little difference between legumes (0.40%) and grasses (0.50%) or between fresh grass and hay. Pastures invariably provide sufficient chloride for grazing animals. Cereals provide more chlorine than sodium, with maize providing the least (0.50%) and barley the most (0.18%). Cereal straws contain 3 to 6 time more chlorine than the grains. Vegetable protein supplements are also low in chlorine (0.03 – 0.08%). Ruminants not give sufficient access to roughages could possibly receive inadequate dietary chloride intakes and may be supplemented as common salt.

VI. MAGNESIUM

Magnesium (Mg) is closely associated with calcium and phosphorus. About 70% of body magnesium is present in selection, the remainder is found mainly in the liver and striated muscles, with detectable quantities in the kidney, brain, and red blood cells.

Functions

Magnesium is a key element in cellular reactions as a commonest enzymes activator for carbohydrate and lipid metabolism. Calves on low magnesium milk diet may develop low serum Mg, depleted lone Mg, tetany and death. The term "grass tetany" grass staggers, lactation tetany, magnesium tetany is known for long in many parts of the world. Cattle affected with grass tetany usually succumb. In a herd of cows, almost one-half can die within 24hr, even before the owner is aware of the condition.

A normal serum magnesium level is 2-3 mg/100ml. When dietary magnesium levels are low, serum magnesium levels in cattle may drop to as low as 1 mg %. Urine magnesium is better indicator of deficiency as it reflects the deficiency sooner. The symptoms of tetany are nervousness, tremorus, twitching of facial muscles, staggering gait and convulsion. Heavy dressings of nitrogenous and potassium fertilizers through cation – anion imbalance has also been related with tetany occurrence. A daily intake of 18-20 g of magnesium oxide (MgO) per head will protect the cattle from this condition. Cattle apparently dislike the taste of MgO, and its incorporation into a free-choice mineral mixture may result in almost complete refusal of such a mixture.

Magnesium sulfate commonly known as Epsom salt plays at least two roles: (1) it is a well-recognized laxative, and (2) it has a sedative effect.

Requirement and Sources

Ruminants require rations containing 0.16 – 0.20% magnesium and monogastric species only 0.05-0.07%. Most of commonly fed diets for farm animals contain sufficient amount of magnesium to meet the requirement. The magnesium content of forage plants also varies with the species, and the soil and climatic conditions in which the plants are grown. In general, leguminous species are richer than grasses in magnesium, as they are in calcium. As far as roughages are concerned, straws are low in magnesium, while root and leafy brasssica crops are similar to grasses (0.18-0.36%). Cereal grains contain 0.11 - 0.19 % magnesium, oilseed meals/cakes 0.30 – 0.58% and fish meal (medium quality) 0.17 – 0.25%. Thus concentrates used in animal feeding very in magnesium contents. Diets too low in magnesium are supplemented with magnesium oxide or carbonate.

VII. SULPHUR

Most of the sulphur is the body is present in form of organic compounds. The organic sulfur compounds include the amino acids methionine, cysteine the vitamins biotin and thiamin, certain mucopolysaccharides, insulen, glutathione, heparin, and coenzyme A. On the other hand, the ruminants can utilize inorganic sulfur, for synthesizing sulphur containing amino acids in rumen.

Function

Generally, the sulfur containing amino acids are present at least in minimal quantities to meet the dietary needs of monogastric animals. However, constant vigilance must be maintained to monitor natural dietary sources. Crystalline methionine is added routinely to many poultry diets, but many swine researchers report that supplementing typical corn-soy diets with methionine produces no benefit. Use of NPN compounds in ruminant rations, is being done by commercial feed manufacturers in India. Under such conditions sulphur supplementation in form of sodium sulphate is recommended in ratio of N: S (10: 1), though it could be reduced to 15: 1 also, without much problem.

Consideration of sulphur in animal feeding should be done in areas of intensive livestock production, where sulphur in soils is not replaced regularly by fertilizer application.

Requirement and source

The estimated sulphur requirement for cattle is 0.20% the maximum should be limited to 0.35% of the diets with no more than 0.20% from the inorganic supplements. Since the excessive amount of sulphates will reduce the feed intake and adversely affect the animals by reducing the availability of other minerals like zinc and manganese. Most diets for farm animals are adequate in sulphur containing proteins. Sulphates can be used by rumen microorganism for protein biosynthesis. Ruminants depending largely on non-protein nitrogen source (urea) may need supplemental inorganic sulphates to correct this deficiency. In case of shortage of sulphur containing amino acids in the diet of monogastric species, part of the sulphur requirement may be met by inorganic sulphates. The concentrations of sulphur in pasture and conserved forage range from 0.05 to 0.50%. Cereal gains tend to be low in sulphur (0.10%), cereal straws slightly richer (0.14 – 0.40%) and protein supplements are rich (0.50%).

VIII. IODINE

The concentration of iodine in animal body is very low. Early three-quarters of the available iodine are concentrated in the thyroid gland. Interestingly, there is a fairly high concentration in the hair and in the ovaries, skeleton, muscle, bile, pituitary and salivary gland.

Iodine Storage

Iodine in the body is found mainly in the thyroid gland, existing as inorganic iodine, as well as in organic forms, including thyroxin, thyroglobulin, triodothyronine, and mono and diodotyrosin. Iodinated amino acids join together via a peptide linkage to form thyrogloublin, the storage form of the thyroxin.

Functions

The function of iodine is largely that of thyroxin synthesis of which it is a major component. thyroxine controls the metabolic rate of the body, or the rate at which energy is utilized. An overactive thyroid produces large quantities of thyroxin, which results in increased energy utilization. Animals effected by this condition produce a lot of body heat, are quite active, and do not fatten easily. An under active thyroid produces reverse effect.

Thyroxine exerts a profound effect on growth and differentiation. Young animals deprived of thyroxine fail to grow because of delayed bone formation. This deficiency early in life results in a type of dwarfism. Thyroxine also affects the nervous system. If excess thyroxine is produced, animals become nervous and restless and shows markedly increased breathing rates and excess sweating. A thyroxine deficit results in slow, lazy, and lethargic animals.

Maternal iodine deficiency in mammals can result in a condition known as hairlessness or woollessness in the young animal. Such animals are usually born dead and exhibit "big neck" or goiter, Reproductive abnormal ties are one of the important consequences of reduced thyroid functions. The goitrogens present in brassicas has been identified as goitrin (L – 5 – vinyl – 2 – oxazolidine – 2 – thione), which inhibits the iodination and thus inhibit synthesis of thyroxine. Therefore, it cannot be overcome by adding Iodine. Thyocynate, problen could be solved by supplemental iodine. Goiter can be either prevented or cured by administration of iodine. Although simple goiter may occur universally, geographical location has some effect on this condition. Simple goiter rarely occurs in animals living near the oceans, but it frequently found in those far from the seacoasts where little iodine occurs in the soil.

Metabolism

Iodine is absorbed from the small intestine, either in the in organic or the organic form. It is absorbed easily and almost completely in the amino acid form. Iodinated casein administered orally maintains its effectiveness as a pseudo-thyroxine. Excess iodine is removed from the bloodstream by the kidneys and is excreted in the urine.

Toxicity

Iodine toxicity has been developed in experimental animal since body is so designed that it can excrete very high levels of ingested iodine.

Goitrogens

Thyroid depressants or goitrogens are found in many feedstuffs, For example, thiouracil, which has a thyroid-depressing effect, cause symptoms of iodine or thyroxine deficiency. Certain feedstuff groups are also classed as goitrogenic, such as those in the brassica family, including cabbage, rape, and turnips.

Requirement and Sources

Dairy cattle require 0.25-0.50 ppm of iodine depending upon the different physiological stage like growing, lactation and pregnancy. In fact, iodine deficiency is a regional problem. The existence of iodine low soils can be attribute to the small accumulation of iodine of marine origin due to the long distance of such areas from the sea. Animals feedstuff very widely in iodine concentration. Cereals (0.04 –0.09ppm) and oil seed meals (0.10-0.20ppm) are poor source of iodine and fishmeal very rich sources. The concentration on iodine in seaweed may be as high as 0.40 – 0.60%. However, legume forage (0.12-0.50ppm) is rich source than grasses.

IX. IRON

Iron exists in the body largely (is 90%) in the heme portion of a hemoglobin and transferrin, although it is also present as a vital constituent of oxidation-reduction enzymes. The liver and spleen have the greatest tissue stores of iron, followed, respectively, by the kidney, heart, pancreas, brain, and muscles.

Functions

Heamoglobin combines readily with oxygen to form oxyhemoglobin, which is turn, gives up its oxygen equally readily at the cellular level. Enzymes containing or activated by iron are catalare, peroxidares, phenylalanine, hydroxilars and TCA cycle enzymes.

The daily requirement of iron in the adult is low since the body recycles it. The iron requirements are much greater in growing animals. A growing pig requirements about 10 mg of iron per day; calves require about 25 mg.

Metabolism

Iron is primarily absorbed in the ferrous form through the wall of the duodenum. Ingested ferric iron is reduced to ferrous iron in the digestive tract. Unabsorbed iron is excreted via the feces, with only trace amount in the urine. The presence of higher levels of phosphorus lowers the digestibility of iron by forming insoluble ferric phosphate.

Deficiency

Iron deficiency characterized primarily by a condition known as anaemia (without blood). Because most feedstuffs, except milk, contain a reasonable amount of iron, under most feeding situations. There are several other types of nutritional deficiency anemias (cobalt, copper, iron, vitamin B_{12}), toxicity anemias (excesses of molybdenum, zinc, or selenium), metabolic disorder anemias, and anemias involving severe blood losses typical of lambs with a heavy infestation of internal parasites that literally suck the blood. Pronounced iron deficiency may be most common in the case of the suckling pig. Anaemia in piglets is characterized by poor appetite and growth. Breathing becomes laboured and spasmodic, hence the descriptive term “thumps” for this condition.

Iron Therapy

Iron therapy is actually quite simple; it consists mainly of providing sufficient quantities of iron to the animal. Thus, pigs that have access to soil rarely develop anemia providing them with a ferrous iron solution is beneficial. Swabbing the sows udder with a ferrous sulfate solution has also proved successful. Intramuscular injection of 100mg of iron as iron-dextran is therapeutic. Iron from ferric oxide is not available nutritionally.

X. ZINC

Zinc deficient soils have been identified in many parts of India. It is often related with sole use of excessive inorganic fertilizers. Zinc relationship with reproduction of male animals has been documented for long. Skin condition in swine known as parakeratosis was observed in swine. Intensive research has related this condition to a combination of excess calcium (1%) and low levels of zinc in swine diets.

A. Functions

The functions of zinc are associated primarily, to those of enzymes. These include carbonic anhydrase, pancreatic carboxypeptidase, lactate dehydrogenase, alcohol dehydrogenase, alkaline phosphatase and thymidine kinase. In diabetes mellitus the zinc content of the pancreas is about one–half of normal, which suggests that zinc, may be involved in storage and/or use of insulin. It is found in carbonic anhydrase, (an enzyme that is involved in respiration). During thiamin deficiency the zinc levels of blood, nails, and skin are lowered. A deficiency of zinc causes depressed phosphatase and catalase activity of liver and kidneys.

Deficiency

Symptoms of zinc deficiency in calves include inflammation of nose and mouth, stiffness of joints, swollen feet and parakeratosis. In deficiency there is decreased functioning of the testes, resulting in poor quality sperm and impaired development of sex organs in both male and females. Bone formation is also impaired. The long bones are shorter and thicker and contain less zinc, which suggests that calcification of bones involves this element. A zinc deficiency in pigs causes many symptoms. The skin over many parts of the body, but particularly the underside, becomes thickened and rough, and a straw-colored liquid exude from open sores. In zinc-deficient chickens feathering is poor. This condition is accompanied by dermatitis and a degeneration of feather follicles.

The adverse effect of soybean protein on zinc utilization has been documented for swine and poultry diets. It was demonstrated that a corn soy diet containing 12 ppm of zinc was not adequate for swine. However, milk protein-supplemented diet containing 10pm zinc was closer to the actual requirements. Apparently, the soy protein diet chelates at least a portion of dietary zinc, making it unavailable. By lowering the calcium level in swine corn-soy diets or raising zinc levels, zinc deficiency symptoms were alleviated. This was also observed in poultry.

Metabolism

Zinc absorption occurs in the duodenum, which is affected by calcium and copper levels in diets. Also, since soy protein diets require higher dietary levels of zinc, it is assumed that soy protein causes a chelating effect probably phytate, rendering zinc less available. Zinc is excreted from the body primarily via the faeces, representing both undigested and endogenous zinc. Zinc is stored mainly in the pancreas, liver, kidney, and spleen. However, it is also found in a large number of other tissues, including hair, bones and the nervous system.

Requirements and sources

Most practical diets supply adequate zinc for farm animals (30 – 40ppm). The zinc concentration in pastures varies considerably (20 – 50ppm). But the legume forages are rich sources (20 – 60ppm) of zinc than grasses (10 – 30 ppm). Cereals grains also contain 20 – 40 ppm zinc. But zinc resembles manganese in being highly concentrated in the outer layer of the grains (wheat bran 49 ppm). Cereals straws (< 12 ppm) contain only one-third of the concentration found in the grain. Protein sources of plant origin like soybean, cotton seed and groundnut meals contain less zinc (50 – 70 ppm) than animal protein sources like fish meals and meat meals (80 – 120 ppm).

XI. MANGANESE

Manganese occurs in all living tissues and in the blood is present at a level of 2-mg/100 ml. It is found in the greatest concentrations in the bones liver, kidneys and pancreas. Manganese is required by most farm animals, but its role in preventing “slipped tendon or perosis” in poultry was first observed in the early 1930s. Since then its role in other species has also been identified.

A. Functions

Manganese is required as cofactor/coenzymes in (activator) many enzyme systems, namely, in blood and bone phosphatase activity, which in turn, is related to the bone calcification process. Manganese also appears to be involved in cartilage. There is a decrease in cartilage chondriotin sulphate, due to manganese deficiency.

Its presence in pyruvate carboxylase indicates that manganese plays a role in carbohydrate metabolism. It may also be involved in insulin synthesis as well as lipid metabolism. The addition of manganese to a manganese-deficit diet result in lowered backfat deposition in pigs, which seems to be related to a relationship involving manganese-choline. Other enzymes are hydrolases, kinases, arginase and manganese sueperoxide dismutase.

B. Deficiency

Deficiency symptoms include retarded growth, skeletal abnormalities, ataxia of new born and reproductive failure. The deficiency may not be common in grazing ruminants but reproductive performance of Australian sheep improved by Manganese supplementation for two consecutive years. Low manganese diets for cows and goats led to depressed or delayed oestrus, conception and increased abortions. A deficiency affects fertility in poultry. Embryos from manganese – deficient hens usually die during the final stages of incubation; in those that do hatch, there is chronic congenital ataxia, usually in the form of tetanic spasms. A manganese deficiency after hatching results in the slipped tendon syndrome, Egg production is greatly diminished and breaking strength of the shells is reduced. Pigs fed manganese-deficient diets develop lameness that is characterized by pain on rising. In addition to stiffness, the hocks become enlarged and the legs appear crooked.

Metabolism

Metabolism of manganese begins with limited absorption from the small intestine perhaps 3-4% of that ingested. Manganese is excreted through urine. Some manganese, however, is excreted in the pancreatic juice. The presence of increased levels of calcium results in decreased body levels of manganese.

Requirements and Sources

The manganese requirement of poultry (50 ppm) is higher than other farm animals (40 ppm). It has become common practice to increase the manganese content of poultry diets by addition of manganese salts. Manganese content of forages vary considerably (20 – 86 ppm). However, the manganese contents of common ruminant diets neither produce deficiencies nor surpass the maximum tolerable levels. Rich sources are wheat bran (115) and rice bran (200 ppm), since manganese is concentrated in the outer layers of the cereal grains. Protein sources of animal origin (0.2 – 20 ppm) contain less manganese than protein supplements of plant origin (35 – 55 ppm).

XII. COPPER

Copper is involved in the incorporation of iron into the hemoglobin structure. Although a limited amount of copper is found in the body, it is part of all tissues, with the largest amount in the liver, spleen and kidneys.

A. Function

Copper plays a role in hemoglobin formation. In addition, copper is needed for the absorption of iron, by apparently converting ferrous into ferric iron. It is involved in red cell maturation. Copper is also involved in bone formation. A deficiency adversely affects collagen formation in bone.

Swayback in lambs has long been associated with copper-deficient pastures (2 – 4 ppm) typical of those found in Australia. Such lambs develop a nervous disorder and a loss of muscular coordination. Although this condition has been identified by a number of names, they all fall under the category of enzootic neonatal ataxia. This is apparently related to a deficiency of copper in the brain that results in a deficiency of cytochrome oxidase in motor neurons.

Pigs fed up to 250 ppm copper (nutritional level is 5-10 ppm) gained 8-10% more rapidly and required 4-5% less feed per unit gain. The effect is probably because of a bacteriostatic or bactericidal effect on intestinal microorganism, which is harmful to the pig.

Copper toxicity is a frequent occurrence in sheep and is characterized by loss of hemoglobin. Which is excreted in the urine. There is a yellowing of the membranes resulting in jaundice. Copper toxicity symptoms in lambs may occur when dietary levels of copper are not too high, but when dietary molybdenum levels are very low.

Deficiency Symptoms

These include anaemia, poor growth, scouring, infertility, depigmentation of hair and wool, gastro-intestinal disturbances and lesions in the brain stem and spinal cord. An early sign of copper deficiency in sheep is a loss of "crimp" or waviness in wool, referred to as stringy or steely, accompanied by deterioration in quality, as losing its elasticity and affinity for dyes. The disulphide linkages in keratin formation involve enzyme, in presence of copper. Pigmentation of wool and hair (melanin) is also affected. The wool fibres produced are brittle. A condition in cattle known as "teart" or "peat scours" which is characterized by unthriftiness and scouring, occur on pastures having very high molybdenum. The animals recover by copper sulphate supplementation. Copper thiomolybdate is formed by rumen, microorganisms due to reaction between molybdenum, sulphur and copper thereby limiting the absorption of dietary copper. Animal body contains a number of inorganic elements rather because present in the food. Nutritionally essential elements are only those, which are known to have metabolic role in the body and the deficiency symptoms develop in their absence. As many as forty or more elements may have metabolic roles in mammalian tissues. The essential minerals are classified into two groups, major and trace elements. Trace elements are present in animal body in a concentration not greater than 50 mg/ kg and are required at less than 100 mg/ kg diet. Major elements namely calcium, magnesium, phosphorous, potassium, sodium, chlorine and sulphur are present in higher concentration than trace elements. The later group includes iron, zinc, copper, molybdenum, selenium, iodine, fluorine, manganese and cobalt. Elements chromium, arsenic, boron, lead, lithium, nickel, silicon, tin, vanadium, rubidium and aluminium have also been included in the list of essential, but these are required in so minute quantities that they can possibly be not required to be supplemented in natural diets.

Requirements and Sources

The requirement of copper for dairy cattle is 10 ppm and for beef cattle and sheep 5 ppm. But it is considerably higher in the presence of molybdenum and sulphur. Pigs and poultry need 5 – 6 ppm of copper in their diets and deficiency does not occur normally in monogastric species fed on diets of natural feedstuffs. The copper content varies with the species of plants. Leguminous forages are richer in copper and then produce feedstuffs, which reflect the deficiency.

XIII. COBALT

Ruminants suffering from diseases as coast disease and wasting disease respond to cobalt administration. The synthesis of the vitamin B_{12} in presence of cobalt indicated that it was an integral part of the vitamin B_{12} molecule. Ruminant animals cannot synthesize vitamin B_{12} if their diets are deficient in cobalt.

A. Deficiency

Cobalt deficiency symptoms are same as that of vitamin B_{12}. Such symptoms become apparent only when ruminants have been subjected to cobalt-deficient regimen for a period of time because vitamin B_{12} can be stored in the liver unlike other B vitamins. Ruminants fed cobalt-deficient diets show loss of appetite, reduced gains and even loss of weight to the point of emaciation. Such animals show all the characteristic of anaemia, such as or other deficiencies. These conditions bleached out areas around the eyes and lips. Milder cases of cobalt deficiency are difficult to identify since this condition is manifested as general unthriftiness, which could easily be confused with a mild case of internal parasite infestation.

B. Supplementation

Supplementation could be achieved by using cobalt-containing mineral mixtures or by top-dressing deficient pastures with cobalt-rich fertilizers (cobalt sulphate 2 kg/ hectare). Since most ruminants do not store large quantities of vitamin B_{12}, there is need for cobalt supplementation. Cobalt bullets (900 g cobaltic oxide/ kg) have been known as "pining", "salt sick", bush sickness and wasting disease, which are a combination of iron and cobalt, are administered with a balling gun. They settle to the bottom of the rumen and slowly dissolve, releasing cobalt and iron over many months. They have been used successfully in cobalt-deficient areas.

Requirements and Sources

The requirement for cobalt is only 1% that of copper. The critical level below which cobalt deficiency signs appear in dairy cattle and sheep is 0.07 ppm. Beef cattle and lambs are more susceptible to cobalt deficiency and shows deficiency symptoms with food containing 0.10 ppm of cobalt. Pasture plants and forages contain cobalt in concentration that vary widely with the species and soil concentrations, though not with advancing maturity. Legumes (0.12 – 0.35 ppm) are usually richer in cobalt than grasses (0.03 – 0.18 ppm). Cereal grains are poor sources of cobalt (0.01 – 0.06 ppm). Feeds of animal origin, except liver meal, are also poor sources of cobalt.

XIV. SELENIUM

The role of selenium in animal nutrition is critical. Although it is essential, it can cause toxic side effects. The toxic aspect of selenium was recognized first, and will be discussed below.

Toxicity

Selenosis, or selenium toxicity, has been recognized in farm animals for many years. The diseases known as blind staggers and alkali disease were identified as being caused by excessive intakes of selenium. The condition arises from consumption of feedstuffs grown on soils containing 5 or 6 ppm of water-soluble selenium. Such selenium is absorbed by plant, which they store at a toxic level o 5 ppm. Seleniferous soils occur through out the world, but are particularly prevalent in certain areas of north and Southern India. Deg Nala disease has been ascribed as selenium toxicity.

The most severe toxicity results in death, which is preceded by typical blind staggers, blindness, grinding of the teeth, and excessive loss of saliva. Such animals exhibit breathing problems and death usually results from suffocation. Necropsy of such animals shows destruction of the heart muscle and cirrhosis of the liver. The joints of the long bones show marked erosion. Such animals become so lame that they are unable to walk to water or food supplies. Sulphur excess in mineral mixture could recover the animals, at early stages.

Requirement

The requirement of selenium has been established for most farm animals. Since the land in most areas of the world contains either too much or too little selenium, the addition of selenium to the diet be considered systematically.

White muscle diseases in lambs and calves are recognized first by a stiff, stilted walking gait followed quickly by death unless the condition is treated. Necropsy of such animals shows chalky white striations of the muscle overlying the cervicular vertebrae. Because the hear muscles are also involve in this disease, death probably results from hear failure. Supplemental vitamin E provides some relief from white muscle disease, as does selenium. However, a combination of the two is more effective that either one alone.

Exudative diathesis is a condition in poultry characterized by edema in the breast and wing area, hemorrhaging through the capillaries, and eventual massive fluid accumulation throughout the body. The growth rate is greatly impaired and the incidence of death is high. The administration of vitamin E in the diet greatly reduces the incidence of this condition, but it is not nearly as effective as supplemental selenium, which is 100% effective in alleviating this condition. Selenium deficiency in swine is recognized as a practical field problem, except in limited cases where feedstuffs grown on seleniferous soils are available in many parts of the world.

Metabolism

Selenium is found in milk and reflects dietary levels in the cattle ration. Similarly, selenium is found in eggs at a normal level of about 0.1 ppm. However, the level found in eggs varies in accordance with dietary selenium levels.

Selenium administered intramuscularly is rapidly and readily absorbed from the intestinal tract. It is stored throughout the body, particularly in the kidney. Selenium is excreted in both feces and urine. Although selenium can function as an antioxidant, this does not appear to be its role in the body. It is found in the enzyme glutathione peroxidase, which is essential for removal of hydrogen thereby protecting cell membranes from oxidative damage. Glutathione peroxidase contains four selenium atoms and forms a second line of defence after vitamin E. Selenium has a sparing effect on vitamin E by ensuring normal absorption of the vitamin. Conversely, vitamin E spares selenium by maintaining the element in its loss. The other major role selenium is in the production of the thyroid hormones, as a component of enzyme. The condition known as "ill thrift" in lambs, cattle at passive can be prevented by selenium. It is necessary for the production of lipase in the pancreas and plays a role in the absorption of fats and vitamin E from the intestine.

Requirements and Sources

The dietary requirements of selenium vary with the form of selenium ingested and other constituents of the diet, particularly the vitamin E level. The recommended selenium requirement for most species is 0.10 ppm. Mineral salts containing sodium/ selenite slow release capsules may be employed. The cereals grains and other seeds vary widely in selenium contents between the locations, with levels as low as 0.006 ppm to 3.06 ppm in some areas. The selenium concentrations in forages also vary, depending upon the plant species and selenium status of the soils on which they grow.

XV. FLUORINE

Fluorine, like selenium, falls into the category of both a potentially toxic and a nutritionally required mineral element.

Toxicity

Fluorine toxicity is a result of intake of feedstuffs produced on soils high in fluorides, the intake of water containing high levels of fluorides or the intake of contaminating fluoride in almost any form.

Ruminants being more susceptible than non-ruminants. Levels of more than 20 ppm in the diet of cattle become toxic. Animals may consume relatively high dosages of fluorine for a relatively long time without exhibiting any symptoms of fluorosis because the kidney removes excess fluorine and because bones have a great affinity and a great storage capacity for fluorine. Excess fluorine beyond bone capacity is deposited in soft tissue, and at this stage, death will ensue. The toxicity symptoms include dental pitting and wear. Further increases cause depression of appetite, lameness and reduced production. Bone and joint abnormalities also occur. Fatty acid utilization is perhaps impaired due to the fact that fatty acid oxidase activity is inhibited and alkaline phosphate activity is increased under condition of fluorosis.

Young animals given high levels of fluorine prior to the appearance of permanent teeth experience excessive tooth damage characterized by severe mottling, pitting and whitening. Such teeth wear down until the pulp cavities are exposed. At this stage, either eating or drinking cold water is so painful that intake of both is greatly reduced. Such dental defects do not occur in the first teeth, indicating that toxic levels of fluorine do not cross the placental barrier. Nor do these defects occur in the permanent teeth if toxic levels of fluoride intake are delayed until after the permanent teeth have erupted.

Requirement

Most plants have a limited capacity to absorb fluorine from soil and normal level in pasture herbage range from 2 to 20 ppm. Cereals and other grains usually contain about 1 – 10 ppm. The benefits from the intake of optimal levels of fluorine are related mainly to protection against dental cavities. Mass population studies revealed that children reared in communities in which the drinking water contained 1–2 ppm fluoride had fewer cavities than those reared where the water had either higher or lower level of fluorine. The manner in which fluoride protects teeth has not been demonstrated. One theory suggests that its presence in the structure of the teeth lower their solubility. The other suggests that fluoride act to depress enzymatic activity in the mouth, which in effect dissolves enamel.

XVI. MOLYBDENUM

This element is found in very low concentration in many tissues of the body (liver, kidney, spleen, lung, brain, muscle) but in greatest proportions in the liver and kidney.

The function of molybdenum as a required nutrient is not known, but it has been identified as a constituent of xanthine oxidase.

Perhaps the toxic aspect of molybdenum is equal in significance to its nutritional need. Swine appear to be highly tolerant to excessive intake of molybdenum, but other species manifest certain toxic symptoms. It appears that molybdenum toxicity, when it exists, is actually a manifestation of an imbalance among copper, sulfur, and molybdenum.

Table 1: Inorganic sources of minerals

Element	Source compound	% element in compound	Bioavailability
Calcium	Steamed bonemial	29.0 (23-37)	High
	Deflourinated rock phosphate	29.2 (19.9-35.7)	Intermediate
	Calcium carbonate	40.0	Intermediate
	Soft phosphate	18.0	Low
	Ground limestone	38.5	Intermediate
	Dolomitic limestone	22.3	Intermediate
	Monocalcium phosphate	16.2	High
	Tricalcium phosphate	31.0-34.0	-
	Dicalcium phosphate	23.2	High
	Hay sources	23.3	Low
Cobalt	Cobalt carbonate	46.0-55.0	- [b]
	Cobalt sulphate	21.0	- [b]
	Cobalt chloride	24.7	- [b]
Copper	Cupric sulphate	25.0	High
	Cupric carbonate	53.0	Intermediate
	Cupric chloride	37.2	High
	Cupric oxide	80.0	Low
	Cupric nitrate	33.9	Intermediate
Iodine	Calcium iodate	63.5	High
	Ethylenediamine dihydroiodide	80.0	High
	Potassium iodide, stabilized	69.0	High
	Cuprous iodide	66.6	High
Iron	Iron oxide	46.0-60.0	Unavailable
	Ferrous carbonate	36.0-42.0	Low
	Ferrous sulphate	20.0-30.0	High
Magnesium	Magnesium carbonate	21.0-28.0	High
	Magnesium chloride	12.0	High
	Magnesium oxide	54.0-60.0	High
	Magnesium sulphate	9.8-17.0	High
	Potassium and magnesium sulphate	11.0	High
Manganese	Manganese sulphate	27.0	High
	Manganese oxide	52.0-62.0	High
Phosphorous	Defluorinated rock phosphate	13.1 (8.7-21.0)	Intermediate
	Calcium phosphate	18.6-21.0	High
	Dicalcium phosphate	18.5	Intermediate
	Tricalcium phosphate	18.0	-
	Phosphoric acid	23.0-25.0	High
	Sodium phosphate	21.0-25.0	High
	Potassium phosphate	22.8	-
	Soft phosphate	9.0	Low
	Steamed bonemeal	12.6 (8-18)	High

Element	Source compound	% element in compound	Bioavailability
Potassium	Potassium chloride	50.0	High
	Potassium sulphate	41.0	High
	Potassium and magnesium sulphate	18.0	High
Selenium	Sodium selenate	40.0	High
	Sodium selenite	45.6	High
Sulphur	Calcium sulphate (gypsum)	12.0-20.0	Intermediate
	Potassium suplphate	28.0	High
	Potassium and magnesium sulphate	22.0	High
	Sodium sulphate	10.0	Intermediate
	Anhydrous sodium sulphate	22.0	-
	Flowers of sulphur	96.0	Low
Zinc	Zinc carbonate	52.0	High
	Zinc chloride	48.0	Intermediate
	Zinc sulphate	22.0-36.0	High
	Zinc oxide	46.0-73.0	High

Source : McDowell (1996)

Mineral Mixtures for Supplementing Feeds

The ingredients like bone-meal, calcined bone-meal, cobalt chloride, cobalt sulphate, common salt, copper sulphate, cicalcium phosphate, ferrous sulphate, limestone, manganese sulphate, oyster shell, potassium iodide or potassium iodate, zinc chloride and zinc sulphate may be used for the preparation of mineral mixtures. All ingredients used shall be of a quality suitable for animal consumption and shall contain no substances harmful to cattle.

Mineral mixtures should be in the form of a free flowing, powder, thoroughly mixed and completely homogeneous. It should be ground to such fitness that in case of Type 1 material, not less than 70 per cent by mass of the material and in case of Type 2 material, not less than 90 per cent by mass of the material shall pass through 106 – micron IS sieve. Mineral mixtures should be free from adulterants, insect or visible fungus infestation and undesirable odour.

Mineral mixtures should also be free from spores of Bacillus anthracis, Clostridium sp. etc. In case of Type 2 mineral mixtures, salt shall be added at the time of mixing the feed by the user at the rate of 22 per cent by mass of the material. The percentage of chlorine as sodium chloride in the mixture after the addition of salt should be at least 22 per cent by mass.

Table 2: Mineral Mixtures containing salt (Type 1) for supplementing feeds

S. No.	Characteristics	Requirement
1	2	3
i)	Moisture, per cent by mass, *Max.*	5
ii)	Calcium, per cent by mass, *Min.*	22
iii)	Phosphorous, per cent by mass, *Min.*	9
iv)	Salt (chlorine as sodium chloride), per cent by mass, *Min.*	22
v)	Magnesium, per cent by mass,	5%
vi)	Iron, per cent by mass, *Min.*	0.4
vii)	Iodine (as KI), per cent by mass, *Min.*	0.02
viii)	Copper, per cent by mass, *Min.*	0.06
ix)	Manganese, per cent by mass, *Min.*	0.09 - 0.12
x)	Cobalt, per cent by mass, *Min.*	0.01
xi)	Fluorine, per cent by mass, *Max.*	0.03 – 0.05
xii)	Zinc, per cent by mass, *Min.*	0.15 - .3
xiii)	Acid – insoluble ash, per cent by mass, *Max.*	3
xiv)	Sulphur, per cent by mass,	0.4

Table 3: Mineral mixtures without salt (Type 2) for supplementing feeds

S. No.	Characteristics	Requirement
1	2	3
i)	Moisture, per cent by mass, *Max.*	5
ii)	Calcium, per cent by mass, *Min.*	23
	Magnesium, per cent by mass,	6.5
iii)	Phosphorous, per cent by mass, *Min.*	12
iv)	Iron, per cent by mass, *Min.*	0.50
v)	Iodine (as KI), per cent by mass, *Min.*	0.026
vi)	Copper, per cent by mass, *Min.*	0.077
vii)	Manganese, per cent by mass, *Min.*	0.12
viii)	Cobalt, per cent by mass, *Min.*	0.013
ix)	Fluorine, per cent by mass, *Max.*	0.04 – 0.07
x)	Zinc, per cent by mass, *Min.*	0.18 - 0.38
xi)	Acid – insoluble ash, per cent by mass, *Max.*	2.5
xii)	Sulphur, per cent by mass,	0.5

Summary

Minerals fulfill physiological, structural and regulatory functions. Mineral supplements take various forms: minerals salts, rumen boluses, 'organic' compounds, pasture applications. The roles of individual mineral elements and the effects of their deficiencies are summarized below:

Mineral Element	Function	Symptoms Deficiency
Calcium	Bone and teeth, transmission of nerve impulses	Rickets, osteomalacia, thin eggshells, milk fever
Phosphorous	Bone and teeth, energy metabolism	Rickets, osteomalacia, depraved appetite, poor fertility
Potassium	Osmoregulation, acid-base balance, nerve and muscle excitation	Retarded growth, weakness
Sodium	Acid-base balance, osmoregulation poor egg production	Dehydration, poor growth,
Chlorine	Acid-base balance, osmoregulation, gastric secretion	Alkalosis
Sulphur	Structure of amino acids, vitamins and harmones, chondroitin	Equivalent to protein deficiency
Magnesium	Bone, activator of enzymes for carbohydrate and lipid metabolism hypomagnesaemia	Nervous irritability and convulsions,
Iron	Haemoglobin, enzymes of electron transport chain	Anaemia, poor growth
Copper	Haemoglobin synthesis, enzyme systems, pigments	Anaemia, poor growth, depigmentation of hair and wool, swayback, reproduction problems
Cobalt	Component of vitamin B_{12}	Pining (emaciation, anaemia, listlessness)
Iodine	Thyroid hormones	Goitre; hairless, weak or dead young
Manganese	Enzyme activation	Retarded growth, skeletal abnormality, ataxia, perosis
Zinc	Enzyme component and activator	Parakeratosis, poor growth, depressed appetite, infertility in males
Selenium	Component of glutathione peroxidase, iodine metabolism, immune function	Myopathy, exudative diathesis

Bibliography

Agricultural and Food Research Council (1991). Technical Committee on Responses to Nutrients, Report No. 6. A Reappraisal of the Calcium and Phosphorous Requirements of sheep and cattle. *Nutrition Abstracts and Reviews, Series B* **61**: 573 – 612

Ammerman, C.B; Henry, P.R. and Miles, R.D. (1998). Supplemented organically bound mineral compounds in livestock nutrition. In: Garnsworthy P C and Wiseman J (eds). *Recent Advances in Animal Nutrition,* Nottingham, Nottingham University Press, pp 67 – 91.

BIS. (1979). Indian Standard: Specification for Mineral Mixture for Cattle. Bureau of Indian Standards, New Delhi.

Cheeke P.R. (1991). Applied Animal Nutrition; Feeds and feeding, Prentice- Hall of India Private Limited, New Delhi.

Fredeen, A.H., DePeters, E.J. and Baldwin, R.L., (1988). Effects of acid base disturbances caused by differences in dietary fixed ion balance on Kinetics of calcium metabolism in ruminants with high calcium demand. J. Anim. Sci. 66: 174-184.

Gawthorne, J. .M; Howell, J. McC and White, C.L. (eds) 1982. *Trace Element Metabolism in Man and Animals,* Berlin, Springer-Verlag.

Georgievskii, V.I; Annenkov, B.N. and Samokhim, V.I. (1982). *Mineral Nutrition of Animals,* London, Butterworth.

McDonald, P. Edwards, R.A., Greenhalgh, JFD and Morgan, CA (2002). Animal Nutrition, Pearson Education, Singapore.

McDowell, L.R. (1992). *Minerals in Animal and Human Nutrition,* New York, Academic Press.

McDowell, L.R., (1985). Nutrition of grazing ruminants in warm climates. Orlando, F.L; Academy Press.

McDowell, R.E. (1996). Feeding minerals to cattle and pasture. Anim. Feed Sci. Technol. 60: 247-271.

Mertz, W. (ed.) 1988. Trace Elements in Human and Animal Nutrition, Vols. I & II. Orlando, FL: Academy Press.

National Research Council (1980). *Mineral Tolerances of Domestic Animals,* National Academy of Sciences. Washington, DC.

National Research Council (2001). Nutrient Requirements of Dairy Cattle. National Academy Press, Washington, DC.

Spears, J.W., (1989). Zinc methionine for ruminants: Relative bioavailability of zinc in lambs and effects on growth and performance of growing heifers. J. Anim. Sci. 67: 835-843.

Thompson, J.K. and Fowler, V.R. (1990). The evaluation of minerals in the diets of farm animals. In: Wiseman J and Cole D J A (eds). *Feedstuff Evaluation,* London, Butterworth, pp 235 – 59.

Underwood, E.J. (1977). *Trace Elements in Human and Animal Nutrition,* New York, Academic Press.

Underwood, E.J. and Suttle, N.F. (1999). *The Mineral Nutrition of Livestock,* 3rd edn., Wallingford, CABI Publishing.

VITAMINS

Vitamins are organic substances required by animals in very small amounts for regulating various body processes towards normal health, growth, production and reproduction. These are classified as either fat-soluble or water-soluble. Vitamins A, D, E and K are fat-soluble and B vitamins and C are water-soluble. Vitamins have diverse functions in body as in metabolic pathways, immunity and gene regulation. In deficiency particular symptoms or disease may occur but in sub clinical cases health performance is less than optimal.

Under ordinary circumstances rumen microbes synthesize all of B vitamins and vitamin C is synthesized within body tissues. Ruminants under tropical conditions will synthesize sufficient vitamin D, during exposure to direct sunlight. Most ruminant rations are considered to be more than sufficient in vitamin E and carotene a precursor of vitamin A is present in leafy pasture grass and forage materials. Vitamin K is present in diets and rumen microbes synthesize K_2. Vitamin A storage in liver could be large enough to meet requirements up to 6 months. It is commonly believed that exposure to stress increases vitamin requirements.

Table 1: Vitamins and their synonyms

Vitamin	Accepted name
Vitamin A	Retinol
	Retinal
	Retinoic acid
Vitamin D_2	Ergocalciferol
Vitamin D_3	Cholecalciferol
Vitamin E	Tocopherol
Vitamin K	Phylloquinone
	Menaquinone
	Menadione
Vitamin B	
Vitamin B_1	Thiamin
Vitamin B_2	Riboflavin
Nicotinamide	Nicotinic acid
Vitamin B_6	Pyridoxine
Pantothenic acid	Pantothenic acid
Biotin	Biotin
Folic acid	Monopteroylglutamic acid
Vitamin B_{12}	Cyanocobalamin
Choline	Choline
Vitamin C	Ascorbic acid

Vitamin A: vitamin A, chemically known as retinol with the following structural formula.

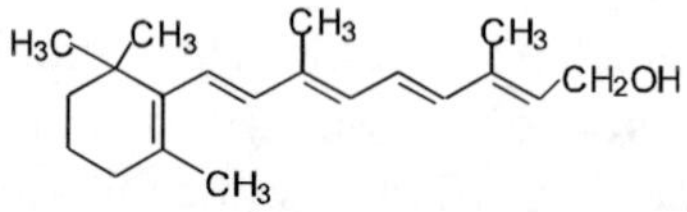

Vitamin A

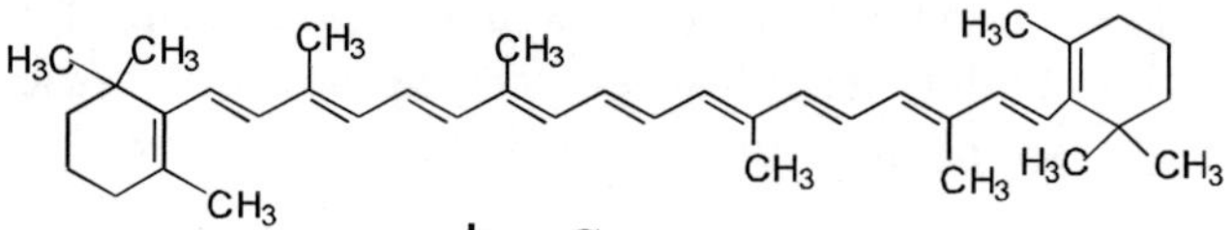

b - Carotene

It is a pale yellow crystalline solid, soluble in fats. It is readily destroyed by oxidation on exposure to air and light. Weathered and matured forage thus are deficient in this vitamin.

Deficiency symptoms

One of the earliest symptoms of vitamin A deficiency is known as night blindness. Various forms of the vitamin have essential roles in vision, i.e. component of Rhodopsin a vision pigment that trap light in eye and trigger nerve impulses via optic nerve to brain. Cattle may develop scaly skin and hair appears coarse in mild deficiency. In prolonged cases excessive watering of eyes, softening and cloudiness of the cornea and drying of conjuctiva i.e. xerophthalmia may occur. Constriction of the optic nerve canal may cause blindness in case of calves. Infertility in breeding male's and abortion or resorption of faetus or birth of dead, weak or blind calves takes place.

Grazing animals are likely to get their requirements, but penned animals receiving high concentrate diets with straw over long periods as in peri urban dairies may suffer vitamin A deficiency. This vitamin also plays an important role in combating infection. Keratinization of the epithelial tissues and defective immune system activity may also occur.

Vitamin supplements

The β – carotene molecule consists of two vitamin A molecules joined together. β – carotene, independent of its provitamin a function, is an antioxidant and can enhance the killing ability of neutrophils. Enzymes in the intestinal mucosa convert β – carotene to vitamin A by splitting. The efficiency of this conversion varies among species. Buffaloes are more efficient than cattle in this respect. Retionic acid serves all metabolic functions of vitamin A except the visual role. Vitamin A activity is usually expressed in international units (IU):

Destruction of vitamin A occurs in pelleting and during feed storage. The loss of vitamin A activity during pelleting varies from 6 to 30 percent, depending on presence of an antioxidant, temperature reached during pelleting, etc. Rancidity of feeds and presence of trace elements, such as copper, accelerate destruction of vitamin A. Vitamin A is degraded by rumen fermentation. Cattle fed high-concentrate diets have higher vitamin A requirements than those fed forage. Common source of carotene in livestock diets are green feeds, forage and yellow corn. Supplemental vitamin A is usually esterified vitamin, which is less sensitive to oxidation. Feed grade vitamin sources, consisting of bead lets with a protective gelatin covering and containing an antioxidant, are available. Requirement for dry as well as lactating cattle was same (110 1U/ kg of BW) as per NRC (2001).

Among all the vitamins, vitamin A is the most likely to present toxicity problems. Signs of vitamin A toxicity are in many respects similar to those of deficiency. The presumed safe limits for this vitamin is 66,000 IU/ kg of diet for lactating and non-lactating cattle.

1 IU = 0.3 μg retinol	1 IU = 0.55 μg retinyl palmitate
1 IU = 0.344 μg retinyl acetate	IU = 0.6 μg beta-carotene

Vitamin D

Vitamin D is fat-soluble. The two most important form of the vitamin are ergocalciferol (D_2) and Cholecalciferol (D_3). In fact, this is a pro-hormone, a necessary precursor of production of the calcium regulating hormone 1, 25-dihydroxy cholecalciferol. The structure of the D_2 and D_3 are:

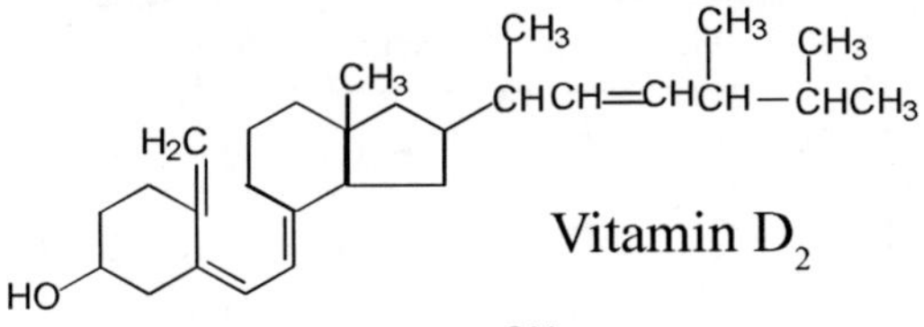

CH_3 CH_3 CH_3
CH_3 $CHCH{=}CHCH-CHCH_3$
H_2C
HO

Vitamin D_2

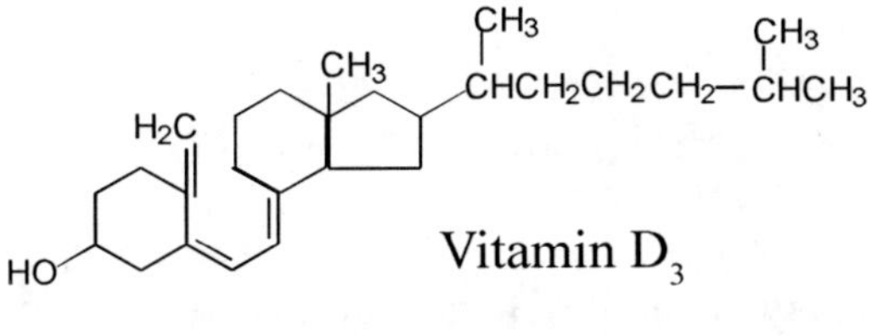

CH_3 CH_3
CH_3 $CHCH_2CH_2CH_2-CHCH_3$
H_2C
HO

Vitamin D_3

Functions

Ergocalciferol is activated plant sterol; cholecalciferol is activated animal sterol, differing slightly in chemical structure. Animals not exposed to sunlight will require a dietary source of vitamin D. Grazing animals exposed to sunlight normally do not require supplementation. Vitamin D_2 is converted to D_3, with D_3 the latter being metabolically active form. Vitamin D plays an important role in calcium metabolism, being involved in the regulation of calcium absorption and bone mineralization. The metabolically active form of the vitamin is 1, 25-dihydroxycholecalciferol (1, 25-OHD_3). This is formed by the addition of two hydroxyl (-OH) groups to D_3. The first hydroxylation occurs in the liver to produce 25 – OHD_3. This compound is then secreted in to the blood. In the kidney, it is converted to 1, 25 – OHD_3, which in turn is secreted into the blood. It acts like a steroid hormone in regulating calcium absorption and bone mineralization. The rate of formation of 1, 25 – OHD_3 is regulated by the parathyroid hormone (PTH) which, in turn, is regulated by the serum calcium level. In dairy cattle, administration of vitamin D metabolite prepartum has protective activity against milk fever. The presence of mycotoxin such as aflatoxin in the diet increases the vitamin D requirement. This has been of practical importance in poultry. Aflatoxin impairs vitamin D absorption, and its hepatotoxic effects may interfere with the formation of 25-OHD_3 in the liver.

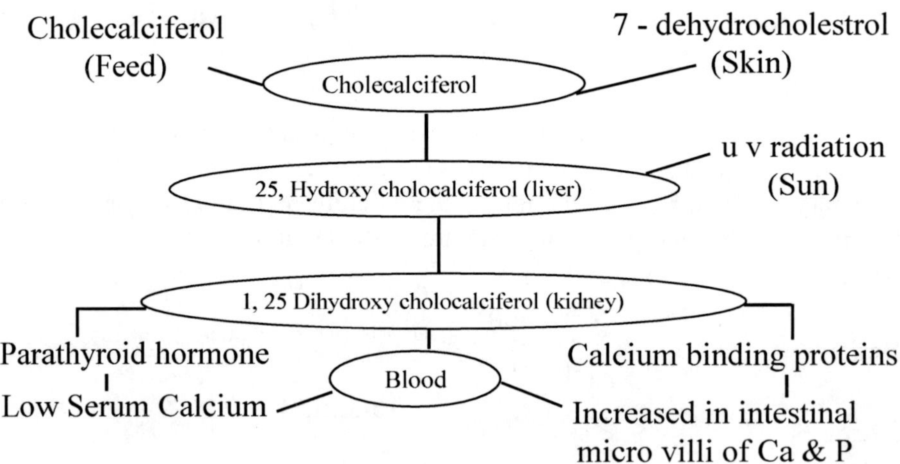

Metabolism of Vitamin D

Deficiency symptoms

Young calves may develop rickets, a disease of growing bones, resulting weak bones prone to fracture swollen knees and hocks and arching of the back may occur. In older animals the disease is known as osteomalecia, these diseases are also caused by the deficiency of calcium and phosphorus minerals.

Sources

Fresh green cereals and yeast have been shown to have rachitogenic (ricket-causing) properties. Insolated soybean proteins have a similar effect on poultry. Supplementation up to ten times of requirement may be tried to overcome the effect. Heating destroys the rachitogenic activity. Dietary sources of vitamin D include hay and other sun-cured forage. In contrast to their vitamin A activity, plants contain vitamin D activity only after the plant cells have died. Milk is not a rich source of vitamin, but colostrums usually contain six to ten times then the amount present in milk.

VITAMIN E

Vitamin E is a generic name for a series of lipid soluble compounds called tocopherols and tocotrienols depending on presence of saturated and unsaturated side chains, respectively. α –tocopherol, the most common and active of eight forms of the compound. Vitamin E, is not stored in body in large amounts. Losses during haymaking are very high, but silage remains good source.

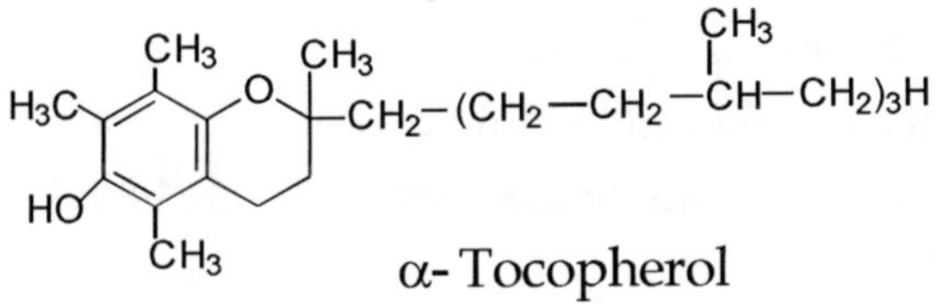

H_3C, CH_3, CH_3, O, $-CH_2-(CH_2-CH_2-\underset{|}{\overset{CH_3}{|}}CH-CH_2)_3H$, HO, CH_3

α- Tocopherol

The best understood function of Vitamin E is as antioxidant, it is also involved in maintenance of cellular membranes, arachidonic acid metabolism, immunity and reproductive function.

Deficiency Symptoms

White muscle disease in calves is a classic sign of a clinical deficiency of vitamin E. Wherein, muscle degeneration results into weak leg muscles and ultimately animal has difficulty in standing. The name white muscle disease is given owing to the presence of pale patches or white streaks in the muscles. Dietary or parenteral supplementation to dairy cows during the prepartum period has consistently improved the function of neutraphils and macrophages. Vitamin E along with selenium has been reported to decrease the cases of retained placenta, duration of intramammary gland infection and clinical mastitis. Encephalomalacia, brain degeneration is examples of vitamin E deficiency.

Sources

Vitamin E is very widely distributed in feeds. Green forages are good source of vitamin E. One IU of vitamin E is equivalent to 1 mg of *dl*-α- tocopherol acetate; *dl*-α- tocopherol has a potency of 1.1 IU per mg. Activities of *d*-α - tocopherol acetate and *d*-α- tocopherol are 1.36 and 1.49 IU per mg, respectively. Commercially, vitamin E is available as *dl*-α- tocopherol acetate. The esterified acetate group renders the vitamin resistant to oxidation and stable in feeds but these forms have antioxidant activity only after they are hydrolyzed in the digestive tract, so they do not protect feeds against rancidity. Good source of vitamin E in animal diets include green forage, good-quality hay, alfalfa meal, and cereal grains. The recommendation of total vitamin E is approximately 2.61U/ kg of body weight during lactation and gestation.

Supplementation of vitamin E increases with under following situations:

1. When diets have less forage, as concentrates are poor source of vitamin E in general, possible exceptions are raw whole soybean and cottonseeds. Heat treatment of soybeans destroys almost all of α – tocopherol.
2. Cows in sub optimal selenium status.
3. Milk is not a major excretion route for α – tocopherol.
4. When intake of polyunsaturated fatty acids are higher in non-ruminants.

VITAMIN K

Vitamin K is a generic term used to describe a group of quinone compounds exhibiting antihemorrhagic effects. The basic form of vitamin K is 2 methyl – 1, 4 – naphthoquinone. Three most common isomers or vitamins of K are Phylloquinone (K_1), Menaquinone (K_2) and Menadiones (K_3). The phylloquinones are commonly present in the chloroplast of green plants. Menaquinones (K_2) are synthesized by bacteria flora. Whereas K_3 menadiones, are the synthetic forms of vitamin K used in feed supplements, and do not exist naturally and is water-soluble. In contrast to the other fat-soluble vitamins, vitamin K is synthesized by ruminal and intestinal microbes.

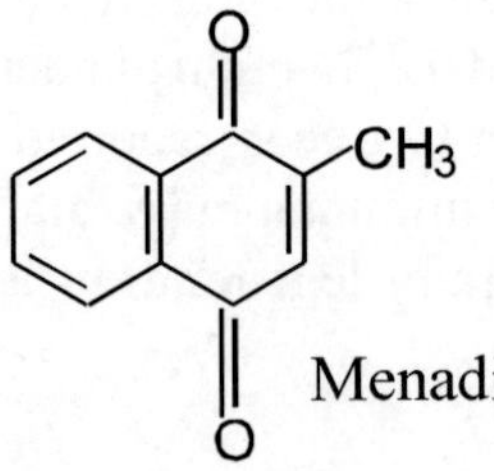

CH_3

Menadione

Cattle require vitamin K for synthesis of at least a dozen proteins. Among these are four blood clotting factors, Prothrombin (factor II) and factors VII, IX and X. These factors ultimately cause clot formation. Vitamin K is essential for the activation of prothrombin (a plasma protein) by catalyzing the addition of carbon dioxide to glutamic acid residues in prothrombin, creating calcium-binding sites. During the clotting process, prothrombin is activated to thrombin by addition of calcium; thrombin converts the soluble protein fibrinogen to insoluble fibrin, forming the clot. Cattle under natural pastures or forages are not likely to produce deficiency. The only reported deficiencies have occurred when mouldy sweet clover hay was fed. Dicoumarol is a fungal metabolite produced from substances in sweet clover that inhibits synthesis of clotting factors. The early signs of vitamin K deficiency include stiffness or lameness and hematoma of tissues. Prolonged feeding of dicoumarol leads to uncontrolled bleeding. Dicoumarol can pass placental barriers resulting in the fetus or newborn animals being affected. Vitamin K_3 was ineffective in preventing the anticoagulant effects of dicoumarol.

Sources

Vitamin K is rarely necessary to supplement ruminants. Swine and poultry diets should be supplemented with vitamin K. aflatoxin and other mycotoxins increase the vitamin K requirement. Coccidiosis infection increases the vitamin K requirement of poultry. The principle commercial forms of vitamin K used in the feed industry are derivatives of menadione, including menadione sodium bisulfite and menadione dimethyl-pyrimidinol bisulfite.

B Vitamins The B vitamins function as cofactors (Table 2) of enzymes involved in energy metabolism, some have additional roles as well. The B-complex vitamins are synthesized in adequate amounts in the ruminant animals. For swine and poultry, a complete vitamin mixture containing all known vitamins is routinely used. Toxicity is rarely of concern because the water-soluble vitamins are readily excreted.

Table 2: Vitamin Related to Coenzymes and Enzyme Prosthetic Groups or Activated Carriers Important in Metabolism.

B Vitamin	Coenzymes or Prosthetic Groups	other functions
Thiamin	Thiamin pyrophosphate (TPP)	Oxidative decarboxylation
Riboflavin	FMN, FAD	Hydrogen carrier
Nicotiamide	NAD, NADP	Hydrogen carrier
Pyridoxine	Pyridoxal phosphate (pp)	Transaminases Decarboxylases

Pantothenic acid	Coenzyme A (CoA) Acyl carrier protein	Acyl transfer
Folic acid	Tetrahydrofolic acid (THFA)	One carbon transfer
Biotin	Biotin	Carbon dioxide transfer
Cyanocobalamin	Methylcobalamin	Isomerases, dehydrases

Thiamin (Vitamin B_1) Thiamin in pure form is white in colour and has sulphurous odor. It contains a pyrimidine ring joined to a thiazole ring. In animal tissues it is mostly present as thiamin pyrophosphate (TPP).

Thiamin plays important role as co-enzyme decarboxylation (removal of CO_2) reactions such as the conversion of pyruvate to acetate in carbohydrate metabolism. Thiamin deficiency results in impaired carbohydrate metabolism, with an accumulation of pyruvate in the blood. Elevated pyruvate affects the nervous system, causing loss of appetite (anorexia) and neurological signs such as convulsions, (head retraction, "star gazing posture"), paralysis, reduced heart rate, and subnormal body temperature. Deficiency signs are very quickly reversed when the vitamin is administrated.

Most feedstuffs contain adequate levels of thiamin, so thiamin deficiency is rare. Most instances of deficiency are caused by destruction of thiamin in the diet or in the digestive tract by thiaminase. Thiaminase enzymes that spilt the thiamin molecule and destroy its activity. Various types of fish as carp, bracken fern, *Equisetum* spp., and bacteria produce thiaminases. Bracken fern poisoning of horses causes thiamin deficiency.

Feeding diets high in sulphate or those which cause a sudden drop in ruminal pH as rich in soluble carbohydrates can result thiamine deficiency called as Polioencephalomalacia (PEM) or cerebrocortical necrosis. A number of thiaminase-producing microorganisms, such as *Clostridium sporogenes,* proliferate when high-concentrate diets are fed. Lactic acidosis is suggested as a predisposing factor for the development of PEM. Approximately 1 g thiamin per animal per day is necessary to prevent PEM in the presence of rumen thiaminase. Feeding ionophores (rumensin, lasalocid), which inhibit lactic acid production, may have beneficial effects in prevention of PEM.

Sources

Soybean meal, grain, grain byproducts, Brewer's yeast and fermentation products are rich sources. Thiamin sources include the hydrochloride and mononitrate form is more stable, particularly under humid conditions.

Riboflavin (Vitamin B_2) It is a yellow crystalline compound and is only sparingly soluble in water. It is heat stable but unstable to light, particularly ultra violet light.

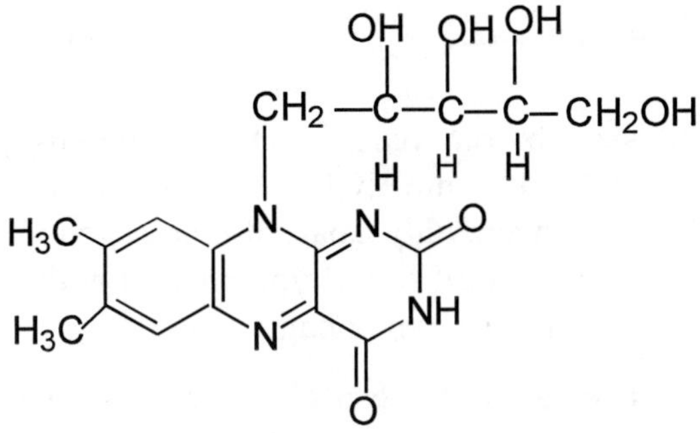

Riboflavin

Riboflavin functions in coenzymes such as flavin adenine dinucleotide (FAD) and flavin mononuceotide (FMN), which are involved in energy metabolism. Dietary riboflavin is destroyed in rumen, but much more is synthesized. A rumen flow of 15.2 mg/ kg of digestible organic matter consumed has bean observed, by only 25 per cent is absorbed from intestine. Riboflavin is one of the vitamins most likely to be deficient in swine and poultry diets because grains and plant protein supplements are poor sources. Alfalfa meal is a good diets but its use in swine and poultry diets is less common than formerly. Signs of riboflavin deficiency include curled toe paralysis in chicks, reduced egg production and hatchability in poultry and skin lesions, reduced growth, reproductive failure and high neonatal mortality in pigs. Riboflavin is commercially available as the crystalline vitamin. It is stable to heat but sensitive to degradation by light.

Sources

Riboflavin occurs in all biological materials. Rich sources are leaf meals, yeast, milk and other green leafy plants. Cereal grains are poor sources of the vitamin B_2.

Niacin Niacin is generic name for pyridine 3 – carboxylic acids (nicotinic acid) and their derivatives that demonstrate activity similar to the amide form i.e. nicotinamide. Niacin functions a coenzyme for the pyridine nucleotide electron carriers NAD(H) and NADP(H). Consequently it plays a critical role in mitochondrial respiration and the metabolism as whole.

Nicotinic acid   Nicotinamide

Niacin is synthesized by ruminal microbes and on unsupplemented diets the net supply to duodenum, is mostly higher than intake, but supplemental niacin is degraded in rumen and only a part of it passes to lower tract as such. Nicotinic acid can be synthesized from tryptophan in body tissues but maize contains very little of vitamin or tryptophan.

Nicotinic acid in cereal grains (especially corn) is bound to amino acids and has low bioavailability. Corn-based diets in particular require niacin supplementation. Niacin deficiency in human is responsible for pellagra, a condition characterized by a fiery red tongue, dermatitis, dementia and diarrhoea. Deficiency signs in pigs include poor growth, anorexia, dermatitis, diarrhoea and entiritis. Niacin deficiency in poultry results in leg problems as enlarged hock joint, poor feathering, and dermatitis. Niacin is available commercially in two forms niacinamide and nicotinic acid. Niacin is stable in the presence of heat, oxygen, moisture and light.

Sources

Niacin requirements for ruminants remain to be established. Supplemental niacin may be required for calves fed milk replacer, but not for growing animals. Most of the literature neither support routine use of niacin to enhance lactation performance, nor it minimize the risk of ketosis and fatty liver; though, there are reports claiming these benefits.

Pyridoxine (Vitamin B_6) The vitamin exists in three forms, which are interconvertible. They are pyridoxal, pyridoxine, and pyridoximine. Pyrodoxil phosphate participates in the catalysis of several reactions of protein metabolism, as transamination, decarboxylation and racemization.

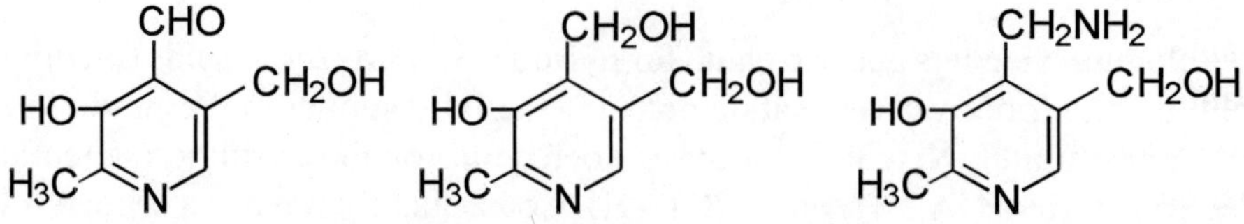

Pyridoxal   Pyridoxine   Pyridoxamine

It is found in adequate quantities for swine and poultry in grains and plant protein sources, so is seldom deficient in diets. Requirements are increased when linseed meal is used. Linseed meal contains linatine, a pyridoxine antagonist. Signs of pyridoxine deficiency include convulsions, dermatitis and impaired reproduction. Chicks on a deficient diet develop jerky movements, reduced hatchability and egg production. Vitamin B_6 is available commercially as crystalline pyridoxine hydrochloride. It is fairly stable in mixed feeds.

Sources

Vitamin B_6 are widely distributed in plant and animal, cereal grains are especially rich sources, yeast, pulses, liver and milk are also rich in pyridoxin.

Pantothenic Acid: The name is derived from the Greek word "*Pantothen*" means from everywhere, indicating its wide distribution. It is a constituent of coenzyme A and is therefore essential for several fundamental reactions such as fatty acid oxidation, amino acid catabolism and acetylcholine synthesis. Pantothenic acid is synthesized by ruminal microorganisms and quantity reaching intestine may be many times more than dietary amounts.

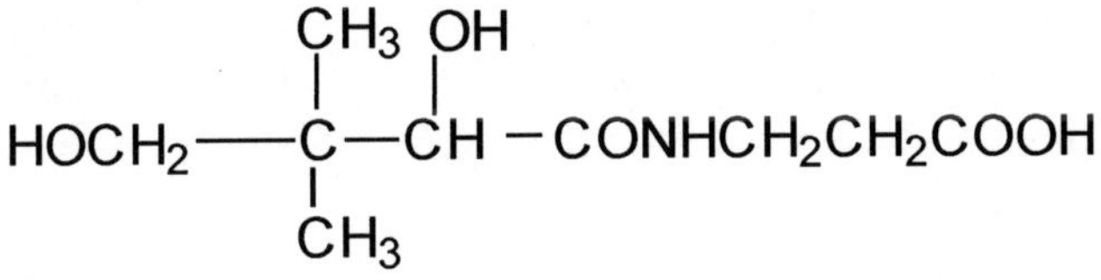

$$HOCH_2-\overset{\displaystyle CH_3}{\underset{\displaystyle CH_3}{\overset{|}{\underset{|}{C}}}}-\overset{\displaystyle OH}{\overset{|}{C}H}-CONHCH_2CH_2COOH$$

Pantothenic acid

Its bioavailability from barley, wheat and soybean meal is high but low in corn and sorghum. Signs of deficiency include reduced growth, dermatitis, decreased feed intake, skin lesions, change in hair coat and neurological defects. Pigs exhibit a characteristics goose-stepping gait (leg extended without bending while walking).

Sources

Pantothenic acid is available commercially as *d*- and *dl*-calcium pantothenate. Only the *d*-isomer is biologically active. One gram of *d*-calcium pantothenate is equivalent to 0.92g *d*-pantothenic acid activity, whereas 1 g of *dl*-calcium pantothenate has a potency of 0.46 g *d*-pantothenic acid. Calcium pantothenate has hygroscopic and electrostatic properties that cause in to pick up moisture and stick to metal, causing handling problems. It is usually combined with a carrier that prevents these problems. Requirements in ruminants are yet to be established.

Biotin Vitamin has also been called as "Anti-egg white injury factor". The term refers to observation that the deficiency could be induced by feeding large quantities of avian egg white. The latter contains a protein *Avidin*, which has very strong binding ability with biotin.

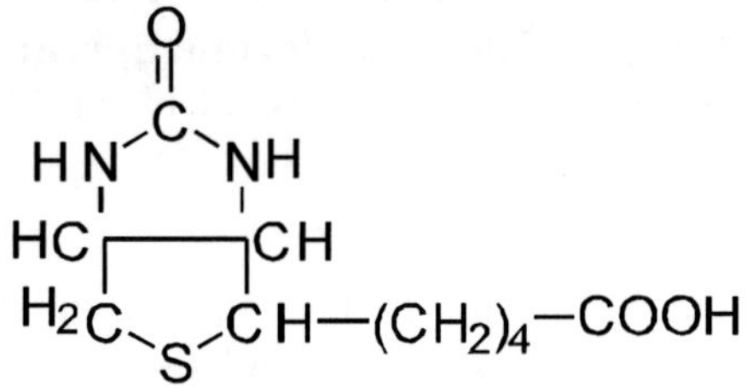

O
C
HN NH
HC CH
H_2C CH—$(CH_2)_4$—COOH
S

Biotin

Biotin is important in carbohydrate and fat metabolism being essential for transfer of carbon dioxide groups in decarboxylations and fatty acid synthesis. Ruminal bacteria normally synthesize biotin and its degradation is lower, thereby resulting efficient transfer of supplemented biotin to serum and milk. It is widely distributed in common feedstuffs but its bioavailability in some feeds (e.g., wheat, barley, sorghum and oat) is low. Biotin deficiency of poultry fed wheat-based diets has been a common problem. Biotin deficiency causes dermatitis and cracks in the feet (foot pad dermatitis), poor growth, loss of hair or feathers, and poor reproduction.

Synthetic biotin is quite expensive. It is available as the synthetic vitamin, either at 100 percent potency or in various dilutions. As it is readily destroyed in feeds by rancidity, it should be protected by the addition of an antioxidant to the feed. Twenty mg/d of supplemented biotine has been reported to improve hoof health in dairy cattle.

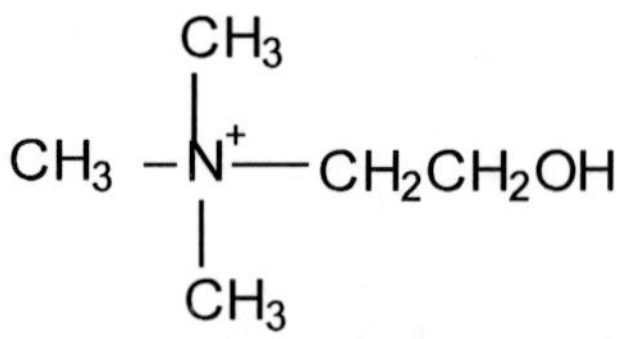

CH_3
$CH_3 - N^+ - CH_2CH_2OH$
CH_3

Choline

Choline is often considered a vitamin, although it is not, strictly speaking, a dietary essential, and not a metabolic catalyst it can be synthesized in the body using methyl groups from other sources such as methionine. Deficiency symptoms are difficult to produce with practical diets. The predominant deficiency symptom is fatty liver. In calves, muscular weakness, renal hemorrhage and fatty infiltration of the liver are observed. Cholines in feeds occur in phospholipids (lecithin), and supplemented choline chloride is extensively degraded by ruminal microflora. Supplementation of choline, in unprotected form may not be of much benefit. Methionine methyl groups are transferred to choline, thereby suggesting that methionine could be spared with addition of methyl group containing metabolites supplementation of choline led to increased milk yield, when intestinally observed methionine is limiting. Because of the relationship between fatty liver and ketosis, choline may probably he evaluate for ketosis treatment. Choline is a constituent of phospholipids in cell membrane and it essential for nerve function as a component of acetylcholine. Green forage, yeasts, egg yolk and cereals are rich source of choline. Choline is added to feeds as choline, either as a liquid or a dry powder. The liquid form is highly corrosive and requires special handling care and equipments. Although Choline can be synthesized from methionine, it is less expensive to provide it as supplementary Choline rather than by providing extra mithionine. Choline supplementation is often recommended in pig rations and the save may improve reproductive performance.

Folic Acid (folacin)

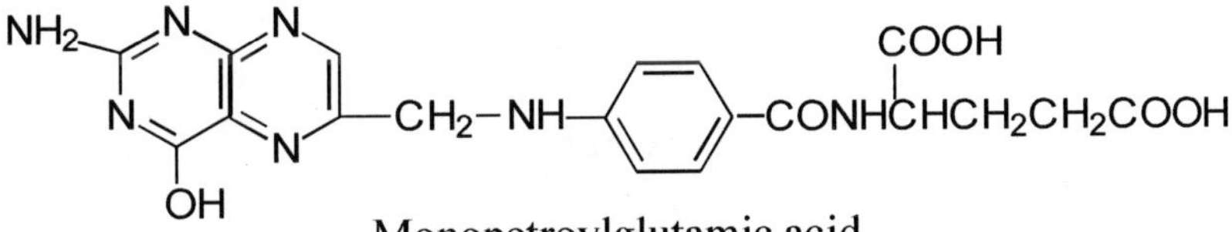

Monopetroylglutamic acid

Folic acid has an important metabolic role, along with vitamin B_{12} in the synthesis of hemoglobin. The principal deficiency sign is anaemia. Several reports indicate that folic acid supplementation of practical diets enhances reproductive performance of swine with an increase in litter size.

Sources

Folic acid is available commercially as the crystalline vitamin and is often used in form of the sodium salt. It is sensitive to heat with moisture and light but is not affected by oxidation. Green leafy materials, cereals and extracted oil seed meal are good sources.

Vitamin B_{12} (Cyanocobalamin) Because vitamin B_{12} does not occur in plant tissues for the dietary purpose. It is a dietary essential for swine and poultry. In ruminants and other herbivores, it is synthesized in adequate amounts by gut microflora. Cobalt is required in the diet. Vitamin B_{12} and folic acid have a dual role in formation of erythrocytes. The classic deficiency symptom in humans is pernicious anemia, caused by a genetic lack of the "intrinsic factor" necessary for vitamin B_{12} absorption. Pig and poultry on diets deficient in the vitamin result reduced growth, anemia, and poor reproduction. Ruminants have a high metabolic requirement for vitamin B_{12}, because it is involved in the metabolism of propionic acid and methionine synthesis. This requirement is met by ruminal synthesis, but adequate dietary cobalt is required. Vitamin B_{12} is available in crystalline form as cyanocobalamin. It has good stability in feeds. Synthesis is also higher on of forage-based diets.

Ascorbic Acid

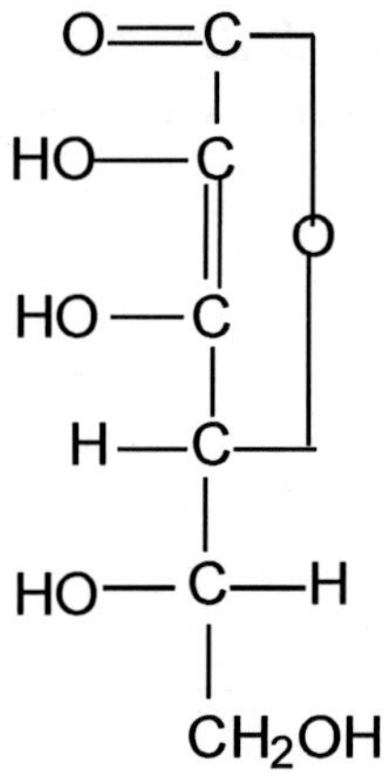

O=C
HO—C
O
HO—C
H—C
HO—C—H
CH_2OH

L -Ascorbic acid

Ascorbic acid (vitamin C) has traditionally been regarded as a dietary essential only for primates and guinea pig. These species lack an enzyme, *L* – gulonolactone oxidase, necessary for synthesis of ascorbic acid from glucose. Calves cannot synthesize ascorbic acid until approximately 3 weeks of age. The vitamin functions as a water-soluble cellular antioxidant. Specifically, it is through to be involved in regulation of steroid synthesis. The classic condition 'scurvy' characterized by oedema, emaciation and diarrhoea. Failure in collagen formation results in problems with teeth and bone.

Although livestock and poultry are considered to be able to synthesize adequate vitamin C, there are numerous reports of positive responses to supplementation with the vitamin. Stress has been suggested as a factor for increasing requirements. The effects of dietary ascorbic acid on egg cell quality and resistance to heat and cold stress have been suggested. However, the conditions under which vitamin C supplementation may be beneficial have not yet been adequately defined, and the vitamin is little used in the feed industry. It is primarily used in fish feeds. Citrus foods, green materials are good sources of this vitamin.

Table 3: Vitamin in livestock nutrition

Names	Basic Functions	Deficiency Symptoms	Good Farm Sources	Additional Information
Vitamin A	Essential for health of epithelial cells. Functions in eyesight-also in bone formation	Slow growth night blindness. Reproductive disorder. Rough coat. Stiff or swollen joints. Total blindness. Low level of vitamin A.	Whole milk (carotene) Berseem, green forages	Not found in plants. Formed in animals form carotene. Formed in wall of intestine, liver and certain other tissues. Rapidly destroyed by oxygen in heat and light. Reserves stored in liver and body fat.
Carotene	As a precursor of vitamin A	Possible vitamin A deficiency.	Green pastures good hay, alfalfa meal. Silage, yellow corn, whole milk.	Formed in plants. Used by animal to from vitamin A. rapidly destroyed by oxygen in heat and light. Reserves stored in body fat. Yellow in color. Vitamin A + Carotene = Vitamin A value.
Vitamin D	Function in Ca absorption and in Ca and P metabolism	Poor growth, rickets, osteomalacia, osteoporosis.	Sun-cured hays, whole milk, (sunlight)	Formed by irradiation of sterols with ultra violet light. Ergo sterol + UVL - D_2. 7-dehydrocholostrol + UVL - D_3.
Vitamin E	As an anti-oxidant. As a metabolic regulator of the cell nucleus.	Poor growth. Muscular distrophy - "white muscle" and "stiff-lamb" disease. Reproductive failures	Cereal grains, (especially the gram), green forage, good hay, oil seeds.	a-ocopherol most active, Utilization dependent on adequate selenium. Some storage in liver and other tissues. Fairly stable but rapidly destroyed in presence of rancid fat. Seldom deficient if selenium is adequate.
Vitamin K	Essential for prothrombin formation and blood clotting.	Prolonged clotting time of blood. Multiple hemorrhages.	Green forage, good hay, most seeds.	Fairly stable under normal conditions. Considerable intestinal synthesis. Seldom deficient for livestock.
Thaimine (vitamib B_1)	As a coenzyme in energy metabolism.	Poor appetite. Slow growth. Weakness. Hyperirritability.	Whole grains, germ meals, barns, green forage, good hay,	Fairly stable under normal conditions, Limited intestinal synthesis. Seldom deficient for livestock.
Riboflavin	In several enzyme systems related to energy and protein metabolism	Slow growth. Dermatitis. Eye abnormalities. Diarrhea and leg troubles in pigs.	Green forage, good hay - especially alfalfa.	Fairly stable under normal conditions. Most swine rations not containing milk and or/alfalfa meal wil be low in this vitamin.
Niacin	In enzyme systems related to glycolysis tissue respiration.	Digestive disorders dermatitis, and retarded growth in pigs.	Some present in most feeds, but that in corn, wheat and grain sorghum is largely unavilable.	Quite stable under normal conditions. Can be synth-esized in body tissues from surplus tryptophan. Swine rations sometimes deficient in available niacin.

Names	Basic Functions	Deficiency Symptoms	Good Farm Sources	Additional Information
Vitamin B_6	As a coenzyme in protein metabolism.	Poor growth, anemia, and convulsions in pigs.	Most common feeds are fair to good sources.	Rations for livestock normally require no B_6 supplementation.
Pantothenic acid	As a part of coenzyme A. in other metabolic reactions.	"Goose-stepping" in pigs-also digestive disorders and unhealthy appearance.	Widely distributed but it is low in corn.	Quite stable under normal conditions. Can be deficient for swine.
Biotin	In enzyme systems related to carbon dioxide fixation and de-carboxylation.	Dermatitis, loss of hair, and related growth in all species.	Widely distributed.	Quite stable. Considerable intestinal synthesis. Not likely to be deficient for livestock. Rendered unavailable by raw egg white.
Choline	In transmethylation. In fat metabolism in the liver. In cell structure (not a true vitamin function).	Unthriftiness. In coordination, fatty livers, and poor reproduction in swine.	Most commonly used feeds are fair to good sources.	Some synthesis in body tissues. Not likely to be deficient. Surplus may be used as a methyl donor to spare methionine.
Folic acid	Functions in the transfer to single carbon units in various biochemical reactions.	Poor growth, and various blood disorders in some species.	Most commonly used feeds are fair to good sources.	Some intestinal synthesis. Not likely to be deficient for livestock.
Vitamin B_{12}	As a coenzyme in several biochemical reactions.	Poor growth, inco-ordination, and poor reproduction.	Proteins feeds of animal origin. Fermentation products.	Originally animal protein factor. Some intestinal synthesis. Cobalt essential for B12 synthesis. Can be deficient for swine and earlyweaned calves.
Vitamin C	Functions in reactions are related to the formation and maintenance of collagenous inter-cellular material.	None demonstrated in livestock.	Not required.	Livestock synthesize vitamin C in body tissues. Never deficient with livestock under normal conditions.

Bibliography

Bieber-Wlaschny, M. (1988). Vitamin requirements of the dairy cow. In: Garnsworthy, P.C. (ed.) *Nutrition and Lactation in the Dairy Cow,* London, Butterworth, pp 135 – 56.

Brent, B.E. and Bartley, E.E. (1984). Thiamin and niacin in the rumen. J. Animal Sci. 59: 813 – 822.

Cheek, P.R. (1991). Applied Animal Nutrition: Feeds and Feeding. Prentice – Hall, Inc.

Lehninger, A.I. (1978). Biochemistry. 2nd edition. Kalyani Publisher, New Delhi.

Maynard, L.A; Loosli, J.K; Hintz, H.F. and Warner, R.G. (1979). Animal Nutrition. 7th edition. Tata McGraw-Hill Publishing Co. Ltd., New Delhi.

McDonald, P; Edwards, R.A; Greenhalhg, J.F.D. and Morgan, C.A. (2002). Animal Nutrition. Pearson Education, Singapore, Pte, Ltd., Indian Branch, Delhi, India.

McDowell, L.R. (1989). Vitamins in Animal Nutrition, San Diego: Academic Press.

NRC (2001). Nutrient requirements of dairy cattle. 7th revised edition. National Academy Press, Washington, D.C.

Olson, J.A. (1989). Provitamin A function of carotenoids: The conversion of â – carotene into vitamin A. F. Nutri. 119: 105 – 108.

Preston, T.R. and Leng, R.A. (1987). Matching Ruminant Production Systems with Available Resources in the Tropics and Sub-tropics. Armidale, N.S.N., Australia: Penambul Books.

Wang, J.Y; Owen, F.G. and Larsen, L.L. (1988). Effect of â – carotene supplementation on reproductive performance of lactating Holstein cows. J. Dairy Sci. 71: 181 – 186.

FORAGE PRODUCTION

The demand of the day is to make animal husbandry as an economically viable and ecofriendly sustainable farming that can boost overall economy of the country. It is known that for any livestock farm, feeding cost alone represents about 60 to 75% of the total recurring expenditure. Forages are economical source of nutrient than concentrates therefore livestock feeding programme should build around maximum use of forages. The forage based feeding system maintains the normal health and reproduction and increases the longevity as well as productivity of the livestock. The lack of adequate supply of quality fodders alone is the main constraint to the production potential of livestock. Therefore, whenever there is an opportunity to produce fodder, intensive forage production system may be practiced. In multiple cropping systems, 3-4 high yielding forage crops are grown in a particular piece of land in a calendar year. The crop sequences are tailored with an objective of achieving higher nutritious and digestible green forage yield and maintaining soil fertility. The system assures regular supply of green forage when staggered sowing and harvesting schedules are followed. The concept of overlapping cropping system may also be practiced.

I. Non leguminous forage

JOWAR (*Sorghum bicolor*)

Jowar is one of the most important grain and fodder crop of low rainfall areas of India and world. The crop has ability to grow over a very wide range of climatic and soil conditions. It can withstand heat and drought, also tolerates water logging for short periods to a greater extent than maize. The other advantage is that it can give two or three cuts and sometimes even four as in the case of M.P. Chari. The plant puts forth tillers particularly in Sudan type. Varieties may be as short as one meter and as tall as 4 to 5 meters. The leaves are generally waxy, possess saw like teeth on their margins and are comparatively smaller than those of maize.

Geographical distribution: Although jowar is believed to have originated from Africa, it has been grown in India since prehistoric times. It is grown in Maharashtra, Madhya Pardesh, Andhra Pradesh, Karnataka, Gujarat, Rajasthan, Tamil Nadu, Uttar Pradesh, Haryana, Punjab and Gujrat.

Climatic requirement: It is a tropical crop ideally suited to moderate rainfall areas. Minimum rainfall of 350 to 400 mm is required to bring the crop to flowering stage although it can tolerate a rainfall up to 600 mm. Mean temperature varying from 20 to 25^{0}C are congenial for its growth. The growth is not seriously affected even at mean temperature of 35^{0}C and humidity as low as 20 to 30 %. Winter varieties can tolerate minimum temperature up to 15^{0}C.

Soil and land preparation: The crop can be grown on a variety of soils ranging from sandy loams to clay loams. Comparatively, it is more tolerant to saline conditions than maize. It performs best under irrigated conditions on heavy and fertile soils.

The crop itself being shallow rooted, does not require deep ploughing. One mould board ploughing followed by one *desi* ploughing or 2 to 3 operations by *desi* plough generally sufficient. Two to three cultivations or harrowing by tractor also meet the requirements.

Time and method of sowing: In the northern states of India sowing time extended from March to August. If the irrigation facilities are not adequate and the crop is likely to wilt in the field before flowering, sowings should be commenced with the first showers of rain during June to July. In the south, four crops can be taken from February to November provided shoot fly is kept under control.

Drilling is the most common method. The seeds should not be sown deeper than 3 to 4 cm. During the rainy season, seeds can be broadcasted at a little higher seed rate and mixed with desi ploughs or harrows to complete the sowings well in time. Jowar seedling can also be transplanted in small areas. Whenever gaps occur in the field, seedling from thick patches can be uprooted and transplanted in the evening or in light drizzle on a cloudy day. Spacing of 25 to 30 cm between the rows and 10 to 15 cm between plants within the rows is recommended.

The seed rate is 40 to 50 kg per hectare. In mixtures with cowpea or guar, 25 kg jowar and 20 kg cowpea seeds may be used per hectare. In places where jowar fodder is not chaffed before feeding or where it is intended to be converted into hay, seed rate as high as 100 kg is used to produce soft and succulent stems.

Manuring: Jowar crop should be manured with 25 tones of farmyard manure (FYM) about a fortnight before sowing. In the rain fed or low rainfall areas, drill 20 kg of N (45 kg urea) per acre in rows at sowing. In high rainfall or irrigated areas, apply 20 kg of N (45 kg urea) and 8 kg phosphorus (50 kg single super phosphate) per acre at the time of sowing and another 20 kg N (45 kg urea) per acre about one month later. Add potassium to the crop on the soil test basis. For improving the protein content of the forage, 40 kg urea may be dissolved in 2,000 liters of water and sprayed on the crop 15 to 20 days before harvesting.

Under dry farming conditions, a fortnight before sowing, 10 to 15 tones of FYM may be incorporated deep into the soil with a mould board plough followed by 50 kg urea, 150 kg super phosphate and 40 kg muriate of potash at sowing. It will be better plan to spray 15 to 20 kg urea dissolved in 750 to 1,000 liters of water every fortnight after the establishment of the crop for quicker absorption of the fertilizer and better succulence of the crop.

Irrigation: The summer crop sown during the period of March to May, requires irrigation every fortnight in the hot and dry season. In the rainy season, water should not be allowed to stagnate but drained out quickly to keep the root system well aerated. In case of failure of rains and during post monsoon season, irrigation may be given every 20 to 25 days.

Harvesting: The crop should be cut immediately after flowering and harvesting should be completed with in 50% of flowering. Delay in harvesting results in sharp decline in crude protein and digestibility of nutrients and increase in crude fiber. During the time of harvesting stems should be soft and tender and not hard and woody. In case of multicut varieties, the first cut may be taken after 2 months of sowing and subsequent cuts at an interval of 35 to 40 days.

Yield: The average crop yields varies from 400 to 450 quintals per hectare. With excellent management yields as high as 850 quintals have been reported from 3 to 4 cuttings. The important thing to remember is that the crop should not be allowed to mature beyond 50%flowering stage. With advancing maturity, any increase in yield is at the cost of the quality of the fodder, particularly the digestible crude protein and subsequent loss in milk yields by animals.

Nutritive Value: It is maintenance type of fodder. When the crop harvest at dough stage it contains DCP 2.97% and TDN 63.3%.

Anti-nutrient: The young plants contain a glucoside called *dhurrin*, which breaks down into prussic or hydro cyanic acid in the rumen of the animal. Young plants up to 30 days growth and new shoots near soil surface are known to contain higher contents of the acid. The damaged crop is also toxic to the animals in the young stage. The crop should preferably be harvested after flowering but never before 45 to 50 days from sowing. Young plants, which have wilted in the field for want of water, cause instantaneous death of animals due to very high content of the glucoside in the leaves.

The hay and silage are generally free from prussic acid poisoning as the acid contents are reduced during the curing process or escape in the form of gas from silage while being carted.

The leaves contain higher amount of the glucoside than the stems. If the sorghum is manured with heavy doses of nitrogenous fertilizers at sowing, large quantities of nitrates may gets accumulate in the young plants, which in turn increases the dhurrin content. At 30-day growth, the accumulation of the glucoside is so high that a cow can die by consuming 4 to 5 kg of young plants.

The treatment against prussic acid poisoning consists of immediate administration of intravenous injection of 15 ml of 20% sodium nitrate followed by 75 ml of 20% sodium thiosulphate or a mixture of 3g of sodium nitrate and 15g of sodium thiosulphate in 200 ml of water.

Conservation: The jowar can be conserved best way for silage, harvest at milk to soft dough stage with moisture content of 65 to 70%. Chaff the plants to a size of 2-3 cm and quickly transfer to silo pit, trench or tower silo. Mix 4.0 kg urea per tone of green-chaffed material at filling to increase protein content. The protein content can also be increased by mixing leguminous crop like cowpea, guar, sunhemp, green gram, etc., in the proportion of ½ to 1/3rd of jowar fodder. Fill the silo fast and press it well to eliminate air. If jowar fodder is cut before the 'milk' stage, it may be sprinkled over with molasses at the rate of 40 to 50 kg per tone of fodder at the time of filling the silo, to increase the content of soluble carbohydrates. Application of 1% urea + 0.5 kg sodium chloride combination has been recommended as chemical preservative if the moisture content is higher (75%). Wilting for a day may also decrease moisture content significantly.

Crop rotations:

1. Cowpea-Jowar + Cowpea – Berseem
2. Jowar + Cowpea – Berseem +Sarson – Jowar + Cowpea.
3. Maize + Cowpea + Jowar + Cowpea – Berseem + Sarson
4. Jowar + Cowpea – Maize + Cowpea – Maize + Cowpea.

Recommended Varieties:

Areas	Varieties
All India (Early to medium, Single cut)	PC-6, PC-9, HC-171, HC-260, HC-308, Pant Chari-5 (UPFS-32)
North India (Multi cut)	PC-23, PC-106 (Hybrid)
North India (Single cut)	SL-44, M.P.Chari, U.P.Chari-2 (Sel.-278)
Uttar Pradesh, Maharashtra, Tamil Nadu, Andhra Pradesh (Single cut)	U.P.Chari-1 (1S-4776)
Madhya Pradesh (Single cut)	JC-6

Areas	Varieties
Madhya Pradesh (Multi cut)	JC-69
Maharastra (Single cut)	CSV-15 and csh-13 9 (Hybrid) as dual purpose
Maharastra (Medium duration)	Parbani Sweta (PVK 801/SPV 13333), SPH-981,(Mahabeej-7) RSLG-262 (Maulee, rabi season)
Rabi sorghum growing areas under irrigated conditions	SPH-1010 (CSH-19R; medium) SPV-1359 (CSV-216 ; Late)
All India	HC-136
(Late ,Single cut)	Raj- Chari-1 (SU-52)
Tamil Nadu (Two cuts)	CO-27
Tamil Nadu (Five cuts)	TNFS-9602 (C0FS-29)
Uttar Pradesh (Single cuts)	Pant Chari- (UPFS-23)
Gujarat Forage Sorghaum-1 (AS-16)	Gujarat (Two cuts)
Meethi Sudan (SSG-59-3), Pro-agro Chari ,SSG-988, Safed Moti (FSH- 92079) MFSH-3, Hara Sona (Hybrid), 855 F	All India (Multi cuts)
Punjab Sudex (Hybrid), LX-250	Punjab (Multi cuts)

MAIZE *(Zea mays)*

This crop is very much popular among the farmers due to its dual-purpose use for grains and fodder. Plant height varies from 1.5 to 6.0 m. Leaf blades are 30-150 cm long and 5-15 cm wide. The green portion of the plant left after removing cobs is used for forage source. The maize when grown for forage is most nutritious at silk stage.

Geographical distribution: Its origin is in Mexico and from pre-historic times it has been cultivated in America. It is one of the earliest introductions from America.

Climatic requirement: It requires warm and humid climate and well-drained soil.

Soil and land preparation: Maize requires well-drained clay loam soils with high humus. The soil pH should be near to neutral (6.5 – 7.0). Under good management and fertilization it can be grown in sandy loam to loam soils. The soil should be pulverized with a ploughing followed by 2-3 harrowing and planking.

Time and method of sowing: In the northern part of the country, maize should be sown from mid-February to August, while it may be grown throughout the year in most part of south India. The seed is placed in rows maintaining row-to-row distance 25-30 cm, at the rate of 45-50 kg seed per hectare.

Manuring: The application of nitrogen should be in 2 split doses. About 100 - 120 kg nitrogen per hectare is required when grown as a sole crop. The nitrogen requirement is reduced to 60-80 kg/ha when crop is sown in mixture with leguminous crop like cowpea. If 20-25 tones farmyard manure (FYM) is applied per hectare, about 50 kg additional nitrogen is required per hectare. The soil deficient in phosphorus and zinc should be fertilized with 20-30 kg phosphate (P_2O_5) and 10-15 kg zinc sulphate ($ZnSO_4$) according to the soil testing report and requirements of the maize variety.

Irrigation The water requirement of the crop is very high. Water is required throughout the growth period. The rainy season crop hardly needs any irrigation. The crop grown during summer, or if scanty rain falls, may require 4-6 irrigation. Water logging is highly undesirable and rainwater should drain out of the field quickly. Water logging damages the maize crop extensively.

Harvesting and yield: Maize should be harvested at silk stage, which has been found to be ideal for optimum harvesting of fresh fodder as well as nutrients. The yields depend on the stage of harvest, variety and management. The average yield may range between 250-300 q/ha, with good management 400 - 600 q/ha green forage can be obtained.

Nutritive value: The nutritive value of the green forage in terms of DCP 4.11% and TDN 53.3%. It can support 3-4 kg of milk production, above maintenance.

Conservation: For the conservation of maize fodder in the form of silage, it should be harvested at milk stage due to lower moisture and higher content of soluble carbohydrates.

Crop rotation:

Maize+Cowpea-Maize+Cowpea-Oats-Maize+Cowpea

Bajra+Cowpea-Maize+Cowpea-Oats

Maize+Cowpea-Sorghum+Cowpea-Berseem+Mustard

Maize+Cowpea-Dinanath grass-Oats

Maize+Ricebean-Berseem-Mustard

Sorghum+Cowpea-Maize+Cowpea-Maize+Cowpea

Recommended Varieties:

Areas	Varieties
All India	African tall, Vijay Composite, Moti Composite, Manjri Composite, Jawahar
Punjab	J-1006
North East India	A-de-cuba

BAJRA (*Pennisetum americanum*)

Bajra or Pearl Millet is grown as a dual purpose crop. It is a quick growing short duration crop has high forage value for the dry farming areas for its potential for high tillering, draught and heat tolerant.

Geographical distribution: Bajra is grown tropical agro climatic zones of the country particularly in Gujrat, Rajasthan, Maharastra and central India.

Climatic requirement: Bajra is grown during summer and rainy season under tropical climates. It is more popular in areas with 400-500 mm annual rainfall. Bajra is considerably tolerance to drought conditions and can also be grown in low rainfall (120-150 mm) areas. High temperature is required for its rapid growth, especially at later stage of plant development.

Soil and land preparation: Sandy loam soil is suitable for bajra cultivation. It gives a good yield in fertile soils and requires less water. The soil should be well drained for its cultivation. Land became ready for sowing after 2-3 harrowing followed by planking.

Time and method of sowing: Under rain fed condition it is usually sown in the northern part of the country on the onset of monsoon (June-July). Under irrigated condition, it can be sown in March as a summer crop. In South India, bajra is sown during May and under irrigated condition during January and February.

Seed should be sown in rows (25-30cm row to row distance), about 1-1.5 cm deep at, the rate of 15-20 kg/ha. Transplantation favors increased tillering in bajra, which is desired for more forage yield.

Manuring and irrigation: Nitrogen requirement of the crop varies according to type of crop, soil and soil moisture. About 90 - 100 kg nitrogen/hectare is applied in 2-4 split doses depending on rainfall and type of soil. The variety having regrowth tendency may require additional nitrogen dose (25-30 kg/ha) after cutting. Phosphorus and potash should be applied at the time of sowing according to soil test results.

Nitrogen dose is reduced to 60-80 kg under un-irrigated and low rainfall conditions. On the application of 200-250 quintals FYM per hectare of land, nitrogen dose can be reduce to 30-40 kg/ha in split doses. In desert, bajra may require iron, zinc, and manganese for proper growth.

Bajra requires low moisture and is able to resist draught for a longer period. It may not need any irrigation during monsoon season, while during summer; irrigation at an interval of 15-20 days may enhance the forage production.

Harvesting and yield: The first cutting should be done after 50-80 days after sowing, and subsequent cutting after 35-40 days. For good regrowth the cutting height should be 15 cm from the ground. The average green matter production varies from 400-600 q/ha. The kabdi production of 50-60 q/ha may be harvested from grain crop.

Nutritive value: It is a maintenance type of fodder and contains DCP 4.3% and TDN 59.2%.

Anti-nutrient: The forage contains 1.19-2.16 % oxalates.

Crop rotation:

Bajra+Cowpea-Maize+Cowpea-Oats

Recommended varieties:

Areas	Varieties
All India	Giant Bajra (Multi cut)
Maharastra, Central India	Giant Bajra (Single cut)
Entire bajra growing tract	Raj-Bajra-Chari-2 (UUJ-IV-M), Fodder Cumbu-8 (TNSC-1), L-72, L-74, WCC-75
Gujarat & Rajasthan	Rajko

OAT (*Avena sativa*)

Oat is the important cereal crop grown in winter season in the northern, western, and central India. It is usually an erect annual crop with a fairly good tillering habit, attains a height from 1 to 2 meters. It is quick growing, palatable, succulent and nutritious crop and forms an excellent combination when fed along with other legumes like berseem, lucerne, senji. On dairy farm, oats fodder is a must, as it can be fed green and surplus converted in to silage or hay for use during the scarcity period.

Geographical distribution: The crop appears to have been cultivated in western Europe as a grain crop. In India, it is grown in Punjab, Haryana, Jammu and Kashmir, Himachal Pradesh, Uttar Pradesh, Madhya Pradesh, Rajasthan, Maharashtra and West Bengal.

Climatic requirement: Cool and moist conditions as prevail in temperate regions, are congenial for its growth. A well-distributed rainfall of 400mm during the four months duration is sufficient to meet its requirement as a fodder crop. Frost and severe cold conditions retard its growth where as hot and dry conditions hasten its maturity resulting in poor yield and inferior quality of fodder.

Soil and land preparation: Fertile and well drained loam to clay loam soils are the best. It can tolerate slight acid or saline conditions.

The land should be opened with a bullock-draw soil inversion plough followed by two operations of *desi* ploughs.

Time and method of sowing: Sowing should continue from the first week of October up to end of November to supply green fodder from mid December to end of April.

Seed should preferably be drilled. Alternatively sowing can be done either by pora or kera methods. Transplanting of four weeks old seedling can also be done successfully. In case of low tillering varieties, the spacing between rows may be kept between 20 to 25 cm but increased up to 30 cm in heavy tillering types.

Small seeded varieties can be sown at the rate of 75 to 80 kg per hectare. A bold seeded varieties like Kent requires a seed rate of about 100 to 125 kg per hectare.

Manuring: The crop may be manured with 25 tonnes of FYM of 10 to 15 days before sowing. In addition a mixture consisting of 65 kg urea, 200 kg super phosphate and 50 kg muriate of potash should be applied at sowing.

Irrigation: A good pre sowing irrigation (7.5 to 8.00 cm depth) may be given 5 to 7 days before sowing and seeds drilled as soon as land is fit for cultivation. The first irrigation should be given after 30 to 40 days depending up on the soil texture. Subsequent irrigation may be given every 25 to 30 days. Irrigation at boot stage is essential as the water requirement at flowering stage is high.

Harvesting & yield: It should be harvested immediately after the initiation of flowering up to 50% bloom. Delay in harvesting will result in sharp decline in digestible crude protein. In a two cut management, the first cut is taken by middle of December. Normally, first cut can be taken after 60 to 70 days.

The average yields vary from 350 to 400 quintals per hectare. With top management, yields of 550 to 600 quintals can be obtained.

Nutritive value: The oat fodder contain 6.61% DCP and 63.4% TDN.

Conservation: An excellent crop for making hay or silage. For hay, the crop may be harvested from mid flowering to milk stage and for silage, when the grains are in soft dough stage. Best hay is made in the months of March and April when the days warm up and there is not much dewfall at night.

Since oats for making silage at the soft dough stage will be poor in protein, it is better to mix it with berseem or lucerne in the proportion of 3 parts of oats with one part of legume. Berseem and lucerne should be wilted in the field to 65% before chaffing and mixing.

Recommended varieties:

Areas	**Varieties**
Whole India	Kent,OS-6,OS-7 (single/double cut)
North, North west India and Southern hills	OL-9 single/double cut)
Central India	JHO-822 (Bundel Jai-822) (Multicut)
North India, Andhra Pradesh & Southern hills	UPO-212, OS-96 (Multicut)
North east & Central India	OS-108
Whole India	UPO-94, PO-3 (Multicut)
North west & Central India	OL – 125 (Single cut)

TEOSINTE (*Euchlaena Mexicana*)

Teosinte or Buffalo grass is a coarse, tall growing (2.5-4 m), annual, exotic grass grown as fodder crop in different states of India. It is an inter-generic cross between maize and chari and is capable to give multiple cuts. It can grow in the areas suitable for maize. Unlike maize, it tillers freely. This can resist water logging for a short period, and it does not lodge like maize.

Geographical distribution: Native of Central America and Mexico and is considered to be the wild ancestor of maize.

Climatic requirement: It grows well in warm regions with high humidity and rainfall.

Soil and land preparation: The well drained loam or sandy loam soils (pH 6.5-7.0) are suitable for cultivation. The land preparation depends on previous crops, however, one ploughing followed by 2-3 harrowing is enough.

Time and method of sowing: It can be sown at any time of the year. To get green forage during summer this can be sown during March-April. The crop sown in October-November can supply greens during winter but the growth is poor in this season.

The healthy seeds can be separated through salt water. The healthy seeds can settled in the bottom of the pot are removed and washed 2-3 times before sowing.

The seed can be planted in rows (25-30 cm distance) or broadcasted at the rate of 35-40 kg/ha.

Manuring and irrigation: To harvest a good yield from crop about 200-250 quintals FYM and 60-70 kg nitrogen per hectare is required which can be applied in 3 split doses (15-20 kg at the time of sowing, 20 kg after 30 days and rest after the first cutting). The phosphorus, potash and other elements can be applied after soil testing.

The crop may not need any irrigation during rainy season, however, if required the irrigation at 15-20 days interval would be beneficial.

Harvesting and yield: The crop sown in the month of February-March is ready for first cut in May-June. The second and third cutting can be taken in August and November, respectively. The crop sown in June-July can be harvested in September –October. The forage should be harvested at 175-200 cm height. The cutting height should be 10.15 cm from the ground for profuse and quick re-growth. In 2-3 cuttings it produces 600-700 q/ha green fodders.

Nutritive value: It is a maintenance type of fodder and it contains 6.00% DCP and 57.9% TDN when cut at pre-flowering stage.

Recommended varieties: The local varieties are more popular, however, some new strains have been evolved at various Institutes collaborating in All India Coordinated Research Project on forage crop, and they are Teosinte-I IGFRI, Teosinte-II Rahuri, and Maizinte Rahuri.

SUDAN GRASS *(Sorghum sudanese)*

Sudan grass is a profuse tillering leafy, tall (3m) annual, exotic grass. It is a draught and heat registant summer crop, serves as an emergency forage crop due to its high growth rate.

Geographical distribution: The grass has been grown in Egypt from very early times and the value of this grass was first reported in Sudan in 1909. In the same time it was also introduced in America. In India it is grown in various states.

Climatic requirement: This requires 30-34^{0}C temperatures. The crop is susceptible to draught so either irrigation should be assured or it should be grown in the areas with 40-75 cm rainfall.

Soil and land preparation: Sandy loam or clay loam soils (pH 6.5 to 7.0) are suitable for cultivation. The land can be prepared with the help of 2-3 harrowing.

Time and method for sowing: Sudan grass can be sown from February to June. The crop sown during February-March can produce good forage for summer season. This can be sown in lines (25 to 30 cm row to row distance) at the rate of 10-15 kg per hectare depending on the type of seed. Sowing can be done through broadcasting with 50% higher seed rate.

Manuring and irrigation: The phosphorus and potash should be added if soils are deficit. Nitrogen requirement depends on the type of previous crop and harvesting management. If 200-250 quintal FYM is put per hectare of land, 30 kg nitrogen can be added at the time of sowing, 30 kg nitrogen can be dressed after each cutting for maximum yield. In the absence of FYM manure, 60-80 kg of nitrogen can be added at the time of sowing.

The spring or summer crop requires irrigation at 10-15 days interval. The rainy season crop may not require irrigation but during failure of rain the irrigation may be required.

Harvesting and yield: The crop is ready for harvest after 50-60 days of sowing the subsequent cutting can be taken at 30-40 days interval. The average yield of green forage can be varied from 450 to 550 q/ha according to management.

Nutritive value: It contains DCP 4.13% and TDN 52.7%.

Anti-nutrient: Prussic acid poisoning in livestock has been reported when the forage is harvested and fed at the young stage. For this reason, animal should not be allowed to graze on Sudan grass in the young stage.

Recommended varieties: Sweet sudan, piper sudan and S.S.G. 59-3.

DINANATH GRASS *(Pennisetum Pedicellatum)*

Dinanath grass is an erect, annual grass of 0.7-0.9 m height. The leaves are 40-60 cm long and light to dark green in colour.

Geographical distribution: It is grown in West Africa, Australia and in India particularly in Bihar, West Bengal, Orissa, North region of Maharashtra and A.P., Haryana, Punjab, Uttar Pradesh and Madhya Pradesh.

Climatic requirement: It prefers warm climate with high humidity and rainfall ranging from 80-125 cm.

Soil and land preparation: The loam soils are best suited for its cultivation. The grass can be grown in less fertile land. With sufficient manure application field become ready for sowing after 2-3 harrowing.

Time and method of sowing: The grass can be grown any time from March to August. Sowing in the months of July to August is advisable to meet the green forage need in the months of October-November.

The sowing can be done in rows at 25 cm distance or broadcast. Seed rate vary from 15-20 kg per hectare depending on the method of sowing. From the transplantation of forage during rainy season about 20 days old seedlings are more suitable.

Manuring and irrigation: Farmyard manure (100 quintal per hectare) is supplied before 4-6 weeks of sowing. Application of nitrogen per hectare in split doses may be beneficial for growth. If FYM is not available, 60-80 kg nitrogen per hectare should be applied in two split doses (half at the time of sowing and rest after 30 days of sowing). The phosphorus and potash should be applied according to soil testing results.

Efficient water drainage system is advisable for cultivation. During rainy season the crop may not need any irrigation, but during summer and after monsoon, irrigation help in increase production of forage.

Harvesting and yield: The forage can give two cuttings under good management. First cutting can be taken after 60-70 days after sowing and second at the flowering stage. The yield depends on varieties, stage of harvest, and managemental conditions. The total green forage production may vary from 400-800 quintals per hectare.

Nutritive values: It contains 7.4 % CP, 0.42 % Ca and 0.21 % P.

Recommended varieties: Apart from local strains few cultivars like IGFRI-3808, P.S.-2, Bundel-1 (IGFRI-43-1), Bundel-2 (IGFRI-4-2-1), TNDN-1 are high yielding varieties.

GUINEA GRASS (*Panicum maximum*)

Guinea grass is most suited to the southern region of the India as it is well adapted to the tropical areas with moderate humid conditions and warm temperatures. The grass is leafy, tall tufted perennial and grows normally to a height of about 2.75 meters, rich in calcium and does not become coarse quickly. In palatability, it compares well with jowar and maize. Guinea grass tolerates shade of the trees and comes up well in the mango groves, orchards and forest areas.

Geographical distribution: The grass is native to tropical and subtropical Africa. Now it is grown throughout tropical and subtropical regions of the world. In India it is generally grown in southern states of Karnataka, Andhra Pradesh, Tamil nadu and Kerala.

Climatic requirements: Warm and humid conditions with partial cloudiness and light showers of rain favours the rapid growth of this grass. The minimum temperature should remain above 15°C but maximum temperature beyond 38°C do not favour this grass. As a pasture grass, it requires a minimum rainfall of about 600 to 1000 mm and a relative humidity of 70%. It can tolerate rainfall up to 1,200 to 1,500 mm.

Soil and land preparation: Rich, fertile, deep, loam to clay loam soils are the best. Sandy loam soils can be used provided large doses of FYM are incorporated to improve the fertility and water holding capacity of these soils.

Since the grass is of perennial nature and remains on the same land for three to four years, it should be given a thorough cultivation to prepare a firm and fine seed bed. One mould board ploughing should be followed by three ploughings by *desi* ploughs.

Sowing time and method: In the southern region, the best time however, appears to be the month of June, July and August when cool and humid weather prevails during the monsoon season. In the north, plantings can be done from April to August under irrigated condition.

The grass can be propagated by rooted slips, stem cutting and seeds. In summer season, it would be preferable and safer to use and rootstock. The seeds may also be sown in the nursery and seedlings transplanted in the fields with the break of monsoon. The old stands of 3 to 4 years can be renovated by burning.

Ridges can be opened one meter apart and planting on the ridges done with a spacing of half a meter between the plants.

The stem cutting or rooted slips may be planted in the north at the rate of 500 kg per hectare. In the south, the planting material required for spacing of 1.0 m x 0.5m x 0.75 m x 0.30 m x 0.75 m x 0.4 m x 0.6 m x 0.6m will be 1,000 kg, 2,240 kg and 1,400 kg per hectare, respectively. When the seeds are drilled directly in the field the seed rate may vary from 4 to 6 kg per hectare.

Manuring & irrigation: FYM may be applied at the rate of 40 tones per hectare2 to 3 weeks before planting. At planting 250 kg superphosphate and 65 to 70 kg muriate of potash may be drilled in to the soil. After taking each cutting, urea may be side dressed at the rate of 90 kg per hectare to boost the yield and to improve the protein content of fodder.

After planting, the first two irrigation should be given every 7 to 10 days in the summer season and 15 to 20 days in the post monsoon season.

Harvesting & yield: The first cut is generally ready after 9 to 10 weeks from planting. Subsequently, the grass can be harvested every 6 to 7 weeks. With sewage irrigation, a cut can be taken every 5 to 6 weeks in the monsoon season.

On an average about 150 quintals of grass can be obtained per hectare per cut. Yields can be improved to 200 to 250 quintals per cut. Thus, total yield of about 2,000 quintals can be obtained per hectare per year in the southern and eastern regions. In the North, yields of 1,200 to 1,300 quintals can be obtained in five cuts.

Nutritive values: It contains 5-14 % CP and 52-60 % TDN depending on the stage of clipping.

Recommended varieties:

Areas	Varieties
Kerala	Macuenii
Karala, Tamil Nadu, A.P. West Bengal, Bihar, and North- Eastern States	Hamil
North-West	PGG-1
Punjab	PGG- 19
Punjab	Pgg-101
Northern, North-West and Central India	PGG-3
Norther, North-West and Central India	PGG-9

NAPIER- BAJRA-HYBRID *(Pennisetum purpureum* X *P. americanum)*

It is vegetatively propagated bajra like grass. It is perennial but maximum fodder yield between March and November. The crop once planted gives fodder for 2-3 years.

Geographical distribution: The grass is native of tropical Africa. It is grown under warm and humid conditions. In India, the hybrid are now grown practically throughout India except in very high rainfall areas or cold temperate regions.

Climatic requirement: This hybrid requires hot and moist climate as prevails in the monsoon season is most congenial for the vigorous growth. Intermittent showers followed by bright sunshine result in lush growth. Its minimum rainfall requirements vary from 800 to 1000 mm. The mean optimal temperature varies from 24 to 28 °C.

Soil and land preparation: This hybrid can be grown on variety of soils, but heavy soils are best for getting a high tonnage of green fodder.

The land should be prepared well and should be free from weeds. Give the first ploughing with a mould- board plough and subsequent two ploughings with a cultivator. Planking should follow every ploughing.

Time and method of sowing: This variety can be planted from last week of February to May in north India. The planting should be completed by mid April to avoid high mortality of root slips. In central, southern and eastern region, planting can be done throughout the year except in the cold months of December and January and hot summer months of May and June.

It can be propagated vegetatively from root slips or stem cuttings. Root slips 30 cm long, or stem cutting with two or three nodes are used for planting. About 11,000 slips or cuttings are required per acre for plantation. A small part of the root or shoot slip is allowed to remain exposed and the rest of the slip is buried in the soil. The stem cuttings can also be planted like sugarcane sets in 7 to 8 cm deep furrows which are afterwards filled with soil. The crop should be planted at 90x 40 cm or 60x 60 cm apart in lines under good conditions of soil moisture.

Manuring and irrigation: A minimum of 40 tonnes of FYM may be applied 2 to 3 weeks before planting. At planting, 250 kg superphosphate and 65 to 70 kg muriate of potash may be drilled into the soil. This dose may be repeated every year with the first monsoon showers. After taking each cut, urea may be applied at the rate of 110 kg per hectare.

The crop needs frequent irrigation during hot and dry months. In the mid-season, however, irrigate after every 10 to 14 days. Drain away excess water from the field during the rainy season.

Harvesting and yield: The first cutting is ready in about 50 days after planting. Take subsequent cuttings when the crop is about one meter high. If the crop is allow growing beyond this height its nutritive value falls drastically.

Average yields from each cutting vary from 250 to 350 q per hectare. High average yields of 450 to 500 q per cutting per hectare can be obtained with liberal application of fertilizers and irrigation. On an average yields of 1500 and 2500 q per hectare are reported from the North and South, respectively.

Nutritive value: Its contains about 3 % DCP and 65% TDN on dry matter basis.

Recommended varieties:

Areas	Varieties
Whole of India and tropics abroad	Pusa Giant
Whole of tropical humid part	NB-21
Tamilnadu, and Karnataka	CO- 1
U.P., M.P., NE hills, Punjab and hills of North India	Swetika –1(IGFRI-3)
U.P., H.P., NE hills, Punjab and hills of North India (intercropping)	IGFRI-6
Punjab	PBN-83
Tamilnadu and southern part	CO-2, CO-3
Whole of India (acid soils sub-temperate regions)	IGFRI- 7, IGFRI-10
Whole of India	Yeshwant (RBN- 9)

II. LEGUMINOUS FORAGE

BERSEEM (*Trifolium alexandrium*)

Berseem also called Egyptian clover, is succulent annual growing to a height of 80 to 90 cm. Cow yielding 7 to 8 kg of milk can easily be maintain on green berseem alone without any concentrate feeding. With increasing availability of irrigation water the slow growing and low yielding winter fodders like senji, metha, pea and shafatl are being replaced by berseem in North India. Apart from supplying high protein fodder, berseem built up the fertility of soil as a results in the economy of fertilizer use for the next crop in the rotation.

Geographical Distribution: Berseem appears to have originated from Asia. The crop was introduced in India in 1904 and now it is grown in Himanchal Pradesh, Punjab, Haryana, Delhi, Rajasthan, Uttar Pradesh, Bihar and parts of Madhya Pradesh, Maharashtra, West Bengal, Assam, Orissa and Jammu and Kashmir.

Climatic requirements: Cool to moderately cold conditions as prevail in North India during the winter and spring seasons. Temperature, 4 to 5°C below freezing kill the plants. Warm and humid climate as prevails in the eastern and southern states is not congenial for the growth of this crop.

Soil and land preparation: Well drained loam to clay loam soil, rich in phosphorous, potash and calcium with neutal to slightly alkaline reaction is ideally suited to cultivation of berseem. The land should be very well leveled as the young plant die in stagnant water. Saline soils due to excess amount of sodium, potassium and magnesium salts adversely effect the berseem seeds, but once established, the crops can tolerate fair amount of salts due to dilution and leaching by frequent irrigation.

The land should be opened with a soil inversion plough followed by two or three operations by a *desi* plough. In case of tractor implements, three cultivators or disc harrowings generally suffice. Sandy loam soils should preferably be puddle before broadcasting the seeds to prevent excessive downward movement of water and leaching of soluble nutrients.

Time and method of sowing: In the North the sowing can commence from the middle of September after the end of the monsoon seasons as the temperature cool down earlier than in the Eastern and Central parts of India. Sowing can be continued up to the end of October in the North and first week of December in the Eastern region

The earliest and quickest method is by letting in water to a depth of 4 to 5cm in a well leveled field and broadcasting soaked seeds of berseem uniformly in a crosswise direction. The seeds should be inoculated with *Rhizobium trifolii* bacteria for fixation of atmospheric nitrogen before sowing.

Manuring and irrigation: Twenty-five tones of farmyard manure may be applied and thoroughly mixed with the soil before sowing. At sowing, 375 to 400 kg super phosphate, 50kg muriate of potash and 55kg urea should be added.

First two irrigations should be very light and given after 5 to 7 days for quicker germination of hard seeds and establishment of young plants. Afterwards, irrigations may be given every 15 days in winter, 12 days in the spring and 10 days from middle of April to mid-May. About 16 to 18 irrigations in all are given and the total requirement of water work out to about 100 cm.

Harvesting and yield: First cutting is generally taken after 60 to 70 days of sowing, second, third and fourth in 30 days intervals and fifth and sixth after 20 days intervals. The dry matter in the first cut averages about 10 to 11 percent and increases by about 1½ percent in every subsequent cut.

The total yields per hectare average about a thousand quintals. The yields are generally lower in the first cut (175 q) and highest in the second and third cuts (225 q) with a steady fall from the 4th cut (175 q) to 5th cut (125 q) and a very small yield (75 q) from the 6th cut.

Nutritive Value: Its green fodder contains 13.99 to 23.10% crude protein and 62.3 to 70.2% TDN on dry matter basis. The *in vitro* dry matter digestibility is recorded 70%. It is also rich in minerals.

Anti-nutrient: Berseem shows fairly high oestrogenic activity but continuous feeding all through the winter and spring does not seem to bring about any significant disorders or changes.

Excessive feeding of berseem in the young stage or when it is wet with dew in the morning, causes bloat called tympanitis.

Rotation:

a) Maize+Cowpea-Berseem+ Mustard/Oats-Maize+Cowpea
b) Bajra+Cowpea-Berseem+Mustard/Oats- Maize+Cowpea
c) Jowar+ Cowpea/Guar-Berseem+Mustard/Oats- Maize+Cowpea
d) Sweet sudan (3 cuts)-Berseem+Mustard/Oats.
e) Rice bean-Berseem+Mustard- Maize+Cowpea
f) Hybrid Napier interplanted with Berseem+Mustard in the winter season and Cowpea+Maize in the summer season.

Recommended Varieties:

Area	Varieties
Whole India	Wardan (S-99-1)
Whole India	Mescavi
Punjab, Himanchal Pradesh & Jammu	BL-1

Area	Varieties
Hills & sub-temperate region	BL-2
Punjab, Himanchal Pradesh & Jammu	BL-10 (long duration)
Hills & sub-temperate regions, North West India	BL-22
North & Central India, Andhra Pradesh	UPB-103
Southern India	UPB-110
Central India	JB-1, JB-2 & JB-3

LUCERNE *(Medicago sativa)*

Lucerne is fondly called Alfalfa. It is very valuable leguminous forage and hay crop for the dry and irrigated areas of the world. It is perennial, persistent, productive, heat and drought tolerant and provides better seasonal distribution. Supplementation with lucerne either in the green or hay form will make up the protein, calcium and other mineral deficiencies. Lucerne is also an excellent source of protein for the poultry and a constituent of mixed cattle feeds.

Gergraphical distribution: Lucerne is native of the Mediterranean region. It is believed to have originated in South West Asia. In India, it is grown in Jammu and Kashmir, Himachal Pradesh, Punjab, Haryana, Uttar Pradesh, Madhya Pradesh, Rajasthan, Gujarat and Maharashtra.

Climatic requirements: Lucerne persists in a wide range of climatic conditions from extreme hot as prevails in Rajasthan to cold temperatures in Ladakh and Canada. Lucerne performs better during the cooler, drier years than cloudy, humid and wet seasons. It can tolerate short spells of drought and remain dormant. Fresh growth starts as soon as moisture becomes available. Moist but not wet conditions are conductive for vigorous growth. It does not tolerate water logging and high atmospheric humidity and gets over run with grassy weeds in the rainy seasons.

Soil and land preparation: Well-drained loam soils are the best although lucerne can grow even on sandy loam to clay types of soil. Drainage is very necessary as the crop is intolerant to water logging tolerate slight salinity but not sonic conditions. Very fertile soils rich in calcium, phosphorous and potash produce extremely high yield as lucerne is a heavy user of these three nutrients.

A deep fine and a firm seed bed should be prepared by one mould board ploughing followed by about three desi ploughings. If prepared by tractor the land may be opened by a mould board or a disc plough followed by two cultivators or disc harrowings. Perfect leveling is a must for proper drainage and uniform irrigations in rectangular plots. Heavy rolling is recommended for clay soil before and after sowing.

Sowing time and method: Seeds are generally sown both in autumn and spring. In North India, sowing are commenced from early October and completed by the end of November. The spring sowing are done in March. In the Eastern region, lucerne is sown up to the late December. In the southern region also, sowing can be done from October to December.

The seed is broadcast at the rate of 25 to 30 kg per hectare and mixed in to the soil by shallow discing or cultivation. The seed can also drilled in rows 30 to 35 cm apart.

Manuring and irrigation: In the first year, farmyard manure may be applied at the rate of 25 tonnes per hectare before sowing and well mixed for quicker decomposition. At sowing a fertilizer mixture, consisting of 40 kg urea, 400kg superphosphate, 75kg muriate of potash and 150gm of sodium molybdate may be applied. In the subsequent years, farmyard manure may be applied at the rate of 10 tonnes per hectare.

Two light irrigations may be given every week after sowing for quicker germination of the hard seeds. In summer months of May and June when hot and desicating winds blow, irrigations may be given every 10 to 12 days. In winter, irrigations may be spaced every 15 to 20 days.

Harvesting and yield: The first cutting is generally taken after about 9 to 10 weeks from sowing. In the cold winter season, a cutting may be taken at the bud stage at an interval of 40 to 45 days in a spring season at 10 percent bloom and in the summer season at 20 days interval. Shorter days in the winter season are more conducive to higher crude protein accumulation. With advancing maturity there is reduction in digestibility and voluntary intake. The leafiness decreases resulting in the higher crude fibre and less protein contents. In all, seven to eight cuts can be taken during the year.

Average yields vary from 850 to 900 quintals per hectare. Under high fertility, improved agronomic practices, superior germplasm and assured irrigation, 1,200 to 1,500 quintals of green yields can be obtained in 9 to 10 cuttings in a year.

Nutritive value: The fodder contains about 16 % DCP and 60% TDN. It can support not only the growth but milk production up to 8 kg per day when fed *ad libitum*.

Anti-nutrient: Cows grazing on lucerne pastures may develop bloat. Like most of the legumes, lucerne fodder contains oestrogen, which may create reproductive disorders if fed in large quantities. It may also cause photosensitization in which case cattle, sheep, horses and pigs become hypersensitive to light. Saponins are glycosides, which are bitter in taste, foam forming and hemolyze red blood cells, are toxic to animals and are present in lucerne.

Rotations: Where lucerne is taken as an annual crop from October to June, it can be rotated with jowar, maize, teosinte or bajra in the rainy season.

Recommended varieties:

Areas	**Varieties**
Gujarat, Rajasthan, Uttar Pradesh & Madhya Pradesh	Anand-2 (Annual)
Gujarat & Himachal Pradesh	Anand-3 (Annual)
Whole India	LLC-3 (Annual)
North West & Central India	LLC-5 (Annual)
Hills, North & Central India	Chetak (S-244)
Tamil Nadu & Karnataka	CO-1(Perennial)
Whole India	T-9 (Perennial)
Whole India	RL-88

COWPEA*(Vigna sinensis or Vigna unguiculata*)

Cowpea is a warm season annual leguminous fodder crop. Cowpea fodder is rich in protein and forms an excellent mixture with maize, jowar, bajra and teosinte for increasing the flow of milk. It can be fed green, converted in to good quality hay or conserved in the form of silage when mixed with cereal fodders.

Geographical distribution: It appears to have originated from W. Africa, India or Iran. Now it is grown in the warm tropical and sub-tropical Asia, Africa, South America, Australia and southern countries of Europe. In India, it is mainly concentrated in Punjab, Haryana, Delhi, Uttar Pradesh, Rajasthan and some parts of Madhya Pradesh.

Climatic requirement: The crop flourishes in warm and humid conditions. Intermittent showers of rain followed by bright sunshine are conducive to its growth. Excellent yields can be obtained under dry and warm conditions during the summer season with 3 to 4 irrigations. It can easily be grown in the Eastern and Southern region in the winter season as the minimum temperature hardly ever drop below 15^{0}C. It can withstand moderate shade and drought but succumbs to frost and prolonged waterlogged conditions.

Soil and land preparation: Cowpea can grow on a variety of soils from clay loam to sandy loam. Deep fertile and well-drained soil is the best.

One ploughing with soil inversion plough to a depth of about 15 cm followed by two desi ploughings and plankings generally sufficient. In case of tractor cultivation, two or three operations by cultivators or disc harrow meet the requirements. A gentle slope may be given to the land to drain away excess water. This is particularly necessary in case of monsoon crop, which is likely to have waterlogged.

Sowing time and method: The best time to sow this crop in North India is from middle of March to end of August. In South India and Eastern regions, it can be grown throughout the year.

In the first two to three years, the seed should be inoculated with Rhizobium bacteria for vigorous growth. Sowing should preferably be done in close rows 30 to 40 cm apart when the seeds are sown by *Pora, kera* should be increased to 50 kg when the seeds are broadcast. A much higher seed rate up to 100 kg may be used in soils badly infested with weeds. In mixtures with cereal fodders, the seed rate may be reduced to 20 to 25 kg.

Manuring and irrigation: FYM may be applied at the rate of 10 tonnes per hectare about a fortnight before sowing. At sowing fertilizer mixture consisting of 25 kg urea, 200 kg super phosphate and 35 kg muriate of potash may be incorporated. When grown in mixtures with cereal fodders, the dose of urea may be doubled.

The first irrigation may be given about 20 to 25 days after sowing in the summer season. Subsequent irrigation may be given every 20-25 days in summer and at monthly intervals in the post monsoon period. The land should be kept well drained during the monsoon season.

Harvesting and yield: Cowpea gets ready for a cut after about 70 to 75 days when the first pods have just been formed. Delay in harvesting will sharply bring down the protein content and increase the crude fibre. In a two cut management the first cut can be made after about 50 to 55 days and clipped about 15 cm above the ground level. The second cutting can be taken after 40 to 45 days. In case three cuttings are attempted, the first cut should be taken after 45 to 50 days and subsequent two cuttings after every 25 to 30 days.

The average yields of cowpea fodder generally do not exceed 275 to 300 quintals per hectare. A yield of 350 to 400 quintals can be obtained with better management and multicut system.

Nutritive Value: It contains about 12.4% DCP and 58.9% TDN on dry matter basis and is very palatable. The fodder can support a growth rate of 400 to 450 g per day and milk production of 6 to 7 kg, when fed exclusively.

Recommended varieties:

Areas	Varieties
North India	Russian gaint, UPC-5287
North, West & Central India	EC-4216, Bundel Lobia-2 (IFC-8508)
UPC-287, UPC-5286, UPC-5287, C-30, Bundel Lobia-1, (IFC-8401), UPC-8705, GFC-1, GFC-2, GFC-3, GFC-4,UPC-4200	All India North-East India
Charodi, GFC-1, GFC-2, GFC-3	Gujarat
CO-1, FCP-2	South India
Haryana Lobia-88	North west India
IGFRI-5-450	All India (for summer season)
Shweta (No. 9998),DFC-1	Naharashtra
C-88	Punjab
UPC-9202	Subtropical to temperate conditions in Central zone, i.e., Madhya Pradesh, Maharashtra, Gujarat and Southern Central Uttar Pradesh.

GUAR (*Cyamopsis tetragonoloba*)

Guar or cluster bean is hardy leguminous summer annual fodder crop that can tolerate high moisture stress, one to one and a half meter tall. It may be spreading or non spreading, single stemmed to much branching. It is adapted to the arid and semi arid region of India. Besides, it has multiple uses, as the crop can be raised for the fodder production.

Geographical distribution: Guar appears to have originated from India. In India, 80 percent of the area is concentrated in the states of Rajasthan, Gujarat, Harayana, Punjab and West Uttar Pradesh. It is also cultivated to some extent in parts of Maharashtra, Madhya Pradesh and AndraPradesh.

Climatic requirements: Arid to semi arid condition characterized by high temperature and low humidity. Intermittent showers of rain followed by clear skies help in the vigorous growth of the legume. A well distributed rainfall to about 150 to 200 mm during the growing meet its water requirements adequately.

Soil and land preparation: Sandy loams to deep alluvial clay loam soils is the best. The soil should be well drained as guar does not withstand waterlogging. Moderately saline soils with a soil reaction up to 8.5 pH can be used without much loss in yield. The soil should be rich in phosphorous, potassium and calcium.

Two operations by Desi ploughs, harrows or Bakhars are enough. Cultivation deeper than 10 to 2 cm is not necessary except in dry farming areas where a depth of 20 cm once in three year is more helpful in conserving moisture.

Sowing time and method: Summer crop in April to May. The Kharif crop can be raised any time from June to August with the onset of monsoon.

The seeds are generally broad cast and mixed with desi plough and harrows. Under irrigated conditions, the inter-row spacing may be kept at 40 cm while under dry conditions it may be increased to 50 to 60 cm. The following rates per hectare may be used. Broadcasting (45 to 50 kg),pora or kera (45 to 50 kg), drilling (35 to 40 kg) and for dry farming areas (30 to 35 kg).

Manuring and irrigation: The land should be dressed with FYM at the rate of 10 tonnes per hectare 10 to 15 days before sowing. A fertilizer mixture consisting of 25 kg urea, 200 kg super phosphate and 35 kg muriate of potash may be placed 4 to 5 cm below the seed line at the time of sowing.

The crop may be irrigated once in 20 to 25 days during the summer season. In the monsoon season the field should be kept well drained. In the post monsoon period, the interval may be increased to one month.

Harvesting and yield: The crop is ready for harvesting from flowering to pod initiation stage in about 70 to 80 days from sowing. It should not be allowed to mature, as it would become woody and fibrous resulting in the loss of digestible nutrients and palatability.

The tall varieties yield about 120 to 150 quintals per hectare under dry farming conditions and 250 to 300 quintals per hectare under irrigated and high fertile conditions.

Nutritive Value: The fodder at flowering stage contain about 13.2% DCP and 61.3% TDN.

Anti-nutrient: There is a feeling with some farmers that guar fodder is not suitable for milch animals. By preference, it is fed to draught animals. The protease inhibitor is found only in seeds.

Conservation: Guar makes excellent and is relished by the animals. For making silage, it should be mixed with jowar, maize, teosinte, bajara, sudan grass, Hybrid Napierin the proportion of one part of guar with two to three parts of the cereal fodder.

Recommended varieties:

Areas	Varieties
Punjab, Haryana	Guar 80, No.2, FS-277, HFG-119, HFG-277, HG-365, HG-75
Entire guar growing tract, especially Rajasthan & Gujarat.	HFG-156, HFG-258, Bundel Guar-1 (IGFRI-212-1), Bundel Guar-2 (IGFRI-2395-2), RGC-471, RGC-936, RGC-197, RGC-986, RGC-1002, RGC-1003
Entire guar growing tract (relatively drier with low to medium rainfall areas in arid and semi-arid zone)	Bundel Guar-3 (IGFRI-1019-1)
Rajasthan	Maru Guar
Gujarat	GUAG-34
Haryana	HG-182

RICE BEAN (*Phaseolus calaratus*)

Rice bean is popularly called red bean. It is annual, hairy and twining legume and can easily be grown in mixtures with maize, jowar, bajra, teosinte and sudan grass. The crop is more drought resistant than cowpea and grows quite well even on poor soils. It remains available during the acute scarcity period when Kharif fodders are generally over and Rabi fodders are yet not ready for harvest.

Geographical distribution: It is typical tropical legume and is considered to be native of India. It is generally cultivated in West Bengal and Assam.

Climatic requirement: Warm and dry with intermittent showers of rain. It can also grow in high rainfall area during the monsoon season provided the crop is raised in uplands where excess water is freely drained out.

Soil and land preparation: Loam to clay loam soils rich in calcium and phosphorous are the best. Sandy loam soils can be used successfully as the crop grows quite well on such types of soils with higher level of organic manure and fertilizer. Crop cultivation with desi ploughs or harrows is sufficient.

Sowing time and method: The best sowing period from middle of August to middle of September. With irrigation facilities, the sowing of this crop can be easily taken up right from the middle of April.

Generally the seeds are broadcast and mixed with disc harrows, cultivators or *desi* ploughs. The seed should preferably be sown by drill. Spacing of 40 cm may be kept between the rows. About 60 kg seed rate per hectare should be used when the seeds are broadcast. When sown in lines the seed rate may be reduced to 45kg.

Manuring and irrigation: The land may be manured with 10 tonnes of FYM per hectare 10 to 15 days before sowing. At sowing the fertilizer mixture consisting of 25 kg urea, 200kg superphosphste and 35 kg muriate of potash may be placed 4 to 5 cm below the seed line.

The summer crop may be irrigated once in 20 to 25 days. In the monsoon season the field should be kept well drained and water should not be allow to stagnate.

Harvesting and yield: The crop generally takes about 70 to 80 days to be ready for cut. The best time to harvest appears to be early pod stage. If cut after 45 to 50 days, the second cut can be taken in another 45 to 50 days.

An average crop yields about 200 to 250 quintals per hectare. Higher yields up to 300 quintals can be obtained in fertile soil.

Nutritive Value: It is most palatable tropical legume contain about 16.9% crude protein. The dry matter and crude protein digestibility are reported to be 55.05 and 63.08%, respectively.

Conservation: The crop can be conserved in the form of hay and should preferably be harvested at the early pod stage. It can be contribute in improving the crude protein and mineral content of silage from Kharif silage fodder when it is mixed in the proportion of one part with two to three part of silage fodders.

Recommended varieties:

Areas	Varieties
All India subtropical to tropical region, particularly west Bengal, Orissa, Assam, Manipur, Mazoram and Nagaland (Single cut)	BIDHAN-1
Konkan region of Maharashtra	Konkan Rice bean-1

SENJI(*Melilotus indica*)

Among the sweet clovers, *senji* or Indian clover,. is the most popular among the north India. It is yellow flowered annual. The flowering shoot reaches a height of about 80 to 100 cm and seldom up to 1.25 to 1.50 meters. It does not require seedbed preparation, high soil fertility, frequent irrigation or much aftercare.

Geographical distribution: Senji is indigenous to India and was the most popular winter forage of the Punjab.

Climatic requirement: The crop grows best in the temperate to subtropical regions with cool and milder winters. Intermittent showers of rain from 150 to 200 mm suffice to bring the crop to flowering stage. Cool nights and bright sunny weather during the day with occasional showers are most conducive for the vigorous growth of senji.

Soil and land preparation: Loam to sandy loam soils with neutral to alkaline reaction and rich in lime, phosphorous and potash are the best. It is tolerant to saline conditions and is therefore useful in the reclamation of such soils. Free drainage is essential, as the crop cannot tolerate waterlogged conditions.

One operation by *desi* plough, or disc harrowing by tractor should be considered as optimum beyond which further cultivation is not necessary.

Sowing time and method: The sowing generally starts by late September and continue through October to about middle of November.

Sowing can be done neither by drilling in close rows after land preparation or seeds broadcast in dry land and mixed with a light followed by irrigation. The spacing between rows should be kept between 25 to 30 cm. In case of broadcasting the seed rate of 50 kg per hectare is used.

Manuring and irrigation: Generally senji crop is not mannured. For obtaining higher yields and higher nutritive value of the forage it should be fertilized with 20 to 25 kg urea and 125 kg of super phosphate.

The seeds may either be sown in high moisture or the fields irrigated immediately after sowing. Afterwards, winter rains of 100 to 125 mm in 2 to 3 showers generally sufficient.

Harvesting and yield: In pure stand, the crop is generally harvested after 15 to 20 percent plants have flowered. In association with oats or barley, the mixture may be harvested after 90 to 100 days in a single and 60 to 65 days in a two cut management.

An average crop yields about 200 to 250 quintals per hectare. With better management the yields can be pushed to about 300 to 350 quintals per hectare.

Recommended varieties:

Areas	Varieties
Punjab, Harayana and Himachal Pradesh	Safed-76, FOS-1, YSL-106

Mixed cropping

To cope up with the conditions of malnutrition and low productivity in livestock, and malnutrition in human due to tremendous increase in their population, it is most important to make the best utilization of available land for the cropping of food crops and forages. Fertility of land should be maintained for prolong productivity. Under intensive cropping system generally the land loses its fertility at a much faster rate if sufficient care is not taken. There is ample scope of increasing crop production through rational management practices and mixed cropping. Mixed cropping can be done in various ways.

(a) Forage mixed with forage

(b) Forage mixed with grain/cash/oil seed crop

(c) Forage interplanted with trees of economic importance including fruit trees.

(a) Forage mixed with forage: On the basis of compatibility, the forages of different nature are sown together to harvest more nutrients through a mixture of forage crops. The legume forage may be mixed with non-legume and a deep rooted crop can be mixed with shallow rooted crop. The annual forage is interplanted between the rows of perennial forage crop. There are various advantage of mixed cropping over sole cropping as follows.

1. Best utilization of the nutrients applied or available in soil.
2. Due to intercropping of legume with non-legume, the growth vigoure and protein content of non-legume forage increases.
3. The intensity of erosion is reduced due to perfect coverage of land.
4. Encroachment of weed is reduced.
5. More nutritionally balanced forage mixture is harvested.

Examples

1. Maize/jowar/bajra + cow-pea/guar.
2. Cow-pea + Hybrid napier (3 year)

3. Maize/jowar + subabul (perennial fodder tree)
4. Lucerne (Perennial) + subabul (Perennial)

(b) Forage mixed with grain/cash/oil seed crop: With the increasing need of grain, cash and oil seed crops, it is not always possible to spare the land exclusively for fodder production. The marginal farmers and small farmers may not be interested to leave land for the fodder crop. Under such conditions the fodder crop should be mixed with the grain crop so that some amount of green fodder may be produced without affecting the production of main crop. The choice of the fodder will, however, depend on the relationship of fodder with main crop, price and cost ratio and variability of prices and yield.

Examples

(a) Senji may be interplanted with cotton.
(b) Rice bean may be sown with rice.
(c) Oat may be sown in furrows while potatoes on bunds.
(d) Cowpea and other legumes may be sown in the sugarcane field.

(c) Forage interplanted with trees of economic importance including fruit trees: The importance of trees in human life has never been secondary. Trees are responsible for maintaining suitable agroclimate, minimizing soil erosion, maintaining oxygen pool of environment apart from providing various valuable products. The plantations are, therefore, essential to maintain ecosystem. Forages can, however, be produced using suitable types of crops capable to grow successfully in between the plantations. The production of *Cenchrus ciliaris* and cassava, cow-pea, green gram is usually practiced in between coconut, rubber and other plantation crops in Kerala. The production of legume crops like cowpea, soyabean, and non-legume like maize and oat is usually practiced in the orchards of U.P. Hills, Himachal Pradesh, etc.

In the adopting of mixed crop system the selection of crops should be according to the facilities available with the farmers.

CROP ROTATIONS

In the areas of assured irrigation facilities, intensive cropping system should be followed in which 3 or more crop are taken in a year, after selecting suitable crop mixture and applying required quantities of fertilizers. On the basis of the type of crop taken into consideration, the rotation may be discussed into the following two heads.

(A) All Forage Rotations

For the intensive dairy industry areas (milk shed areas) round the year around supply of good quality green fodder is essential to reduce the cost of production, through the saving of costlier concentrate feed supplements. Being the main occupation, farmers may also spare sufficient land for the forage production. The fodders may, therefore, be cultivated using intensive rotations. The high intensity rotations, however, require assured water supply, availability of good seed of high germination value, nutritious and short duration varieties and higher investments on fertilizers. On the basis of area and other variables affecting the crop yield, some of the following rotations may be followed in different agroclimates. However, it is possible to layout a new rotation using the fodder crop suitable for cultivation in a particular area.

North zone:

1. Maize + Cowpea – Sorghum + Cowpea (two cuts) – Berseem + Mustard.
2. Sudan grass+ Coepea - Maize + Cowpea - Turnip - Oat (two cuts).
3. Hybrid napier or Setaria inter-planted with cowpea in summer and berseem in winter (9to 10 cuts per year).
4. Teosinte + Cowpea (two cuts) – Carrot – Oats + Mustard/ Senji (two cuts).

Western and Central zone:

1. Bajra + Guar (two cuts) - Annual Lucerne (6 cuts).
2. M.P. Chari + Cowpea (two cuts) – Maize + Cowpea – Teosinte + Cowpea (two cuts).
3. Hybrid napier or Guinea grass or Setaria interplanted with Lucerne (8-9 cuts/ year).

Southern Zone:

1. Sorghum + Cowpea (3 cuts) – Maize + Cowpea – Maize + Cowpea.
2. Hybrid napier or Guinea or Setaria grass interplanted with Lucerne (8-9 cuts) or Hybrid napier + Subabul/ Sesbania (9-11 cuts year).
3. Sudan grass + Cowpea (3 cuts)-M.P. Chari + Cowpea (three cuts).
4. Para grass + Centrosema pubescens (9-11 cuts/year).

Eastern Zone:

1. Maize + Cowpea- Teosinte + Rice bean (two cuts) – Berseem + Mustard (3cuts).

2. M.P. Chari + Cowpea – Dinanath grass (two cuts) – Berseem + Mustard (3 cuts).
3. Para grass + Centrosema pubescens (8-9 cuts/year).
4. Hybrid napier or Setaria inter-planted Subabul or Sesbania sesban (9-10 cuts per year).

(B) Mixed Rotation

The rotation applicable to all farm size should essentially be formulated in such a manner that it may provide grain, cash and fodder crop from the same field. In the small holding of the Indian sub continent, the need of mixed farming is always beneficial. The idea of mixed farming is as ancient as Indian agriculture. Most of the traditional tribes and unexploited areas follow mixed farming although they could not utilize the efficiency due to one reason or other. Many types of rotations may be formulated for different areas. For example in U.P. and Himachal hills, the land is better utilized if there are orchards. In between fruit plants, legumes like Soyabean, Urd are usually grown during the rainy season and wheat in the winter. Several other examples may also be cited for different areas and nature of main crop. In the areas where only crops are grown, the forage may be fitted in between to increase the total return from the land. Some of the rotations suggested are as follows:

***(i) Forage.* Grain crop/cash crop/root crop (2 year rotation):**

1. Wheat – chari + guar – gram – cotton + senji
2. Cotton + berseem – chari + cow-pea – wheat – chari + cow-pea
3. Maize + cow-pea – wheat + mustard – maize + cow-pea
4. Paddy – wheat + mustard – ragi + urd – fallow (2 year rotation in upland, rainfed, Garhwal hills)
5. Paddy – wheat + mustard – ragi + urd – oat + mustard
6. Bajra + bean – wheat + mustard – paddy – barley
7. Wheat + mustard – maize + cow-pea – paddy – potato + oat – maize + cow-pea
8. Wheat – maize + cow-pea – cheena – paddy- cabbage + potato + cauliflower.

(ii) Rotations in Plantations.

(a) Coconut/Rubber plantations (south):

1. Cassava – cow-pea + linseed – ginger/turmeric

2. Cow-pea + cassava
3. Green gram + cassava
4. Black bram + cassava
5. Groundnut – cassava.

(b) In Apple/Plum/Walnut and other fruits of Hills.

1. Soyabean – wheat + mustard – potatoes
2. Maize/jowar + cow-pea – ragi + pulses – oat
3. Potatoes – upland paddy + pulses – tomatoes
4. Tobacco – tomatoes – oat + mustard.

(c) In the orchards of Mango, Litchi and other fruits of Northern plains:

1. Soyabean (summer) – maize + cow-pea – wheat + mustard
2. Maize + cow-pea – maize + cow-pea – wheat + mustard
3. Soyabean – maize + cow-pea – potato + oat.

Bibliography

Choubey R.N Forage crops for different regions of the country, their productivity and feeding value in buffalo production under different climatic region (ed. S.S. Kundu, A.K. Mishra, P.S. Pathak, ICAR, IGFRI, Jhansi) pp 351- 369.

Hazra, C.R. (1993). Improved varieties of forage crops and their production potential. Souvenir. IGFRI, Jhansi, pp. 21 – 25.

Hazra, C.R. (1995). Advances in Forage Production Technology. Project coordinator (Forage crops), IGFRI, Jhansi, India.

L.L.Relwani (1979). Fodder Crops and Grasses ICAR, New Delhi

Melkania .N.P (1999) production technology for potential rabi forages. Technical Bulletin ICAR, New Delhi.

Melkania .N.P (2001) Production technology for potential kharif forages. Technical Bulletin ICAR, New Delhi.

Menhi Lal Shukla, N.P., Tripathi, S.N., Sriniwasan, B.S., Niranjan, K.P., Gupta, S.D. and Arya, R.L. (1994). Forage Production Technology, IGFRI, Jhansi.

Menhi Lal, Singh, R.P. and Tripathi, S.N. (1995). Important fodder crops of India and their cultivation. In: Forage production and utilization, IGFRI, Jhansi. pp. 44 – 59.

Pathak, N.N. and Jakhmola, R.C. (1983). Forage and livestock production. Vikas Publishing House Pvt. Ltd., New Delhi

Singh. R.P; Menhi Lal and Vinod Shanker (1995). Forage production in India-past, present and future. In: Forage production and utilization, IGFRI, Jhansi. pp. 1-18.

Tripathi, S.N. (1990). Principles and practices of forage crop production. Lecture notes, Summer Institute on 'Forage Production Technology', organized at IGFRI, Jhansi, 2 – 21 May, 1990, pp: 26 – 30.

PASTURE DEVELOPMENT AND UTILIZATION

Animal husbandry was based on Grasslands and pastures in past. The forage production from grassland and pastures depends on various factors, which directly or indirectly influence the plant growth. The purpose of a grassland or pasture establishment may be (1) to improve soil fertility of degraded lands (2) to harvest nutrients for livestock and (3) to minimize or avoid the further erosion of soil through perfect soil conservation measures. While fulfilling the basic purposes it is always desirable to explore the possibility of forage production from the grasslands at a higher rate through proper management practices. In India, livestock production is still based on rangeland grazing in many parts. The grazing activity is mainly dependent on the availability of the grazing resources from pastures and other grazing lands viz. forests, miscellaneous tree crops and groves, cultivable wastelands and fallow land. India contributes 15% of world livestock population while it has only 2% of world's geographical area. The cropping area under fodder production is about 8.3 million ha (4.4%) and there is no scope for expansion of fodder cultivation. The grazing intensity in our country is very high 12.6 ACU/ ha as against 0.8 ACU/ha in developed countries. Therefore, the task is two fold viz. improvement of pasture and judicious implementation of grazing management.

The management practices of grassland and pasture include the following steps:

1. Selection of plant species.
2. Establishment and renovation of pasture and grasslands.
3. Management of grasslands and pastures.

1. Selection of Plant Species

The plant species vary in their physical nature and nutrient up take. The selection of plant species for the pasture will not only depend on its nutritive value but also on several other factors like purpose of the grassland and pasture, soil and climate of the area and the system of grassland and pasture which is being followed.

Purpose

The grasslands chiefly aimed to be established for soil conservation, would have quite different types of species than the grassland established with the purpose to improve the soil fertility. These objectives can be fulfilled with the selection of species capable of good forage production also, under these circumstances. The deep rooted, dense rooted, grasses and spreading type of the legumes are suitable for soil conservation because the deep and dense roots hold the soil while cover crop protect soil from surface run off and wind erosion. The legumes with well-developed root system are generally preferred over grasses for the grassland and pasture to be established with an aim to improve the soil fertility. A combination of grass and legume herbage may provide an almost balanced feed for livestock and may also fulfill the purpose of soil improvement.

Soil

On the basis of soil conditions the species of grass, legume, bush and tree can be selected. The selection lays in the compatibility of plant species of co-existence and at the same time their tolerance to specific soil conditions.

Climate

The climate in India varies from tropical to temperate. Some herbage species suggested for various ecological conditions are dealt elsewhere is this chapter.

2. Establishment of Grassland

Area which has been under continuous grazing without any management generally contains poor quality grasses and the bushes of inedible plant species. The establishment of fresh grassland for high quality forage production, along with improvement of soil fertility and conservation, is a tough job and includes a series of operations. The operational step depends on the quality and quantity of previous vegetation, type of land and quality of desired grassland. The establishment after the complete eradication of previous vegetation is as follows:

i. Land Clearing

Although clearing of land is recommended for the development of a nutritious pasture, but the trees should not be destroyed. Since they to protect the riverbanks from erosion and work as a windbreaker and also provided shade for the animals. The bushes of inedible and poisonous plant type should be destroyed completely before the establishment of the grassland afresh. Such type of bushes and plant not only restrict the growth of the required biomass but may also cause health hazards to livestock. However, the leaf yielding shrubs may be retained in appropriate proportion. It is suggested that in rangelands of Indian deserts about 14 percent of land area, covered by Zizyphus nummularia shrub canopy, is optimum for high forage as well as leaf fodder production. The clearing may be done by mechanical, chemical or other methods.

a. Mechanical

The big bushes can be removed either using hand tools or machines. The use of machines like specially equipped tractors is limited to the plains. In the hills or stony areas, the land clearing operations are less efficient. The uprooting of poisonous and other undesirable plant species should also be done simultaneously.

b. Controlled burning

The burning, which is the most common and practical method of clearing, has got its own advantages and disadvantages. The beneficial effects are (i) removal of old growth, (ii) partial control of unwanted bushes, diseases and pests. Some of the serious limitations of the burning are (i) increased erosion, (ii) injury to the other required vegetation like trees (iii) loss of organic matter and nitrogen and (iv) exposure of soil. Therefore, burning should be used with caution. Controlled burning can be used and should be limited to the burning of bushes and undesired plants only. It should be avoided during the dry-hot season because the other vegetation and thatch houses may catch fire.

c. Chemical application

Thus method includes the use of chemicals to kill the vegetation. Some of the herbicides used for the destruction of bushes are 2, 4-D; 2,4, 5-T and dicamba. The concentrated mineral acids may also be used to kill the undesirable biomass. However, it should be used with caution as these chemicals are corrosive.

ii. Seeding and Planting of desired species

After the land is practically cleaned and free from undesired vegetation it needs planting or seeding of desired species. The season of the seeding or planting vary from region to region depending on the climatic factors. In most areas of the northern plains the seeding can be done during the pre-monsoon period. The quantity of grass seed and legume should be calculated keeping in view the desired grass: legume ratio in pasture and weight of the seed. While seeding, care should be taken that the seeds are not dormant. This problem is more associated with the legume seeds, therefore, seed treatment should be done before seeding. The seed treatment should be done with care to avoid the spoilage of seeds. The treatment against various insects, pests and diseases may also be done prior to sowing. The rhizobium treatment of legumes is beneficial as it enhances the growth of a plant due to early formation of rhizobium nodules.

The seeding can be done either through broadcasting or with the help of a seed drill. The seed should be covered with soil as the exposed seeds may be picked up by the birds. Moreover, the exposed seeds will not get suitable environment for germination. In the areas where the land is not uniform and stony or gravelly some other suitable methods can be applied for seeding of the grass legume. The grass seed is usually broadcasted while the legume seeds are sowed in rows at the distance of 20-25 cm.

Although seeding is easier and widely followed, the transplantation of the seedling of grass/legume is also practiced due to some advantages over direct seeding method. The transplanting of the seedling during rainy season is much effective. In the hills transplanting method is more applicable. For the production of seedling of grass and legume a nursery is prepared during May-June. The seeds of grasses and legumes are grown in a nicely prepared and well-fertilized seedbed. At the age of 25-30 days the seedlings are uprooted carefully avoiding the breakage of the roots. These seedling are then planted while the soil is wet. The operation is preferably done in a cloudy weather. Under the sunshine larger percentage of seedling may die due to excessive loss of moisture through faster transpiration rate. Some species of grass/ legume can be propagated by root slips or stem cuttings. On the basis of the facilities available, the form of germplasm available (seed, root slip/cutting) and the type of topography, etc. any of the above mentioned methods may be applied.

iii. Application of fertilizer

Fertilizer application is required in the eroded and low fertility soil and is necessary for the boosting of initial growth of the grass-legume mixture, at the same time to correct the acute deficiencies of nutrients in the soil. The randomly taken soil samples can be analyzed for the mineral status of soil. The required amount of fertilizer nutrients may be added to soil by broadcasting them uniformly just before sowing so that the fertilizer get mixed with the soil during the sowing operation. The fertilizer mixing with soil is a problem in the steep stony land. The fertilizer may be broadcasted as usual and the mixing should be done while planting the legume seeds.

At first, a basal application of 5 tones of Farm Yard Manure (FYM) along with 40 kg N and 20 kg P_2O_5 per hectare land is to be mixed thoroughly in the soil for land preparation. After one month of establishment a top dressing with another 20 kg N/ha is given. In subsequent years mixture of 20 kg N + 20 kg P_2O_5/ ha is beneficial after first shower of rains.

iv. Care of Newly Established Grassland/Pasture

The newly seeded or planted forage species are very much susceptible to the environmental changes. These are usually dominated by the precursors of earlier biomass at the initial phase. The weeding is, therefore, necessary at the initial stage. The grasses of undesired plant type should be removed as early as possible to give sufficient opportunity for growth to the desire grass species. The regrowth occurring as harmful bush should also be removed. The legumes although do not face much competition with grasses may suffer some degree of depression during germination in a newly sown pasture. The birds usually eat the cotyledon coming out of the soil and thus affect the ratio of grass-legume. A watch on the birds during morning and evening is required. This problem does not arise where the legumes are planted as seedling or root slip.

The entry of the animals and grazing on the establishing pasture should be avoided. The grazing should not be allowed during the first year of the growth, since it badly affects the future production of the pasture. For this purpose protection through fencing can be provided. In the area where irrigation facilities are available, the pasture should be irrigated during the winter and summer seasons depending on the spread and intensity of rains in the area.

v. Establishing Pasture with Partial Cleaning

In the area where natural grass cover is present in a bad condition, the removal of grass is not required. The thorny or harmful bushes and undesirable plant species can only be removed to reduce the cost of establishment. Such conditions generally occur due to over-grazing in unmanaged pasture. Thus, the area should first be fenced to avoid the uncontrolled entry of the animals. For the establishment under such conditions the following measures may be adopted.

1. Removal of the thorny and harmful bushes and undesired toxic plant species.
2. Seeding or planting of the desired species of the legumes.
3. Application of the fertilizers.
4. The proper care of the pasture to maintain suitable grass-legume ratio of the herbage.
5. Regular removal of the undesired plant species, after a suitable span.
6. In case the legumes are not present in suitable ratio (1:1), reseeding or re-planting of legume in pasture is required.

A careful management of the newly establishing pasture is always required for harvesting the optimum nutrients from it for a longer time.

3. Management of the Pasture/Grasslands

The management of the grasslands and pasture is a science as well as an art of the planning and directing grassland/pasture in a way to obtain maximum livestock production consistently and at the same time conserving the fertility of the land alongwith soil conservation. The longevity and productivity of the grassland/pasture depends on various factors like grazing intensity of animals, exhaustion of soil nutrients and soil conservation practices, etc.

I. Harvesting Schedule of Grass-Legume Mixture

Depending on the agro-climatic and the type of grass legume mixture present in the pasture, the growth of the pasture plants differs. The harvesting schedule of pasture may be fixed on the basis of agro-climatic conditions and the type of vegetation in the pasture. The intensity of harvesting also influences the total yield of the forage mixture and the longevity of the pasture. In the monsoon fed area of northern India, 2-3 cutting of the grass-legume mixture can be taken during rainy season. Once the legumes start bearing flowers after monsoon, the harvesting should be avoided to allow the seed production and subsequent spreading. This practice is more desirable when pasture contains annual legumes, which need seeding every year for the maintenance of ratio. Once the seed gets spread, the residue can be harvested during the month of February and this may be utilized for feeding during the summer. The areas in which rainfall is distributed throughout the year, 6 - 8 cutting can be obtained. In the areas having irrigation facilities the grass-legume mixture can be harvested 8-10 times in a year. Irrigation after each harvest followed by application of fertilizer mixture will improve its production potential.

ii. Renovation of the Pasture/Grassland

The grasses of pasture are perennial in nature. Their multiplication with seed as well as with the developing root system is possible. The profuse growing legumes are usually spread in the pasture through seed. The legumes are also more preferred over the grasses by the grazing animals, which causes slow depletion of the legume in the pasture. As the ratio of legume decreases the nutritive value of pasture also reduces. Although, regular care should be taken to correct the grass-legume ratio still after every 3 years the renovation programme should be implemented.

1. Unwanted bushes and trees should be removed.
2. Seeding of the legume in the pasture should be done.
3. Fertilization of the pasture.
4. Keep the grassland/pasture protected from the animal grazing till the legume establishes in the grassland/pasture.

iii. Managing Livestock on Grasslands

The herbage production of the entire grassland and the intake of herbage by the animal together determine the capacity of the grassland. For example, a block of Cenchrus-siratro pasture is expected to yield 24 quintals of dry matter and if an adult animal of about 400 kg body weight consumes the grass legume mixture at the rate of 2.0 kg per 100 kg body weight daily, the said block will last for 300 days. In other words 30 cows or 20 cows and 40 sheep (considering that one cow consumes as much forage as consumed by four sheep), or 20 cows, 10 heifers and 15 sheep (considering a heifer consumes 2/3 of forage consumed by a cow) can be allowed to graze on this pasture for 10 days. In practice, however, safety margin should always be kept to avoid under feeding by certain group of animals. The said pasture can, therefore, be allowed for grazing for 8-9 days. The consumption of forage differs from animal to animal. The feeding habit of buffalo, cattle, sheep and goat, is quite different from each other. So the type of animals in the stock will be helpful to regulate the pasture in efficient manner. The mixing of animals of different feeding habit together while grazing, results in the efficient utilization of the herbage. The practice of grazing buffaloes, cows, sheep and goat together in the pasture/grassland is very common in India. The cows which clip spontaneously are followed by buffaloes in the pasture. The sheep graze at ground level generally follow the cattle and buffaloes, while goat browse on the bushes. The high live weight gain in sheep and cattle has been observed in mixed grazing. It has been considered that better utilization of herbage in mixed grazing is attributed by more complete herbage utilization, better re-growth of grass and lower parasitic burden. The crowding or overstocking in the grasslands should be avoided as it causes social problems in the herd. The large and strong animals generally do not allow to graze weaker or younger animal, freely. This causes poor growth rate and production in such animals. At the same time fighting among the herd some time poses serious problems due to injuries. The number should, therefore, be adjusted in such a way that there is sufficient space for an animal to walk according to its will in the grasslands.

The high milk yielder should not generally be mixed with dry animals and they should be allowed to graze on a nutritious pasture. In addition sufficient mineral mixture should be provided along with the concentrate mixture when animals return for milking. If possible, there should be provision of water in grassland, so that the animals can drink water as and when required. This is more important in the case of buffaloes and crossbred cows and especially during summer season. The rainwater can be utilized by collecting it in a pond. The natural water pond or stream can be used very efficiently after slight modifications.

The grasslands should be free from the infections and parasites to check disease in the herd. The animals should be sprayed with suitable insecticide/ germicide when there is parasitic infestation. The deworming and other suitable measures to minimize the internal parasitic load increase the efficiency of the grassland utilization.

Grazing on the Grassland/Pasture

The grazing and browsing is a natural feeding habit of animals. The longevity and productivity of the grassland are very much affected by grazing. A poorly managed pasture is usually overgrazed. Overgrazing has some of the following important limitations:

1. Overgrazing causes depletion of the legume in the pasture followed by complete eradication because the animals prefer to eat succulent legume over grass.
2. The grasses do not get sufficient time for re-growth, thus the edible grasses deplete in the pasture.
3. The unpalatable grasses or bushes dominate the pasture, which reduces the efficiency of utilization of the pastures.
4. Due to continuous grazing the soil erosion increases, which causes the loss of fertility apart from the other problems arising out of soil erosion.
5. The return from pasture in terms of forage for animal production, maintenance of soil fertility and conservation of the soil badly reduces.

The grazing practices should, therefore, be determined according to climate and other factors affecting the growth of pasture. In the first year of the pasture establishment the grazing should either be prohibited or strictly controlled. The grazing on the grassland should be regulated on the basis of following factors:

1. Carrying capacity of the grassland.
2. Growth rate of the herbage.
3. Type of animals.

Once the forage is grazed, it should be left for re-growth. Several factors including plant type, soil and climate, etc. determine the re-growth of the herbage. If the growth of herbage is slow, it will take more time to achieve the height of about 50-60 cm, which is ideal for allowing the animals to graze. A quick growing herbage mixture is generally ready for the grazing after 20-30 days. The growth rate will, therefore, determine the resting period of the grassland.

Grazing systems

Various grazing systems viz. continuous, rotational, deferred rotation are practiced throughout the world. However, continuous mixed herd grazing on degraded community land is the widely existing grazing system practiced in our country. Every grazing system has its own merits and demerits. However, the superior grazing system like rotational and deferred rotational system may be adopted to replace the existing continuous grazing system for better livestock productivity.

Continuous system

It is the day old practice of small and marginal Indian farmers to graze their livestock in community land. The pasture from such land cannot fulfill the feed requirement of livestock due to high stocking rate and high grazing pressure without any rest, resulting poor selectivity of the available palatable vegetation and gradually replaced by the biotic composition, which are not selected by animals. The rangeland is mainly composed of natural vegetation, which undergo seasonal changes. Therefore, continuous type grazing system cannot sustain the livestock productivity round the year. Under the prevailing socio-political and administrative structure in the country neither the administrator nor the beneficiaries are responsible for maintenance of these rangelands.

Rotational system

It is one of the improved techniques of grazing management by which livestock owners can sustain the productivity of their livestock for longer period in a year. For the maximum utilization, the grasslands should be divided into blocks. The animals should be rotated from one block to another. This method, not only gives sufficient time for the re-growth of the grass-legume mixture, but also provides quality grass-legume for continuous grazing. This system has a particular advantage in arid and semi-arid ecology where drought is a recurring phenomenon. This system is also widely practiced to control parasitic infestations in intensively grazed locations. However, practices of such type of grazing system in abundant grazing land resources are the prerequisite, which is generally not common in our system of land utilization. Even due to seasonal changes of pasture quality and availability, this system of grazing management can also not support the productivity of livestock round the year.

Deferred rotational system

Under this improved grazing management system part of the pasture land is protected from livestock grazing for a particular period and the area is then made open for grazing in the subsequent years. The pasture from the protected area harvested during appropriate time and stage of maturity and conserved for the feeding of livestock during lean period. Under such type of grazing management livestock productivity can be sustained round the year with little to no supplementation of the top feed/ concentrate feed during lean period. However, for practical execution of this system, stocking rate should be judiciously fixed so that area under grazing should not be over grazed as well as quantity of forage harvested from protected area can sustain the livestock productivity during lean period. The prime limitation to this system of grazing management is higher input due to harvesting of pasture as well as protection of pasture. However, under judicious use this system is cost effective as well as sustainable for pasture and livestock production over the other systems of grazing management.

Stocking Rates and Carrying Capacity

Stocking rate as defined by the Society for Range Management is the amount of land allocated to each animal unit for the entire grazable period of the year. Selection of the correct stocking rate is the most important of all grazing management decisions from the standpoint of vegetation, livestock, wildlife and economic return. It is a basic problem confronting ranchers and range managers. Although, it is generally agreed that there is no substitute for experience in stocking rate decisions on specific ranges. The Society of Range Management defines an animal unit as one mature (450 kg) cow. Based on the recent research, this animal would be expected to consume 9.1 kg of forage per day, 273 kg per month, and 3318 kg per year. An animal unit month is the amount of feed or forage (273 kg) required by one animal unit for one month.

The terminology, carrying capacity, is commonly used when discussing stocking rate. Carrying capacity is defined as the maximum number of animals that an area of land can support on a sustainable basis. A basic technique for determining carrying capacity is to calculate the total amount of forage at the end of the growing season multiplied by a correction factor and then divided by the average yearly feed requirements of an animal unit. Carrying capacity is usually expressed as a stocking rate in ha/animal unit.

Livestock cannot use all the range forage as some are not accessible to the animals and some are unpalatable. Further losses occur due to trampling or fouling by the animals. Most importantly after the harvestable portion of forage has been taken by grazing or browsing animals, a residue should be left on sustainable basis. This is known as 'proper use'. Three correction factors are thus needed if the range is to continue producing.

1. Grazing inefficiency
2. Losses
3. Proper use

With little or no research substantiating these correction factors estimates remain subjective. In most studies a single multiplier has been used which combines corrections for all the three. As a result of differences in the multiplier uses, carrying capacity estimates vary.

The success behind any grazing management practice depends on the accurate stocking rates considering the condition of the range. This principle cannot be over emphasized. Many stocking rate experiments have shown that moderate and conservative stocking rates give greater long-term returns than does a heavy stocking rate (Table 1). With moderate and light stocking rates there are improved animal condition, more wool, greater percentage of calf crop, higher weaning weight, fatter cull animals, less death loss, less supplementary feeding and higher selling prices. Generally, as stocking rate is increased, productivity per animal declines. Differences in animal productivity between light and moderate stocking rates are much less than between moderate and heavy stocking rates. Although productivity per animal unit declines as stocking rate increases, productivity per unit area increases upto a point. It then decreases as scarcity of forage reduces nutrient intake by livestock. During drought periods, heavy stocking can be economically disastrous because the complete lack of forage will necessitate that all livestock be removed from the range and fed hay or sold at low prices.

Basically, livestock producers strive to maintain a stable enterprise, therefore generally a constant stocking rate is maintained. However, the flexible stocking rates should be practiced during following conditions:

1. The pasture growth pattern is markedly periodic.
2. Quality and quantity of forage cannot be maintained for long period if forage is not consumed.

Therefore, following strategies for the fixing of stocking rates should be taken :

a) Stock conservation so that the range is not overstocked, even in fairly severe drought.
b) Stock at a high rate commensurate with forage available during favourable years.
c) Stock at the average forage supply.
d) Vary stocking to meet supplies for the year.

Table 1: Live weight gain per unit area and per animal unit from various studies involving grazing intensity.

Animal	Grazing intensity	Level of use	Weight gain/unit areea (kg/ha)	Weight gain/ (kg)	Reference
Cattle	Heavy	-	37.1	199	Kothmann et al., 1971
	Moderate	-	23.7	200	
	Light	-	18.1	210	
	Heavy (Residue, 218 kg/ ha)		7.0	62	Piper et al., 1978
	Moderate (Residue, 305 kg/ha		6.4	81	
	Heavy	34%	8.9	74	Skovlin et al., 1976
	Moderate	28%	8.1	95	
	Light	17%	6.9	105	
	Heavy (Residue, 166 kg/ ha)	15.5	57	Bement, 1969	
	Moderate (Residue, 333 kg/ ha)	16.5	106		
	Light (Residue, 550 kg/ ha)	12.9	118		
	Very Heavy	81%	132[a]	164[a]	Willms et al., 1986
	Heavy	47%	82[a]	205[a]	
	Moderate	36%	62[a]	230[a]	
	Light	26%	45[a]	224[a]	
Sheep	Heavy	68%	18.2	58	Smoliak, 1974
	Moderate	53%	15.2	60	
	Light	45%	12.3	62	
	Heavy	-	52[b]	28[b]	Lang et al., 1956
	Moderate	-	42[b]	33[b]	
	Light	-	27[b]	35[b]	

[a] Cow + Calf

[b] Ewe + Lamb

Diet Selectivity

Forage selected by grazing animal is determined by a complex interaction of multiple factors including both animal and plant characteristics. Some obvious factors affecting animal diet selection include forage quality and availability, animal prehensile, grazing ability, animal anatomy, secondary plant metabolites, topography, animal agility, physical plant properties and inter-specific and intra-specific animal species competition. Many studies correlate palatability with various plant chemical components. It is commonly noted that forage high in crude protein, phosphorous, gross energy and low crude fiber is highly accepted by cattle and sheep. Forage high in sugar and high fat content usually correlated with high palatability. Livestock accepted the grass cultivars highest in phosphorus and potassium before those with low contents of these minerals. Percentage of lignin and crude fiber are as common as positive relationships with other compounds.

Manure-affected plants in a pasture were not grazed by cattle but were grazed by sheep, although they were higher than unaffected plants in CP, calcium, phosphorus, tannin, chloride and sugars, Cattle exhibit preferences not only for certain plant species, but for the same species at different stages of growth and even for individual plants within a species. The climatic factors, altitude, soil properties, level of water table and application of fertilizer also affect the palatability of pasture species.

Most grazing animals select leaf over stem and green material over dry. Leafy and green materials are preferred because they contain more CP, soluble carbohydrates and P and less CF. With the maturity, the plant becomes higher in fiber and the leaf/ stem/ fruit ratio changes towards a higher proportion of stems.

Diet selectivity also affected by the presence of awns, spines, excessive hairs, stickiness, coarseness of texture and unfavourable odor from external glands of the plant. Seasons have significant effect on forage nutritive quality in most ranges. Levels of cell solubles, CP and P are highest in actively growing forages and show substantial declines as plants become dormant. Annual grasses show much greater declines in nutritive quality than do perennial grasses.

Cattle and sheep diet generally have CP levels of 10 – 12% and 12 – 16%, respectively during active forage growth period. During forage dormancy, CP levels drop upto 4 – 7% for cattle and 6 – 12% for sheep. The feeding habits of goat indicate that goat consumes huge amount of browse species as compared to grasses and weeds (Table 2). In a three-tier silvopastoral system during rainy season Jamunapari goats spent 72% of the feeding time on browse and only

28% on ground flora. In Barbari goats the value was reverse, 24% and 76% respectively. However, nutrient composition, dry matter intake and digestibility of nutrients did not vary between the breeds. The higher values for goats are explained by greater selectivity by goats and the fact that goats are generally grazed on ranges with a much higher forb and shrub component. The diet of sheep and goats composed primarily of fallen tree leaves are of relatively good quality during the dry season. These diets contained CP levels of 8 – 12%, IVDMD of 40 – 50% and sustained organic matter intake levels of 2 – 3% of body weight. DM digestibility of forages ingested by livestock during active forage growth period is generally over 50%. Goats select pasture which is higher rate of degradability as well as faster fractional rate of passage of digesta than the diet selected by sheep. Again goat prefers to browse on small species of thorny plants. Goats changed their diet composition frequently and exhibited a strong preference for plant species least available in particular zones. As goats can change the botanical composition of diet, the problem of increasing goat production and survivability would be solved best by examining nutritional aspects of intake rather than botanical components of the diets.

Goats are useful companion in sheep and cattle grazing system, while grazing with cattle, goats consume different sward components and with sheep, then prefer grass over white clover. However, the same preference are not seen when goats graze tropical grass – legume pasture. For Brachiara decumbens pasture in summer and autumn, both goats and cattle preferentially select legume from the sward while sheep select grass over legume (Table 3). However, in Paspalum plicatulum sward, grass was preferred by all species in summer but in autumn only sheep continued to select grass and cattle and goats preferred legumes. These observations suggest a number of management options for using sheep, goats and cattle to maintain productive balances of grass and legume in tropical pastures.

Table 2: Grazing choice of range plants by domestic livestock (per cent intake)

Plant	Horse	Cattle	Sheep	Goats
Grass	90	70	60	20
Weeds	4	20	30	20
Browse	6	10	10	60

Table 3: Selection of legume and grass by cattle, sheep and goats under grazing

Species/ Component		Summer		Autumn	
		BD	PP	BD	PP
Cattle	Grass	27	71	40	35
	Legume	64	13	50	51
Sheep	Grass	79	82	66	66
	Legume	11	6	25	20
Goat	Grass	11	51	34	7
	Legume	80	37	58	78

BD, Brachiara decumbens, PP, Paspalum plicatulum

Forage Intake

Intake from a given vegetation type is assumed to depend on the area covered by the vegetation type and the potential daily intake of digestible DM available from it in relation to those for all vegetation type present. Potential daily intake of digestible DM is calculated as the product of diet digestibility and daily DM intake. Daily DM intake is assumed to be limited either by diet digestibility or by a maximum grazing time, which ever is lower. There are also indications that low forage availability may seriously limit intake. Forage intakes of the grazing animal also vary with body weight as well as with forage quality. Daily DM consumption by most range ruminants is about 2% of their body weight when values for different seasons are averaged across the year. However, considerable seasonal variation occurs with values as low as 1% during periods when forage is low in quality and availability, to over 2.5% when quality and availability are high.

Considerable evidence from penned ruminants offered various forage in different physical forms shows that intake is more affected by particle passage rate through the digestive system than by palatability. Limited research indicated that leafy materials from range forbs and shrubs are consumed at higher levels than grasses for a given levels of digestibility had a 8% higher intake than a grass dominated diet with 43% digestibility. The browse dominated diet had a shorter retention time than the grass dominated diet.

Lower levels of cell wall can result in higher intakes of leafy materials from palatable forbs and shrubs compared to grasses. Diets high in cell wall content occupy a large volume in the rumen, limiting gut capacity. The rate of break down of this material determines the potential intake. Forbs and shrubs leaves generally have thinner cell wall than those of woody material or grass leaves and stems. Forage with thin cell wall tends to break down and pass more quickly out of the rumen than do those with thick cell wall.

Grazing Behaviour

The ability of a grazing system to enhance the quantity of livestock products produced on rangelands is determined in part by the efficiency with which livestock harvest forage produced under a grazing system. One measure of efficiency of harvest is the ratio of the amount of energy obtained from the forage. Information pertaining to the amount of time animals spend in different activities is necessary to determine energy budgets and efficiency of harvest.

Cattle graze during 4 or 5 periods in each 24 hours of grazing. The major periods of grazing during the day occur just before dawn, during mid morning, in the early afternoon and near sunset. Grazing peaks during the middle of the day are variable in number and extent from day to day. During hot weather grazing animals are likely to be more active at night and less so in cold weather. Cows spent a similar time grazing per day in summer and winter though significantly more grazing was done between afternoon and morning milking in summer than in winter, i.e. day time grazing was reduced in summer when compared with grazing time in winter. Cattle graze tropical pastures for a longer time each day than temperate pasture even when large quantities of herbage are available for grazing. Under conditions where leafy material is difficult to harvest, cows compensate for the small size of bite eaten by increasing the time spent for grazing and also by increasing the number of bites taken.

Grazing patterns follow closely the periods of daylight when the quantity and quality of feed is good but under adverse conditions cattle extend grazing into hours of darkness and high producing lactating cows can be identified by the extent of foraging during the night. Since more than 50% of grazing can occur between 6.00 p.m. and 6.00 a.m. it is highly desirable to have good night paddocks to obtain optimum production. During inclement weather grazing time is reduced but cows compensate for these shortages by increasing in subsequent periods. Bos indicus cattle can graze during hot weather, whereas Bos taurus cattle seek shade.

Concentrate supplementation change the grazing time in cattle. Daily feeding of 4.5 kg crushed maize to cow reduced grazing time per day by approximately 15%. Concentrate supplementation also change the grazing behaviour in sheep. The supplemented groups graze less time than the unsupplemented one when two groups graze separately, however, grazing and resting times for both groups were similar when two groups grazed together.

Bite frequency of all grazing animals was higher during morning and evening hours of the day. Again the average bite frequency (bite/ hour) was highest for goats (962 ± 25) followed by oxen (726 ± 30) and lowest for buffalo (569 ± 36). The time spent for grazing in sheep was higher than cattle. Grazing system also directly effect the grazing behaviour of cattle. Cattle in continuous grazing system consistently grazed 1 hr per day longer than cattle in the rotational system. Grazing time also varied seasonally in a rather consistent manner. Grazing time was longest when forage quality was highest and shortens when forage quality was lowest. The behaviour of sheep was studied and noted that Rambouillet sheep traveled greater distances and spent more time resting while Columbias traveled the least distances, rested least and grazed longer than the other breeds. All breeds traveled further in the morning and grazed longer in the afternoon.

Supplementation

Animal production is primarily a function of quantity and quality of forage consumed by the grazing livestock but because quantity and quality of available forage vary among range sites, season, year and management strategy, available forage seldom meet the nutrient requirements of livestock for and entire year. To correct these deficiencies such practices as supplemental feeding is often employed. During the period of December to June when good quality pasture is not available, supplementation by more DCP and TDN doubled the growth rate of heifers. Body weight gain in grazing heifers was declined form August to November when supplementation not given, however, with the supplementation of 2 kg concentrate mixture (16% CP) daily and ad lib Lasiurus sindicus leaf improve the weight gain. Nutritional stress on goats and sheep during the dry season is responsible for high weight losses and periodic mortality. With poor quality diet, supplementation of only urea or molasses has no beneficial effect in goats, however, supplementation of both urea and molasses increased the live weight gain of animals.

Period of Supplementation

Identification of deficient nutrients time by and the severity of the deficiency are major concerns of range livestock producers. Clipped samples of the available forage provide inaccurate measures of diet quality because grazing animal show high selectivity for particular plant part and species. Esophageal fistulation have been useful for studying nutrient levels and trends in livestock diets. Studies have shown that nutritional quality of livestock diets varies substantially between years on most range type. The period when supplementation can be advantageous, vary substantially within and between years. Although fistulated animals are a good research tool in recent year's faecal analysis, has shown potential as a quick, practical means for ranchers to detect periods when energy, protein and phosphorous supplementation will be most advantageous economically.

The concentration of nitrogen in the faecal organic matter of ruminants shows strong linear association with livestock weight changes, forage intake, diet digestibility, and diet crude protein content. Research indicates that when total nitrogen concentration of the fecal organic matter drops below 1.40 %, cattle undergo weight losses due to inadequate crude protein in the diet. Severe weight losses (over 0.25 kg/day) can be expected if fecal nitrogen concentrations drop below 1.30%. A faecal nitrogen concentration below 1.40 % for cattle indicates that diet crude protein levels are probably lower than 6%. Once this level is reached, the forage intake declines precipitously because rumen microbial needs for nitrogen are not satisfied. Increases in range forage intake (20 to 60%) and digestibility (5 to 15%) can be expected from supplemented protein when diet crude protein levels are below 6%. When diet crude protein is above 6%, supplemented protein has had small to no effect on range forage digestibility and intake.

Protein Supplementation

The two basic types of protein supplements provided to range livestock are non-protein nitrogen (urea and biuret) and high protein natural feeds (alfalfa hay, cottenseed meal, soybean meal). Non protein nitrogen source can substitute for costlier feed protein because rumen microorganism can convert nitrogenous compound into proteins. Under the right conditions about one third of the total protein requirement can be met by non-protein nitrogen. For non-protein nitrogen supplement to be utilized effectively, a good energy source must be available. This approach is used with feedlot animals consuming grain but seldom with animals consuming dormant range forages.

Cottonseed meal is probably the most heavily used plant protein supplement and 0.1 to 0.2 kg, per day for ewes have given the best economic returns. These levels allow animals to meet their protein requirement and often improve intake of the range forage. Regarding the frequency of protein supplementation, no real differences was observed in livestock performance between daily, alternate day, every third day and even weekly feeding of cottonseed meal or alfalfa hay.

Energy Supplementation

Barley and cracked corn are two of the more common energy supplement used under range conditions. These feed usually depress the intakes and digestibility of range forage and serve primarily as substitute for range forage.

Rams fed on leaves of F. albida, L. leucocephala, S. sesban and lucerne hay gained 49 g/ d with maize and 23 g/d without maize supplementation. Alfalfa hay can be a good source of energy as well as protein. If protein as well as energy is deficient in the range animal diets, it is particularly advantageous. When lush forage is with high protein levels (over 15%) one of the grain supplements would be advisable.

The time of day when energy supplement is provided to cattle influenced livestock performance. Two groups of steers were fed the same diet of cracked corn daily. One group was fed in the early morning while the other group was fed in the middle of the afternoon. Steers fed in the afternoon outgained steers fed in the morning by about 0.2 kg. This difference is explained by reduced grazing of the group fed in the early morning compared to those fed in the late afternoon. The early fed group tended to remain near the feed grounds waiting for the feed truck instead of grazing in the morning. The other group grazed in the morning and waited for feed in the afternoon. Cattle typically grazed in the morning and evening and water and loaf in the afternoon. Therefore, normal grazing activities of the afternoon fed group were unaffected. Steers fed in the late afternoon had a higher total intake of feed, which probably accounts for their weight gain.

Grazing Resources in India

In India, about 40% of total area is available for grazing of livestock. In some states viz. Himachal Pradesh, Jammu & Kashmir, Meghalaya, Nagaland and Arunachal Pradesh the grazing land available is even over 70%. The grazing activity depends on the availability of grazing resources like pasture lands, forest lands, cultivated wastelands, fallow lands, nonagricultural lands, miscellaneous tree crops and groves, etc.

1. Pasture land

This is the main grazing resources of the country available over area of 12 million ha and about 3.94% of the geographical area of the country. The distribution of pasture lands are mostly noticed in the states like Himachal Pradesh (36.44%), Sikkim (13.31%), Karnataka (6.54%), Madhya Pradesh (6.35%), Rajasthan (5.39%), Maharastra (5.11%) and Gujarat (4.49%). The distribution of pasture in India as follows:

(i) Pasture of northern region

The pasture lands of Jammu & Kashmir, Himachal Pradesh and Uttar Pradesh provide good quality pasture for livestock grazing. This region has potential resources in the form of green meadows and pasture, which at some places are mixed with the forests. The alpine meadows have an important economic value, which provide pasture for the sheep and goats of migratory livestock owners. Both tropical pasture in lower hull and temperate pasture at higher altitude is a common phenomenon in the pasture of this region. According to different situation the grass cover varies from 16.8 to 88.9% and biomass production has been estimated from 44 to 1077 kg dry matter per hectare.

(ii) Pasture of western region

It includes the pasturelands of Rajasthan and Gujarat. In this region a vast area of pasture land are available for grazing of livestock. In Rajasthan the available grazing land is over 40% and in Gujarat the area is about 30%. The land under permanent pasture is about 5.4% in Rajasthan and 3.5% in Gujarat, which provide good quality fodder for livestock.

(iii) Pasture of Peninsular India

Quantity and quality of pasture from the Deccan plateau in the state of Maharastra, Karnataka, Andhra Pradesh and Tamilnadu can not sustain the mixed herd grazing. The tract is rocky, strewn with boulders and gravels and receives annual rainfall less than 50 cm. Sheep rearing is the major occupation in this region due to limited forage availability.

(iv) Pasture of eastern region

The availability of pasture land in this region is limited. Most of the states under this region viz. Manipur, Tripura, West Bengal and Bihar have pasture land less than 1%. Therefore, livestock grazing is not a popular practice in this region.

2. Forest land

Total 22.1% of India's geographical area i.e. about 37.33 million ha is covered by forest and provide valuable grazing resources for livestock. Several states have a sizable area under forests viz. Arunachal Pradesh (93.88%), Jammu & Kashmir (61.23%), Tripura (55.15%), Orissa (42.7%), Sikkim (32.35%), Meghalaya (36.10%), Madhya Pradesh (31.68%), Kerala (27.82%), Maniupur (27.22%), Himachal Pradesh (27.06%), Nagaland (25.90%), and Assam (25.28%). The forest of northeastern hilly region does not provide sufficient fodder for livestock due to its dense nature which is beyond the reach of livestock. The forest land of Jammu and Kashmir, Himachal Pradesh, Uttar Pradesh, Madhya Pradesh have a fairly wide spread pasture to support the grazing of large number of livestock.

3. Cultivated wasteland

India's 5.1% of geographical area is under this category. The large area of cultivated wasteland belongs to Meghalaya (20.19%), Rajasthan (16.77%) and Gujarat (10.56%) and it is about 5% in other states. The large proportion of cultivated wasteland viz. ravine areas, water logged areas, saline and alkaline lands, shrubs and bushes infested lands and riverine lands has a great potential to provide valuable fodder for livestock.

4. Fallow lands

About 22.9 million ha of land constituting 7.5% of the total land area of India is under fallow land. Majority of this land was situated in Nagaland (32.28%), Bihar (18.26%), Tamil Nadu (16.14%), Meghalaya (13.87%), Andhra Pradesh (11.72%) and Rajasthan (11.10%). The fallow land actually is the cultivated land not utilized for cultivation due to scanty rainfall in Tamil Nadu, Andhra Pradesh, Rajasthan or due to heavy rain in Meghalaya, Nagaland and Bihar.

5. Nonagricultural land

The land which is earmarked for nonagricultural uses viz. the side of railway tract, roads and canals, dams or bunds and river banks provide considerable amount of pasture land for the grazing of livestock.

6. Miscellaneous land

The total 1.14% geographical area of India, comprising 3.46 million ha is under this category. The maximum of such type of grazing resources situated in northeastern states viz. Nagaland (17.99%), Tripura (9.35%), Meghalaya (6.45%) and Assam (3.18%).

Future Studies

There is little to no scope for increase the land under fodder production due to increased demand of food grain production for ever increasing human population of the country. Livestock industry is therefore, facing major challenges for increasing the productivity using available resources in scientific manner. The available technology should be properly implemented to increase forage production from the available land so that it can provide quality fodder round the year. Regarding the livestock component, grazing behaviour of different species of animals should be thoroughly studied. The grazing resources should be developed as per the grazing habits and pasture requirements of particular animal species. For the browser species like goats, grazing land can be predominated with nutritionally superior shrubs and tree species. For grazing livestock like sheep and cattle proper balances of superior quality, grasses and legumes may be the major component of the pasture. According to the carrying capacity and availability of forage species, grazing of mixed herd like cattle, sheep and goats should be adjusted intermittently so that quality forage can be available round the year. The improved grazing management systems like, rotational, deferred rotational systems can be practiced to improve the quantity and quality of pasture from degraded land. Supplementation of energy, protein and mineral during lean period at individual level or in combination of various proportions may be exploited for economic livestock production under grazing condition.

Bibliography

Bement, R.E. (1969). A stocking rate guide for beef production on blue grama range. J. Range Manage, 22: 83-86.

Kothmann, M.M., Mathis, G.W. and Waldrip, W.J. (1971). Cow-calf response to stocking rates and grazing systems on native range. J. Range Manage. 24: 100-105.

Lang, R.L., Barnes, O.K. and Rauzi, F. (1956). Shortgrass range: Grazing effects on vegetation and sheep gains. Wyo. Agric. Exp. Stn. Bull. 343.

McDowell, L.R. (1985). Nutrition of grazing ruminants in warm climates. Academic Press, INC, London.

Pathak P. S. (2003). Strategy to maximize production despite poor rainfall: A focus on forages. Short Term Training Course on Forage Development and Seed Production, February 25 – March 26, IGFRI, Jhansi. pp 1-7.

Pathak, P.S. and Sunil Kumar, (2004). Forage and grazing resources in different agro-climatic regions of India. In: Buffalo Production Under Different Climatic Regions (S.S. Kundu, A. K. Mishra and P. S. Pathak, eds.), IGFRI, Jhansi. pp 1-41.

Pieper, R.D., Nelson, A.B., Smith, G.S., Parker, E.E., Boggino, E.J. and Hatch, S.L. (1978). Chemical composition and digestibility of important range grass species in South-central New Mexico. N.Mex.Agric.Exp. Stn. Bull. 662.

Rai, P. (1990). Establishment and management of tropical pastures in India. In: Silvipastoral System in India (P.S. Pathak, Panjab Singh, eds.). Range Management Society of India, IGFRI, Jhansi. pp 131-146.

Singh, R.P. (ed.) (1995). Forage Production and Utilization. IGFRI, Jhansi.

Skovlin, J.M., Harris, R.W., Strickler, G.S. and Garrison, G.A. (1976). Effect of cattle grazing methods on ponderosa pine-bunchgrass range in the Pacific northwest. U.S.Dept. Agric. Tech. Bull. 1531.

Smoliak, S. (1974). Range vegetation and sheep production at three stocking rates on Stipa-Bouteloua prairie. J. Range Manage. 27: 23-26.

Tyagi, R.K. and Singh, Panjab (1987). Grazing resources and grazing systems in India. Pasture and forage crops research. A state of knowledge report. Third International Rangeland Congress. RMSI, IGFRI, Jhansi.pp. 17-34.

Willms, W.D., Smoliak, S. and Schaalje, G.B. (1986). Cattle weight gains in relation to stocking rate on fescue grassland. J. Range Manage. 39: 182-187.

SILVOPASTURE FOR RUMINANTS

Silvopasture is an agroforestry practice that is specifically designed and managed for production of tree, tree products, forage and livestock. Introduction of forage crops in a tree production system or trees into a forage production system or re-vegetation of degraded lands not suitable for agriculture through tree and forage production system results into a silvopasture system. In a typical nature rainfed silvopasture the upper most strata constitutes the trees, in the middle the bushes of different nature and the ground is covered of grass-legume mixture.

There are some areas, which need focused attaintion other than the forests and community pasture which can be utilized for the development of silvopasture. They are:

1. River bank and nalas
2. Tank bunds
3. Road sides
4. Premises of railway tract
5. Canal bunds.

The terrace bund in the hills can also be very successfully utilized for the development of silvopasture system. The field bund, unreclaimed land, follow land are useful for the establishment of silvopasture.

The plant species of silvopasture system are selected on the basis of relative competence and their capability for the production of good quality forages for the livestock. The selection of the herbage for middle and ground strata depends on the type of tree required and climatic conditions.

Raising and Planting Desirable Species of Trees: The method generally followed in establishing the silvopasture is the planting of the young plants. The high cost involved in the raising of sapling in the nursery is usually nullified while considering the advantages. There are lower deaths, faster growth and quicker establishment in the planted stock. The site of the nursery, management of the nursery and the method followed for planting, determine the efficiency of the system.

Various nursery raising practices required for trees are discussed briefly.

Raising of the nursery: There are different methods for raising saplings suitable for planting. The sapling may be raised in polythene pots, soil balls, soil blocks, under cut beds or boxes depending on the facilities available and cost involved. The polythene pots are now most commonly used in most of the nurseries dealing with the tree plantation. Polythene pots can be used for the raising of sapling, striplings (nursery grown saplings containing a few apical leaves and sometimes pruned roots or stumps (root or shoot cuttings) depending on the type of species and facilities available. The care during the process of nursery raising is essential as it is an important factor for the proper establishment. The sapling should be protected from direct sunlight during the summers and frost during the winter. Proper shade and shelter, and timely watering required to give to protect from heat/frost damage and drought. The pruning of root is required to prevent the development of deep tap root and to encourage the growth of lateral root system, as well. The control of insect-pest especially the termites is much more necessary. The soil fungi responsible for the early deaths of sapling should be treated with suitable fungicide prior to the nursery raising. The bird and rodents may also create problems while raising the nursery, so necessary arrangements should be made either using repellants or other protecting measures.

Planting Operation: The land is first cleaned using various techniques like burning, harrowing, uprooting, cutting and grazing depending on the density of the undesired plant species, cost involved and topography of the land. The chemical methods may also be used to clean the vegetation. Once the land is free from undesirable vegetation it is ready for planting. A small pit of about 50 cm diameter and 50-70 cm deep is needed for planting. The pit is refilled with a mixture of soil, organic manure, insecticides (to protect sapling from termite), and fertilizer mixture (depending on the soil status for the minerals). The sapling is then planted in the pit. A small portion of plastic bag should be above the soil surface. The side gaps are thoroughly filled using the mixture in a way to avoid air gap. The distance between tree to tree and bush to tree is dependent on the type of species and purpose of plantation. The season of planting is also important. The suitable time of planting is after the onset of the monsoon in the Indian sub-continent, which may vary from place to place depending on the rainfall. Off-monsoon planting results in high mortality due to lack of water and too late planting is also not beneficial as the total span of rainwater available to plants is very much reduced.

Pasture Establishment

Immediately after completion of planting of tree seedlings, grass seeds may be sown/transplanted in lines 90-100 cm from row to row and 45-60 cm apart within the row. In between the two lines grasses 1-2 lines legumes may be sown. For quick establishment and also to boost up production, application of

40 kg P_2O_5 and 20 kg N/ha at the time of land preparation followed by 20-40 kg N/ha annually depending on the rain.

Because of low nutritive values of tropical grasses, introduction of legume in pure pastures is important. In IGFRI, Jhansi both stylo (Stylosanthes hamata) and siratro (Macroptilium artopurpureum) have successfully been established under silviopastoral systems. It has been found that sowing of one line of stylo and one line of siratro in between grasses keep the pasture less weed infested and thereby help in better establishment and yield of the pasture.

In general, during first two years grazing should not be allowed. The grasses may be harvested during November and fed to livestock. Seeds of grasses and legumes may also be collected during this period.

Care and Management of newly established silvopasture: The young trees and shrubs are much tender to withstand the various anti-factors. The high mortality at the initial stage increases the cost of production.

The weeding of the pit and surrounding area is necessary to minimize the competition of the young tree for the nutrients. The weeding is required during the first 2-3 years. Once the tree surpasses the height of weeds and develops its canopy, the weeding is not required. The tolerance against weed varies among the species. Nearly all the saplings are vulnerable in the early phase of establishment. Acacia and Prosopis species however, rapidly penetrate down deep tap roots. Once established, they tolerate a certain amount of surface weed growth. The weeding can be performed by manual, mechanical or chemical method depending on the cost involved and other factors. The weeding should be done as and when required.

The watering is not required when there is rain. In a planting season if there is drought, watering is required for the survival of the young trees.

The trailing type of legume should be avoided during the early phase of the establishment. If planted, regular cutting of the creepers is desirable for the survival of young tree. The thinning and pruning depending on the condition may be done when necessary.

The grazing on the newly established silvopastoral system should be avoided till the trees grow to inaccessible height.

The fire hazard should be avoided as far as possible. The firebreak, roads and clean surface belt may be used to reduce chance of the spread of the fire from one place to another. The protection of early planted trees from the insect, pest and diseases is also necessary to exploit the benefit in the long run.

Care and management of forest based silvopasture system: The management practices are derived to harvest maximum benefits in terms of forage production without adversely affecting the quantity and quality of the timber, industrial raw material and fruits, etc. The scientific management for these purposes includes the quick growing good quality grass-legume mixture, a fair compatibility among the species and the harvesting and/or grazing management in the forest. The good quality legume shrubs can also be planted in the forest for such purposes. The proper ratio and the distance between trees and bushes should be maintained through the planting of new sapling of desired species, thinning to reduce the density of the trees and loping of de-shaped, dry or diseased branches. The purpose of the forest should however be kept in mind while determining the tree-bush ratio and the distance.

In India the forage production is a part of most of the forest. The system is quite variable. In some forest areas harvesting of grasses is done once in a year after monsoon season, while other forest areas are open for grass cutting round the year. The free and controlled grazing in forest can also be observed at some places. Keeping in view the potency of the forest in terms of the forage production, a suitable harvesting/grazing schedule should be adopted for individual forest. Grazing in the forest, although not much popular in the protected forests, has several advantages. The grazing avoids the competition of undesirable tree species by consuming them. This is more important where pine, decodar or similar species dominate which are not generally consumed by the animals. The grazing helps in the regeneration of such species. The animals graze on the annual or desiduous type of the species, and thus reduce the chances of fire. Over grazing is a serious limitation and should be avoided for deriving more benefits from the forests. The following points should be considered before allowing the grazing of livestock:

1. Economic value of the forest resources.
2. Density of the grass-legume cover.
3. Number of the livestock per hectare of forest land.
4. Grazing habits of the livestock induced in the forest.
5. Distribution of the livestock within the forest.

The growth of the forage species and season of grazing are also important for the optimum forage production. The harvesting schedule should be chalked out keeping in view the aforesaid and other factors influencing the plant growth. The intensity of the harvest should be such that the sufficient time may be available to the plant species for its growth. The weeding, fertilization and plant protection measures also increase the productivity.

Care and management of silvopasture system (forage/fuel type): The trees planted for the fuel/fodder purpose should also be maintained in the manner described for the forest based silvopasture. The trees should be preserved as a buffer of nutrients. For maximum benefit, lopping be done during the scarcity periods. The lopping during rainy season or just before rainy season increases the chance of soil erosion. The thick branches, which provide canopy to the tree, should not be lopped. The dry branches and the left out branches after removal of the leaves from a lop are used for fuel purpose. Tree can also be sacrificed for fuel purpose provided younger follower plant has grown enough to replace it. For the fodder production the intensity of lopping should be adjusted in a way to provide sufficient time for regeneration of the leaves and new branches. It, therefore, depends on the rate of regrowth of the plant species and several environmental factors affecting the growth of the plant. The maintenance of bushes in the multi-tier system is most important. Browsing and lopping of the bushes should be controlled to provide sufficient regeneration time to the bushes when goats are allowed with cattle. This can be done by following rotational grazing.

The grass legume mixture can either be harvested or the animals may be allowed for grazing on it. During the early stage of the establishment the harvesting of the grass is advantageous over the grazing. During grazing the animals may destroy the growing tree and bushes. The carefully chalked grazing scheduled should be followed to avoid the losses incurred due to over grazing. The grazing should be avoided on the land prone to soil erosion. The grazing on such land is possible only when it has developed a good tree, shrubs and grass-legume cover.

With similar managemental practices the forage production is even possible in orchards. The riverbanks, barren-land, eroded sloped, terraced land, roadsides, etc. can serve a potent source of the forage. However, for the full and economic exploitation of the various forage resources the seasonal availability of forages, production of the forages and the requirement of the various categories of the animals at different levels of production/reproduction should also be considered. The forages should be harvested at the most nutritious stage without loosing the yield. The surplus forages should be preserved or processed using suitable methods keeping in view the limitations of economy, transport facilities and season, etc.

The Trees

The silvopasture system when developed, the tree species should be selected on the basis of its immediate use other than fulfilling forage/fuel requirements. For example, a silvopasture system proposed to be developed along the roadside should generally be selected for shade/fruit or shade/fodder trees. The dual purpose trees fulfilling fodder or fuel need other than soil conservation are preferred. In the community barrenland, community pasture or terrace bund, preference should be given to fodder and fuel trees.

The principal aim of any forage improvement programme is to maximize the use of basic feed resources that in mostly are crop residues or low quality roughage. One approach could be the addition of leguminous forages that will act as a nitrogen supplement to quality roughage material. Many tree and shrub species have been documented as useful fodder. For a sole feed source a good quality forage should possess certain attributes viz., (i) high feed digestibility (i.e. 65%); (ii) soluble protein in the diet producing about 30 kg N/ha digestible dry matter (DDM); (iii) by pass protein providing about same amount of protein as arises from microbes (i.e. 200g protein/kg DDM intake); (iv) a level of fat in diet approaching that of temperate pastures (i.e. 4-8%).

Thus for improving the quality of low quality roughage nutrient rich tree or shrub fodder is important. The palatability, digestibility and nutritive value of leaves varies with different stocks. Also, there is extreme variability in the nutritive value of shrub and tree species used as livestock feed. This variation due to greatly varying inherent nutritive values between species as well as the variation found within species because of differences in plant parts, age of tissue, soil and climate in which plant is growing.

Scarcity of good quality roughages and high cost of protein supplement necessitates the strategic supplementation of shrubs and tree leaves to achieve maximum feed utilization and in turn efficient animal productivity.

Trees can be grown either in combinations with agriculture crops under agroforestry system or on separate lands usually not fit for common agriculture. Since, trees have the capacity of growing under different conditions and in areas where it may not be possible to grow conventional crops, such as saline and waterlogged soils, steep and rocky mountain slopes and areas with severe climatic conditions. The ability of trees to tap water from deep underground layers and to with stand drought is outstanding advantage over crops. Thus, tree sustained the supply of fodder over long periods without any climatic risk, therefore, deserve due attention for the production of green nutritious fodder for small ruminants especially in the lean period. The fodder production potential of some trees at Jhansi are presented in Table 1.

Table 1: Fodder production potential of some trees at Jhansi

Species	Year of growth	Dry fodder yield (kg/tree)	Pod yield (kg/tree)
Acacia tortilis	7	3.2	-
Albizia amara	7	13.5	-
Albizia lebbek	6	5.4	4.5
Albizia procera	6	11.3	3.2
Dichrostachys cinerea	6	0.8	0.7
Hardwickia binata	10	3.7	-
Leucaena leucocephala	3	0.8	-
Leucaena leucocephala	4	1.9	-
Leucaena leucocephala (K8)	1.5	1.5	-
Sesbania grandiflora	3.5	0.3	-
Sesbania sesban	3.5	0.3	-

Source: Deb Roy et al, 1997, Gupta et al, 1983, Pathak and Patil, 1980, Rai et al, 1984, Roy et al, 1984, Roy, (1991).

The tree leaves are not only classed as emergency fodders but also considered nutrient rich fodder for livestock in general. Top feeds form an integral part of sheep and goats feeding. About 60% of the total feed available to sheep and goat comes form top feeds only. This is more in goat since they spend more than 90% time in browsing on top feeds and hardly 10% to graze on surface vegetation that too when sufficient top feeds are not available. Chemical composition and nutritive value of some important tree leaves used as top feed are given in Table 2.

Table : 2 Chemical Composition and nutritive value of some important tree leaves.

S.No	Species	Local Name	CP	CF	NFE	EE	Total Ash	Ca	P	DCP	TDN
1.	*Acacia catechu*	Khair	13.0-18.7	21.9-22.5	46.7-51.0	3.0-2.7	9.7-9.8	2.5-2.7	0.1-0.2	24	46.33
2.	*Acacia nilotica*	Babul	13.9	9.2	69.8	-	7.1	2.6	0.1	-	-
3.	*Adina cordifolia*	Haldu	8.7-12.5	12.1-13.8	-	3.1-5.6	7.7-8.2	2.3-3.1	0.2-0.5	28	50.90
4.	*Aegle marmelos*	Bael	15.1-15.3	16.4-18.1	48.4-52.8	1.4-1.5	14.0-16.8	4.2-4.8	0.1-0.3	-	56-67
5.	*Ailanthus excelsa*	Maharukh	19.9	12.8	51.8	3.5	12.0	2.1	0.2	13.07-16.24	63.03-63.80
6.	*Albizia lebbek*	Siris	16.8-26.5	26.5-37.5	36.0-42.8	2.8-4.7	7.1-11.5	1.1-2.7	0.1-0.2	-	-
7.	*Amaranthus spinosus*	Goja	16.1	20.3	42.1	2.5	-	2.28	0.28	11.6	-
8.	*Azadirachta indica*	Neem	12.4-18.3	11.4-23.1	43.3-66.6	2.3-6.2	7.7-18.9	0.9-4.0	0.1-0.3	-	-
9.	*Bauhinia racimosa*	Jinja	15.4	18.5	54.6	3.3	-	4.1	0.14	-	-
10.	*Bauhinia variegata*	Kachnar	10.7-15.9	25.3-33.0	40.9-51.8	1.3-3.9	6.3-12.3	1.8-4.1	0.2-0.4	4.98	47.92-55.54
11.	*Careya arborea*	Kumbi	10.4	25.9	42.6	7.7	7.5	1.6	0.3	-	-
12.	*Carthamus tinctorius*	Kaar	12.4	10.9	58.3	1.6	-	1.7	0.34	12.4	-
13.	*Celtis tetranda*	Kharik	14.0	16.6	44.0	4.5	-	4.0	0.13	-	-
14.	*Cichorium intybust*	Kasni	21.7	10.9	32.1	1.4	-	-	-	15.7	-
15.	*Cordia dichotoma*	Buhal	12.4-15.1	16.4-26.8	41.9-52.8	1.5-2.9	12.6-17.4	2.4-4.2	0.2-0.3	-	-
16.	*Dalbergia sissoo*	Shisham	11.3	19.6	51.6	3.2	-	3.17	0.14	-	-
17.	*Dendrocalamus strictus*	Bamboo	14.1	27.6	44.6	1.7	-	1.22	0.26	9.3	-
18.	*Ehretia leavis*	-	13.5	18.0	51.2	6.0	11.3	1.5	0.3	-	-
19.	*Ficus bengalensis*	Bargad	7.7-11.2	13.9-29.9	37.6-51.6	2.5-5.1	9.3-16.9	2.5-4.1	0.1-0.4	-	57.05
20.	*Ficus glomerata*	Gular	11.2-15.2	12.3-16.5	53.5-59.0	2.2-2.9	11.9-18.4	1.7-3.0	0.2-0.4	6.69	53.82
21.	*Ficus lacor*	Pakur	8.4-15.7	12.8-25.9	44.8-61.1	1.8-3.4	-	2.0-3.5	0.2-0.4	-	48.11
22.	*Ficus religiosa*	Aswatha	13.8-13.9	19.5-21.4	46.7-50.5	2.2-3.8	14.0-14.2	2.2-3.9	0.2-0.4	7.05	38.27-48.22

S.No	Species	Local Name	CP	CF	NFE	EE	Total Ash	Ca	P	DCP	TDN
23.	*Ficus roxburghis*	Timla	13.4	7.7	64.9	4.7	-	1.31	0.17	-	-
24.	*Grewia optiva*	Bhimal	19.0	22.1	43.9	2.2	12.8	3.3	0.2	12.3-12.41	61.35-62.1
25.	*Gymnospora spinosa*	Kankera	9.4	-	-	-	-	-	-	2.7	-
26.	*Hardwickia binata*	Anjan	9.0	30.4	49.7	4.1	-	2.28	0.13	2.0	-
27.	*Kydia calycina*	Baranga	11.1-13.6	14.5-23.7	46.1	3.3-6.1	15.3-19.7	3.1-5.2	0.3-0.6	63	45.18
28.	*Lannea coromandalica*	Jhingan	11.4-16.2	16.1-19.8	53.1-59.0	2.5-4.2	8.5-9.9	1.7-2.4	0.2-0.3	43	55.15
29.	*Leucaena leucocephala*	Subabul	18.9-27.6	10.1-17.2	46.7-52.6	2.6-5.9	7.5-10.9	-	-	12.64-16.37	57.08-70.22
30.	*Mallotus philippinensis*	-	14.5	21.6	50.4	3.9	9.7	1.3	0.2	7.70	36.65
31.	*Moringa oleifera*	Sahijna	15.3-20.7	7.1-17.9	48.7	4.3-11.1	11.8-14.2	3.2-3.8	0.3-0.6	-	-
32.	*Morus alba*	Mulberry	15.0-27.6	9.1-15.3	48.0-49.7	2.3-8.0	14.3-22.9	2.5-4.7	0.2-1.0	7.84	48.35
33.	*Ougeinia oojeinensis*	Bondhan	11.6-15.1	21.9-26.5	47.6	1.4-4.3	10.0-14.9	2.4-3.4	0.2-0.4	-	-
34.	*Prosopis cineraria*	Kheri	14.2	22.1	53.2	4.3	6.1	-	-	4.49	40.99
35.	*Prosopis juliflora*	Vilayati khejri	21.4	20.8	50.0	-	-	-	0.20	1.5	-
36.	*Robinia pseudoacacia*	False acacia	25.5	17.2	46.5	3.3	7.5	1.5	0.3	-	-
37.	*Schleichera oleosa*	Kusum	10.4	32.3	49.2	1.9	6.1	1.7	0.3	-	47.53
38.	*Shorea robusta*	Sal	10.1	27.4	55.4	3.2	3.9	0.8	0.1	-	42.70
39.	*Syzygium cumini*	Jamun	10.8	20.1	60.4	1.9	6.9	1.8	0.1	-	-
40.	*Tamarindus indica*	Imli	13.8	15.9	60.5	1.9	7.94	1.6	0.1	-	-
41.	*Terminalia alata*	Sain	8.3-12.7	13.1-21.8	54.1	1.4-4.9	9.1-12.1	2.7-3.2	0.2-0.4	-	-
42.	*Tribullus terrestris*	Gokhru	18.5	36.9	23.9	2.9	-	3.5	0.52	14.8	-
43.	*Zizyphus mauritiana*	Beri	16.9	13.5	56.1	2.2	11.5	1.4	0.3	-	30.65
44.	*Zizyphus nummularia*	Pala/ Jharber	14.3	15.7	57.1	4.3	-	2.4	0.10	3.5	-

Some of the important fodder trees suitable for planting in silvopastoral system are presented in the following text.

Anjan *(Hardwickia binata)*: Anjan is a tree of dry hot climate with scanty to moderate climate and porous soils with good drainage are favourable sites for natural regeneration. The tree pollards well and coppices indifferently. The leaves are classed as good fodder, but mature leaves are not much palatable to sheep and it cannot form the sole feed.

Ardu *(Ailanthus excelsa)*: Leaf fodder from Ailanthus excelsa is an excellent fodder with high palatability and nutritious for sheep and goats. It is relatively a fast growing tree. An average full-grown tree yields about 5-7 quintals of green leaves lopped twice a year in the months of November-December and May-June. It is the native of India and grows almost throughout the tropical and subtropical part of the country especially in the dry tracts.

Babool *(Acacia arabica)* pods: The pods are also used as fodder. An average tree yields about 18-40 kg of pods in a year. Pods of Acacia species are relished by sheep and goats they possess good nutritive value (Ravi and Natanam, 1995). Adult sheep could be maintained on Acacia pods alone while for growth, lambs require supplementation with leguminous fodder or grazing. It was also observed that Milled Acacia pods with seeds provided supplement similar to maize bran (Gohl, 1981). Acacia pod as sole feed can meet the maintenance requirements of N in sheep. However, Ca and P have to be supplemented. Higher tannin content might have resulted in low protein digestibility. The DCP and TDN of Acacia pods are 5.14 and 60.69% respectively (Madhavilatha, 1999).

Babul *(Acacia nilotica)*: The leaves of this are excellent fodder and are extensively lopped for fodder. The palatability rating of the leaves is reported to be good. The leaves are fairly rich in crude protein and ether extract and have low crude fiber. It is indigenous to the northern part of the deccan plateau and the indo gangetic plain. It is a tree of dry and warm regions and thrives best in areas with annual rainfall of about 500 to 1250mm.

Bargad *(Ficus bengalensis)*: Bargad is a tree of subtropical and tropical climate grows on a variety of soils. It is not a tree of choice for silvopastoral system, however, leaves are important top feed resources for livestock. Goats relish more than other animals the leaves of bargad. The crude protein, calcium and phosphorous contents generally increase, as the leaves mature and crude protein content is the highest during monsoon season. This refers that its lopping during winter season have better fodder value. Goats can easily be maintained on its leaves alone.

Ber ***(Zizyphus mauratiana)*****:** It grows in dry tropical and subtropical climate on wide variety of soils, but best growth is obtained on deep sandy loam soil neutral to slight alkaline. The tree coppices and suckers well. The leaves are considered to be good fodder for goats and in some parts, it is the sole green fodder available to the animals. The leaves have fairly high nutritive value.

Bhimal ***(Grewia optiva)*****:** Bhimal is a tree of subtropical climate. Sandy loam soil with adequate supply of soil moisture supports good growth. The tree pollards and coppices well. The leaves are rated as good fodder. The leaves are fairly rich in protein and other nutrients and do not contain any tannin. The composition of Bhimal leaves compares favourably with berseem, alfalfa/ cowpea hay in respect of chemical composition. Crude protein is high in young leaves, gradually decreases through rainy season and increases again in winter. The leaves are fairly palatable. The DCP value of the leaf fodder is comparable with that concentrates like bajra and maize, thus can be a substitute of concentrate when fed with straws and other roughage.

Kardhai ***(Anogeissus pendula)*****:** It is a tree of dry hot region and its good growth is obtained where the soil is deep. Leaves are considered to be good fodder and are as such extensively lopped, which are moderately palatable.

Gliricidia ***(Gliricidia maculata/Gliricidia sipium)*****:** Gliricidia is medium size, quick growing and deep-rooted tree legume mostly grown as live fence support plant for the vines of black pepper and also as green manure crop. Leaves of Gliricidia sipium were found to be very nutritive value for ruminants. Goats utilized Gliricidia leaves more efficiently as compared to sheep. Both sheep and goat could not relish Gliricidia maculata leaves as sole feed but when mixed with sorghum strover in equal parts (DM basis), the animal could meet DM requirement.

Heens ***(Capparis horrida)*****:** Heens plant is a thorny climbing shrub, which is available in abundance under range conditions. It is green in rainy and winter season, whereas its growth becomes dormant during summer. Its fruits are prickled. It is also relished by goats. The balance of N, Ca and P are positive indicating the availability of these nutrients to the body in sufficient amount. The heens plant is having DCP 11.54% and TDN 48.31%.

Vilayati babul ***(Prosopis juliflora)*** **pods:** Iron wood (Prosopis juliflora) pods are also utilized in concentrate feed for cattle and sheep at 40% level in place of wheat bran. Prosopis juliflora post itself could not supply the required quantity of DCP and hence needed supplementation.

It is readily browsed by goats. The leaves are highly nutritive and palatable. It is much valued as a fodder tree that way lopped heavily. On an average a full-grown tree is expected to yield about 60 kg of green forage popularly called "loong". Pods have a sweetish pulp and are used as a fodder for livestock.

Kachnar ***(Bauhinia variegata)*****:** It grows in the sub Himalayan region and outer Himalayan valley up to an elevation of about 1500m. Tropical and subtropical climate is preferred and is capable of growing on wide variety of soils ranging from gravelly soils on mountain slopes to sandy loamy soil in the valleys. Natural reproduction takes place from the seed. The leaves are regarded as good fodder and its average fodder yield is reported to be 15-20 kg/tree. The chemical composition of the leaves varies with locality and season. The crude protein and crude fibre contents generally increase as the leaves mature. The leaves are fairly palatable.

Katai ***(Flacourtia indica)*****:** It grows mostly in dry open places and rocky hills as a large shrub. It coppices well and readily browsed. It is a favourite for sheep and goats.

Khair ***(Acacia catechu)*****:** It is tree of warm and dry regions. Acacia catechu is considered to be a good fodder tree and is extensively lopped to feed goats. It grows throughout the greater part of India excerpt temperate and very humid regions. It grows on a wide variety of soils such as sandy, gravelly, alluvium and loam. Seeds have no dormancy and germinate readily under favorable conditions and it reproduces naturally through coppice and seeds.

Khejri ***(Prosopis cineraria)*****:** Khejri is grown in dry and arid regions on coarse sandy soils. It can survive on soils poor in organic matter and nitrogen. It is the only indigenous tree species, which can withstand well the rigorous and exacting conditions of the Rajasthan desert. Young tree coppices well, but old trees fail to coppice. Feeding of these leaves during winter when no other green fodder is generally available under rain fed areas has been found to be profitable.

Lasura ***(Cordia dichotoma)*****:** The tree of tropical and subtropical climate grows on a wide variety of soils. It coppices and pollards well. The leaves yield good fodder and lopped for this purpose.

Neem ***(Azadirachta indica)*****:** Neem leaf fodder is not browsed by cattle, but goat and camel browse it well. The leaves contain negligible quantities of phytin-P and no anhydrous oxalic acid and consequently calcium utilization from them is very good. The leaves thus provide enough nutrients and can very well serve the purpose of green fodder supplement. It grows throughout the greater part of India. It can not withstand excessive cold and does not as such grow in Himalayas above an elevation of about 1000m. it can be raised through direct sowing/planting.

Nutan ***(Dichrostachys cinerea)*****:** Dichro is a thorny shrub of arid and sub-arid region, survived well on sandy loam soils under moisture stress conditions. It is mostly propagated through suckers, its leaf fodder is considered to be of good quality for small ruminants and greedily browsed by goats.

Pipal ***(Ficus religiosa)*****:** It is a tree of subtropical humid climate prefers deep alluvial sandy loam soil with good drainage. Pipal is epiphytic when young, drought resistant and frost hardy. Its leaves are classed as good fodder. Chemical composition of leaves varies with season. Goats relish the leaves and palatability for other animals is also fairly good. The leaves can serve as maintenance ration for goats. The silage prepared from the leaves is richer in CP, EE, Ca and P than Jowar silage. The silage is palatable and fairly digestible.

Reonja ***(Acacia leucophloea)*****:** It is a thorny deciduous tree with a spready crown. The leaves and pods of this tree are toxic during June-July due to higher HCN content (5-12 mg/100g material) when the leaves are usually thin and tender. Low content (1.12 mg/100g material) of HCN was reported in matured leaves .The chemical composition of reonja leaves compared well with those of species of Acacia, Albizia, Bauhinia, and Anjan tree.

Sanjna ***(Moringa oleifera)*****:** It is a tree of subtropical climate and prefers alluvial sandy soil with sufficient moisture supply. It is susceptible to drought and reproduces through seeds and coppice. Leaves yield good quality fodder. Total digestible nutrient content is 61.49-kg/100 kg of dry material and nutritive ratio is 4.5. Tender pods are used as vegetable and may also be fed to animals.

Jayanti ***(Sesbania sesban)*****:** it is fast growing fodder tree which is fed both as a green fodder and hay but hay is preferred by the animals than green leaves. Sesbania sesban hay has a good source of protein and Ca and its feeding in combination of poor quality cellulosic material in the ratio of 1:3 can meet the maintenance requirement of adult sheep.

Shahtut ***(Morus alba)*****:** Shahtut (Mulberry) grows in areas with subtropical to mid temperate climate and prefers alluvial, deep, loamy soil with sufficient moisture supply for its best growth. It coppices and pollards very well. Its leaf fodder is of good quality and can be profitably utilized as a supplement to poor quality roughage. Leaves are highly palatable to sheep.

Shisham ***(Dalbergia sissoo)*****:** It is a tree of subtropical climate, prefers well-drained sandy loam soil with good moisture supply. It has good coppice power and produces root suckers. Leaves are classed as medium to good fodder. These are more palatable when succulent and the palatability decreases as the leaves mature. Wide seasonal variation also exists in chemical composition of leaves. Feeding of green leaves sometimes causes digestive disorders in animals.

Siris *(Albizia lebbek)*: The leaves of Albizia lebbek are a good fodder. It is a strong light demander and requires complete overhead light for its optimum growth. It grows wild in the sub Himalayan region and is a tree of mixed deciduous forest. Natural reproduction is through seed. Its leaves are rich source of protein and contain lower level of crude fiber. Supplementation of 40-45% siris leaves seems to be optimum for adult goat fed on mature dry grass.

Subabul *(Leucaena leucocephala)*: Though, it is a tree of the tropics and subtropics, it is capable of withstanding large variations in rainfall, temperature, wind and drought. The tree coppices well and with stands browsing also. Leaf fodder of Subabul is highly palatable, however, high mimosine content limits its use as a sole feed. Now a days low mimosine varieties are available. Subabul leaf forage makes more or less complete ruminant feed, pretty much comparable to Lucerne. Since its leaf meal is specially valuable because its of amino acids composition which is superior to many other plant protein sources and is almost as good as Lucerne.

Zinja *(Bauhinia racemasa)*: Zinja is a small to moderate size tree with drooping branches and crooked bole. Its natural habitat is broadly subtropical and tropical. The leaves are lopped for fodder in Rajasthan, M.P. Mahrashtra, U.P. and Orissa and considered to be good fodder. Sheep and goats may be fed Zinja as supplementary ration during lean period.

Other trees and shrubs also have good fodder quality for sheep and goats are Acacia Senegal (Kumta), Acacia suma (Kantia), Albizia amara (Kala siras), Betula alnoides (Kath bhay), Cassia siamea (Beati), Cotoneaster bacillaris (Reosh), Dalbergia latifolia (Rose wood), Dracaena angustifolia (Bakripatti), Populus euphratica (Bahan), Prunus cornuta (Jamoi), Pyrus pashia (Kaint), Rododendron arboreum (Chiu), Rhus cotinus (Tunga), Rhus pariflora (Jurma), Salix wallichiana (Bhains), Syringa emodi (Ghia), Ulmus villosa (Marinu) and Zizyphus ruqosa (Suran).

Some important tropical tree species available in different regions are given in Table 3.

Table 3: Important tropical tree species

I. Arid and Semi-arid Tropics	*Acacia albida, A. nilotica, , A. Senegal, Albizia lebbek, Inga dulcis, Leucaena leucocephala, Prosopis cineraria, Sesbania aegyptica, Hardwickia binata, Dalbergia sissoo, Melia azaderach, Azadirachta indica.*
II. Humid Tropics	*Gliricidia sepium, Leucaena leucocephala, Sesbania grandiflora, Acacia mangium, Albizia falcataria,* *A. lebbek, A. procera, Calliandra calothyrsus*
III. Highland Tropics	*Acacia farnesiana, A. mearnsii, Leucaena diversifolia, Albizia falcataria*

The Legumes

The ground level in silvopastoral system is usually the grass-legume mixture. Leguminous fodders contain higher level of protein, which is often deficient in low quality fodder. They can be incorporated in small ruminant rations to improve intake and productivity. Inclusion of leguminous fodder in complete rations of ruminants is the best strategy for maximizing animal productivity. The yield of the grass is also higher when mixed with the legumes. They not only provide nutritious forage but also improve land fertility by adding nitrogen equivalence to tune of 40-60 kg/ha.

The legumes should be selected keeping in view the following characteristics:

1. The legume species should be able to fix nitrogen in an efficient manner, so that sufficient quantity of nitrogen and protein are made available for the production of pasture and animals, respectively.
2. They should have high persistency in the pasture in spite of the competition with grasses for certain nutrients during various climatic variations.
3. The legume species should have the ability to adopt to the variable soil conditions.
4. The legumes for their suitability in silvopasture system should be as much resistance as possible from insect-pests and diseases because the monsoon season generally provides very congenial atmosphere for their growth. The growth and nutritive value of the forage is thus very much reduced, if a species is susceptible to the insect-pest and diseases to a high degree.

5. The legume should have high feeding value. This will be possible through selecting a species which has high nutritive value and palatable to animals.
6. The legumes should not have toxic principles.

Some important tropical pastures legumes suitable for different agroclimatic zones also given in Table 4.

Table 4: Important tropical pasture legume

Zones	Common name	Botanical name
	Arid	
	Shrubby stylo	*Stylosanthes scabra*
	Atylosia	*Atylosia scarabaeoides*
	Siratro	*Macroptilium atropurpureum*
	Semi arid	
	Desmodium (green leaf)	*Desmodium intortum*
	Glycine	*Glycine wightii*
	Stylo	*Stylosanthes hamata*
	Lablab beans	*Lablab purpureus*
	Sub humid and Humid	
	Desmodium (silver leaf)	*Desmodium uncinatum*
	Centro	*Centrosema pubescens*
	Stylo	*Stylosanthes guianensis*
	Glycine	*Glycine wightii*

Some of the range legume may be used for pasture component are discussed as follow.

Archer dolichos *(Macrotyloma axillares)*: It is a summer growing, trailing and twining perennial legume. Stem is cylindrical and hairy. Leaves are trifoliate. This legume is well adapted to subtropical or tropical climate. It grows on variety of soils from sands to clays with good drainage. Its dry forage production is 15.5 t/ha. It contains about 12.0 % CP in full growth which declines to 6.2 per cent at seed formation stage.

Bankulthi *(Atylosia scarabaeoides)*: It is herbaceous perennial twiner with slender stem. Leaves are tri-foliate. It is native of India and found in all parts of the country. Bankulthi grows in 250-1000 mm rainfall zones of arid and semi-arid climates. It prefers a wide range of soils ranging from light to heavy even clayey soils with adequate drainage. Green forage production ranges from 8 to 12.5 t/ha. Its C.P content ranges from 11 to 14%.

Butterfly pea (*Clitoria ternatea*): It is a perennial pasture legume which once established can resist the grazing for a long period. It is suitable for grass legume mixed pasture from soil and moisture conservation point of view and provides more nutritious fodder. It can be planted in 6-8 alternate rows with pasture grasses. Dry matter yield varies from 1.1 to 3.3 t/ha in first year under rainfed condition while under irrigated condition it yields around 13.3 t/ha. The CP content ranges from 10.5 – 15.5% on DM basis.

Calopo *(Calopogonium mucunoides)*: It is an annual creeper of 0.3-0.5m height. Leaves are trifoliate and hairy on both surfaces. It is grown preferably under hot, humid tropics with an annual rainfall in excess of 1525mm. It is moderately drought and shade tolerant. The forage yield of calopo is 56 t/ha (green) and 14.5 t/ha (dry) in 3-4 cuttings. It contains 16 % CP on dry matter basis. In new mixed pasture, the Calopo can be sown after the grasses get established.

Centro *(Centrosema pubescens)*: Centro is a perennial vine of trailing habit. In India it was introduced and well suited to west coast of the peninsula. It is found in hot humid climate. It thrives well in alluvial to medium fertile soils. Forage yield ranges from 15 to 20 t/ha (green). The CP content of the species varies from 12 to 15 % in winter and reaches up to 23% in summer.

Dalrymple vigna *(Vigna luteola)*: It is a short lived, summer growing, leafy perennial with climbing and twining habit. The optimum temperature for its growth ranges from 20 to 30^{0}C. A minimum annual rainfall of 1525 mm is required. The green forage yield varies from 12 to 18 t/ha while its dry matter production is from 1.9 to 2.5 t/ha. It contains 17.4 per cent CP.

Dolichos *(Lablab purpureus)*: It is a biannual pasture legume, which is drought resistant and provides nutritious and palatable fodder. It attains a height of 0.9 to 1.8. The stems are robust and well branched. Leaves are large trifoliate. Pods are 3 – 12 cm long, curved with 2 – 4 big seeds. In well managed pasture it produces 5.4 t/ha dry forage. It contains about 12% CP. It can be planted in alternate strips of 6-8 rows with pasture grasses.

Faba bean *(Vicia faba)*: Faba bean is, known as bakla and Anhari, a leguminous plant which mature in 90-120 days and its seed possessed fairly high content of protein (CP=23-24%) and there is good yield (3.1-4.3t/ha) on reasonably fertile soil. It possesses a high nutritive value for cattle, horse, swine and sheep. The methionine supplementation is necessary because faba beans are deficient in sulphur amino acids. It also contain some anti-nutritional factors like tannins, convicine and trypsin inhibitor which limits the availability of energy and protein.

Lotononis ***(Lotononis bainesii)*****:** It is a creeping perennial and forms a dense pasture. It is adapted to tropical and sub-tropical climate with an annual rainfall in excess of 900mm. Lotononis thrives best to well drained acid sandy soils. Its dry matter yield varies from 1.8 to 5.0 t/ha whiles under irrigation the average dry forage yield of 8.2 t/ha has been recorded. Lotononis is one of the most palatable tropical legumes and provides high yields of protein rich fodder to cattle. It contains about 18.3 % CP.

Phasey bean ***(Macroptilium lathyroides)*****:** Phasey bean or Wild pea bean is an erect annual or biennial legume and attains a height of 0.6 to 0.9 m. Leaves are trifoliate and leaflets are smooth. It is native to India but has now spread to many tropical and sub-tropical countries. It is adapted to sub-tropical areas with 25 to 30^{0}C optimum temperature and 760 to 2030 mm annual rainfall. Its green forage yield ranges from 15 to 20 t/ha. It has good nutritive value and contains 19.2 % CP.

Puero ***(Pueraria phaseoloides)*****:** Puero also called Tropical Kudzu is a deep-rooted perennial legume of twining and climbing nature. It is suited to humid zones of tropics and grows well in the areas of annual rainfall above 1500 mm. It has adaptability to a wide range of soils from sand to clay. The average dry matter yield was found to be 4.1 t/ha. In a mixed pasture of molasses, grass, puero contributed 9.1 t/ha (dry matter) of the total (22.9 t/ha). It contains 19.0 per cent CP.

Scabra stylo ***(Stylosanthes scabra)*****:** It is an erect and woody perennial. It is dark in colour and slightly sticky/ viscous in feeling. It thrives well in warm tropical climate with 325 to 1200 mm annual rainfall. It is grown on all types of soils with good drainage. It contains 9.0 to 12.0 per cent crude protein. It is used for grazing and gives high quality hay and silage.

Siratro ***(Macroptilium atropurpureum)*****:** Siratro is a very important summer grown range legume having high drought resistance and good palatability. It is a deep rooted perennial herb and has trailing habit. Stem is hairy and root readily at the nodes. Leaves are trifoliate, dark green. It thrives well on wide range of soils from light textured sandy soils to heavy clay with good drainage. The average green fodder yield ranges from 15 to 20 t/ ha and dry fodder from 3 to 5 t/ ha. The inclusion of siratro in the mixture considerably enhances the quality of herbages during the dry season. The CP content of some of the siratro genotypes is comparable to berseem and lucerne, which makes it an important legume crop of silvopastoral system.

Stylo *(Stylosanthes hamata)*: Stylo is a commonly known as schofield. It is much branched perennial with trifoliate leaves. It is remarkable in its hardiness and adaptability to various climate and soil conditions. It can be grown on poor soils. It is easily established, both under arable cultivation and in natural grasslands, where it extends grazing in dry season. The green forage production ranges from 20 to 30 t/ha while the dry forage is 10 t/ha. It contains 10-14 percent CP. Dry matter (DM) and TDN intakes are sufficient for maintenance need of the animals. Study revealed that Stylosanthes hamata straw was nutritionally viable feed resource that could cater the maintenance need of lambs and it should be supplemented with appropriate amount of concentrate or mixed with good quality green fodder to meet the production requirements.

Velvet Bean *(Stizolobium deeringianum)*: Velvet Bean or Makhmali sem is a fast growing annual climber or runner and spreads the branches upto 10m. It thrives well in warm, moist conditions in the zones of 650 – 2500mm annual rainfall. It grows in all types of soils but prefers sandy loam. The green forage yield ranges from 15 to 25 t/ha. It is very nutritious fodder and contains 15.6 per cent CP.

The Grasses

A grass in the pasture performs many functions including soil conservation and production of forage. Grasses exhibit a wider range of adaptability; especially in humid to arid areas and alpine peaks. They have the advantage of reproducing fresh shoots through tillers, recovering from grazing or cutting. The root system bind the soil particles together the forming a sod. Thus surface layer nutrients are utilized that otherwise leach into subsoil through rainfall. Many species having the capacity to spread by rhizomes or stolons, provide rapid ground coverage. These grasses in a suitable mixture with legumes provide much balanced diet to the grazing animals. These are perennial in most of the conditions and dominate the pasture under continuous grazing or unmanaged conditions. Seed production is high and seeds are versatile. The seed weight much less than that of legume, so that generally confuse while calculating the seed rate for grass and legume mixture. The tolerance of most grass species against natural calamities makes them a key factor while establishing the pasture. The yield, palatability and the nutritive value should be considered while selecting a grass species for the pasture in addition to the adaptability in the different ecological conditions. A few of the important grass species suitable for different ecological conditions are mentioned below:

Anjan grass *(Cenchrus ciliaris)*: Cenchrus ciliaris (Buffel grass) commonly known, as Anjan is one of the dominant pasture species of arid and semi-arid zones of India. It grows on sandy silty or stony soils. It is drought resistant, hardy, perennial, nutritious and palatable to all kinds of livestock. The dry matter yield ranged from 6.0 – 11.4 t/ha depending on the rainfall. Cenchrus based diets can be fed to sheep and goats for maintenance. The grass is relished by all classes of livestock. It contains as high as about 11.0 % CP at young stage.

Bahia grass *(Paspalum notatum)*: It is a perennial, densely tufted grass. It prefers warm and moist climate. It grows well on poorly drained and swampy soils. At early stage it gives 11-12% CP. It is used for grazing as well as stall-feeding.

Blue Panic *(Panicum antidotale)*: It is locally called Ghamur, Gunara, Bansi or Parwari. Blue panic is an erect, thin stemmed, tufted perennial grass with solid culms, thickened nodes and blue green coloured internodes. Besides being an excellent sand binder, it is drought resistant, tolerates moderate salinity and known for its fodder value in dry regions. The grass is also famous for its medicinal antidote. Its production potential is 150-250 q/ha of green fodder in 4-6 cuttings per year. The grass contains 7.5 per cent crude protein and a satisfactory balance of calcium and phosphorus.

Congo signal grass *(Brachiaria brizantha)*: It is an exotic grass. In India it is found in coastal regions. It requires warm and humid climate. This grass thrives well on loamy soils. It responds well to fertilizer application. It contains 9.6 and 8.1 % crude protein in July and October, respectively. The grass remains green for major part of the year even under the rainfed conditions. It is much relished and is quite palatable.

Dallis grass *(Paspalum dilatatum)*: It is tall, tufted, tussocky, leafy at base, perennial grass and attains a height of 1to 1.5m. It is a grass of semi-arid climate. It is drought resistant grass. It grows well in rich moist soils of low-lying areas. The green forage yield varies from 50 to 60 t/ha. It is a good fodder grass, contains 6.7% CP and utilized for grazing and hay purpose.

Dhaman *(Cenchrus setigerus)*: Dhaman is another cenchrus species dominant in semi-arid and arid areas. It grows well under extreme climatic conditions. It grows well on well drained sandy-loam soils. Under rainfed conditions dry matter yield varied from 3.9 – 7.9 t/ha. It is hardier than anjan and also drought resistant. Palatability and nutrition is as good as anjan grass and can be grazed heavily.

Dhwalu *(Chrysopogon fulvus)*: It is highly variable, tufted, perennial and 1.8 m tall grass. It is found in arid to semi-arid regions with 250 to 850 mm rainfall. Dry forage yield is 4.2 t/ha. Fertilized pasture produces about 10.0 t/ha forage dry matter. It contains about 4.6 to 5.1% CP during the growth period

(Aug-Sept.). If it is cut before flowering the grass makes good green fodder or may be grazed directly from 2nd year onwards up to 6th year.

Doob *(Cynodon dactylon)*: It is a profusely much branched, leafy perennial stoloniferous with a large number of runners. Doob is native to India. It grows well in semi-arid conditions between 300-200mm rainfalls. This grass thrives best on heavier silt and clayey soils. The average green forage yield ranges from 15 to 16 t/ha (4-5 t/ha dry matter) under rain fed conditions. It contains about 11.0% CP in young stage which reduced to 7.0% at maturity.

Fulkara *(Bothriochola intermedia)*: It is a grass of semi-arid regions and prefers 900 mm average annual rainfall. It grows mainly on heavy clay loam to clay loam soils and heavier alluvial soils. In normal conditions the grass produces 2.4 t/ha dry matter and with the application of fertilizer the yield is 7.8 t/ha. It contains 6.0% crude protein at flowering stage. In India it is found widely distributed in U.P. and other northern and southern states. It prefer drier habitat in arid and semi arid climatic conditions. It grows well on coarse sand to fine texture sandy loam to loamy soils. Under rainfed and unmanaged condition it provides good forage in two cuts at 60 days intervals i.e. end of July and September.

Johnson grass *(Sorghum halepense)*: Johnson grass (Sorghum halepense) also known as Barru grass or Banchari is a tall, coarse, warm season perennial grass forming about 20% of the grass cover of India. This grass at the preflowering stage can be solely used as a maintenance ration for sheep.

Karnal grass *(Diplachne fusca)*: It is a perennial, tuft and erect grass growing 0.9 to 1.5 M tall. It is distributed in upper Gangetic plains, Bengal, Orissa, A.P., Tamilnadu and Kerala along coasts. It gives good growth in alkali and saline soils. In highly alkali soils this grass gives 30 to 40 tones green forage in one hectare without fertilizer but with the application of recommended doses of fertilizer the green forage yield reaches up to 60 t/ha. It contains 8.5% CP. Animals relish this grass in young stage.

Kikuyu grass *(Pennisetum clandenstinum)*: It is a perennial grass forming dense mats on ground. It prefers cool and moist climate. It grows well in rich loamy and acidic soils. The average total green forage yield is 35-40 t/ha. At early stage i.e. 30 days after transplanting it is relished by all kind of animals. The average CP of the crop is over 12%. Being highly nutritive and palatable it is a very good fodder for dairy cattle. They relish it well due to its high leaf to stem ratio and succulent nature. It can stand close grazing.

Lampa *(Heteropogon contortus)*: It is densely tufted, perennial and highly palatable, 0.9 to 1.0 m tall, erect or decumbent grass. It grows well in arid and semi-arid conditions in the rainfall zones ranging from 180 to 1200 mm. It is adaptable to all types of soils. At early stage it is a highly nutritive grass and even 10 % CP is found in July which decreases regularly and only 3 % is found in December. It is very good grass but due to presence of sharp awns at maturity may be cut at pre-flowering stage.

Machhauri ***(Iseilema laxum)*****:**It is short erect to decumbent perennial grass. It is found in the semi-arid regions of 450-1500 mm rainfall zones. It thrives best in black clayey soils and heavy loams. Under natural conditions the dry matter forage yield varied from 3 to 4 t/ha. The grass is most nutritious and contains over 9.6% CP at preflowering stage (July) which comes down to 5.2% in September. The grass being nutritious is highly acceptable and palatable at all stages of its growth for grazing animals.

Marvel grass ***(Dichanthium annulatum)*****:** It is an erect tufted, fine stemmed, perennial grass and 1.2 m tall. It grows well in arid and semi-arid regions. It can grow on well drained, medium black or red alluvial soils. Under rain fed condition its dry matter yield ranges from 2.8 to 4.5 t/ha in single cut. It can give 3 times more forage under irrigation. Its CP content varies from 5 to 7 %.

Molasses grass ***(Melinis minutiflora)*****:** It is tufted perennial grass of about 1.5 to 1.8 m high. This grass grows well in moist tropical climate. It is found on light sandy to loamy soils. The average dry matter yield of molasses grass is 4.8 t/ha. At the initiation of flowers it contains 8.9% CP. Being nutritive and palatable it is used for hay and silage purposes. It tolerates high grazing but heavy grazing thins it out.

Nandi ***(Setaria sphacelata)*****:** It is an erect, bunchy perennial grass. This is a grass of warm climate. It thrives well on fertile loamy soils. The pasture of Nandi grass produced 22.8 to 23.8 t/ha green forage under rainfed condition. The grass is very leafy and quite palatable and highly nutritive. It contains 5.3% CP. The grass can be used for soiling, hay or grazing particularly due to its good winter growth ensures round the year forage supply.

Pangola grass ***(Digitaria decumbens)*****:** Pangola grass is of creeping habit. It is adaptable to wide range of soils from extremely poor sands to heavy clays of low fertility. Rotational grazing with one-week rest during growth period and 2 to 3 weeks rest in summer period is recommended for higher quality forage. Its analysis at floral stage revealed 11.8% CP. At young and vigorous stage it is most palatable and is used for hay, silage and grazing purposes.

Para grass ***(Brachiaria mutica)*****:** It is also called buffalo grass. It is an exotic, hairy and perennial grass. It grows well in warm humid situations climate of high rainfall areas. The grass contains 7 % crude protein. It can withstand moderate grazing.

Rhodes grass *(Chloris gayana)*: It is fine stemmed, leafy, perennial, erect grass. It grows well in warm-moist conditions. It prefers loamy to sandy loams. At preflowering stage CP is nearly 5%. Even after severe trampling it provides grazing up to September and could be utilized for hay thereafter.

Sain *(Sehima nervosum)*: It is a perennial grass forming dense tufts with numerous tillers. It prefers hot and dry climate. Eroded, red gravelly/stony to medium sandy loamy soils are common of its availability. Maximum forage (green) obtained from an unfertilized pasture is 16 t/ha if harvested in September. Thereafter, the dry forage yield as well as nutritive value declines due to leaf fall. It contains maximum CP (6.9%) in July and minimum (2.3%) in December. It is a good forage grass and may be utilized for grazing.

Sewan *(Lasiurus sindicus)*: In dry areas among fodders sewan is called a 'king' of desert. It gives good production under climatic severity and soil stress. It can sustain in pasture for a long period of 10-15 years. Average dry matter yield of this grass is 2.5 – 3.5 t/ha in rainfed situation but improved strains under well management contributes yields about 7 – 8.5 t/ha. At flowering stage it contains 12.5 per cent CP. Its palatability is also high for most of the livestock. It is the only grass that can survive better under desert condition.

Sola grass *(Aexhynomene indica)*: Locally known as sola grass, it is an erect suffruticose deciduous membranous herb or under shrub. It is widely distributed in Chhatishgarh plain and fed to goats. The peduncle and pedicels of plants are usually vision and rich in protein. Goats can be reared under stall fed conditions on sole feeding of sola grass without any deleterious effect on the health of the animals. Its DCP and TDN content are comparable to some leguminous green fodder with high Ca (1.5%) and P (0.16%) levels.

In continuation some of the important tropical grasses are also presented in Table 5.

Table 5: Important tropical grasses

Zones	Common name	Botanical name
	Arid	
	Sewan grass	*Lasiurus sindicus*
	Anjan grass	*Cenchrus ciliaris*
	Dhaman grass	*C. setigerus*
	Blue panic	*Panicum antidotale*

Zones	Common name	Botanical name
	Semi arid	
	Anjan grass	*C. ciliaris*
	Blue anjan grass	*C. glaucus*
	Sain grass	*Sehima nervosum*
	Marvel grass	*Dichanthium annulatum*
	Dhwalu grass	*Chrysopogon fulvus*
	Machhauri grass	*Iseilema laxum*
	Sub-humid and Humid	
	Guinea grass	*Panicum maximum*
	Nandi grass	*Setaria sphacelata*
	Signal grass	*Brachiaria brizantha*
	Para grass	*B. mutica*
	Rhodes grass	*Chloris gayana*
	Gautemala grass	*Tripsacum dactyloides*

Bibliography

Deb Roy, R., Patil, B.D. and Gupta, S.K. (1979). Silvicultural studies on fuel cum fodder trees. Annual Report, IGFRI, Jhansi.

Gupta, S.K., Pathak, P.S. and Deb Roy, R. (1983). Biomass production of Sesbania sesban in different habitats. My Forest, 19: 145-155.

Mandal, L. (1997). Nutritive values of tree leaves of some tropical species for goats. Small Ruminant Research, 24: 95-105.

Pathak, P.S. and Patil, B.D. (1980). Fuel wood and forage production from Leucaena leucocephala. Leucaena Research Report, 1: 11.

Pathak, P.S. and Roy, M.M. (1992). Fodder trees in agroforestry: their selection and management. Range Management & Agroforestry, 13: 63-87.

Pathak, P.S. and Roy, M.M. (1994). Silvipastoral system of production. A Research Blletin, IGFRI, Jhansi.

Pathak, P.S., Pateria, H.M., Solanki, K.R. (2000). Agroforestry systems in India: A diagnosis & design approach. ICAR, New Delhi.

Rai, M.P., Pathak, P.S. and Deb Roy, R. (1983). Biomass production of Sesbania sesban in different habitats. Intl. J. Ecol. Environ. Sci., 9: 21-27.

Roy, M.M., Pathak, P.S. and Deb Roy, R. (1984). Dichrostachys cinerea (L.) wight & Arn with special reference to its potential on wastelands. My Forest, 20: 213-224.

Roy, M.M. (1991). Some tropical fodder trees for sustained fodder and firewood availability during lean period. J. Trop. For., 7: 196-205.

Sharma, N.K., Singh, R.P., Singh Manjit and Harsh, L.N. (2001). Silvopasture management in hot arid and semi-arid ecosystem. Agrotech Publishing Academy, Udaipur, India.

Singh, Panjab and Pathak, P.S. (1990). Establishment and growth of MPTS species in natural Sehima-Dichanthium grass cover of Bundelkhand region. In: MPTS for Agroforestry Systems (P.S.Pathak, R. Deb Roy and P. Singh, eds.), RMSI, Jhansi, pp. 127-132.

Thakur, S.S. and Kundu, S.S. (1987). Tree leaves as livestock fodder. Indian Dairyman, 39 (8):369-371.

Trivedi, B.K. (2002). Grasses and legumes for tropical pastures. Technical Bulletin, IGFRI, Jhansi.

AGRO –INDUSTRIAL BYPRODUCTS

India has a very large livestock population with more than 204.58 million cattle and 84.20 million buffaloes. The annual requirements of feeds and fodder estimated to be, 25.4 million tonnes of concentrates, 353.0 million tones of dry fodder and 308.1 million tonnes of green fodder. However, only 16.5 million tonnes of concentrate, 300.5 million tonnes of dry fodder and 261.0 million tonnes of green fodder are available. The gap between the availability and requirement of concentrates and green fodder is wide and with there is a shortage of about 44% of concentrate and 34% of green fodder. This shortage of feed and fodder is a chronic problem, which will likely to continue in years to come. Crop residues (mainly straws and stovers) are likely to continue to make significant proportion of bovine diets. These residues usually have energy-protein depression effect due to low digestibility besides being deficient in many essential minerals and vitamins and may lead to productive and reproductive disorders. In order to narrow the gap between demand and availability of feed resources, unconventional feeds and fodder are being widely used to meet the requirements of our large animal population. If all crop residues and other non-conventional materials are properly put into the use as animal feeds, the requirement of the animals can probably be largely meet. It would therefore be appropriate to give more emphasis on the use of agro-industrial byproducts that are available in huge tonnage in our country. This chapter highlights the use of various agro-industrial byproducts as feed substitute in the diets of animals.

Byproducts of sugar industry: Sugarcane production in the country is of the order of 300 MT, and by-product availability is also huge.

Sugarcane molasses: molasses is one of the main byproducts of sugar industry that has long been investigated as a source of energy in animals feed. The production of molasses is about 9 MT or1/3rd of the quantity of sugar produced. Molasses when added in the diet of the animals serves the following purposes:

a) Improves the palatability of the diet as it acts as sweetener for increasing the voluntary intake of unpalatable feeds.
b) As a binder for making pellets and block of diet.
c) As a medium for adding certain nutrients such as minerals to diets.
d) To reduce dustiness of feed material.

Molasses contains about 61% sugar and 3% protein and has been used in animal feeding at 5-10 per cent.

Sugarcane Bagasse: Bagasse is considered as one of the main byproducts of sugar industry. The quantity of bagasse produced is equal to 44-45% of cane crushed. Chemical analysis of bagasse shows that it has a high fibre content and low protein percentage. Thus, it is a source of cellulose that is poor in protein and rich in lignin. As such it is not very much relished by animals, therefore, its use as animal feed requires certain treatments. Cattle and swamp buffaloes can consume about 80% of their expected dry matter intake through bagasse impregnated with 1 to 2 % urea. Buffaloes appeared to have significantly higher apparent digestibility of crude protein of rations consisting of 60% bagasse and 1% urea, than cattle. In India, bagasse has been used as famine roughage for ruminants. About 60 kg rations consisting of 20 kg bagasse, 6 kg molasses, 0.34 kg common salt, 0.34 kg mineral mixture and 65 liters of water forms a maintenance ration along with 800 g of groundnut cake for adult animals weighing 400 kg. The optimum steam pressure for treatment of bagasse is 7 kg per cm^3 for 30 minutes and fortification with 1 percent urea and molasses. The steam treated bagasse +urea + molasses cost compares favourably with conventional roughage. Sodium chlorite treatment of bagasse has decreased the lignin content and increased *in situ* dry matter disappearance. It has been reported that bagasse could be used for fattening at the rate of 30% of the total diet offered.

Tea industry wastes: India produces nearly 400 million kg of tea every year and main tea growing states are West Bengal and Assam. Tea wastes in the form of fluffs, stalks and sweepings become available during production, storage, handling of tea in garden, factories, warehouses, packing factories and shipment shed at ports. Tea wastes DCP and TDN value are 9.0 and 43.0% respectively. It contains about 1.9% tannic acid. The material can be used up to 20% level in the concentrate mixture of crossbred calves to attain a weight gain of 350 g per day. Caffeine can be isolated from waste tea leaves. The decaffeinated tea waste is also available during tea processing. The material is not very palatable and should be mixed with more palatable feed ingredients. It is a moderate source of protein and energy containing 7.5% DCP and 58% TDN. This material can be fed to pigs at 15% level in the concentrate mixtures to give daily weight gain of 415g. Spent tealeaves (STL) containing 25% CP, can be used upto 20% in the concentrate mixture.

Apple wastes *(Malus pumila)*: Apple is a premier table fruit and is cultivated in Jammu and Kashmir, Himmachal Pardesh and Uttranchal. The damaged apple, broken or injured during plucking and not suitable for packing, are available in plenty during the apple season, and after slicing, drying and grinding these can be incorporated as energy source upto 30% level in the concentrate mixture of crossbred calves replacing 100% maize and supporting daily weight gain upto 427 g. This can also be incorporated as energy source at 12.5 % in broiler mash replacing maize. Apple pomace is a main byproduct obtained after juice extraction. Apple pomace and damaged apples are high in energy (TDN 62-65%) but low in protein (DCP 1.6-2.6 %). This material can be preserved as silage which can be safely fed to buffaloes and cattle from 15 to 30 % of their total DM intake. Urea treated apple pomace replaced 40 % maize grain without affecting dry matter intake, feed efficiency, nutrients digestibility and milk production.

Salseed meal ***(Shorea robusta):*** Potential yield of salseeds in our country is about 6 million tonnes annually, but about 70,000 tonnes of the seeds are actually collected. The seeds contain about 10-15 percent oil. The availability of cakes is about 50-60 thousand tonnes annually. Thus, a large quantity of deoiled salseed meal is available for livestock feeding. Salseed meal contains about 8-10 % CP but the digestibility of protein is almost negligible due to the presence of tannins (8-13 %). The TDN value is about 45%, hence it is only a source of energy. Salseed meal significantly, decreased CP digestibility due to the presence of tannins, however ammoniated salseed meal has improved protein digestibility. It such as is not palatable but when added to a level of 20-30%, the feed intake was not significantly affected. The general recommendations are to use deoiled salseed meal to a maximum level of 10% in the rations of growing, dry and lactating cattle and buffaloes. About 30% of the tannins of deoiled salseed meal can be made inactive by treating meal with NH_3 and this appears to be the practicable way of using salseed meal.

Tapioca/cassava starch waste ***(Manihot esculenta):*** The availability of tapioca starch waste is estimated at 40,000 tonnes. Most of it is used as animal feed. Cassava feeds and roots are poor in CP (2 %) but rich in soluble carbohydrates (about 80 %) where as tapioca starch waste has 2% DCP and 64% TDN. This material can be included upto 25% level in the concentrate mixture of crossbreed calves for sustaining about 370 g /day body weight gain. It does not contain any toxic factor.

Mango seed kernel ***(Mangifera indica)*****:** It is rich in the starch content i.e. 92% of its total carbohydrates. It has about 68-70 % neutral lipids of the total lipids. Stearic and oleic acids are the main components and saturated and unsaturated fatty acids are 40 and 60 %, respectively. Histidine and cystine are in greater amount as compared to other amino acids. The kernels can be incorporated upto 10% level in the concentrate mixture of milch cattle producing 8 kg milk daily. These are very good source of energy (TDN 70%) and moderate source of protein (DCP 6.0 %). The kernels are rich in the tannins (5-6 %). They are recommended for use in the rations of ruminants only and at a level not exceeding 10 % and that also with other feed ingredients which do not contain high levels of tannins.

Tomato pomace: Tomato pomace consists of skin, pulp and seeds and accounts for about 20 % of the total production. Dried and ground pomace contains 22% CP and 28 % CF with a nutritive value of 14 to 16% DCP and 61 to 73 % TDN. Tomato waste can be incorporated up to 35-40 % of the concentrate mixture in the diet of cattle and buffaloes.

Potato waste: Potato waste is the byproduct remaining after extraction of starch with cold water. About 15-20 % of starch is produced and 3-3.5 % dried potato waste is obtained for feeding to cattle and buffaloes.

Castor bean meal *(Ricinus communis)*

The castor production, during the year 1990, was 5,00,000 MT. The castor bean meal (CBM) is one among inedible protein rich byproduct of oil industry which contains 30-40% CP but it has toxic principles viz. Ricin, ricinine, allergen and chlorogenic acid. Incorporation of processed CBM (after suitable detoxification) as protein source in the diet of ruminants had no ill effect. The CBM contains 20% DCP and 57% TDN.

Karanj cake *(Ponamia glabra)*

More than 111,000 tonnes of karanj seeds are available in India under wide range of agro climatic conditions. The seed contains on an average 28% oil. The cake left after oil extraction was found to be unpalatable and toxic even at lower level (4%) of dietary incorporation. The toxic principles karanjine, pongaglabrone and pongapin are associated with oil fraction. The deoiled (solvent extracted) karanj cake is a rich source of protein (33% CP, 25% DCP), low fibre (5.4%) and sufficient TDN (62-71%). This cake can be incorporated in the concentrate mixture of cattle and buffaloes.

Kosum cake (Schleichara oleosa)

Kosum is available widely in Madhya Pradesh, where more than 100,000 tones of cake is available (Singh, 1987). It contains cyanogenic glucosides. Autoclaving, boiling, deoiling or even storing reduces cyanide content by 65%. Its DCP, and TDN values are 16 and 78%, respectively.

Tobacco seed cake *(Nicotiana tobaccum)*

After USA and China, India (Orissa and Gujarat) ranks third in world tobacco production and produces about 9,000 tonnes of tobacco seed cake. It contains about 26% CP and 20% CF. It can partly replace conventional concentrate in the ration of cattle and buffaloes, if replaced gradually.

Citrus wastes

Large ruminants can be fed citrus waste upto 40 kg/head/day without any harmful effects. After pressing oranges, lemons or grape fruit for juice, the left over which comprises the pod seeds and rag etc amounts to 40-60% by weight of the fruit and could be very palatable to buffaloes. Citrus pulp contains a relatively high amount of pectin and soluble carbohydrates, and for this reason dried citrus pulp has been used to replace cereals in ruminants diet.

Coconut meal (*Cocos mucifera*)

Coconut is cultivated mainly in Karnataka, Kerala, Tamil Nadu and Andhra Pradesh. Coconut availability in Asia is about 3.75 MT and 35-40% left as meal. Its DCP content is 21% and ME/Kg is 1.9 Mcal. The storing quality is poor as rancidity develops due to its oil content (3.8%) which may lead to diarrhoea in animals.

Coffee waste (*Coffea arabica* and *C. robusta*)

Coffee is cultivated commercially in Karataka, Tamil Nadu, Kerala and Andhra Pradesh. In these states, coffee is available about 70,000 tonnes, which yield about 45,000 tonnes of coffee husk. It has about 9% CP, 40% CF, 1.0% EE and 5% ash. It has 3.4% DCP and 42% TDN, and can be incorporated in buffaloes ration.

Kapok seed meal (*Ceiba pentandra*)

Kapok or silk cotton seed meal is grown in Madhya Pradesh and Tamil Nadu. It has 20% CP and 1.3 Mcal ME/kg. It has cyclopropenoid acid (18%) and tannins, even then during scarcity it is incorporated in the diet of ruminants.

Rubber seed cake (*Hevea brasiliensis*)

In India, rubber trees are grown over an area of 224,428 hectares of which 202,320 hectares are in Kerala with an estimate potential of 150,000 tonnes of rubber seed cake. The DCP is about 18% and TDN is 54%. Rubber seed protein is poor in methionine but rich in lysine and cystine. Feeding of this cake upto 30% level in the concentrate ration of lactating cows showed no significant effect on milk yield, butter fat, melting point, iodine and saponification values.

Neem seed cake (*Azadirachta indica*)

Neem is a large evergreen tree, which is native of India. Neem seed cake is produced from seeds after extraction of oil. Neem seed kernel cake has 35-40% CP, but found unsuitable for animal feeding due to presence of bitter and toxic triterpenoids (azadirachtin, salanin, nimbin, nimbidiol etc). Neem seed cake besides being unpalatable is also adversely affecting the growth, male reproductive system and even sometimes leading to haematuria. The cake in alkaline medium without water washing was tried either by soaking in water (1:5 w/v) containing either NaOH for 24 h or by ensiling with 2.5% urea (w/w) for 5-6 days. The sun dried and ground alkali treated and urea ammoniated neem seed kernel cake was found suitable in the feeding of cattle and buffalo calves without affecting their growth, nutrient utilization, blood profile, rumen fermentation pattern indicating that this cake after treatment/ammoniation can be substituted as wholesome protein in the buffalo feeds.

Mahua seed cake *(Madhuca indica)*

Mahua is widely distributed in the deciduous forests of India, producing 0.12 million tonnes of cake from 0.21 million tonnes of seeds collected annually. Mahua seed cake is a moderate source of protein (15-24% CP) and contains reasonably good amount of NFE (53 – 60%). It contains saponin (mowrin) varying from 4 – 20% and hydrolysable tannins upto 6% which renders the cake unpalatable and toxic. Ensiling the cake with fertilizer grade urea @ 6.7% (w/w) for 30 days converts the cake into a palatable product and improvs feed intake and nutrient utilization in large ruminants. Groundnut cake N of concentrate mixture at 25, 50% replaced with processed mahua seed cake and fed to buffaloes milk yield and milk composition remain normal.

Pineapple wastes (*Ananas camosus*)

From pineapple cultivation and canning of its fruits, large amount of waste materials of nutritional value are available for livestock feeding. Pineapple wastes are of two types, viz cannery wastes and plant residues. Pineapple cannery waste consists of outer peels (shell), the crown and bud ends of the fruit, fruit trimmings, the inner core and the pomace of the fruit from which the juice has been extracted. The pineapple waste contains 80-95% moisture and 5% CP. Buffaloes can consume about 2 % of dry matter through this waste.

Jackfruit and other fruit wastes

The waste from ripen fruits is more palatable than the waste from raw fruits. It contains 13 % CP, 5 % EE, 12 %CF, 0.2 % Ca and 0.35% P. It is a rich source of energy having 60% NFE. It can be incorporated in buffaloes ration up to 10-15%. Banana peels, orange peels, mango peels, lemon peels have 6-13% CP and 55-74% NFE and can be included in the ration of buffaloes. Cauliflower leaves, papaya leaves, cabbage leaves and pea pods have 12-3 % CP, 38-55% NFE, 1.0-4.50% Ca, 0.05-0.60% P and can be included in the ration of ruminants.

Panewar seed (*Cassia tora*)

The seeds after grinding can be incorporated at 15% level in the concentrate mixture of lactating cows (daily milk yield 6kg). The seeds contain 16% DCP and 66% TDN, thus these are good source of energy and protein. The seeds contain 0.083% crysophanic acid. Most of the toxic factor is removed during fresh water treatment. The tannic acid content is very low, i.e. 0.8%.

Cocoa pods (*Theobroma cacao*)

Cocoa pods contain 6% DCP and 64% TDN. The dried and ground pods can be included in the concentrate mixture of growing calves at 20% level (daily gain 522 g) without any deleterious effect.

Warai bran (*Panicum miliaceum*): Warai bran contains 4.5% DCP and 57% TDN. It can be incorporated upto 30% level in the concentrate mixtures of crossbred cows (daily gain 418 g) and lactating cows (daily milk yield 12.9 Kg). The bran can replace wheat bran/rice bran by suitably adjusting the protein contents of the ration. It has not been reported to contain any toxic effect and is quite palatable.

Rice bran: Rice bran consists of the pericarp of bran layer, germs of the rice and some grain fragments remained during milling rice for human food, and it should contain only such quantities of hull fragment as are unavailable in the regular milling process. From this type of rice bran, oil is removed and a product of better keeping quality, known as solvent extracted or deoiled rice bran, is obtained. Rice bran is fairly palatable to animal. It should have crude fiber between 8-14% and acid insoluble ash between 4-8.5%. Deoiled rice bran forms an indispensable item for feed manufactures and its judicious use can save upto 10% of feed cost.

Ambadi cake (*Hibiscus cannabinus*): It is a good source of energy and protein (TDN 64%, DCP 18%). The cake does not contain any toxic factor and is palatable to the cattle. It can be incorporated at 20% level in the concentrate mixture of crossbred calves (daily gain 418 g) and lactating cow yielding daily 13 kg milk.

Poultry droppings: The poultry excreta is a rich source of nitrogen and other digestible nutrients. Its NPN fraction consists of purines, uric acid and allantoin. The rumen microorganisms use these NPN fractions for the synthesis of microbial protein available to the host animal. Dried poultry droppings from caged birds, either untreated (15% level) or pleurotus treated (20% level) can be used in broiler rations replacing 7.5 and 10% of groundnut cake on nitrogen basis, respectively.

Fish wastes: In the sea food canning industry, prawns and shrimps are utilized for manufacturing frozen headless fish, peeled and deveined fish, cooked/frozen and semidried fish, dry fish pulp, etc. In processing, large quantities of shells are left out as wastes. A scientifically processed shrimp-meal has an average of 47% protein and 27% ash. A process of fish ensiling using surplus fish as animal feed was developed at the CIFT, Cochin. Fish meal is one of the most valuable byproducts of the fish industry. Fish meal contains about 60% protein and very high acid-insoluble ash.

Frog meal: Frog meal is the leftover of the frog leg industry. It can replace fish meal twice by weight in poultry rations for the growth and egg production. It contains about 60% protein.

Prawn wastes: Shrimp shell powder, a waste product of the shrimp processing industry, consists largely of discarded portions of shrimp like heads and crustacean stills. These wastes can be incorporated up to 15% in the broiler rations when fish meal is not available.

Bijada cake (*Citrullus vulgaris*): Bijada cake contains 20% CP and only 38% TDN. Being poor in TDN, it can replace only partly the protein rich feeds in concentrate mixture of the livestock and that too after increasing the percentage of the energy rich feeds like maize, barley and gram.

Meat meal: Meat meal is prepared by sterilizing and rapidly reducing the moisture content of the meat of fallen animals in wet-rendering plants. Meat meal has a high nutritional value. A good meat meal contains 60 to 70 % CP. It is a good source of essential amino acids, vitamins of B group. It is generally incorporated at 8 to 10% level in the rations of poultry and swine to make up the deficiency of essential amino acids.

Blood meal: This is a major slaughterhouse byproduct. It contains over 80% of protein. It is a good quality protein source, but is poor in Ca and P. It is unpalatable to the animals. It can be fed to poultry upto 10%. Blood can be soaked on wheat bran which could form an ingredient in the poultry ration. Supplementation of the basal cereal free poultry ration with

3.6 and 9% of blood meal gave a similar performance as the control ration containing maize, groundnut cake, and 10% fish meal.

Feather meal: Feathers are available in large quantity in the poultry-processing plants. They can be effectively utilized in the preparations of feather meal. If properly processed, feather meal can replace a part of the ordinary protein supplement in the poultry rations. Satisfactory results have been reported with 25% incorporation in the broiler mashes.

Thus, agro-industrial byproducts have great potential in animal feeding. Therefore, there is need to explore the possibilities of more byproducts that can be incorporated in the ration of animals. If all the agro-industrial byproducts are used judiciously, the gap between availability and demand of concentrate feed could be narrowed and there is an urgent need to explore new feed resources in view of the present status of feed and fodder availability.

Table: Availability, chemical composition and nutritive value of agro-industrial byproducts as livestock feed

Name	Availability (lakhmetric tonnes)	Chemical composition (% DM basis)				Nutritive value (% DM basis)		Level of feeding	Toxic/ Incriminating factor (s)
		CP.	Ash	Ca	P	DCP	TDN		
1.	2.	3.	4.	5.	6.	7.	8.	9.	10.
Ambadi-cake (*Hibiscus cannabinus*)	0.30	23.4	10.4	-	-	18.7	63.8	20% level in CM of crossbred calves	Nil
Babul seeds (*Acacia nilotica*)	0.60	17.6	4.4	0.9	0.3	13.8	59.0	30% level in CM of crossbred calves	Tannins (5%)
Castor-bean-meal	0.60	34.8	10.4	-	-	-	-	Detoxified-meal at 10% in CM of sheep	Ricin (0.22%)
Cocoa pods (*Theobroma cacao*)	0.30	7.8	7.2	-	-	6.6	63.5	25% in the CM of growing calves	Nil
Coconut pith (coir waste) (*Cocos nucifera*)	2.00	1.8	5.4	0.3	0.1	0.0	62.7	20% level in CM of crossbred calves	Lignin (35-40%)
Damaged apple waste (*Malus sylvestris*)	0.20	12.0	3.2	-	-	2.0	70.0	30% level in CM of crossbred calves and 12.5% level in broiler mash.	Nil
Decaffeilnated tea waste (*Camellia assamica*)	0.05	17.9	9.1	5.6	0.8	7.5	58.0	15% in CM of pigs 10% in broiler feed.	Tannins
Karanj-cake (*Pongamia glabra*) (Solvent extract)	1.3	34.1	3.3	-	-	25.5	62.0	15% level in CM of crossbred calves and lactating cows	Karanjine and some unidentified factors (10-15 mg/ 100g cake)
Kokam cake (*Garcinia indica*)	0.15	16.6	7.4	-	-	9.3	80.0	15% level in CM of crossbred calves	Nil

Name	Availability (lakhmetric tonnes)	Chemical composition (% DM basis)				Nutritive value (% DM basis)		Level of feeding	Toxic/ Incriminating factor (s)
		CP.	Ash	Ca	P	DCP	TDN		
1.	2.	3.	4.	5.	6.	7.	8.	9.	10.
Kosum cake (*Schleichera oleosa*) (Solvent extracted)	0.30	22.11	7.80	0.49	1.09	14.73	79.62	35% level in CM of crossbred calves and 17 % level in broiler mash.	HCN (2.4mg/100g)
Mango seed kernels (*Mangifera indica*)	10.0	8.7	3.6	0.3	0.3	6.1	70.0	10% in CM of milch cattle	Tannins (5-6%)
Mahua seed-cake (*Madhuca indica*)	3.0	20.4	6.5	0.4	0.2	9.3	49.8	20% level in CM of crossbred calves	Morwin (19%)
Niger seed-cake (*Guizotia abyssinica*)	1.0	34.2	9.5	-	-	32.7	49.4	57% level in CM of crossbred calves	Nil
Panewar seeds (*Cassia tora*)	0.30	18.6	9.4	0.9	0.6	15.9	66.0	15% in CM of milch cattle	Crysophanic acid (0.083%)
Rice straw (*Oryza sativa*)	2000	4.0	11.0	-	-	0.8	41.5	Major roughage	Lignin oxalate, selenium
Rubber seed-cake (*Hevea brasiliensis*)	1.5	35.1	10.5	1.0	0.8	18.6	66.0	25-30% level in CM of crossbred calves and milch cattle	HCN (9mg/100g)
Sal seed cake (*Shorea robusta*)	0.70	8.6	3.1	0.2	0.2	0.1	57.8	10% in CM of milch cattle	Tannic acid (8-10%)
Spent annatto seed (*Bixa orellana*)	0.30	11.8	5.1	-	-	7.9	67.2	20% level in CM of crossbred calves	Tannic acid (10%)
Spent Brewer s grains	0.50	28.9	25.4	5.7	3.4	-	-	Air-dried material at 50% in CM of buffalo calves & lactating buffalo	-

Name	Availability (lakhmetric tonnes)	Chemical composition (% DM basis)				Nutritive value (% DM basis)		Level of feeding	Toxic/ Incriminating factor (s)
		CP.	Ash	Ca	P	DCP	TDN		
1.	2.	3.	4.	5.	6.	7.	8.	9.	10.
Sugarcane bagasse i) Untreated ii) Steam treated	 200 -	 2.9 3.2	 4.9 4.4	 - -	 - -	 - -	 - -	 Low intake Intake & dugesti-bility improves	 Lignin (8-10%) Liginin inter-ference decreased
Sunflower straw (*Helianthus annuus*)	2.0	2.8	12.0	0.43	0.09	-	-	50% in complete ration for ruminants	Tannins (1.5%)
Sunflower heads (*Helianthus annuus*)	2.0	7.2	10.6	1.40	0.12	-	-	50% in complete ration for ruminants	-
Tapioca starch waste (*Manihot essulenta*)	0.40	4.9	5.6	-	-	1.8	64.0	25% in CM of crossbred calves	-
Tamarind seed (decorticated) (*Tamarindus indica*)	87.0	19.8	3.1	-	-	11.26	63.92	25% in calf starters	Nil
Tea waste (*Camellia sinensis*)	0.15	28.7	5.6	-	-	9.7	43.3	20% level in CM of crossbred calves	Tannins (1.9%)
Vilayati babul pods (*Prosopis juliflora*)	10.0	12.5	5.1	0.4	0.2	7.0	75.0	20-30 in CM of growing calves and lactating cows	Tannins (0.74-1.5%)
Warai bran (*Panicum miliaceum*)	0.20	6.2	12.1	-	-	4.5	57.4	30% in CM of growing calves and milch cattle	Nil
Wheat bhusa (*Triticum vulgare*)	1500	3.1	11.4	0.26	0.09	0.0	40.1	Major roughage	Lignin (8-11%)
Wheat bhusa (urea treated) (*Triticum vulgare*)	-	6.4	11.6	-	-	4.0	57.0	Improved	Lignin inter-ference decreased
Water hyacinth *(Eichomia crassipes)*	2000	19.2	16.4	1.4	0.6	4.2	53.1	Feeding mixed with dry roughages to ruminants	Oxalic acid (2.98%)

Bibliography

Barman, K. and Rai, S.N. (2003). Potential of mango seed kernel (*Mangifera indica*) as an animal feed. Indian Dairyman, 55 (8) : 59-62.

Boucque, C.H.V. and Fiems, L.O. (1988). Vegetable byproducts of agro industrial origin. Livestock Product Sci. 19: 97-135.

Dhingra, S. and Kapoor, A.C. (1985). Nutritive value of mango seed kernel. J. Sci. Food Agric. 36: 752-756.

Garewal, N.S. 1969. Conference on the Intensive Approval to Animal Production. pp. 12-15. Bombay.

Maheshwari, P.K., Gupta, B.S., Sinha, R.R.P. and Gupta, R. (1984). Symp on feeding system for maximizing livestock production. HAU, Hisar. pp – 31.

Pathak, N.N. (2003). Buffalo production systems in India. In: Proc. 4th Asian Buffalo Congress on Buffalo for Food Security and Rural Employment. Vol. I, New Delhi, India, Feb 25-28. pp- 36-49.

Punj M.L. (1988). Agriculture Byproducts and Industrial Wastes for Livestock and Poultry Feeding. All India Co-ordination Research Project, ICAR, New Delhi.

Ranjhan, S.K. (1990). Agroindustrial Byproducts and Nonconventional Feeds for Livestock Feeding. ICAR, New Delhi, pp- 7-91.

Sengar, S.S., Tomar, S.K. Lall, D. and Mehra, U.R. (1993). Effect of sodium chlorite treatment of straws on their cell wall constituents and *in situ* dry matter disappearance. Indian J. Dairy Sci. 46 : 130-132.

Tomar, S.K. (2004). Unconventional Feed Resources for Buffaloes in various Agro-climatic Regions. In: Buffalo Production under Different Climate Regions. Eds, S.S. Kundu, A.K. Mishra, and P.S. Pathak, International publications Lucknow. pp – 299-320

Tomar, S.K. and Arora, S.P. (1982). Effect of plane of nutrition on estrous cycle length and progesterone profile during estrous cycle in crossbred cows. J. Nuclear Agric. Biol. 11: 29-30.

CATTLE FEEDING

India is gifted by nature with a diverse livestock germplasm suited to different agro ecosystems of the country. Cattle population of India is 186 million (2004) against 176 million in 1961. During last 4 decades there is invariable increase in cattle population. There are 30 cattle breds distributed in different habitats of the country. Some of the important indigenous and cross breeds along with their characteristics and purpose are given in Table 1. The livestock enterprise in India is largely dependent for feed on agricultural crop residues or by products, which by themselves cannot maintain animals even for normal physiological functions. It is known that the quality of feeds affects the efficiency of the nutrient utilization and in term, affects the quantities of feeds required to meet the nutrient needs. Also, the quality of feeds coupled with physical environment factors may influence the nutrient requirements of dairy animals.

Table 1: Cattle breeds of India with their habitat, characteristics and purpose

Sl.No.	Name Breed	Habitat/main state	Characteristics	Purpose
A. Indigenous				
1	Amritmahal	Erstwhile Mysore State, now part of Karnataka	Well-shaped and narrow head. Forehead is deeply furrowed. Barrel is long and round.	Draught
2.	Dangi	Maharastra and Gujart	Small head, bulging forehead long face, short thick horns.	Draught
3.	Deoni	Andhra Pradesh	Medium size head, pendulous ears, slightly prominent head, thick horns curving outwards and backwards.	Draught
4.	Gir	Gir Hills and forest of south Kathiawar	Extremely bulging forehead, narrow face, long and pendulous ears.	Dairy
5.	Hallikar	Karnataka	Long and bulging head with a distinct furrow in middle.	Draught
6.	Hariana	Haryana and Delhi, Punjab, Rajasthan	Light head with a long narrow face, horns are short and horizontal, white or light grey colour.	Dual
7.	Kangayam	Tamil Nadu	Head is short and forehead is broad, colour is white or grey with black marking on the knees and above the fetlocks on legs.	Draught

Sl.No.	Name Breed	Habitat/main state	Characteristics	Purpose
8.	Kankrej	Gujarat	Narrow face and broad forehead which is slightly dished, barrel is long, which is deep and compact, silver or iron grey in colour.	Draught
9.	Khillari	Maharashtra	Animals have a massive head with a deep furrow in the centre of the forehead. The barrel is long and compact, greyish white in colour.	Draught
10.	Krishna Valley	Maharashtra, Andhra Pradesh, Karnataka	Small head, wide forehead, short and thick horns curving upwards and outwards, greyish white colour with darker shed in the fore and hind quarters in the males.	Draught
11.	Malvi	Madhya Pradesh	Small head, dished face, covered horns, emerging forward from the polls in an outward and upward direction, grey colour with black markings over the neck hump shoulders and quarters.	Draught
12.	Nagori or Nagauri	Rajasthan	Small head and flat forehead well developed horns curving upwards and outwards and than inwards, white in colour.	Draught
13.	Ongole	Andhra Pradesh	Medium size head, broad forehead, short and stumpy horns growing outwards and inward, deep barrel.	Dual
14.	Rathi	Rajasthan	Medium size, powerful heavy built, animals with curly horns.	Dairy
15.	Sahiwal	Punjab, Haryana, Delhi, U.P., Bihar, M.P., W.B.	Heavy breed with symmetrical body and loose skin. Colour is reddish dunn or pale red, sometimes flashed with white patches. Whitish ring along the eyes.	Dairy
16.	Red Sindhi	Pakistan, all parts of India	Small in size, compact frame with round dropping quarters, red colour with shades varying from light to dark. Strips of white colour sometimes seen on dewlap and occasionally on the forehead.	Dairy

Sl.No.	Name Breed	Habitat/main state	Characteristics	Purpose
17.	Siri	Sikkim, Bhutan	Colour is mostly black and white with extensive solid black. Head is small and squarely cut. Forehead is white red and flat, position of hump is slightly forward, dewlap is moderate and seath in males is very tight. Legs are strong and hooves are broad.	Dual
18.	Tharparkar	Pakistan (Sind)	Medium size compact breed. Colour is white or light grey with coffin shaped skulls.	Dairy
B. Crossbred cattle				
1.	Karan Swiss	NDRI, Karnal	Cross of Brown Swiss bulls with sahiwal and Red Sindhi cows. Colour is red dun, resemble to Sahiwal in appearance.	Dairy
2.	Karan Fries	NDRI, Karnal	Cross between Tharparkar and Holstein-Friesian. Black patches on body and sometimes completely dark with white patches on the forehead and switch of the poll.	Dairy
3.	Frieswal	Military Farm, India	Cross of Sahiwal and Friesian	Dairy

In India, a large number of feed ingredients of variable chemical composition and digestibility from agriculture, forest, marine and industrial sources are utilized as animal feeds. Inversely, good quality forage and concentrates (primarily maize and oil-cakes) provide major feed ingredients to dairy animals in developed countries, but are not usually available to our animals. Thus, scientific and judicious combinations of various available ingredients to produce balanced ration for meeting the energy, protein, mineral and vitamin requirements of dairy animals for maintenance, growth and milk production have a greater significance under the situation. This is particularly important under feed resource constraints as exits in our country. Balancing of rations for feeding to cattle is possible, if the nutrient composition of feeds and their requirement for different physiological processes of these animals under the environment that they are raised are known. The levels of nutrients required for them should be adequate to prevent deficiencies and to provide for required growth, reproduction and milk production.

Energy value of feeds

In most of Indian experiments, proximate analysis along with total digestible nutrient system is practiced, despite a resolution of early seventies when the national conference of animal nutritionists recommended adoption of ME system. In NRC (2001), NE_l (net energy of lactation) has been adopted for both maintenance and lactation with the utilization efficiency of 62 and 64%, respectively. NE_l is therefore used in all tables to express energy value of feeds. A summative approach has been used to derive the TDN_{IX} as below:

TDN_{IX} (%) = td NFC + tdCP + (td FA x 2.25) + td NDF – 7

Where: td NFC = 0.98 (100 – [(NDF – NDICP) + CP + EE + Ash)] x PAF

td CP (forages) = CP x exp [-1.2 (ADICP/CP)]

td CP (concentrates) = [1-(0.4 x (ADICP/CP))] x CP

td FA = FA

td NDF = 0.5 x (NDF_m - L) x [1-(L/NDF_m)$^{0.667}$]

Where:

td: true digestible

PAF: Processing adjustment factors.

NFC: Non- fibrous carbohydrates

Estimating DE/ME and NEL of feeds

The gross energy of feeds varies with carbohydrates, protein and fat contents. The following equation is used to estimate DE value of feeds.

DE_{IX} (M cal/kg) = (td NFC/100) x 4.2 + (td NDF/100) x 4.2 + (td CP/100)x 5.6 + (FA/100) x

9.4 – 0.3

Since, this energy value is at maintenance level of production, as a consequence of incremental intake above maintenance there will be depression in digestibility with level of intake. Enough data will be needed to apply the equation on feed having 50 to 60% digestibility. The proposed equation is applicable on feeds that have digestibility coefficients of higher than 70%.

The equation

ME (M cal/kg) = 1.01 x DE – 0.45

The above equation needs to be modified where ether extract content of the ration is higher than 3%, as DE of fats will be utilized with 100% efficiency. Similar modifications were also included for NE_l. The NE_l value of lower quality forage, which may include almost all of our forages, will have lower values than ascribed. Composition of the total diet and dry matter intake has marked effect on digestibility and subsequent energy values. The diets that do not support optimal ruminal fermentation will result in an overestimation of energy values. The new equation proposed to estimate NE_{lp} is:

$$NE_{lp} \text{ (Mcal/kg)} = 0.703 \text{ X } ME_p - 0.19 + ([(0.097 \text{ X } ME_p + 0.19)/97] \text{ X EE-3}])$$

Where ME_p= is in M cal and EE is % of DM

Energy

Energy is probably first most important factor to reduce cattle productivity in most parts of the country. Its deficiency is manifested in different forms. A look in feeding standards, recently published (NRC 2001) only again led to think on our database and its present state of information and knowledge vis a vis developments elsewhere in the world.

Carbohydrate and to some extent fat are primary energy sources in a dairy ration. They supply energy to carry on the life processes and help to grow and to produce milk. If the dairy ration does not contain enough energy, then protein that might otherwise be used for maintenance, growth or milk production is used for energy. To avoid this uneconomic use of nutrients, there should be a proper balance between protein and energy in the diet. A major portion of the energy needs of high producing cows and buffaloes go for milk production. Adequate supplies of energy are particularly important during the early part of lactation. Fats present in usual rations appear to be adequate for dairy animals. Unless fats and oils are cheap sources of energy, there is little economic advantage of feeding fats in addition to what is normally present.

In growing calves, insufficient supply of energy result in retarded growth and causes delay in the onset of puberty. In lactating cattle inadequate feeding, results in decline in milk yield and loss of body weight. Severe and prolonged deficiency of energy depresses reproductive functions. Over feeding also results in excessive fattening of dairy animals, affecting their reproduction efficiency and milk producing ability. Hence the feeding rate should be optimum and balanced in order to allow animals for early calving and acceptable mature live weights at minimal cost. Similarly, energy intake of mature bulls should be controlled to avoid excessive fattening and lowered libido, but should be adequate to allow maintenance of good physical condition. However, young bulls may be fed liberally to stimulate early sexual maturity and semen production.

Energy requirement for maintenance of cows depends on their basic metabolic rate and their activity. Lactating cows require 10 to 15% greater amount of energy for maintenance compared to dry, and non-pregnant cows of similar body weight. High producing cows are unable to consume enough feed of low quality in early lactation, when they need the most to prevent some loss of body weight. The losses are minimized by feeding a properly balanced ration during the first 6-8 weeks after calving. The total energy requirement for a pregnant cow increases markedly during the last 4-8 weeks of pregnancy. An additional amount of energy during this period should be provided to meet the total requirements of the foetus and maternal development of the cow and also for body deposition, which would be utilized during early lactation.

Maintenance requirement

The maintenance requirement of 0.080 M cal/kg$^{0.75}$ has been adopted for mature dairy cows in revised recommendations. These data are based on heavy breeds such as Holsteins and it may be even long known that requirements may be higher at higher temperature specifically when it increases more than 35°C and often coupled with higher humidity.

Lactation requirement

Lactation requirement (NE_l) is defined as the energy contained in milk produced. The concentration of various milk constituents is taken into account for determining energy value. The NE_l concentration in the milk is proportional to the heat of combustion of individual milk components (fat, protein and lactose). As per equation of Tyrrell and Reid (1965) NE_l concentration of milk is determined as follows.

NE_l (M cal) = 0.0929X fat% + 0.0547 X CP % + 0.395 X lactose %.

In condition where only fat and protein contents of milk are known then NE_l is measured:

NE_l (M cal/kg) = 0.929 X Fat % + 0.547 X CP % + 0.192

When milk fat is the sole milk constituent NRC (2001) has used the equation of Tyrell and Reid (1965) to calculate the NE_l concentration.

NE_l (M cal/ kg of milk) = 0.360 + [0.0969 (fat %)]

For NE_l estimation only the yield of total energy in milk has been taken into account and does not consider many of the differences in metabolic transactions or the substrates required for synthesis of individual milk components. Though there is a allowance of 10% for activity but for grazing animals there has to be increase in energy requirements for grazing animals and will largely be function of distance walked, topography of the pasture and body weight. Thus, the energy required for excessive walking was set at 0.00045 M cal/kg/km similar to one adopted by ARC (1980). The energetic cost for cows grazing on hilly topography is rather higher ARC (1980) estimated that 0.03 M cal of NE_L/kg BW is required for a cow to walk 1 vertical km. It has been therefore postulated by 50%, on undulated terrains, due to the process of walking as well as food gathering. In an Australian review, it was estimated that grazing increased requirements relative to maintenance by 20% on flat terrain and by as much as 50% on hilly pasture. It could further be stated that above conclusion has been drawn on limited data. Whereas, the Indian data again is very limited, therefore studies on grazing animals under Indian condition should be undertaken to develop database.

Environmental effects

Lactating cows consuming required amounts of feed may not have problem of lowered temperature at least in most part of the country, if the animals are not exposed directly to wind and kept in cold. An average reduction in DM digestibility by 1.8% units was observed for each 10°C reduction in ambient temperature was ascribed to increased rate of passage of feed through the digestive tract. Western Himalayan, Trans-Gangetic and Gangetic regions may face some difficulty. But proper housing provided in most part of this region and supplementation of small quantities of dry roughages such as wheat or paddy straws may well elevate the (mitigate) the situation. Mild to severe heat stress increases maintenance requirements by 7 to 25%. The data is not sufficient to quantify these effects accurately. Heat stress induces behavioral and metabolic changes in cattle. Some changes such as painting, increase energy expenditure, while others such as reduced intake, reduced activity and reduced metabolic rate will result in reduced heat production.

Pregnancy requirements

The energy required for gestation is assumed to be non significant during early pregnancy and only during last 190 days provision should be made for additional energy. An efficiency of ME used by gravid uterus is assumed to be 0.14 and ME to NE efficiency is 0.64, thus following equation is used for NE_l requirement for pregnancy.

NE_l (M cal/d) = (0.00318 x D – 0.0352) x (CBW/45)/0.0218

Where:

D = day of gestation between 190 and 279.

CPW = calf birth weight in kg

Tissue Mobilization and repletion during lactation and dry period

As a normal physiological process, a dairy cow mobilizes energy stores and protein to a lesser extent in early lactation, and replete tissue reserves during mid and late lactation for the purpose. The cows, which are too fat or thin, are at risk for metabolic disorders and diseases, decreased milk yield, low conception rate and difficult calving. Changes in BW of cows may not reflect true changes in stores of tissue energy. The average gut fill in dairy cows is approximately 15% of BW. Each kg of DMI increase has been found to enhance 2.5 to 4 kg increase in gut fill. The energy value per unit weight of true body tissue that is lost or gained depends on the relative proportions of fat and protein in the tissue and their respective heat of combustion. NRC (2001) has used body reserve model (NRC, 1996) with modification (Fox et al., 1999) to predict body composition based on body score (BCS) of cows of varied body sizes and amounts of body reserves. BCS is correlated with body fat and energy contents. BCS on a 1 to 5 scale is commonly used for dairying cows. Here, BCS of 1 mean extremely thin and a score of 5 means extremely fat. Loss of BCS is expected during early lactation due to energy mobilization for milk production. BCS seems to be an important parameter in deciding the energy requirement of animals.

To convert live body weight to empty body (EB) weight use following equation:

EB weight = 0.817 x LW

Here EB weight values are converted to tissue energy by multiplying by 0.855. This means

Tissue energy/ kg of EBW = EBW x 0.855

The above values are then converted to dietary NE_l using an efficiency of 0.82 for converting energy from live weight loss to dietary NE_l, while an efficiency of 1.12 is used to convert dietary NE_l to tissue energy for gain.

Optimum body condition for maximum milk production may be higher than that for optimum milk production (4 – 4.5 vs 3 -3.5). The body condition of the cow should not be below 2.5 – 3.0 even during peak yield. The optimum body condition at calving would be around 4 – 4.5. Animals with BCS more than 4.5 may suffer from fatty liver syndrome during early post partum.

Protein

In order that dairy cattle grows, reproduces, maintains the body tissue, produces milk and stays healthy, protein must be present in desired quantities in dairy rations. Proteins provide amino acid which are essential for various functions in the body i.e. maintenance, growth reproduction and lactation. Amino acid are the building units of all cells and tissues in the body after meeting needs, protein when giving in excess is used as a source of energy but causing excessive urea level in blood and loading the kidneys. Urea synthesis itself is highly energy requiring process. It is, therefore, very important to feed only the requisite amount of protein required for the animal. Cows being a ruminant, unlike poultry and swine, utilize poor quality protein or non-protein in nitrogen compounds, because bacteria and protozoa in the rumen can form essential amino acids that are required by the animal.

Protein deficiency reduces growth rate in both foetus and the calf, resulting in small calves at birth and / or slow-growing young stock. It affects the appetite, milk yield and general health and vigour. Cows those are severely protein-deficient loose more than usual body weights in early lactation. Lowered immunity and transport proteins of blood and reduced hormone secretions may predispose protein-deficient animals to infections and metabolic diseases.

For efficient digestion and fermentation in the rumen, diets for cattle should contain around 8-9% crude protein or more. When diets are too low in protein (5-6%) digestibility of the total diet is reduced which may result in reduced feed intake and poorer efficiency of utilization of the feed energy. Feed proteins-heated, charred, browned or caramelized-are unavailable or less available to animals. Therefore, for calculating protein needs in the diets, it is essential to increase the dietary crude protein level depending on the extent of their availability. Similarly, solubility of protein may affect its nutritive value. The diet should contain protein of different solubility in right proportion so that there is optimum utilization of nitrogen in the rumen to synthesize microbial proteins. Caution is, therefore, necessary in ration formulation to ensure that balancing feed for protein requirements will not be under estimated because of the unavailable protein.

Feedstuffs contain naturally produced non-protein nitrogen (NPN) and soluble nitrogen compounds. Urea and other NPN compounds are also used as a supplemental source of nitrogen in the diets. The NPN compounds contribute to higher production of ammonia in the rumen. The production of ammonia has to be dovetailed with fermentable energy source, in rate and quantity. Excessive levels or poorly mixed urea to feed can cause toxicity due to rapid ammonia production. A source of S such as sodium sulphate with energy source should be supplemented with such diets.

Metabolizable protein (MP) for maintenance

The maintenance requirement comprises of urinary endogenous nitrogen (UEN), metabolic fecal nitrogen (MFN) and scurf nitrogen (skin and hair). The NRC (2001) has utilized different equation given by NRC (1989) to estimate the EUN, MFN and scurf N.

The endogenous urinary protein requirement in metabolisable protein unit is

$4.1 \text{ X } BW^{0.75}$

Scurf protein requirement of cattle is based on following equation:

Scurf protein (MP) = $0.3 \text{ x } BW^{0.60}$

The metabolic fecal protein (MFP) estimation is based on the intake of indigestible dry matter (IDM). There are several approaches put forwarded by different workers to estimate MFN. However, based on data of these studies NRC (2001) chooses the following equation to estimate metabolic fecal protein

MFP (g/d) = 30 x DMI (kg)

The metabolic fecal protein consists of microbial debris (bacteria and protozoa), keratinized cells. The estimation of fecal protein of bacterial origin was invariable with the nature of diet.

Taking this into account following equation is used to estimate the MFP (g/d) requirement

MP = [(DMI x 30) – 0.50 (bacterial MP/ 0.80) – bacterial MP)]

The efficiency of utilization of the absorbed MP for EMP is 0.67. Thus the equation to estimate the MP requirement for maintenance is:

MP_m (g/d) = 4.1 x $BW^{0.50}$ (kg) + 0.3 x $BW^{0.60}$ (kg) + [(DMI (kg) x 30) – 0.50 (bacterial MP/0.8) – bacteria MP)] +EMP/ 0.67

Pregnancy requirement

Requirement for pregnancy includes maintenance, growth of conceptus and growth of dam. Estimation of pregnancy requirement requires the knowledge of the rates of nutrients accretion/ deposition in conceptus tissues (fetus, fetal fluid, placenta and uterus) along with the efficiency of nutrients utilization for growth of conceptus. Estimation of protein needs to sustain pregnancy is the sole function of a day of gestation and calf birth weight (CBW). NRC (2001) has recommended the following equation to estimate MP for pregnancy.

MPpreg = [((0.69 x days preg.) – 69.2) X (CBW/ 45)] / Eff. MP preg.

days preg = pregnant days

CBW = Calf birth weight

Eff preg. = 0.33

The above equation is applicable to the animals that are pregnant between 190 – 279 days.

Lactation requirement

Requirement of protein for lactation depends on the amount of protein secreted in milk. The MP for lactation (milk production) is calculated by following equation.

MP_{Lact} (g/d) = (y protn/ Eff lact) x 1000

Where,

$Eff._{Lact} = 0.67$

Y protn = milk production (kg/d) X (milk true protein/100)

Minerals

Minerals are required for growth, maintenance and milk production in dairy animals. Some of them are needed for bones and teeth and others for use as constituent of the proteins and lipids. Some are also useful in maintenance of osmotic relationship and acid-base equilibrium. Growing and milking cows need large amount of minerals, especially calcium and phosphorous. In addition to calcium and phosphorous, sodium, chloride, iron, iodine, magnesium, copper, manganese, cobalt, sulphur, zinc, molybdenum, fluorine are essential for dairy cows. These should be supplied in sufficient amount in the regular ration. Forages, unlike concentrate, are high in calcium and low in phosphorous content, while high protein feeds (bran and oil cakes) contain more phosphorous.

The mineral deficiency affect growth, production and reproduction while, excessive levels of these elements. In diets may be detrimental or toxic. Although the range between the minimum required level and the toxic level is relatively wide for most of the essential elements. It is important that the safe tolerance level of minerals should not be exceeded in their diets as higher than minimum required levels. Deficiency of one essential mineral may increase requirements for the others elements and vice- versa. The ratio of calcium to phosphorous in bone is about 2: 1 in older animals and somewhat lower in young animals. In milk the ratio is about 1.5: 1. The dietary ratio of calcium to phosphorous is important in view of their utilization and other animal performance such as growth and milk production. Mineral mixtures available in Indian markets should be purchased carefully. Phosphorous level of about 22% should be maintained however excessive fluorine may be a problem if rock phosphate used has not been deflorinated. More over, phosphorous deficiency has a direct effect on reproduction through suffering oestrous cycles. Parturient paresis (milk fever) in cows is caused by a disturbance in calcium metabolism manifested by a marked drop in blood serum calcium at parturition or soon thereafter. It has been reported that the calcium: phosphorous ratio along with Vitamin D in the prepartum ration is critical. Cattle are able to avoid toxicity from high dietary calcium through excretion of the excess via faeces. However, excessive calcium has an antagonistic effect on the metabolism of other elements including phosphorous, manganese and zinc. When a diet severely deficient in salt is fed to dairy cows an intense craving for salt, manifested by licking and chewing various objects, can occur within two to three weeks.

Generally, forages contain more potassium than required by dairy animals however; the potassium content of many concentrates is below the requirement. Symptoms of relatively severe potassium deficiencies (0.06-0.15 percent potassium in the dry matter of the diet) in lactating cows include a marked decrease in feed intake; reduce body weight gain, decrease milk production, pica, and loss of glossiness, lower plasma and milk potassium. However, a borderline potassium deficiency (0.45-0.55 percent potassium in dry matter) causes lower feed consumption by lactating cows. High potassium content of the diet appears to interfere with magnesium metabolism and / or utilization and may be a factor in grass tetany of lactating cattle. While magnesium deficiency may lead to problems relating to growth and production, magnesium toxicity is not known to be a practical problem. The suggested magnesium requirement is 0.07% in the diet of young calves and 0.20% in the diet of lactating cow. The requirement of iodine is primarily for synthesis by thyroid gland of hormones that regulate the rate of energy metabolism. The iodine requirement for growing or non-lactating cattle is 0.25 ppm and that both for lactating cows and cows during last two months of their gestation is 0.5 ppm. The cobalt requirement of dairy cattle is about 0.10 ppm of the ration. Since the required level is more than the amount contained in many forages and some concentrates, supplemental cobalt is needed in many practical situation. Cobalt is the essential component of the Vitamin B_{12}, which is synthesized by rumen micro organism. Copper requirement, is increased, where molybdenum concentration of herbage is high. Addition of 0.5% copper sulphate to salt is often recommended in copper-deficient areas. In addition to Se, the trace elements copper Mo, Mn and, Zn are important in influencing fertility. Similarly, the requirements for trace elements like sulphur, iron, zinc, manganese, copper etc are higher than the level contained in many feedstuffs for dairy cattle and, thus are required to be supplemented. The relative proportion of ingredients to prepare a mineral mixture is given in Table 2

Table 2: Proportion of Various Ingredients in Mineral Mixture

Ingredients	Percent
Di-calcium phosphate	55.0
Sodium chloride	30.0
Chalk	11.0
Magnesium carbonate	3.0
Ferrous sulphate	3.0
Copper sulphate	0.5
Manganese di-oxide	0.08
Cobalt chloride	0.06
Potassium iodide	0.01
Zinc sulphate	0.26
Total	100.0

Vitamins

Cattle meet most of the vitamin requirements from natural feeds, rumen microbial synthesis and tissue synthesis. Vitamin D and E are usually present in significant amount in high quality forage. B-vitamins and vitamin K are synthesized in the rumen, and vitamin C is synthesized in tissues. Supplementary vitamins are not required by cattle when they receive green forage or good quality hay. In other situation when these supplies are limited or milk replacers are used extensively for young calves, it is necessary to supply commercially prepared feed supplements. Even 0.5 kg of green fodder has been found to provide enough quantities of vitamin A.

Water

Water is the most vital element in cattle ration. Dairy animals suffer more quickly and severely from lack of water than from the deficiency of any other nutrient, particularly when they are in milk. A restricted intake of water lowers the feed intake and the retention of nitrogen. Water intake is largely affected by the nature of the diet, ambient temperature, humidity, breed, body size, milk yield and amount of water and protein in the feed. For maximum growth and milk production, dairy animals should have access to clean and fresh water at all times. Supply of water is equally important in winter as in summer. Water is also needed for sprinkling during hot summer months. The water intake of lactating cows at different production level and environment temperature is given in Table 3.

Table 3: Total water Intake of lactating cows (kg/cow/day)

Milk yield (kg/day)	Live weight (kg)	Environmental temperature (° C) - 17 - +10	11 – 15	16 – 20	21 – 25
10	600	78	81	92	102
	350	52	54	61	70
20	600	88	92	104	119
	350	62	65	73	84
30	600	99	103	116	133
	350	73	76	85	98
40	600	109	113	128	147
	350	88	92	104	119

Calf Feeding

Feeding colostrum

The calf after weaning at birth should be fed colostrum within two hours, and it should be continued for 4 days. Colostrum provides antibodies which are absorbed intact in the first few days of the calf's life. After this age, globulins carrying antibodies are broken down by proteolytic enzymes in the process of exogenous digestion. Gamma-globulin level in blood serum of neonatal calves is only 0.97 mg/ml at birth. It increases to 16.55 mg/ml level after first colostrum feeding at 12 hours and subsequently on the second day, shows a peak of 28.18 mg/ml. This level more or less persists till the reticuloendothelial system of the calf starts producing antibodies. High nutritive value and laxative action to remove the meconium are other important functions of colostrum feeding. Where the dam's colostrum is not available, a substitute of equal nutritive value is necessary; this may consist of two eggs and about 30 ml of castor-oil. In addition, it is necessary to inject the calf with the dam's serum for augmenting the antibody titre in the body.

Milk feeding

Milk has a very high nutritive value and should be given to calves after 4 days of age. The calves when fed milk show best feed efficiency, up to 1 kg weight gain /1.39 kg dry matter in milk. For scientific feeding it is possible to stipulate the requirements of the calf from chemical composition of milk.

Milk is more or less a complete feed for the calf, whose gastrointestinal system is equipped for digestion of the nutrients present in milk. The calf must receive sufficient milk during the first 3 months; alternatively a minimum of 110 litres of whole milk should be fed over a period of 4 to 5 weeks, along with a calf-starter having a good quality of protein and low fibre. A minimum of 160 litres of whole milk up to the age of 7 to 10 weeks in addition to colostrum is an alternative suggestion. Economical feeding of restricted milk quantity slows rates of growth which delays maturity age.

Whatever may be the milk-feeding schedule, it is important to consider birth weight, quantity of whole milk consumed, sex and heredity as these usually affect the growth rate. Some calves have poor growth rate in spite of optimum feeding and management; others can not digest normal quantities of the milk and may suffer from nutritional diarrhoea. This is particularly true when calves of the same age but the different weight (20 to 35 kg) are getting the same quantity of milk on age basis. Milk should be given warmed to body temperature and preferably with a trace-mineral supplement to make up for its deficiency of Fe, Cu, Mg, Mn, and Zn. Green fodder up to 100 g dry matter may be offered daily from the age of 15 days onwards to provide a stimulus for the development of the rumen.

Milk feeding schedules

Some of the milk-feeding schedules for calves are suggested below:

Schedule 1

Age (days)	Colostrum (kg)	Milk (kg)
1-4	1/10 of body weight	-
5-90	-	1/10 of body weight

Calves kept on this schedule of feeding give the best performance but it is expensive. A weight of 76 kg can be achieved at 3 months in crossbreds weighing 31 kg at birth.

Schedule 2

Body weight (kg)	Age (days)	Colostrum	Milk
Up to 30	0-4	1/10 of body weight	-
Up to 30	5-90		1/10 of body weight
31-60	5-90		1/20 of body weight above 30 kg

It is possible to raise crossbred calves with feeding schedule 2 from 26 kg birth weight to 54 kg at 3 months.

Schedule 3

Body weight (kg)	Age (days)	Colostrum	Milk	Separated milk
Up to 30	0-5	1/10 of body weight		
Up to 30	6-30		1/10 of body weight	
25-30	31-60		1/15 of body weight	1/25 of body weight
35-70	61-90		1/25 of body weight	1/15 of body weight

Schedule 4 (based on age alone)

Age (Birth weight-30 kg)	Rate of colostrum/or milk feeding (kg)
1-4 days	3.0
5-15 days	3.0
16-28 days	3.5
1-2 months	4.0
2-3 months	4.5

Reduce milk by 1 kg every 4 days afterwards

Feeding milk-replacer to calves

Milk-replacer (MR) diet is fed to calves as early as at 10 days of age to replace milk from economy point of view. The calf at an early age cannot consume large quantity of feed in bulk form, and hence requires MR which can be fed on milk-dry matter-equivalent basis. MR should resemble milk more or less on broad chemical composition especially in term of protein quantity, volatile fatty acids, minerals and vitamins. It should have a biological value equivalent to that of milk, and the ingredients used for constituting MR should be low in crude fibre and free from any anti-metabolite. In addition, milk-replacer may contain butyric acid, citric acid, and some antibiotics as additives to simulate growth and to build up vitality and resistance against diseases. The replacement of milk by milk-replacer should be gradual to facilitate its acceptance and to avoid a drop in growth rate.

Composition of milk-replacer

The ingredients of a milk-replacer and its rate of feeding are given in the Tables 4 and 5

Table 4 Ingredients of milk-replacer (kg)

Wheat	10.0
Fish-meal	12.0
Linseed-meal	40.0
Milk	13.0
Cotton seed oil/coconut oil	7.0
Citric acid	1.5
Molasses	10.0
Mineral mixture[a]	3.0
Linseed-oil[b]	3.0
Total	99.5
Butyric acid	0.3
Antibiotic mixture	0.3
Rovimix A, B_2, D_3	0.015

[a]Mineral mixture per 3 kg: Dicalcium phosphate, 1.6500; sodium chloride, 0.9000; chalk, 0.3312; magnesium carbonate, 0.0900; ferrous sulphate, 0.0150; copper sulphate, 0.0021; manganese dioxide, 0.0021; cobalt chloride, 0.0015; potassium iodide, 0.0003; sodium fluoride, 0.0003; and zinc sulphate, 0.0075 (total 3.0000).

[b]linseed oil will provide adequate linoleic acid.

Table 5 : Rate of feeding milk-replacer

Days	Body weight (kg)	Colostrum (kg)	Milk[a]	Milk-replacer (g)
1-5	-	1/10 of body weight		
6-9	-		1/10 of body weight (kg)	-
10-13	-		As above	50
14-17	-		As above	100
18-21	-		Less by 1/2[a, b] kg	175
22-25	-		Less by 1 kg	250
26-29	-		Less by 1.5 kg	325
30-33	35		Less by 2 kg	375
34-36	40		Less by 2.5 kg	450
-	40		1.5 kg	450
-	45		1.5 kg	525
-	50		1.5 kg	600
-	55		1.0 kg	800
-	60		1.0 kg	900
-	65		1.0 kg	950
-	70		1.0 kg	1000

[a] when the calf reaches 1½ months of age, milk is to be reduced to 1 kg irrespective of weight.

[b]Less from that of milk quantity of the 13th day.

Crossbred calves of 25 kg birth weight can be raised to 51 kg at 13 weeks and 95 kg at 25 weeks on feeding milk replacers. This compares well with controls of the same birth weight given milk the rate of 1/10 of body weight up to 30 kg body weight and thereafter 1/20 kg of body weight up to 60 kg. The respective weights in the control group may be 53 kg at 13 weeks and 96 kg at 25 weeks. With this schedule, the calves can be raised with as little as 114 liters of milk plus milk-replacer when compared with the control given 299.2 liters of milk. These weight gains are at about half the cost of the control group. It has to be borne in mind that a very young calf is equipped to digest only lactose and milk proteins. Since polysaccharides ordinarily are not digested, and the calf requires sufficient adapting time to added carbohydrates, the change from milk to milk-replacer should be very gradual. Normally 10 days old calves to not take MR. It is necessary to include at least 1 liter of milk per day to impart its taste to the milk-replacer and clean water to dilute it to the extent of 1:8.

In a follow-up programme, the calves fed milk-replacer did not show any anomaly in subsequent growth rate or on maturity age and even lactation yield. The maturity age calves was similar in both cases.

Growth Requirements

For the estimation of energy and protein requirement of young calves for maintenance and growth NRC (2001) has used following equations:

$NEm\ (M\ cal) = 0.086\ LW^{0.75}$

Where LW is the live weight in kilograms

$NE_G\ (M\ cal) = (0.84\ LW^{0.355} \times LW_G^{1.2}) \times 0.69$

Where, LW and live weight gain (LW_G) are in kilograms.

$MEm\ (M\ cal) = NEm/\ 0.75$

$ME_G\ (M\ cal) = NE_G\ /\ 0.57$

Thus, sum of $MEm + ME_G$ is the total ME requirement of calf at a particular growth rate.

The equation used to estimate protein is as follows:

$ADP\ (g/d) = 6.25\ [1/BV\ (E + G + M \times D) - M \times D]$

ADP: Apparent digestible protein.

BV = Biological value set at 0.70,

E = Endogenous urinary nitrogen = $0.2\ LW^{0.75}$

G = Nitrogen content of gain, assuming 30 g/kg of gain.

M = is metabolic faecal nitrogen computed as 3.3g/ kg dry matter consumed (D)

Thus, CP requirement is
CP = ADP/ 0.75

Here the total protein requirement has been divided into two components i.e. maintenance and body weight gain (growth rate). Maintenance requirements consist of obligatory losses of N in urine and faeces, while gain is related to tissue N retention (storage).

Formulation of Ration

The ration formulation for dairy animals depends on the type of feeds and fodders available and the requirement of the nutrients for a particular animal (Table 6). The nutrient requirements are largely based on the size of the animal and its physiological status, i.e. growth, pregnancy or milk production. An adequate ration should meet the energy, protein, mineral and vitamin requirement of the animals if consumed in sufficient quantity. Therefore, it is essential to formulate ration having enough flexibility to accommodate diverse situations of feed and fodder availability and their cost.

The first step in formulating a ration is to calculate the nutrient requirements for the cow to be fed. This will depend on the age, size, production and reproduction of the animal. On the basis of this information, the nutrient requirements in terms of energy, protein, calcium and phosphorous can accordingly be obtained from the respective tables. These requirements can be fulfil with much different combination of feed ingredients (Table 7) depending on the cost and availability. The next step consists of finding out, the digestible nutrients contents in the feeds and fodder available for feeding. When these figures are available, the final step is to calculate the quantities of different feeds and fodder, which will require nutrients, calculated in the first step. While calculating the quantity of feedstuff, the capacity of the animal to consume feed should be taken into account. Normally, milch cows may consume dry matter @ 2.0-3.5% of their body weight depending on the quality of feed and fodder.

For example, a mature, non-pregnant cow, weighing 500 kg and producing daily 10 kg of milk containing 4% fat will require the following quantities of nutrients:

Physiological function	DCP (kg)	TDN (kg)	Ca (g)	P (g)
Maintenance	0.278	4.45	31.0	23.0
Milk Product	0.550	3.25	27.0	20.0
Total	**0.828**	**7.70**	**58.0**	**43.0**

The cow must receive O.828 kg DCP, 7.70 kg TDN, 58g calcium and 43g phosphorous to meet her daily requirement. Thus, a cow weighing 500 kg may consume 10-18 kg feed dry matter. It is, therefore, significant that required quantity of nutrients should be provided within this amount of dry matter.

With a limited availability of good quality fodder, it may not be possible for the cow to meet her total requirements from this source. However, her maintenance requirement can be met from one of the following two feed combination:

(a) Wheat bhusa 7.0 kg
Green berseem 10.0 kg

(b) Wheat bhusa 4.0 kg
Green jowar 20.0 kg.
Cotton seed cake 5.0 kg

In addition, 4 kg of concentrate mixture (14 percent DCP) should be added to the maintenance ration in order to provide the needed energy and protein for the production of 10 kg of milk.

Table 6: The requirements of a cow weighing 400 kg and yielding 8 to 10 kg milk:

Nutrient requirement (NRC, 1989)							**Feeds/fodder offered -**	**Qty.(kg)**	**Nutrient available from feeds/fodders (kg)**	
	DM	CP	TDN	ME	Ca	P	I Berseem/Lucerne	60.0	CP	TDN
Maintenance	12.5	0.32	3.10	12.01	16	16	Conc.-mix.	02.0		
									1.3	5.4
Production	-	0.96	3.43	13.20	28	17			0.3	1.4
							II Oat green	50.0	1.66	6.8
							Conc. - mix	03.0	0.77	4.62
									0.54	2.10
							III Maize/ Jowar	35.0	1.34	6.72
							Conc. mix.	04.0	0.93	3.60
									0.72	2.80
									1.65	6.40
Total	12.5	1.28	6.53	25.21	44	28				

Table 7: Composition of concentrate mixture for dairy animals (% in the concentrate mixture)

Feed Ingredients (Kg)	1	2	3	4	5	6	7	8	9	10	11	12	13
GNC/Linseed cake	-	28	20	-	20	17	22	15	-	20	42	25	35
Wheat/bran rice	57	27	25	30	41	20	25	45	40	-	20	24	20
Maize/oils/ barley	-	42	-	27	26	40	10	-	-	20	35	37	32
CSC	-	-	-	-	-	20	-	-	-	20	-	-	-
Mustard Cake	25	-	-	-	-	-	-	-	20	-	-		-
Mung/gram/ urad	-	-	25	20	-	-	-	37	37	17	-		-
Wheat/ rice grit	-	-	27	-	-	-	30	-	-	20	-		-
Guar meal	15	-	-	-	10	-	-	-	-	-	-	-	-
Maize gluten meal	-	-	-	20	-	-	10	-	-	-	-	-	-
Molasses	-	-	-	-	-	-	-	-	-	-	-	10	10
Urea	-	-	-	-	-	-	-	-	-	-	-	1	-
Mineral Mix.	2	2	2	2	2	2	2	2	2	2	2	2	2
Salt	1	1	1	1	1	1	1	1	1	1	1	1	1

Points for Effective Feeding Management

To fully tap the potentials of high-yielding cows the following points of feeding management are of practical importance.

- Feeding dairy animals on home-grown fodder especially leguminous (Berseem, Lucerne, cow pea, guar, stylo etc.) or a mixture of leguminous and non- leguminous would be most economical as compared to feeding them on crop residue and concentrates.
- Feed roughages (dry or hay and silage) twice a day and concentrates or grain mixture at or before milking. Green forages are much more palatable than the hays prepared from them. Through liberal feeding of green or silage, the need to feed concentrates can be cut down by 33%. When leguminous fodder such as lucerne and berseem are fed, the quantity of concentrates can be further cut down. Again, the green especially leguminous fodder should not form the entire roughage ration because of the risk of producing bloat. It should be mixed with green grasses / hay or other non-leguminous fodders / crop residue before feeding. If grass is not readily available, it may be replaced with straw in the proportion of 1 kg of straw for every 4 kg of green grass. Long grasses / other fodder crop residues must be chaffed before feeding to reduce wastage.

- Concentrate must be fed after moistening it with some water to reduce dustiness. The moistened feed should be fed the same day. The concentrate can be soaked overnight.
- The digestive system of a high-producing dairy animal has a limited capacity and so to obtain the required nutrients, the digestibility of the ration on dry matter basis must be 65% or higher. It will enable the cow to assimilate the available nutrients in the feed. Roughages may have to be supplemented "With feed concentrates to ensure nutrients for maximum milk production.
- The cow must be "conditioned" adequately before calving so as to have the needed body reserve for good milk production after calving. For this, it must calve in relatively fat condition and so must be on a fattening ration for the last two months of the dry period. It must be provided energy requirements equal to that of a cow producing about 10 kg of milk a day and gain at least half a kg in body weight every day in the two months of lactation. Feed at least 1 kg of concentrate mixture per day and larger quantities for the high-yielder.
- A dry cow often considered as a cow which is not in 'profit', whereas it should considered as a cow 'preparing for profit' - and the amount of the profit depends on the thoroughness of the preparation. In a herd that is well managed, the dry cow rates high in priority for feed.
- A well-conditioned cow loses weight after calving, but increases her daily milk yield. If it is going to produce 10 to 12 kg or more milk a day for any length of time, it will have to loss some weight. Cows usually can not take enough feed in a day to produce this quantity of milk for any length of time without losing weight, irrespective of how good the feed may be. The weight loss is largely from body fat stored for this purpose when it was dry. A well - grown jersey cow male loss about 50 kg in the two months after calving, while a large Holstein may loss 150 kg over the same, period.
- The cow i.e. poorly conditioned at calving time is hardly likely respond to additional feed inputs. In any case her yield will be mediocre or even low. Some of the additional feed supplied will be diverted to her body conditioning. It is more expensive to fatten cows after calving than it is to fatten them before calving.
- For maximizing income, the dairy animals should be fed individually according to their individual milk yield and nutritional requirements, instead of allowing the same ration to each animal in the herd. Liberal feeding is necessary for continued high milk production and its persistency throughout the lactation.

- Abundant supply of clean drinking water all the time is essential. It is the most essential feed nutrient. Do not permit cattle to drink water from dirty ponds, streams or tanks as it would result in increased incident of diseases, lowered vitality and decreased milk production. The daily water intake for an animal of 300 kg body weight-yielding 5 kg of milk per day will be about 35-40 litres. About the same quantity of water is required for the general cleanliness of the animal and its shed.
- Does not change the ration abruptly; give few days (1 to 2 weeks) for adaptation to a new feed. Harvest the fodders at pre-bloom to half bloom stage.

Bibliography

ICAR, 2002. Hand Book of Animal Husbandry. ICAR Publication, New Delhi.

Kearl, L. C. 1982. Nutrient requirements of ruminants in developing countries. Published by International feedstuffs Institute, Utah State University, Logan, Utah, USA.

NRC, 2001. Nutrient Requirements of Dairy Cattle. National Academy Press, Washington, DC.

Ranjhan, S. K. 1999. Animal Nutrition and Feeding Practises. Vikas Publishing House Pvt. Ltd.

BUFFALO FEEDING

Asian buffaloes (Bubalus bubalis), also called water buffaloes, have been classified into two distinct types swamp and river buffaloes. Cockrill (1976) suggests that water buffaloes were reared as early as 2500 to 2100 BC, whereas Pathak (2004) points outs their rearing during Ramayana period i.e. 5000 BC. River buffaloes, usually found in India and Pakistan, prefer the flowing water of rivers, exhibit superior milking qualities. Whereas swamp buffaloes are common throughout Asia, Europe, Africa and Latin America, their natural habitat being marshlands and swampy areas and used for draft purpose animals. In some areas, buffaloes are known by the name Carabao, which possibly originated in Malaysia from the Malay word phonetically pronounced Krbau. There are over 170 million buffaloes in the world (FAO, 2004) of which 97% i.e. 165 million are found in Asia, 57% in India followed by Pakistan 15%, China 13% and 2% in each of Philippines, Vietnam, Myanmar and 1% each in Indonesia, Thailand and Laos. Thus 95% of all buffaloes in the world are present in ten countries of Asia.

Buffalo is the main dairy animal of the Indian sub continent and about 56% of the total animal milk in India is from buffalo despite being around one third in numbers. The additional advantages of buffalo rearing are the supply of nutritious meat (buffen) and draught power besides the valuable byproducts like hide for leather industry and dung for manure as well as fuel. The economic contribution of buffalo is significant and a large number of families are wholly or partly dependent on dairy buffaloes for sustaining their livelihood. Thus buffalo is main milk as well as meat animal of India.

Future of buffaloes in Indian sub continent

Buffaloes being the native of Indian sub continent are better adapted than the high milk producing taurus cattle under the tropical and subtropical climatic conditions where more fibrous feedstuffs are available. Secondly the males of taurus and crossbred (Bos indicus x Bos taurus) are undesirable whereas, in buffalo males have got good local and export markets as meat animals. Besides, the preference of most of the Indian people in milk shed areas will persist for long time for thick buffalo milk. Quality milk products like ghee and curd are made from the same. Mozzarella cheese is also prepared from buffalo milk. Thus buffaloes have bright future in this subcontinent.

Buffalo Breeds

1. Murrah

Murrah is one of the best breeds of buffaloes (Table 1- 4). Its home tract is Rohtak, Jind, Hisar and Bhiwani district of Haryana and Patiala district of Punjab. However, the animals of this breed can be seen throughout the country. Murrah buffaloes are jet black in coat colour with sparse hairs and large in size with long and deep body. This breed has derived its name from shape of its horns (Murrah means "twisted horns"). Horns are short and tightly curved in a spiral shape. Udder is capacious with prominent milk-veins. Murrah buffaloes produce around 2000 kg of milk, per lactation and are widely used for up-gradation of local buffaloes in many parts of the country. The age at first calving is 45 to 50 months in field but 36 to 40 months under farm condition.

2. Jaffarabadi

The breed is also named as Bhavanagri and Gir. The animals are found in kutch, Junagarh, Bhavanagar, Nreli and Rajkot districts of Gujrat. Jaffrabadi is the heaviest among the Indian breeds. Average weight of a bull is around 1000 kg. The most common colour of Jaffarabadi buffaloes is black but animals of grey and copper colour are also available. Some animals have white spots on forehead, fetlock and tail. Head is drooping and large size with bulging forehead. The rest of the face is small and wide. The bone of forehead is concave and covers the eyes to give "sleepy eye appearance". Horns emerge side ways downwards compressing the head and are pendulous in appearance just falling on both sides of the neck. Body is usually long, massive and fatty but not compact. The buffaloes of this breed produce around 2200 kg of milk in a lactation of 305 days. Exceptional buffaloes are very high producers.

3. Nili Ravi

The buffaloes are found in Firozepur, Gurdaspur and Amritsar districts of Punjab. The breeding tract is spread all along the Sutlej River on the Indo-Pak border. This breed is similar to Murrah in almost all characters except for white markings on extremities and walled eyes. Skin and hairs are normally black-brown, with wall eyes and white markings on forehead, face, muzzle, legs and tail. The most desired character of the female is the possession of white markings due to which the breed is also known as "Panch Kalyani". Moustache is white. Average milk yield is of 1811-2000 kg in 300-days lactation length (1500 to 1850 kg per lactation).

4. Bhadawari

Bhadawari buffaloes are known for higher milk fat content, which may go upto 13 percent. Animals of this breed are found in the ravines of Yamuna, Chambal and Utangan rivers spread over in adjoining areas of Uttar Pradesh and Madhya Pradesh. The breeding tract and natural habitat of this breed are Bhadawar estate comprising Bah tehsil of Agra, Chakarnagar and Barhpura blocks of Etawah in Uttar Pradesh, Ambah and Porsa tehsil of Morena and Mahangaon tehsil of Bhind district in Madhya Pradesh. They produce about 1000-1500 kg of milk in a lactation with an average fat of 8 per cent. Fat may vary from 6- 12 percent as per the stage of lactation. The population of this breed is declining at an alarming rate in the breeding tract. Estimated population in the whole breeding tract is reported between 13500 to 15000 animals (Kushwaha et.al., 2004).

5. Pandharpuri

Pandharpuri buffaloes are spread in Kolhapur, Solapur, Pandharpur, Sangli and Satara distt. of Maharashtra. The animals are medium sized with long and compact body, having long and narrow face, prominent nasal bone, comparatively narrow frontal bone. Hair colour is grey, tan and blackish. The horns are very long, running backwards side, upwards and twisted outward and touching extended beyond shoulder blade. The neck is comparatively longer and thin. The udder is medium in size with well-placed teats. They produce around 1500 kg of milk in a lactation and are considered to be the consumers friendly, by way of allowing door-to-door milking for milk quality satisfaction of consumers.

6. Surti

The animals of this breed are found in Anand, Nadiad, Kaira and Vadodara districts of Gujarat. The animals are also commonly found in South Rajasthan extending over to Udaipur, Bhilwara, Rajsamand, Chittorgarh and Dungarpur districts. The Surti buffaloes are medium in body size. Coat colour of Surti buffaloes varies from black rusty brown to silver grey. There are two white lines on lower side of the neck. The horns are sickle shaped/ moderately with high fat content (8 - 12%). Udder is well developed, finely shaped and squarely placed between the hind legs. Surti buffalo produce around 1500 kg of milk in lactation.

7. Godavari

The breeding tract of Godavari buffaloes lies in the East and West Godavari districts of Andhra Pradesh. This breed has their origin from interbreed crossing followed by grading of local non-descript buffaloes with Murrah. The Animals are medium statured with compact body. Colour is predominantly black with sparse hair coat of coarse brown hairs. The horns are short, flat, curved, slightly downwards, backwards and then forward with a loose ring at the tip. Udder is medium in size, bowl shaped and well placed medium sized teats. Milk yield is around 2050 kg in a lactation of 305 days and animals have shorter calving interval than Murrah and are regular breeders.

8. Swamp

Swamp buffaloes are draft buffalo breed and found in Assam state of India and are spread over entire river bank of Brahmaputra. Animals are reared under semi-domesticated, nomadic (khutis) and settled system of management. Buffaloes grazing on grasses, shrubs and leaves are most commonly found. Body colour of the swamp buffaloes is primarily black with varying shades from dark to slaty black. Albinos are, however, not uncommon in Assam. They produce around 500 kg of milk in lactation with average fat of 8.48 percent. They are primarily reared for draft purpose. The swamp buffaloes have 48 chromosomes.

9. Toda

The Toda buffaloes are reared by Toda tribes of Nilgiri hills and named after them. The breed had been evolved, which could withstand extreme climatic variation of high rainfall and humidity. Toda buffaloes are medium in size and are considered to be quite powerful. Coat colour of the calf is generally fawn at birth which varies from grey, light grey or dark grey. In growing calves, at about 2-3 months, the fawn colour changes to ash grey. The horns are quite large, set wide apart, curving inward, outward, upward to form a characteristic semicircle. They produce around 500 kg of milk in a lactation of 200 days.

10. Mehsana

Mehsana buffaloes are found in Mehsana district of Gujarat from where it derives its name. It is also common in Banaskantha, Sabarkantha, Gandhinagar and Ahmedabad districts. The breed is considered to have originated from regular inter-breeding between Murrah and Surti. The Mehsana buffalo is a medium sized animal. Body colour is mostly black. Horns are generally sickle-shaped with the curve more upward than the Surti breed and less curved than in the Murrah breed. Udder is attached high in the back and carried well. Milk veins are prominent. Mehsana buffaloes produce around 2000 kg of milk in lactation.

11. Nagpuri (Berari)

Nagpuri buffalo is a native of the Vidarbha region of Maharashtra. This breed is commonly found in Nagpur and the adjoining districts of Wardha and Berar. This breed has 4 distinct strains, viz. Purnathadi (Akola distt.), Ellichpuri (Ammravati distt.), Gaulani and Nagpuri (Nagpur distt.). All these strains are commonly known as Nagpuri or Berari buffaloes. Nagpuri is a dual-purpose breed of buffalo. These buffaloes are used for heavy draught purposes. These buffalo produce around 900 to 1,200 kg of milk. The estimated population of Nagpuri buffaloes as 256851 in Nagpur, 46688 in Amravati, 12762 in Akola and 14736 in Yavatmal district. Around 32.75 percent of the total population of Vidarbha region belong to Nagpuri buffaloes.

12. Marathwada

Marathwada buffaloes are medium in size and found in the Marathwada region (Parbhani, Nanded, Beed and Latur) of Maharastra and are named after the name of the region. Coat colours varies from greyish-black to jet black, white markings on forehead and on lower parts of the limbs and white switch are common. Horns are medium in length. Udder is medium and milk yield ranges from 850 to 1000 kg in a lactation.

13. Tarai

The Tarai buffaloes are found in Tarai area of Uttar Pradesh, Uttaranchal from Tanakpur to Ramnagar, but originally are a native breed of hilly areas. The animals are moderate in body size with coarse and slightly convex head and prominent nasal bones. Horns are flat and long with broad base and pointed tip. Back is straight in males, slightly shallow in females. Skin and coat colour varies from black to brown. The animals are commonly used for draught purposes in agricultural operations. Buffaloes produce around 2-3 kg daily averaging around 700 kg in a lactation of 250 days.

Table 1: Buffalo Breeds and their important characteristics/description

S.No.	Name Breed	Habitat/ Main State	Breeding Tract Districts
1.	Bhadawari	Uttar Pradesh, Madhya Pradesh	Bah Tehsil in Agra adjoining areas of Gwalior
2.	Jaffarabadi	Gujarat	Kathiawar and Honreli
3.	Mehsani	Gujarat	Mehsana, Banaskantha, Sabarkantha tract in Gujarat
4.	Murrah	Haryana, Uttar Pradesh, Punjab	Rohtak, Hissar, Karnal, Jind, Gurgaon, Western parts of Uttar Pradesh Nabha & Patiala

S.No.	Name Breed	Habitat/ Main State	Breeding Tract Districts
5.	Nagpuri	Maharastra, Andhra Pradesh	Wardha, Nagpur, Yeotmal, Adilabad and adjoining parts
6.	Nili Ravi	Punjab	Ferozpur (Montgomery Pakistan)
7.	Surti	Gujarat	Kheda, Vadodra (Charottar tract)

Table 2: Average body measurements (cm) of buffalo breeds

Breed		Male			Female	
	Length	Height	Heart girth	Length	Height	Heart girth
Bhadawari	116.9	122.8	184.5	115.0	123.1	184.3
Jaffarabadi	127.7	126.1	207.7	132.6	129.1	200.9
Mehsana	153.7	133.7	200.6	141.7	127.5	189.3
Murrah	150.0	142.0	220	148.0	132.7	202.4
Nagpuri	180.0	140.0	210.0	128.6	122.8	181.8
Nili-Ravi	160.0	140.0	230.0	165.4	134.2	207.7
Pandharpuri				132.9	130	192.8
Surti	142.0	130.0	190.0	118.8	124.9	184.0
Toda				132.7	121.8	180.4

(Source: Nivsarkar et al., 2000)

Table 3: Body weights (kg) of buffalo breeds

Breed		Birth weight		Adult weight	
	Male	Female	Overall	Male	Female
Bhadawari	27	25	25.3	475.0	425.71
Mehsana	29.51	28.51	29.0	565.4	484.2
Murrah	31.7	30.0	30.3	567.0	516.0
Nagpuri	29.0	28.1	28.61	520.0	363.5
Nili-Ravi	35.1	34.5	34.8	567.0	454.0
Pandharpuri	28.0	25.61	26.8	-	416.2
Surti	26.3	24.5	25.2	500.0	382.61
Toda	27.91	28.01	27.91	380.0	380.0

(Source: Nivsarkar et al., 2000)

Table 4: Production performance of buffalo breeds

Breed	Total lactation milk yield (kg)	First lactation milk yield (kg)	Lactation length	Dry period
Bhadawari	903.1	780.0	271.9	190.0
Jaffarabadi	2,238.7	2,151.3	305.1	144.9
Mehsana	1,988.0	1 940.4	316.7	166.7
Murrah	1,751.8	1,678.4	298.7	154.8
Nagpuri	1,055.4	933.3	286.4	129.1
Nili-Ravi	1,850.2	1,483.4	294.2	150.6
Pandharpuri	1,502.3	1,197.3	330.0	144.0
Surti	1,285.43	1,396.5	344.7	205.4±8.79

(Nivsarkar et al., 2000)

FEEDING

Feeding of an animal is based on certain factors, keeping in view the physiological needs for specific functions like maintenance, growth, milk production, reproduction, work etc for which energy, protein, minerals and vitamins must be provided in an adequate quantity to get optimum production. Although, in practice only the protein and energy needs of an animal are met through locally available feeds and fodders.

Factors to be considered

1. Physiological stage of animals
   (a.) Age of active growth
   (b.) Pregnancy
   (c.) Lactation
2. Voluntary feed intake capacity of the animals
3. Nutritional requirements of the animals of different category
4. Nutritional characteristics of the feeds available for the feeding of animals
5. Requirement of supplements and additives to be procured for the balancing of rations of different categories of buffaloes.

Voluntary feed intake

Voluntary feed intake is an important factor for estimating the production potential of the animal, selection of feeds and preparation of ration to meet the requirements of the animals. For quantification of requirements voluntary feed intake is assessed on dry matter basis i.e. dry matter intake (DMI) capacity of the animals and usually expressed as DMI (kg) per 100 kg body weight or DMI (g) per kg metabolic body size ($W^{0.75}$ kg).

1. DMI in calves

The major contribution in DMI in early life upto 3 months of age is from whole milk alone or partly replaced with skim milk and supplemented with calf starter and good quality fodder. Average DMI per kg ($W^{0.75}$ kg) has been found to range between 30 - 40g upto 1 month of age, 40 - 60g from 1 to 2 month and 60 - 75g from 2 to 3 month of age. On such feeding average daily body weight gain (ADG) has been found to vary between 200 to 450g depending on the quantity of milk fed. Frequency of feeding has little advantage. DMI in calves had been summarized as follows: -

A. Dry matter intake capacity of new born calves fed whole milk or milk replacer has been summarized as follows: -

(a) From birth to one month of age 30 - 35g per kg $W^{0.75}$.
(b) From 1 to 2 month of age 35 - 40g per kg $W^{0.75}$.
(c) From 2 to 3 month of age 40 -45g per kg $W^{0.75}$.

B. Dry matter intake on feeding of restricted amount of milk, calf starter and fodder is: -

(a) From birth 1 month of age 30 - 35g per kg $W^{0.75}$.
(b) From 1 to 2 month 50 - 60g per kg $W^{0.75}$.
(c) From 2 to 3 month 60 - 70g per kg $W^{0.75}$.

2. DMI in growing buffaloes

Although rumen becomes functional in weaned calves reared upto 3 month of age on the feeding of limited quantity of milk alongwith calf starter and fodder but still it is in developing stage due to which coarse fodder utilization capacity is limited and calves require nutritious diets for the supply of their nutritional requirements for growth. Body weight gain potential in most of the buffaloes is 400 to 800g during the active growth phase after 3 month of age. The nutritional requirements of calves may be supplied by feeding a concentrate mixture @ 2.0, 1.75, 1.5 and 1.25% of body weight during growth phase of 50 - 100, 100 - 150, 150 - 200 and beyond 200 kg body weight, respectively. The diet is supplemented with either *ad libitum* feeding of a cereal green fodder or a mixture of leguminous green fodder and wheat straw (70:30 in early stage and 60:40 during late cut).

Voluntary DMI per 100 kg body weight and per kg $W^{0.75}$ has been observed to range from 2.27 to 2.65 kg and 75 to 103g respectively in growing buffalo heifers. In growing male buffalo calves mean DMI has been found to range from 2.0 to 3.5 kg per 100 kg or 80 to 132g per kg $W^{0.75}$ at different stages of growth phase.

3. DMI in pregnant buffaloes

Properly fed buffalo heifers may conceive at about 24 - 30 month of age and 350 - 400 kg body weight. The nutritional requirements of the pregnant buffalo heifers are somewhat higher than the adult buffaloes because they are still growing. Additional feeding is generally not required during the first half of the gestation period, which increases gradually to become about 50% higher than the maintenance requirement during the terminal trimester of the pregnancy. The target is that animal gain 50 - 60 kg during last trimester of pregnancy except for over conditioned animals, when buffaloes should maintain the weight. During last week before freshening, bulk feed of laxative nature be offered it may contain feeds such as wheat bran, linseed, molasses etc. The dry period which may be about 6 - 8 week helps the animals to replenish the body and also store nutrients for next lactation. Feeding during last trimester is therefore as important as during lactation.

4. DMI in lactating buffaloes

DMI of lactating buffaloes is highest among all the physiological stages. It is influenced by several factors like body size, age, milk yield, milk fat content, climate and the palatability of feed etc. DMI in lactating buffaloes has been found to range from 2 to 3 kg per 100 kg body weight equivalent to 89 to 155g per kg $W^{0.75}$.

5. DMI in non-lactating and empty adult buffaloes

A significant reduction in DMI of non-producing adult buffaloes is quite common. Since their requirement is limited to maintenance, they consume about 1.5 to 2.0 kg DM per 100 kg body weight equivalent to 65 to 100g DMI per kg $W^{0.75}$. DMI depends on the palatability as well as and protein and energy content the diet.

Nutritional requirements of buffaloes

The average nutritional requirements of buffaloes are generally lower than the requirements of cattle suggested by NRC (1988, 2001), but standards worked out by different workers show great variations. Therefore, nutritional requirements suggested for the dairy cattle, NRC have been used as basis for the preparation of nutritional requirements for buffaloes. The values given in tables (5 to 6) have been used for the preparation of rations and feeding schedules.

Table 5: Non-lactating animals

Animals			**Requirements**		
	DM kg	**CP kg**	**DCP kg**	**TDN kg**	**ME (M cal)**
Calves 6-12 months (150 kg)	3.4	0.55	0.30	2.3	8.3
Young stock 1-2 years (300 kg)	6.9	0.80	0.44	4.2	15.1
Young stock over 2 years (400 kg)	9.7	1.00	0.56	5.1	18.3
Grown Calves (450 kg)	11.2	0.41	0.22	4.1	14.8
Dry animals, last trimester (450 Kg)	11.2	0.93	0.52	4.5	16.2
Breeding bulls (550 kg)	12.3	0.85	0.47	4.6	16.6

Table 6: Lactating Animals (body weight 500 kg)

Milk Yield (kg)	**DM 100 kg B. wt.**			**Requirements**		
		DM kg	**CP kg**	**DCP kg**	**TDN kg**	**ME (M cal)**
Up to 5	2.5	12.5	0.99	0.55	5.9	21.2
6 to 8	2.5	12.5	1.30	0.72	7.1	25.5
9 to 11	3.0	15.0	1.7	0.95	8.6	30.9
12 to 15	3.0	15.0	2.2	1.23	10.4	37.4
16 to 17	3.0	15.0	2.5	1.40	11.3	40.6
18 to 20	3.5	17.5	2.8	1.56	12.7	45.7

NB: 1 kg CP = 0.56 kg DCP, 1 kg TDN = 3.6 M cal ME, 1.0 kg of milk (with 7% fat) requires 0.125 kg CP and 0.450 kg TDN

Common feedstuffs and their nutritive value

Common feedstuffs available for the feeding of buffaloes are listed in Table 7 along with their mean nutritive value. Availability of required quantity of green fodder is a common limiting factor at most of the places and even at the organised farms. Short supply of green fodders is frequently experienced during the month of May, June, October and November in most part of the northern states of India. The common diets of dairy buffaloes are made of a small quantity of chaffed green fodder, wheat straw/ Kadbi/ Paddy straw/ ragi straw and concentrates. The supply of green fodder is highly variable and diet of buffaloes may be devoid of green fodders for several months. However, for the supply of carotenes at least 3 - 4 kg green fodders are required from any edible source and should be made available for the maintenance of normal functions of the body. The fodder planning should be such that at least 30 kg of green fodder (6 kg DM) comes from green fodder like berseem, Lucerne, maize, jowar, lobia, bajra, paragrass, nappier, guinea, setaria.

Table 7: Nutritive values of fodders on raw basis (%)

		CP	DCP	TDN
1.	Jowar	6.2	3.3	52.4
2.	Maize	6.9	4.7	75.0
3.	Bajra	5.3	4.2	59.2
4.	Teosinte	7.3	4.3	58.3
5.	Hybrid Napier	6.8	3.2	55.2
6.	Guinea grass	5.8	3.1	52.4
7.	Para grass	7.9	5.9	45.6
8.	Cowpea (fodder)	18.6	12.3	51.4
9.	Soybean (fodder)	14.4	9.6	49.0
10.	Guar	8.1	6.6	49.0
11.	Lucerne	21.5	16.5	48.8
12.	Berseem	10.2	14.6	64.9
13.	Mustard	10.7	7.1	35.0
14.	Oats	10.5	8.7	82.0
15.	Ragi straw	0.2	0.2	55.1
16.	Jowar Kadbi	1.2	1.0	55.8
17.	Bajra Kadbi	0.9	0.7	52.9
18.	Maize Kadbi	1.0	0.9	53.0
19.	Wheat Bhusa	0.0	0.0	46.2
20.	Gram Bhusa	2.4	1.2	23.5
21.	Oat Straw	0.2	0.2	49.5

Buffaloes require different kinds of diets from birth to production stage of life cycle and for the purpose of feed ingredients selection, ration formulation and rate of feeding the following classification is suggested:

1. From birth to 3 month of age
2. From 3 month to 6 month of age
3. Feeding from 6 month to second trimester of pregnancy
4. Feeding during last 3 months of pregnancy
5. Feeding of lactating buffaloes

1. Calves from birth to 3 month of age

In early life larger part of the diet of calf is made of milk along with highly digestible concentrates i.e calf starter and some good quality fodders. Rumen is non functional and its development is almost completed in about 8-10 weeks in calves fed limited quantity of milk but takes longer time of 4 - 5 months in suckling calves consuming much higher quantity of milk. Just born calves (neonates) should be fed colostrums, preferably within 2 hr of birth, and continued daily for 4 days. Colostrum feeding is essential to provide antibodies, which can be absorbed intact in first few days of calf's life. It is also rich in all nutrients and being laxative helps to remove first faeces. If dam does not give colostrums, a substitute consisting of two eggs with about 30 ml of castor oil, may be given to calf.

Milk is more or less complete food and a calf should receive about one litre of milk in morning and evening up to 30 days. The quantity could be slowly decreased thereafter and beyond the age of 3 to 4 months the calf is used only for let down or the milk intake could be only up to 500 ml or so. The legume forage such as berseem/ lucerne/ cowpea/ saftal/ Methi or succulent cereals as oat, barley, maize, paragrass, guinea grass should be offered to calves after 10 days of age or it may be allowed to nibble forage offered to mother. A good quality concentrate of about 25 per cent crude protein and 75% TDN be fed.

Table 8: Ration Schedule for Buffalo Calves

B. Wt. (kg)	Age (Approx.)	Colostrum	Whole Milk	Skim Milk	Calf Starter
20-35	0-5 days	1/10th of B.W			
25-45	Up to 1 month		1/10th of B. wt.		
35-55	Between 1-2 m		1/15th of B. wt.	1/25th of B. wt.	125 g
50-70	Between 2-3 m		1/25th of B. wt.	1/15th of B. wt.	250 g
	3-4 month			6.5 litre	650 g
	4-5 month			6.5 litre	1.0 kg
	5-6 month			5.0 litre	1.5 kg

The suckling almost upto 6 month of age or even beyond that in certain cases, may be allowed. Suckling is said to be the problem of let down of milk in early weaned buffaloes. Average milk intake through suckling has been estimated to be about 200 - 300 kg during the first 3 months of life. Thus, a tentative feeding schedule of calves from first to 3 months of age with supplementary feeding of concentrate mixture and green fodder or good quality hay (Table 8) along with routine deworming should be followed (Table 9). The calf starter and green fodder should be offered from the second weak of age.

Table 9: Deworming Programme

Body Weight (Kg)			
25- 35	Sulmet	3rd Day	30 ml (2 table spoons)
25-36	Sulmet	4th Day	15 ml (1 table spoon)
26-36	Sulmet	5th Day	15 ml (- do -)
28-40	Piperazine	7th Day	5 g
35-45	Sulmet	1 month	45 ml (3 table spoons)
45-50	Piperazine	11/2 month	10 g in two instalments
55-65	Piperazine	2 1/2 month	- do -
60-75	Phenovis	3 1/2 month	9 g (2 spoons)
65-85	Sulmet	4 month	90 ml (6 spoons)
70-100	Piperazine	5 month	15 g in two instalments
75-110	Phenovis	6 month	13-18 g (3 to 4 spoons)
85-125	Piperazine	7 1/2 month	20 g in two instalments
100-140	Phenovis	9 month	18 g (4 spoons)
130-180	Phenovis	12 month	18 g (4 spoons)

Composition of calf starter

Crushed cereal grain	40 Parts
Crushed groundnut cake	20 Parts
Crushed mustard cake	20 Parts
Wheat bran	17 Parts
Mineral mixture	02 Parts
Common salt	01 Parts

It is not necessary to prepare a separate calf starter for the feeding of suckling calves and concentrate mixture used for the feeding of dams may be offered to calves provided it does not contain very high percentage of deoiled rice bran and mustard cake.

2. Calves from 3 to 6 month of age

Although suckling is allowed to continue beyond 3 month of age to facilitate the let down of milk but calves are allowed to suckle only a small quantity of milk. Thus, the major nutritional requirements of the calves have to be supplied through the feeding of concentrate mixture, green fodder and wheat straw. Since rumeno-reticulum of the calves is still in growing stage and animals generally do not eat large quantity of roughages and the feed intake capacity remains limited. Therefore, for the supply of optimum nutritional requirements a high protein-high energy diet is required. However, there is reasonable scope of compensatory feed intake and calves raised on the feeding milk diet supplemented with calf starter and fodder can eat about 3 to 3.25 kg dry matter (DM) per 100 kg body weight and this may continue upto 150 kg body weight. Thus, a palatable diet containing 13 -14% crude protein (CP) and 60-62% of total digestible nutrients (TDN) may support 500 - 600g average daily grain (ADG) in the body weight. One of the following feeding schedules may be followed for the feeding of growing young calves of 3 to 6 month of age.

(i) Feeding with cereal green fodder

The calves should be offered 1 kg concentrate mixture, 5 kg green cereal fodder and ad libitum wheat straw. Average DM intake is expected to range from 2.8 to 3.25 percent of body weight, which will able to support about 400 to 500g average daily gain in body weight.

(ii.) Feeding with green berseem fodder

The amount of concentrate mixture and wheat straw to be offered to calves depends on the supply of green berseem fodder. Feeding of 1 kg concentrate mixture with 15 kg green berseem and ad libitum wheat straw will be able to support 400 - 500g ADG upto 6 month of age.

(iii.) Feeding with wheat straw / paddy straw/ ragi straw/ jowar kadbi

A concentrate mixture containing about 20 per cent CP in air dry feed should be prepared by mixing crushed cereal grain, groundnut cake or cotton seed cake or soybean cake , mustard cake, wheat bran, mineral mixture and salt in the ratio of 30: 27:10:20:2:1. The diet should be made of 1.5 kg concentrate mixture, 1-3 kg available green fodder (or synthetic vitamin A) and ad libitum wheat straw. On the feeding of such rations, calves may not consume adequate feed. However, intake may be adequate to support about 300 - 400g ADG on such diets.

3. Buffalo heifers from 6 month of age

The ration of growing buffalo heifers may be described in two parts, i.e. feeding upto second trimester of pregnancy and feeding during the last trimester of pregnancy.

(a) Feeding upto second trimester of pregnancy

As usual the roughage part of daily ration will be straw supplemented with available (variable amount) green fodder and suitable concentrate mixture to meet the requirements for supporting at least 500g ADG in body weight. The roughage part of the diet may be variable quantity of cereal or leguminous green fodder and concentrate mixture. In most of the places, availability of green fodder for the feeding of non-lactating stock is limited to 5 to 20 kg per head per day. Thus, staple roughages are straws and stovers. The following feeding schedules may support about 500g ADG in growing buffalo heifers.

(i) A concentrate mixture containing about 20% CP and 63% TDN on air dry basis may be fed @ 1.5, 2.0, 2.5 and 3.0 kg per head daily at 100, 150, 200 and 250 kg body weight alongwith 5-10 kg cereal green fodder and ad libitum straw/ chaffed kadbi.

(ii) A concentrate mixture containing about 15% CP and 63% TDN on air dry basis may be fed along with 25-30 kg green berseem and ad libitum straw/ chaffed kadbi.

(iii) The quantity of concentrate mixture as in (ii) may be reduced to 1.0, 1.5, 2.0 and 2.5 kg at the corresponding body weight when plenty of leguminous green is available for feeding.

(b) Feeding of buffalo heifers during the last trimester of pregnancy

Buffalo heifers and other non-lactating pregnant animals should be fed to gain about 700g ADG during the last 3 months of gestation period. Depending on the quality of roughages the composition and quantity of concentrate mixture may be worked out.

(i) For feeding with straw / kadbi as staple roughage

The pregnant buffaloes should be fed 5 kg concentrate mixture, at least 2 –3 kg available green fodder and ad libitum straw/ kadbis. However, during the last week of pregnancy entire concentrate mixture may be replaced with wheat bran only.

(ii) Feeding with variable amount of cereal green fodder

The quantity of concentrate mixture and green fodder will be as follows -

Amount of green fodder (kg)	Conc. Mixture (kg)	Straw/ kadbi (kg)
05	4.5	ad libitum
10	4.0	-do-
15	3.5	-do-
20	3.0	-do-

(iii) Composition of ration with leguminous green fodder has been shown as follows –

Green legume (kg)	Conc. Mix. (kg)	Straw/ kadbi (kg)
10	4.0	ad libitum
15	3.5	-do-
20	3.0	-do-
25	2.5	-do-
30	2.0	-do-

4. Lactating buffaloes

A lactating buffalo should be allowed to consume adequate feeds and fodders to maintain her body weight to grow, if she has not achieved her mature body weight and also to producing optimum quantity and quality of milk. The diets of buffaloes may be made of following combination of roughages and source of roughages may also be different as follows -

A. Feeding with cereal green fodder and grasses

Green fodder supply per head per day is about (i) 5, (ii) 10, (iii) 20, (iv) 30, (v) 40 or (vi) 50 kg. The quality of cereal fodders changes with the stage of maturity being highest in protein, energy, minerals and vitamins at its vegetative stage, which gradually decreases with the maturity for most of the nutrients. An average value of 20% dry matter (DM), 10% crude protein (CP) and 60% total digestible nutrients (TDN) on DM basis for forages upto dough stage of maturity has been taken in to consideration for the calculation of the requirement of other feeds for balancing the diets for a buffalo (body weight 500 kg) producing 10 kg milk (with 7% fat).

Amount of green fodder (kg)	Conc. Mixture (kg)	Straw/ kadbi (kg)
05	8.0	ad libitum
10	7.5	-do-
20	7.0	-do-
30	6.5	-do-
40	6.0	-do-
50	5.0	-do-

B. Feeding with different quantity of green leguminous fodder

Berseem, Lucerne, cowpea and clusterbeans are the main sources of leguminous fodders and among them berseem is more popular in large part of

India. Therefore, berseem may be used as example. Lush growing berseem alone or mixed with green mustard contains about 10% dry matter and CP and TDN contents in DM are about 18 and 65 per cent, respectively, besides high calcium (1.5 – 2.5%) content. Therefore, a low protein and low calcium concentrate mixture should be prepared to avoid the wastage of valuable nutrients as excess, protein intake will be metabolised to produce ammonia which will give further stress to the liver for its conversion into urea and subsequent elimination from the body through urine.

Amount of legume fodder (kg)	Conc. Mixture (kg)	Straw/ kadbi (kg)
05	8.0	ad libitum
10	7.5	-do-
20	6.5	-do-
30	5.5	-do-
40	4.5	-do-
50	3.5	-do-

Bibliography

Cockril, W.R. (1976). The buffaloes of China. FAO, Rome.

FAOSTAT (2004). www.fao.org

ICAR (2002) Hand Book of Animal Husbandry. Indian Council of Agricultural Research, publication New Delhi, India.

Kearl, L.C. (1982). Nutrient Requirements Ruminants in Developing Countries. International Feedstuff Institute. Utah State University, Logan, Utah, U.S.A.

Kundu, S.S; Misra, A.K. and Pathak, P.S. (2004). Buffalo Production under Different Climatic Conditions. International Book Distributing Co., Lucknow, India.

Kushwaha, B.P., Maity, S.B. and Kundu, S.S. (2004). Present status and future perspectives of Bhadawari buffaloes. In: Buffalo Production under Different Climatic Conditions (S.S. Kundu, A.K. Mishra and P.S. Pathak, Eds). International Book Distributing Co., Lucknow, India.

Nivsarkar, A.E., Vij, P.K. and Tantia, M.S. (2000). Animal Genetic Resources of India: Cattle and Buffalo. ICAR, New Delhi.

NRC (2001). Nutrient Requirement of Dairy Cattle. National Research Council, Washington, D.C.

Pathak, N.N. (1998). Feeding Standards for Buffaloes. Animal Nutrition Division, Indian Veterinary Research Institute, Izatnagar.

Pathak, N.N. (2004). Fundamentals of buffalo nutrition and feeding. In: Buffalo Production under Different Climatic Conditions (S.S. Kundu, A.K. Mishra and P.S. Pathak, Eds). International Book Distributing Co., Lucknow, India.

Pathak, N.N. and Verma, D.N. (1993). Nutrient Requirement of Buffaloes. International Book Distributing Co. Lucknow.

Ranjhan, S.K. and Pathak, N.N. (1999). Management and Feeding of Buffaloes 3rd ed. Vikas Publishing House Pvt. Ltd., New Delhi.

Singh, S., Kundu, S.S., Misra, A.K. and Sachan, C.B. (2004). Nutritional studies on Bhadawari buffaloes. In: Buffalo Production under Different Climatic Conditions (S.S. Kundu, A.K. Mishra and P.S. Pathak, Eds). International Book Distributing Co., Lucknow, India.

SHEEP FEEDING

Sheep production is a traditional occupation of small and marginal farmers and landless labourers in the regions of the country where agricultural production is always a gamble due to unfavourable agro-climatic conditions. Sheep play a pivotal role in agrarian economy especially of semi arid, arid and cold regions of the country, providing economic sustenance of about 5 million families with poor socio-economic back ground. Sheep are also raised in other parts of the country. Garole known for its twining is raised in Sunderban region of West Bengal. A variety of products are produced from the sheep farming, which ranges from carpet to fine wool, leather, meat, manure and also milk to a limited extend. Sheep are traditionally raised under extensive range management on public range land, fallow (waste land) land and forest land. Sheep are able to collect their feeds by close clipping and they follow cattle and buffaloes in rangelands with good pasture cover. Young tender grass blades and weeds are liked by sheep, which they selectively consume from the plants/ pasture, leaving aside the coarse and riper parts. The farmers also extensively use top feed resources to sustain their animals during critical lean seasons. Despite lopping various feed resources, the nomadic farmers from semi-arid and arid zones are forced to migrate seeking greener pasture during lean period and return back to their native tract in favourable season. Thus the animals are exposed to vagaries of nature resulting in loss of production. The nomadic system is also prevalent in Himalayan regions.

Genetic resources

India has 42 descript breeds of sheep. A sizeable population is non-descript due to in-discriminate breeding and intermixing of breeds. In fact, there are no breed societies or agencies to register animals of a particular breed, but most of the breeds are well adapted to harsh climate, long migration, tropical diseases, poor nutrition and shortage of drinking water. The sheep breeds can be classified on the basis of agro-ecological regions (Table-1).

Table 1. Sheep breeds and their major products

North temperate	North-Western Arid and semi- arid	Southern peninsular	Eastern region
Bhakarwal (CW)	Chokla (CW)	Bellary (MCW)	Bollangir (MCW)
Changthangi (CW)	Hissar dale (HW)	Coimbatore (MCW)	Bonpula (MCW)
Gaddi (CW)	Jaisalmeri (MCW)	Deccani (MW)	Chottanagpuri (MCW)
Gurez(CW)	Jalauni (MCW)	Hassan (M)	Ganjam (MCW)
Karnah (AW)	Kheri (MCW)	Kanguri (M)	Garole (M)
Kashmir-Meriono (AW)	Magra (CW)	Kilakarsal (M)	Tibetan (CW)

North temperate	North-Western Arid and semi- arid	Southern peninsular	Eastern region
Poonchi (CW) Rampur Bushair (CW)	Malpura (MCW) Marwari (MCW) Muzaffarnagari (MCW) Nali (CW) Pattanwadi (CW) Pugal (MCW) Sonadi (MCW)	Madras Red (M) Mandya (M) Mecheri (M) Nellore (M) Nilgiri (AW) Rammand white (M) Tiruchy black (M) Vembur (M)	

AW= Apparel wool, CW=Carpet wool, MCW =Meat and Carpet Wool and M= Meat

Among the breeds, Marwari in North West and Deccani in Southern – Central Peninsular are numerically the most important and contribute largely for carpet wool and meat production.

Nutrient requirements

Indian breeds of sheep have much lower body weight and growth rate compared to exotic breeds. Besides the genetic make up, inadequate nutrition is the major factor responsible for low growth and body weight. Thus, there is considerable scope for improvement by feeding animals according to their requirements. Sheep obtain their nutrients from the close clipping of natural herbages from the pastures, which are mostly unmanaged in many tropical countries including India. In fact sheep with their small muzzles and split upper lips, can nibble tiny blades of plants which can not be eaten by large ruminants like cattle and buffalo. The sheep can consume 2 to 5 kg dry matter per 100 kg body weight depending upon the age, sex, production, feed quality and physiological stage. Adult sheep can maintain their health and wool production on a ration/ pasture providing 7-8% CP and 45-50% total digestible nutrients (TDN) on dry matter basis. However, pregnant and lactating ewes require slightly more palatable and nutritious ration/ pasture (CP 10-12% and TDN 52-55%). While growing lambs should be provided with more palatable and nutritious diets containing 12-16% CP and 58-65% TDN. Thus, the ratio between protein and energy should be narrower in early life and may gradually widen during the finishing period. (Table 2-3). High quality of abundant legume hay, silage or legume straw are needed for animals at each physiological stage. If hay is rather stemmy and poor in quality, the chopping may result in enhanced intake. They should be cut early. Under humid areas the stomach worms may be a problem, thus deworming may be adopted as per the observed incidences.

Table 2: Daily nutrient requirement (% of diet dry matter) of sheep

Body Weight (kg)	Dry Matter (as % of BW)	CP (%)	TDN (%)	Ca (%)	P (%)
Adult Sheep (Maintenance)					
20	2.88	7.2	45	0.19	0.14
30	2.58	7.2	45	0.19	0.13
40	2.41	7.3	45	0.20	0.13
50	2.28	7.2	45	0.20	0.13
60	2.18	7.3	45	0.20	0.13
Adult sheep for wool production					
20	3.65	8.0	45	0.21	0.14
30	3.33	8.0	45	0.20	0.13
40	3.08	8.0	45	0.20	0.14
50	2.94	8.0	45	0.20	0.13
60	2.80	7.9	45	0.20	0.13
Pregnant ewes					
30	4.27	10.4	52	0.30	0.21
40	3.98	10.6	52	0.30	0.21
50	3.76	10.6	52	0.30	0.21
60	3.60	10.6	52	0.30	0.21
Lactating ewes					
30	4.70	11.3	54	0.50	0.33
40	4.38	11.4	54	0.50	0.33
50	4.14	11.4	54	0.50	0.33
60	3.95	11.5	54	0.50	0.33

Growing lambs

	ADG (g/d)	DMI	CP	TDN	Ca	P
10	50	3.00	16.0	65.0	0.50	0.33
	100	3.40	16.0	65.0	0.50	0.33
	150	3.55	16.0	65.0	0.50	0.33
15	50	3.00	15.8	64.7	0.50	0.33
	100	3.40	15.8	64.7	0.50	0.33
	150	3.87	15.8	64.7	0.50	0.33
20	50	3.00	13.6	60.0	0.50	0.33
	100	3.40	13.6	60.0	0.50	0.33
	150	4.15	13.6	60.0	0.50	0.33
25	50	3.00	12.7	60.0	0.50	0.33
	100	3.40	12.6	60.0	0.50	0.33
	150	4.15	12.6	60.0	0.50	0.33
30	50	3.00	12.2	58.0	0.50	0.33
	100	3.45	12.2	58.0	0.50	0.33
	150	4.30	12.2	58.0	0.50	0.33

Source: NRC (1985), ICAR (1998)

Table 3: Daily macro & micro mineral requirement (% or ppm of diet dry matter) of sheep

Nutrients	Requirements
Sodium (%)	0.09-0.18
Chlorine (%)	-
Calcium (%)	0.20-0.82
Phosphorus (%)	0.16-0.38
Magnesium (%)	0.12-0.80
Potassium (%)	0.50-0.80
Sulphur (%)	0.14-0.26
Iodine (ppm)	0.10 – 0.80
Iron (ppm)	30 – 50
Copper (ppm)	7 – 11
Molybdenum (ppm)	0.5
Cobalt (ppm)	0.1 – 0.2
Manganese (ppm)	20 – 40
Zinc (ppm)	20 – 33
Selenium (ppm)	0.1 – 0.2

Source: NRC (1985)

Water

Like other animals water is also required to sheep. DM intake decreases, during periods of water deprivation. The water intake is about 13 per cent of body weight during summer months, which could be almost 50% during winter. Growing and fattening sheep require about 2 to 3 kg of water for each kg of DMI. In pregnant sheep water requirement increases by approximately 50 to 100% than growing animals during last 3 months of gestation period. The requirement also higher by about 50 % during early lactation depending on milk yield. Though sheep can tolerate 2 to 3 days of water deprivation, but for optimum results *ad-lib* clean water should be made available to the animals.

Eating habits

It has been observed that sheep, cattle and goats spent 60, 70, 20 per cent time for grazing on grasses, 30, 20, 20% on forbs and 10, 10 and 60% on browse, respectively. Sheep generally prefer to graze in the early morning and late in the afternoon. As such, sheep are good grazers and adapt to many types of management systems, feed resources and environmental conditions including the tropics, arid and sub-arid regions and temperate zones. Sheep are especially adapted to grazing on rough arid hilly lands and therefore, farm flocks are most common where much of the land is unsuited for tillage or taking crops. Sheep are also weed destroyers and fed on nearly all types of weeds.

Feeding

While feeding sheep, their nutritional requirements should be met during the different physiological stages like growth, pre-weaning, post- weaning and finishing, pregnancy, lactation and breeding etc.

(a) Pre-weaned lambs:

The practice of providing supplemental feed to nursing lambs in an area (that can not be entered by their dam) is called creep feeding. Lambs usually start creep feeding around 10-14 days of age and the consumption is inversely proportional to the amount of milk consumed. Upto 10-12 weeks of age suckling lambs should be supplemented with creep rations (Table 4). Again the consumption of creep feed is affected by the palatability of the ration (composition and ration form), the location and environment of the creep area. Initially, lambs prefer ground creep rations to pelleted form. After 4-5 weeks of age, while lambs prefer pelleted ration and thereafter lambs should be given unground grains. For higher weight gain, creep rations must be palatable and high in energy and must contain adequate protein (16 to 18% CP) minerals (especially calcium, since grains are low in calcium) and vitamins. Feeding of creep rations (Table 4) to Muzaffarnagari lambs upto 90 days of age resulted in 110-130 g daily gain in body weight, required to attain 25 kg live weight by 6 months of age.

Table 4: Ingredient composition of creep mixture for lambs

Ingredients	**Composition (%)**		
	A	**B**	**C**
Maize flour	67	50	30
Ground nut cake	10	10	10
Barley flour	-	17	-
Wheat bran	10	10	-
Oat flour	-	-	37
Rice polish	-	-	10
Fish meal	10	10	-
Meat meal	-	-	10
Mineral mixture	2	2	2
Common salt	1	1	1

In 100 kg of the creep mixture, adequate quantity of vitamins and oral antibiotic (powder) should be added. Fish meal and meat meal may be replaced with groundnut cake, soybean cake, linseed cake etc. on equinitrogen basis with slightly lower performances in lambs.

(b) Post weaned lambs

The requirements of CP and TDN for post weaned/ growing lambs have been given in Table 2. A ration providing 12-16 % CP and 58-65% TDN with 3.0 to 4.30 % dry matter consumption meet the requirements for expected average daily gain of 50-150 g. When quality fodders like green oats, cowpea, maize, berseem, lucerne, and their hays are available for *adlib* feeding, a concentrate mixture (cereal grains, groundnut cake, wheat bran, mineral mixture and common salt in a ratio of 27, 30, 40, 2, 1) containing 18-20% CP and 68-70% TDN should be fed @ 50-150 g/ head/ day depending upon the body size to meet the protein and energy requirements. But when there is availability of poor quality forages like mature grasses, straws and stoves, the above concentrate mixture may be fed at an enhanced rate viz. 300-600 g/ head/ day for the lambs weight 10-30 kg.

(c) Pregnant ewes

The last 6-7 weeks of pregnancy is considered to be very crucial for foetal development since 70-80% of gain in foetus mass is achieved during this period. Thus the foetus growth and pregnancy requirements are substantial and average approximately 0.5 times of maintenance requirement for single bearing ewes and 1.0 times of maintenance for twin bearing ewes. Total feed requirements thus increase to about 1.75 times of maintenance for this physiological phase. Ewes in their lambing may be further compensated for body growth through 10% additional nutrients. Under inadequate feeding the body reserves depleted and structural components are diverted to meet the increasing demand of fast growing foetus resulting different deficiency diseases relating to pregnancy and causes a considerable monetary loss to the farmers. A balanced ration containing 10-11 % CP and 50-55% TDN with 3.5- 4.5% DMI to be quite satisfactory. While on a good quality pasture/ browsing, supplementation of about 200 g concentrate mixture will also serve the purpose. DM intake may decrease substantially during last 10 to 15 days of gestation, possibly due to abdomen distension, hormonal changes or both. This may result in some body weight loss during this period. A diet consisting of some laxative ingredients, such as wheat bran or linseed cake may be increased to overcome the intake problem.

(d) Lactating ewes

Sheep have a relatively short lactation period and the actual quantity and composition of milk produced by animals for suckling young are difficult to measure. An ewe nursing twin lambs produces 20-40% more milk than an ewe nursing one lamb. Although, milk production responds to nutrient intake of the ewe and demand for milk by the lamb or lambs. It has been estimated that 65-83% of ME is converted to milk energy during 12 weeks of lactation. The daily feed requirement is about 4 -5% during first half of lactation period. Thus a ration supplying 11-12% CP and 55-60% TDN with sufficient quantity of minerals and vitamins is quite good to maintain the steady milk production. In fact, first 10 days of lactation ewes should be fed on *ad lib* legume hays and then in addition to *ad lib* hays supplementation of concentrate mixture @ 250 g/head/day. For each kg of milk production 5.25 MJ ME and 55g of digestible CP and 9 g calcium and 7 g phosphorous be provided.

(e) Feeding of breeding ewes/ rams

Flushing: The practice of increasing nutrient intake or the dynamic effect that influences body weight changes and conditions prior to and during breeding is called flushing. Its purpose is to increase the rate of ovulation and hence, the lambing rate. But the response to flushing is variable and the explanation for the response is not clear. A high level of hepatic steroid metabolizing enzymes (SME) is thought to be associated with an increased clearance rate of steroids, and decrease in steroids is associated with an increase in gonadotrophins and thus an increase in ovulation. The response to flushing is again affected by the age of ewe (mature ewes show a greater response than yearlings), its breed and the stage of breeding season. Practically about 2-3 weeks before the breeding, the rams are let loose with ewes and the ewes are offered about 200-300 g concentrate mixture in addition to standard feeding of good quality fodders/ rations resulting in more lambs production and lower lamb mortality.

Similarly breeding rams should be fed at the rate of 3 to 3.5% of body weight and rations providing 10-11% CP and 50-55% TDN with sufficient minerals and vitamins can maintain normal health and fertility.

Feeding for higher production

The level of production in sheep and other ruminants depend on the type of the animals, their genetic potential, rearing environment and health, while the over riding determinants are the intakes of digestible energy, protein, minerals vitamins and absence of antinutritional factors in feeds and fodders. Under traditional feeding practices the lambs hardly achieve 16 kg at six months of age and they are marketed for slaughter around 9-12 months of age weighting about 20-22 kg with dressed carcass weight of about 10 kg. Now it is visualized, based on the emerging awareness of quality meat by Indian consumers and export avenue to middle eastern countries that the sheep be raised on intensive feeding or grazing on nutritious pastures having tender grass and high legume with concentrate supplementation to meet the market requirement.

(a) Intensive feeding

The likely changes in agricultural production system in the country and priority of converting excess cereal production in certain parts of the country to animal produce is likely to the change the traditional pattern of small ruminant production favouring intensive system of production in large scale. Although, intensive feeding systems may not be adopted by the farmers in the whole country mainly due to their poor socio-economic background. Limited intensive feeding is practiced by some communities, involving 1 or 2 sheep/ goats per family and raised for sacrifice on the occasion of festivals or feasts on fancy feeding rather than commercial consideration. The crossbred weaner lambs in 90 days of intensive feeding had 180 g ADG compared to 127 g ADG in native lambs. Intensive feeding of concentrate and roughage (50:50) rations resulted in 170, 150 and 128 g ADG in crossbreed (meat), Malpura and Sonadi weaner lambs. Similarly fed native and crossbred fine wools breeds had ADG of 111 to 135 g with 1:12 feed efficiency ratio. Faster gain in intensively fed lambs was associated with higher carcass fat content, some time exceeding 20% hence weaning age was reduced from 90 to 60 days. The 60 days weaner under intensive feeding had 160 and 151 g ADG with 16 and 12% feed efficiency while similarly fed 90 day weaner lambs had 139 and 146 g ADG with 16 and 14 % feed efficiency. Slaughter studies indicated that irrespective of slaughter age and feeding regimen, carcass of finisher lambs weighing around 25 kg provided desirable carcass. This weight could be achieved in 73, 91 and 136 days of intensive feeding with 160, 135 and 112 g ADG and 18 , 16 and 14% feed efficiency in superior mutton breed lambs. The desirable finishing weight was attained by least cost ration formulation with conventional roughage sources, energy and protein supplemented complete feeds.

(b) Grazing with supplementation

Raising of lambs/ kids for meat purposes under grazing with supplementation, although cost effective for the farmer, has not largely been adopted by the farmer owing to their poor economic back ground and age old traditional practices. But it is being realized that with the progressive shrinkage of grazing land and concomitant increase in stocking density on such land, in future, majority of small ruminants may be raised under grazing with supplementation. Studies indicated that in addition to free grazing on range land limited amount of concentrate @ 1.5 to 2.0% of the body would provide 25 and 30 kg finishing weight at 6 and 9 months of age, respectively. However, grazing with supplementation on established pasture has potential of still higher production. On established cenchrus pasture stocked as per carrying capacity with concentrate supplementation @ 1.5% of body weight provided 27.3 kg finishing weight in lambs. Thus there is a scope to increase the slaughter weight of lambs at much lower age.

Feed Utilization

The overall goal of good nutrition is to optimize the efficiency of feed utilization, animal output and finally, the financial return to the farmers. Thus it is important to establish the potential of major feedstuffs and the need for appropriate supplements in order to overcome nutritional deficiency and raise the level of performance. With respect to the animal, the level of performance dictated by the quantity of feed voluntarily consumed and the efficiency of utilization of the major nutrients, namely energy and protein. Again the composition of animal products is equally important, which is affected by the feed utilization. Hence, to meet the requirements, recourse to the feedstuff and their appropriate utilization indices that are thought likely to influence the level and type of performance is necessary.

Energy utilization

Metabolizable energy (ME) as unit of energy is in vogue (AFRC, 1998), representing the total amount of energy available for metabolism. In practice, few direct estimates of ME contents of feedstuff are available and many are derived from digestibility and urine output measurement, conducted maintenance level of feeding with mature animals, and adjusted for estimated of methane production. In general these approaches to determine ME content and hence ME intake are considered satisfactory, but estimates of the efficiency of utilization of ME for maintenance (k_m), growth (k_f) and lactation (k_l) are subjected to doubt (Table 5). A major criticism of current energy system is that they are unable to predict the composition of gain, which is now primary importance to producers.

Table 5: Values for the efficiency of utilization of metabolizable energy of normal diets by ruminants

Function	**Efficiency when metabolizability (q_m) of diet is:**			
	0.4	**0.5**	**0.6**	**0.7**
Maintenance (k_m)	0.643	0.678	0.713	0.748
Pregnancy (k_c)	0.133	0.133	0.133	0.133
Growth and fattening (k_f)				
All diets	0.318	0.475	0.396	0.477
Pelleted diets	0.474	0.479	0.552	-
Lactation (k_l)	0.560	0.595	0.630	0.665

Source: AFRC (1998)

NRC (2001) for dairy animals, has adopted different method to obtain and express feed energy values which can also be adopted for small ruminants with some needed modifications. Earlier feedstuffs were assigned total digestible nutrients (TDN) values that has been determined experimentally using similar feeds. The concentrations of DE, ME and NE_L for each feedstuff were then calculated from TDN values using different equations.

DE (Mcal/ kg) = 0.04409 x TDN (%)

ME (Mcal/ kg) = 1.01 x DE (Mcal/ kg) – 0.45

NE_L (Mcal/ kg) = 0.0245 x TDN (%) – 0.12

Because of certain problems, the TDN values at maintenance level of intake (TDN_{1X}) have been calculated from composition data rather than being experimentally determined. In this summative approach to derive the TDN_{1X}, the concentrations (percent of dry matter) of truly digestible non-fiber carbohydrate (NFC), CP, ether extract (EE), and NDF for each feed are estimated using different equations. Either extract does not represent a nutritionally uniform fraction and therefore, does not have a constant digestibility across feedstuffs. Fatty acids (FA) are a uniform fraction with a true digestibility of 95 to 100 percent when diets contain 3% or less EE. A value of 100% digestibility was chosen, FA content of feed can be estimated as FA = EE -1. A more accurate approach would be to measure FA directly.

Truly digestible NFC (td NFC)

= 0.98 (100 – [(NDF – NDICP)
+ CP + EE + Ash]) x PAF

Truly digestible CP for forages (td CPf)

= CP x exp [- 1.2 x (ADICP/CP)]

Truly digestible CP for concentrates (td CPc)

= [1 – (0.4 x (ADICP/CP))] x CP

Truly digestible FA (td FA)

= FA Note: If EE<1, then FA = 0

Truly digestible NDF (td NDF)

= 0.75 X (NDFn - L) ′ [1 - $(L/NDFn)^{0.667}$]

In above equations, NDICP = neutral detergent insoluble N x 6.25, PAF = processing adjustment factor, ADICP = acid detergent insoluble N x 6.25, FA = fatty acids (i.e., EE – 1), L = acid detergent lignin, and NDFn = NDF – NDICP and all values are expressed as a percent of dry matter (DM).

Above equations are based on true digestibility, but TDN is based on apparent digestibility, therefore, metabolic fecal TDN must be subtracted from the sum of the digestible fractions. It has been determined that, on average, metabolic fecal TDN equally 7. The TDN_{Ix} is then calculated as follows-

$$TDN_{1x} (\%) = td\ NFC + td\ CP + (td\ FA \times 2.25) + td\ NDF - 7$$

While NE_L values are then calculated based on actual intake and the digestibility of the entire diet. Since the digestibility of diets fed to animals is reduced with increasing feed intake. This reduces the energy value of any given diet as feed intake increases. This is particularly important for high producing animals where feed intake even exceeds 4 times of maintenance level of intake.

$$DE_{1X} (Mcal/\ kg) = (td\ NFC/\ 100) \times 4.2 + (td\ NDF/\ 100) \times 4.2 + (td\ CP/\ 100) \times 5.6 + (FA/\ 100) \times 9.4 - 0.3$$

- DE at productive levels of intake (DEp)

  $$DEp (Mcal/\ kg) = DE_{1X} \times \text{Discount factor}$$

Where, Discount factor = $[(TDN_{1X} - [(0.18\ '\ TDN_{1X}) - 10.3]) \times Intake)]/ TDN_{1X}$

Intake = As incremental intake above maintenance

- ME at production level of intake (MEp)

  $$MEp (Mcal/\ kg) = [1.01 \times (DEp) - 0.45] + 0.0046 \times (EE - 3)$$

Where, DEp = Mcal/ kg & EE = % of DM

- NE_L at production level of intake (NE_{LP})

  $$NE_{LP} (Mcal/\ kg) = [0.73 \times MEp (Mcal/\ kg) - 0.19]$$

Protein utilization

The present UK metabolisable protein system for the estimation of protein requirements of ruminants (AFRC, 1993) recognizes that protein, like the carbohydrate present in feed, can be considered as a number of fractions with different degradability characteristics (Figure 1). The immediately soluble protein, akin to WSC, is assumed to be readily utilized by the rumen microflora (QDN, quickly degraded nitrogen), while a second insoluble fraction is degraded more slowly (SDN, slowly degraded nitrogen) leaving a rumen undegradability fraction (UDN, undegradable nitrogen). Assessment of protein quality ideally requires a measure of these fractions coupled with the rate at which the SDN fraction is degraded. Various laboratory methods can be used to measure the immediate soluble fraction, and the amount of nitrogen remaining associated with ADF (ADIN, acid detergent insoluble nitrogen) used to indicate the size of the non-degradable fraction. Subtraction of these values from he total protein will provide an estimate of the amount of SDN present. However, this is a crude approximation, and the various regression equations developed to predicts the needs for growth, lactation or wool production require kinetic data. Typically, values for QDN, SDN, UDN and the necessary rate data are still obtained by the nylon bag method. As a result, ration formulation for ruminants may utilize actual CP values and, occasionally AIDN, but otherwise is based wholly on table values and is unable to take account of variations in raw ingredients.

While NRC (2001) has adopted the model of Cornell Net Carbohydrate and Protein System (CNCPS). In this model, feed CP is divided into five fractions (A, B_1, B_2, B_3, and C) which sum to unity. The five fractions have different rates of ruminal degradation. Fraction A (NPN) is the percentage of CP that is instantaneously solubilized at zero time, which is assumed to have a degradation rate (k_d) of infinity, it is determined chemically as that proportion of CP that is soluble in borate-phosphate buffer but not precipitated with the protein denaturant, trichloroacetic acetic (TCA) (Figure 2). Fraction C is determined chemically as the percentage of total CP recovered with ADF (i.e., ADIN) and is considered to be undegradable. Fraction C contains proteins associated with lignin and tannins and heat-damaged proteins such as the Maillard reaction products. The remaining B fractions represent potentially degradable true protein. The amounts of each of these 3 fractions that are degraded in the rumen are determined by their fractional rates of degradation (k_d) and passage (k_p); a single k_p value is used for all fractions. Fraction B_1 is that percentage of total CP that is soluble in borate-phosphate buffer and precipitated with TCA. Fraction B_3 is calculated as the difference between the portions of total CP recovered with NDF (i.e., NDIN) and ADF (i.e., fraction C). Fraction B_2 is the remaining CP and is calculated as total CP minus the sum of fractions A, B_1, B_3, and C.

Reported ranges for the fractional rates of degradation for the three B fractions are: B_1 (120 – 400%/h), B_2 (3 – 16%/h), and B_3 (0.06 – 0.55%/h). The RDP and RUP values (percent of CP) for a feedstuff using this model are computed using the equations:

$$RDP = A + B_1\ [k_dB_1 / (k_dB_1 + k_p)] + B_2\ [k_dB_2 / (k_dB_2 + k_p)] + B_3\ [k_dB_3 / (k_dB_3 + k_p)]$$

$$RUP = B_1\ [k_p / (k_dB_1 + k_p)] + B_2\ [k_p / (k_dB_2 + k_p)] + B_3\ [k_p / (k_dB_3 + k_p)] + C$$

Although the most commonly used model to describe in situ ruminal protein degradation divides feed CP into three fractions (A, B, and C). Fraction A is the percentage of total CP that is NPN (i.e., assumed to be instantly degraded) and a small amount of true protein that rapidly escapes from the *in situ* bag because of high solubility or very small particle size. Fraction C is the percentage of CP that is completely undegradable; this fraction generally is determined as the feed CP remaining in the bag at a defined end-point of degradation. Fraction B is the rest of the CP and includes the proteins that are potentially degradable.

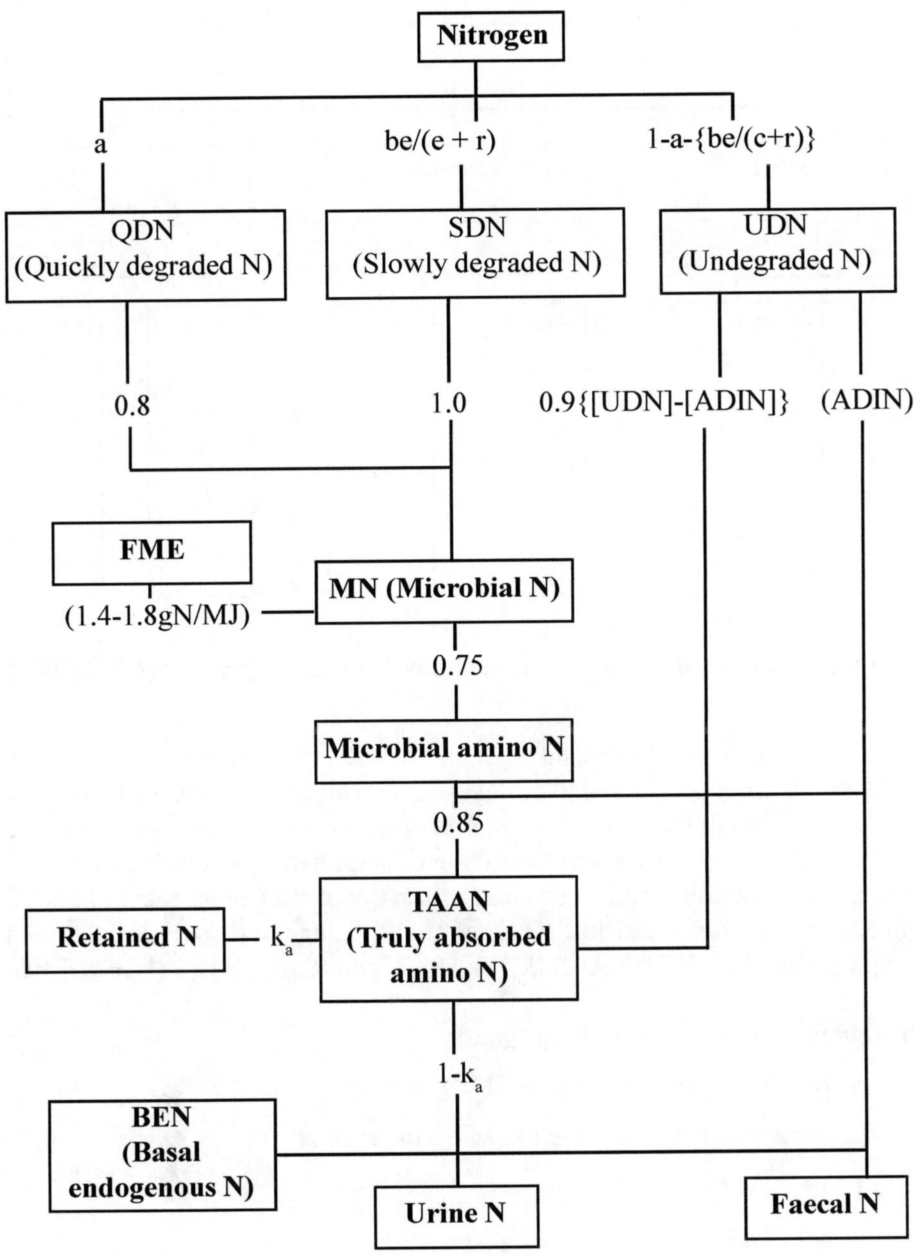

Proportion of N lost at time, t, dg = a + b $(1 - e^{-ct})$

Figure 1: Scheme for the description of protein values in terms of truly absorbed amino nitrogen (AFRC, 1993)

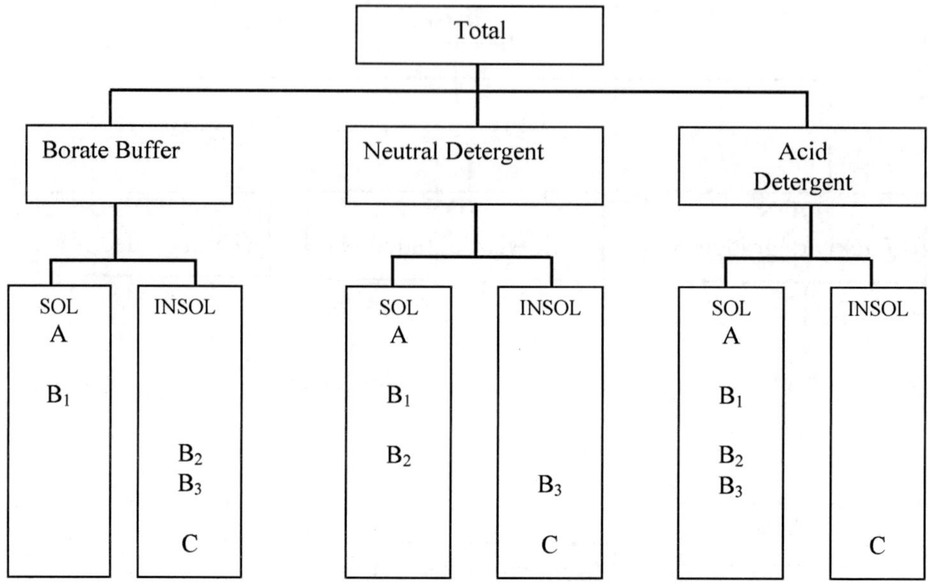

Figure 2: Analysis of crude protein fractions using CNCP System (NRC, 2001)

Only the B fraction is considered to be affected by relative rates of passage; all of fraction A is considered to be degraded and all of fraction C is considered to pass to the small intestine. The amount of fraction B that is degraded in the rumen is determined by the fractional rate of degradation that is determined in the study for fraction B and an estimate of fractional rates of passage. The RDP and RUP values for a feedstuff (percent of CP) using the model are computed using the equations $RDP = A+B\ [k_d / (k_d + k_p)]$ and $RUP = B\ [k_p / (k_d + k_p)] + C$.

Protein requirements for sheep/ goats

AFRC- Metabolizable Protein (MP) concept

(a) Maintenance requirements (MP_m) of goats

$= MP_b + MP_d$

$= 2.1785\ W^{0.75} + 0.1125\ W^{0.75}$

$= 2.30\ W^{0.75}$ g/day

Where MP_b = MP required for basal endogen on nitrogen loss

MP_d = MP required for dermal loss as scarf and hair

For lambs (MP_m)

= MP_b = 2.1785 $W^{0.75}$ g/day (as wool growth is proportional rate of gain and no allowance is made for scurf and hair losses)

For ewes (MP_m)

= MP_b = 2.1785 $W^{0.75}$ + 20.4 g/day (allowance for wool growth)

(b) Requirements for milk (MP_l)

MP_l, g/kg milk = (0.95 x protein content of milk)/ 0.68

= 1.471 x true protein content of milk (48.9 and 29 g/kg in

sheep and goats, respectively)

(c) Allowance for live weight change in ewes and goats

MP_g for live weight gain = 140 g/kg

MP_g for live weight loss = 119 g/kg

In goats MP requirement is reduced by 50 g/day in early lactation

Energy Requirements of sheep

AFRC- Net Energy concept

(a) Energy requirement for maintenance (E_m)

E_m (MJ/d) = 0.226 (W/ 1.08)$^{0.75}$ + 0.0096 W

The activity increment figure 0.0096 W is based on the following assumptions-

Time spent standing = 14 h

No. of positional changes = 14

Distance walked = 50 m

(b) Energy secreted in milk (E_l)

E_l (MJ/d) = 4.6 ´ Y

Where Y = Yield of milk (kg/d)

(c) Energy value of deposited or mobilized body tissue

EV_g = 26 MJ/kg

(d) Efficiency of utilization of the dietary ME for various functions

(i) Maintenance (k_m), $k_m = 0.35\ q_m + 0.503$ (where q_m = ME/GE)

(ii) Milk production (k_l), $k_l = 0.35\ q_m + 0.42$

(iii) Live weight gain (k_f), $k_f = 0.95\ k_l$

(iv) The efficiency of utilization of the energy of mobilized body tissue for milk production assumed to be 0.84

(e) Correction for level of feeding (C_L)

$C_L = 1 + 0.018\ (Y \times EV_l / k_l + W \times EV_g / k_g) \times k_m / E_m$

Where Y = Yield of milk (kg/d)

W = live weight change (kg/d)

EV_g = 27.36 MJ/kg

$k_g = k_l$ when W is positive

$k_g = k_l / 0.80$ when W is negative

Thus the advancement in utilization of feedstuffs in relation to both animal performance and health is highly significant, since our feed resources are varied and location specific. Their effective and economic/ judicious utilization, keeping in view the requirements of different categories of small ruminants, will go a long way in making sheep and goat farming a viable proposition. Accordingly efforts were made to record the utilization of different feedstuff/ total mixed rations following CNCP Systems of analysis. But feeding systems need to be based on the mechanisms that govern the response of animals to nutrients, dealing with quantitative aspects of the digestion and metabolism in the ruminants.

Bibliography

AFRC (1993). *Energy and protein requirements of ruminants*. An adversary manual prepared by the AFRC technical committee on responses to nutrients. CAB international, walling ford, UK.

AFRC (1998). *The Nutrition of Goats*. AFRC technical committee on responses to nutrients. Report No. 10. CAB international, walling ford, UK.

Anonymous (1998). Proceedings of Golden Jubilee Seminar on sheep, goat and rabbit production & utilization during April 24 -26, 1998 at Jaipur, Rajasthan.

Arora, A.L. and Karim, S.A. (1995). Growth response of intensively fed lambs. *Indian J. Anim. Sci.,* 65: 1046-1048.

ICAR (1998). *Nutrient Requirements of Livestock and Poultry*. ICAR, New Delhi.

Kundu, S.S., Mahanta, S.K., Singh, N.P., Singh, S.S. and Gupta, A. (2001). Feed quality assurance for higher productivity of livestock. Proceed. *Xth Anim. Nutr. Conf.* During Nov. 9-11, 2001 at NDRI, Karnal.

Mahanta, S.K., Singh, Sultan, Kumar, A. And Pachauri, V.C. (1999). Subabul leaf meal as a replacement of mustard cake in lamb diets. *Small Rumin. Res.*, 32 : 37-42

NRC (1985). *Nutrient Requirements of Sheep* (6th rev. edn.). National Academy Press, Washington DC.

NRC (2001). *Nutrient Requirements of Dairy cattle* (7th rev. edn.). National Academy Press, Washington DC.

Pathak, N.N. and Kamra, D.N. (1989). *A Text Book of Livestock Feeding in Tropics*. Cosmo Publications, New Delhi.

Pailan, G.H; Verma, N.C; Sharma, D.K. and Gupta, J.N. (2002). Nutrient utilization in sheep fed diet containing different proportion of grass, legume and to feed. *Indian J. Animal Nutrition*, 19 (3): 199 – 203

Pailan, G.H; Shrama, D.K; Verma, N.C. and Gupta, J.N. (2003). Supplemental value of legume and tree leaves in grassland diet for pregnant sheep. *Range Management and Agroforestry*. 24 (2): 135-139

Ranjhan, S.K. (1993). *Animal Nutrition and Feeding Practices* (4th rev. edn). Vikas Publishing House Pvt. Ltd., New Delhi

Thakur, S.S., Kundu, S.S. and Sharma, D.D. (1986). Sheep nutrition research in India. *Inter. J. Anim. Sci.*, 2: 14-48

GOATS FEEDING

Goats rank fourth, after cattle, sheep and pigs and the estimate of their world population is about 450 millions. They are sporadically distributed in all parts of the world, but more densified in dry and humid tropical areas. In industrialized countries where goat husbandry is equated to poverty, goat almost disappeared except in countries, where it was a part of cultural heritage. In India also goats are kept by poor and specially by the traditional castes/tribes. Search for a higher social status involves an increase of cattle and buffaloes rather than goats. Despite of all these odds their number is increasing, since the animal has wide adaptability and able to thrive almost in every agro- ecological niches of the country. Rising of goats from very extensive to intensive situation further indicates their ability to adapt wide range of conditions. The importance of goats does not only lie as protein supplier (meat, milk) or for mohair, but also as a fix bank deposit for poor farmers. It thus leads to more egalitarians social system. Goat husbandry has received poor attention in research and development programmes in past. Moreover, foresters, considered the goat as a destroyer of vegetation, which has been proved a myth and overgrazing by any species, will be equally harmful. Research and development fronts/ issues of goatary have been poorly attended in the past. Due to changing food habits of a large section of society during last quarter of the century, researchers, managers and policy planners have started renewed in goats.

BREEDS

There are about 19 well-known Indian breeds, apart from a number of local non-descriptive scattered throughout the country (Table 1). The breeds are described below based on their locations.

Himalayan (Hilly Tract) Region

This region comprises Jammu and Kashmir, Himachal Pradesh and parts of Uttar Pradesh.

Himalayan breed

The goats of this breed are white-haired and sturdily built. The breed is also known as Chamba, Gaddi and Kashmiri, after the localities where these goats are reared. They inhabit Kangra and Kulu valleys, Chamba, Sirmur and Simla in Himachal Pradesh, and parts of Jammu hills. Castrated bucks are used for transporting merchandise in the hilly tracts.

Pashmina

The goats of this breed are small dainty animals with quick movements. They are raised at elevations above 3,400 m in the Himalayas, Ladakh, and Lahaul and Spiti valleys. They produce the softest and the warmest animal fibre used for high-quality fabrics. The yield of this valuable downy hair varies from 75 to 150 g per goat.

Chegu: This breed is found in the mountainous range of Spiti, Yaksar and Kashmir. The goats of this breed yield pashmina, good meat and a small quantity of milk.

Table: 1 Important goat breeds in India

Name of breed	Home tract	State	Function
Assam Hill goats	Village of Khasi, Naga and Dushai Hills	Assam	Meat
Barbari	Agra, Aligarh, Etah, Mainpuri, Mathura		Dual
	Delhi	Delhi	
	Haryana	Haryana	
Beetal	Gurdaspur	Punjab	Milk
Black Bengal	Villages of Orissa and Bengal	West Bengal and Orissa	Meat
Chegu	High Mountains of Spiti, Kashmir & Tibet	Kashmir	Pashmina
Kutchi	Kutch, Kathiawad	Gujarat	Meat
Gaddi	Hills of Himachal Pradesh	Himachal Pradesh, Jammu and Kashmir	Hair
Ganjan	Orissa and parts of Bengal	Orissa	Meat
Jamunapari	Chambal ravines, Chakanagar area of Etawah district	Uttar Pradesh	Dual
Changhtangi	Hilly tracts of Leh and Ladakh	Kashmir	Pashmina
Bakhrawal	Hilly tracts of Kashmir	Kashmir	Hair
Malabari and Tellicheri	North Malabar in taluka of Kusum, Kottayam, Calicut and Ponnani	Kerala	Milk
Marwari	Jodhpur, Mehsana and Marwar	Rajasthan	Dual
Osmanabadi	Osmanabad	Andhra Pradesh	Dual
Sirohi	Sirohi district, Udaipur	Rajasthan	Dual
Mehsana	Baroda and Mehsana	Gujarat	Milk
Surti	Surat district	Gujarat	Milk
Zalwadi	Zalwad district in Kathiawad and Western Mehsana district	Gujarat	Dual
Zakhrana and Parbatsar	Ajmer and Nagore	Rajasthan	Milk

Northern Region

This region comprises Punjab, Haryana and parts of Uttar Pradesh. In this region there are a few important milch breeds.

Jamunapari

This breed is usually found in the Etawah district in Uttar Pradesh, and in the tracts lying between Jamuna and Chambal rivers. The goats of this breed are large-sized, tall and leggy, with convex face line, large folded pendulous ears and prominent Roman nose. They carry long and thick hair on their hindquarters and possess a glossy coat. The horns are short and flat. The length and height vary, respectively, from 127 to 137 cm and 91 to 102 cm in the buck, and from 116 to 127 cm and 76 to 86 cm in the doe.

The body weight of adult bucks and does varies from 65 to 86 kg and 45 to 61 kg respectively. The average daily milk yield is 2.25 to 2.7 kg. The milk yield in a lactation period of 250 days varies from 250 to 300 kg with 3.5 per cent fat content.

The Jamunapari goats have been used for evolving the famous Anglo-Nubian breed of goats in England. According to the British Goat Society Yearbook (1940) the two Jamunapari bucks, viz. 'Sedgemere Chancellor' and 'Bruchest cross' (the first imported in 1896 and the - second in 1904), used for breeding have greatly influenced the building up of the Anglo-Nubian breed.

Beetal

Mainly found in Punjab and has been evolved from Jamunapari breed. The colour of the goats of this breed is predominantly black, red, tan or black, heavily spotted on white also occur. The length, height and body weight of bucks range, respectively, from 127 to 132 cm, 91 to 99 cm and 65 to 86 kg, and of does from 107 to 122 cm, 76 to 83 cm and 45 to 61 kg. Does yield about 1 kg of milk daily, the maximum yield 591 kg in 177 days lactation period. Bucks generally carry a beard.

Barbari

This breed has its origin from Berboa in Somali Republic in East Africa. The goats are small, short haired and erect-horned, and are popular in urban areas of Delhi, Aligarh, Etah, Etawah, Agra and Mathura in Uttar Pradesh, and in Gurgaon, Karnal, Panipat and Rohtak in Haryana. The colour varies, with white, red and tan spots being common. The length of bucks and does ranges from 96 to 112 cm and 91 to 114 cm, respectively, and the height ranges from

66 to 76 cm and 61 to 71 cm respectively. The adult buck weighs 36 to 45 kg and the doe 27 to 36 kg. They are usually stall-fed and yield 0.90 to 1.25 kg of milk (fat content, 5 per cent) per day in a lactation period of 108 days. They are prolific breeders and usually kid twice in 12 to 15 months. The outstanding quality of this breed is its habit of stall-feeding, which makes it most suitable for cities and towns where grazing facilities are lacking.

Central Region

This region comprises Rajasthan, Madhya Pradesh, Gujarat and Northern parts of Maharashtra.

Marwari, Mehsana and Zelwadi:

These breeds are derived from Jamunapari breed. Commonly found in Rajasthan, Gujarat and Madhya Pradesh. Different colour combinations are found within these breeds. They yield 0.75 to 1.0 kg of milk per day.

Berari

This breed is found in Nagpur and Wardha districts of Maharashtra and Nimar district of Madhya Pradesh. The Berari is tall and dark -colored. The doe yields about 0.6 kg of milk per day.

Kathiawari

This breed is a native of Kutch, Northern Gujarat and Rajasthan. The goats have a black coat and reddish mark on the neck. The doe yields about 1.25 kg of milk per day.

Southern Region

The Southern region comprises parts of Maharashtra, Gujarat, Andhra Pradesh, Tamil Nadu and Kerala. This region possesses distinct breeds, viz. Surti, Deccani, Osmanabadi and Malabari.

Surti

This breed is probably derived from the small Arabian milking goats. Surti goats resemble Berari goats in broad details and possess white, short legs. Surti is popular in Bombay, Nasik and Surat and is known after the locality it inhabits. The does are good milk producers, yielding 2.25 kg per day.

Deccani and Osmanabadi:

These have originated from a mixture of goats of the plains. They are black; mixtures of white and black or red are also found. The milk yield is 1.4 to 2.25 kg per day.

Malabari (Tellicherry)

This breed is found in Northern Kerala. The goats of this breed seem to possess Surti blood, being a mixture of two or more types of goats. They have no uniform color. They yield 0.9 to 2.8 kg of milk per day.

Eastern Region

This region comprises West Bengal, Assam, Tripura, Orissa and part of Bihar. Variations in climate and heavy rainfall are not suitable for rearing high milk-yielding types of goats. The goats in this region manage to thrive on meager food.

Bengal

The goats of this breed are found in 3 colors, viz. black, brown and white. Their legs are short and the body is deep. The meat is regarded as excellent. Bucks weigh 14 to 16 kg and nannies 9 to 14 kg. Nannies kid twice in a year; twins are common. The skin of the Bengal goat is of superior quality and is in great demand both in India and abroad in foot wear industry. The milk yield is low.

Assam hilly breed

This breed is similar to the hilly breeds seen in Kangra and Kulu valleys and in the Himalayan ranges of Uttar Pradesh, but has been described as a distinct type.

There are also other varieties of goats, but it is doubtful whether they represent distinct breeds or are mere types. Their names are, therefore, not mentioned here.

On account of the milk-producing qualities, Jamunapari, Beetal and Barbari goats are in great demand. These breeds are maintained on different government farms for providing bucks of approved quality to the rural areas for improving the local breeds or the non-descript.

Some exotic breeds, known for high genetic potentialities for milk, are being used for evolving new breeds of milch goats, suitable for different climatic conditions and capable of yielding more milk.

Salient Features of A Good Milch Goat

Good milch breeds differ from one another in their breed characters, such as the color, shape of head and ears, size and coat, but certain general features may be recognized as' being constant in the majority of these breeds. These are described below.

Head: The head should be long and of moderate width, with a full and well-developed muzzle and prominent nostrils. It should be naturally hornless or disbudded. The head in the doe should be well carried and should bear a feminine appearance. The eyes should be large and bright, set well apart, indicating docility and some capacity for understanding.

Neck and shoulders: The neck should be long and slim, and of good depth, with the tassels, if present, evenly hung. The withers and the shoulders should be fine in appearance, and connect the neck with the body with little break in continuity. A considerable thickness in the shoulders or a drop immediately between the shoulder blades is undesirable.

Chest: This should be moderately deep and have good width, giving the appearance of strength without coarseness.

Forelegs: The forelegs should be straight, strong and possess good bone.

Feet: The animal should stand well on its legs without the tendency' to 'turn toes' or 'walk on heels'. The hairy growth covering the hoof should be kept trimmed to the 'sole' of the foot, leaving the latter bare underneath.

Body: Good depth is an important feature. It should drop in a gradual curve from the point where the abdomen unites with the chest, i.e. a little way behind the forelegs, and then rise slightly again to meet the udder. The back should be level from the shoulders to the hips and then drop slightly at the tail region. The hips are often slightly higher than the shoulders, but this need not be regarded as a defect. An excessive dip in the back, however, is highly undesirable. Plenty of length from the head to the tail is a desirable feature.

Ribs: The ribs should be well sprung so as to give a barrel effect; flat sides are a common fault. The abdomen should not be protruding beyond the width of the ribs, so that its roundness is not affected.

Hind-quarters: There should be sufficient width across the hips and the rump, and between the pin-bone and the hocks. The hind-legs should face straight forward and not outward; the latter tendency is one of the commonest defects in goats resulting in the so-called 'cow-hocks'. There should be a slight rise from the back to the hips and a gradual drop from the rump to the tail. A sharp drop from the hips to the tail is regarded as a defect

Hind-legs: Bones should give the appearance of strength with the hocks slightly bent. The pastern should be short, its joint 'showing no sign of weakness that might result in dropped pastern.

Udder and teats: The udder should be carried well under the body. When viewed from the side it should be in front of the hind-legs. It should be large, but its size should be proportional to the' size of the goat. The skin of the udder is usually covered partly with fine, soft and pliable hair. The texture of the udder should be reasonably soft. Heavy milk production is associated with the amount of milk-secreting tissue. The udder should undergo considerable shrinkage after milking. Milk ducts and teats should be entirely free from hard lumps, these being sometimes discovered only after milking. Such lumps indicate chronic inflammation, which may be due to several causes. The teats should be quite separate from the udder with a distinct line of demarcation, point downward and slightly forward, and be of moderate length and of suitable size to be conveniently held in hand during milking. Large milk-veins should be present under the belly and lead to the udder. These veins, although usually better developed in old goats, indicate superior milch quality. The shape of the udder in the Indian breeds varies considerably. The occurrence of four milking quarters, all lending themselves to being used in milking, is not uncommon in Jamunapari goats. This trait is found more in certain strains and is probably inherited.

Skin and hair: The skin should be soft, supple and loose. The coat varies in different breeds, but is generally glossy, with fine, short hair.

Judging of goats:

In the following Table (2) standards denoted by numbers have been mentioned for judging the goats.

Table: 2. Standards for judging goats

Character	Doe	Buck
General appearance	5	5
Style and quality	8	10
Temperament	3	5
Head	6	10
Neck	3	3
Fore-quarters	10	15
Body	13	15
Hind-quarters	10	15

Character	Doe	Buck
Legs	5	5
Udder	25	0
Genital organs	0	12
Size and weight	5	5
Polled (naturally)	2	0

Feeding

Despite the fact that the goat was undoubtedly the first ruminant domesticated by man, little work has been carried out to determine the actual nutritional requirements of goats.

Feeding Behavior

The feeding behavior of goat is more complex than other ruminants. They frequently begin by sampling, selecting the fractions it prefer to eat as, leaves and stems. Consequently, the daily time spent in eating by goats generally exceeds than that of sheep. It has been said that they eat round the clock. During stall-feeding, the refusals may be as high as 50%. They specifically choose more leaves than stems. Consequently, the quantity of ration may be higher specially when it is of poor quality. Time spent in ruminating was shorter than sheep during day, and goats ruminate much more during night. Goats generally vary to choose feeds. They may refuse badly conserved feeds, rapeseed meals, urea or high levels of minerals. These problems are much more acute in high yielder. Looking to feeding behavior, goat feeders, with specific design are used to minimize the spoiling and wastage of feeds.

In natural state, given a free hand, majority of the goats prefer browsing rather than grazing or nibbling. Standing on their hind limbs they always like to pluck the tender leaves of plants. Goats have fastidious eating habits. They will accept a wide variety of feed, appreciate it and thrive on it, but what is acceptable to one goat is not always accepted to others. Goats may not relish feeds soiled by other animals. They prefer to select the varieties of feeds and vegetation out of available stuffs. Furthermore the appetite of the goat for any given concentrate or mixture of concentrate fed in quantities over (0.45 – 0.91 kg) per day often decrease within a short period. Goats are largely though not completely omnivorous.

Goats can distinguish between various tastes and they have higher tolerance for bitter tastes than cattle. Therefore, goats can consume a wide range of plant species than either sheep or cattle. Goats may consume certain species at definite stages of maturity and reject them at other times. Higher intakes of grass are normal when goats are grazing in improved pasture, while large percentage of weeds are consumed when they are browsing over wastelands. Over 80% of their feed intake can consist of leave of shrubs and bushes when they are living in a bush country.

There is universal agreement that considerable care in feeding is required if goats are to produce large quantities of milk or are to grow quickly and yield high quality meat. Hay, pasture or concentrates supplements are required for sustained lactation by milking does. At least as much skilled required in formulating adequate rations for high producing dairy goats as is required in the proper feeding of high producing dairy cow. In addition, the successful goat dairyman must always be aware of the feeding idiosyncrasies of this individual animal.

Pasture

Goats are very selective feeders. They choose very carefully, the kind of plant eaten amongst the vegetation, and the parts ingested of each plant, which varies with the season. Moreover goats will discontinue grazing at any minor disturbance like rain or the appearances of an unknown person. Nevertheless, it is possible to make better utilization of pasture by maintaining a high animal density to obtain greater competition between animals.

Rangelands

The choice behavior of goats is generally a disadvantage in intensive pasture or at the trough, but it is favorable in rangelands, particularly in poor quality ranges. Goats move and climb up and down to search the most nutritive plants or parts of grass. They can stand on their hind legs with apparent ease in order to consume foliage, which is unavailable to sheep. In ranges the goat consumes a large number of plants but the species or the plant fractions ingested vary in space and time. Independent of range quality, goats as well as other ruminants prefers vegetative fractions with a high leaf/stem, seed/leaf ratio, low fiber and high protein content.

The more limited the number of plants in an area, the more the diet chosen by the animal corresponds to what is available. The preference of goats for herbaceous strata and especially for grasses becomes evident when grasses are growing. After the disappearance or desiccation of the annual species, the goats go to the low bushes. In the area of CHAPARRAL vegetation, goats may be generalist in their feeding strategies but display some sort of specialized behavior in selecting between the most dominant species. Selection of herbaceous vegetation restricted by availability, corroborates the classification of goats as adaptative mixed feeders rather than browsers. Goats do not prefer woody vegetation but they accept and consume it more easily than sheep. So the behavior of goats, and their ability to search out the most nutritive plants in rangelands partly explains, why goats adapt better than sheep or cattle in arid and even desert areas.

Nutrients Requirements

Dry matter:

Dry matter intake level depends on animal and the diet. The animal factors include live weight, breed, physiological stage, and level of growth/production, while diet factors consist of palatability, protein content, stage of maturity, balance of micro and macronutrients and presence of anti-nutritional factors. One of the most appropriate model proposed to predict probable DMI value, is as follow:

$$DMI = 423.2\ FCMY + 27.8\ (BW)^{0.75} + 440\ CLWG + 6.75\ HP$$

Where,

DMI	=	Dry matter intake (g/d)
FCMY	=	3.5% Fat corrected milk yield (kg/d)
$(BW)^{0.75}$	=	Metabolic body weight (kg)
CLWG	=	Corrected live weight gain (LWG kg- DMI/kg/day) (kg/month)
HP	=	Percent hay in the diet.
		The above model explains 84% of the diet variation.

DMI of adults at maintenance:

To estimate the maintenance dry matter requirement following equation can be used. DMI (kg/d)= 0.522+0.0135 W

The equation for pregnant and lactating goats is:

DMI (kg/d)= 0.062 W 0.75+ 0.305 Y

Where,

W= live weight in kg

Y= milk yield in kg/d at 3.5 % milk fat

Voluntary DMI of goats fed *ad libitum* is generally sufficient to meet their requirement. Intake of Dry Matter may not be sufficient during later part of pregnancy because of reduced rumen volume, probably due to stored fat inside the abdomen and little on the carcass, and also significant volume of foetuses because of the high rate of prolificacy or twinning. Gluconeogenesis may also not be sufficient to meet fetal requirements. Such a situation may lead to ketosis (pregnancy toxemia) also. Therefore it is necessary to supplement roughage-based diets with easily digestible, high energy containing concentrate. FCM yield and DMI may not corroborate during early lactation days because of parturition stress. The DMI capacity remains low during late pregnancy to the onset of lactation, despite the additional drain of nutrients for milk production. Moreover, intake capacity increases rather slowly than the increase in milk yield. This un-synchronization of dietary intake and milk yield results in negative nutrient balance mainly in energy.

Effect of diet quality

The goat has two kinds of appetite regulation, i.e. energy density of the diet or organic matter digestibility due to its fill effect. There could be high variation in intake of forages having the same OM digestibility. Other factors that also influence the dietary intake are such as:

a) Form or state of forage as fresh, dried, ensiled etc.

b) Particle size of the roughage.

c) CP content particularly less than 8 – 9 percent.

d) Supplementation.

Catalytic supplementation is of particularly high significance for animals being fed on poor quality roughages. A marginal nitrogen or mineral correction enhances the DMI, its digestibility and overall production of animals. When there is 15 – 20% addition of concentrate, it may not result in decreased DMI, but higher levels particularly with poor quality roughage may cause depression in digestibility and consequently lowering of DMI. Therefore, substitution rate should be monitored closely to attain optimum dietary intake and its utilization.

Energy

Energy is most common nutrient deficiency in goats, which is also required in higher quantities than all other nutrients. The energy requirements of goats for maintenance seem to be more comparable to the cows than ewes. For similar fat content of goat milk is slightly less energetic than cow milk about by 2.97 MJ per kg of 4% FCM.

Milk energy (MJ/kg) = 1.143 + 0.44 (% Fat content).

Fasting metabolism in lieu of basal metabolism is rather accurate, since state of basal metabolism is difficult to absorb in ruminants. Fasting heat production is more nearly proportional to the surface area of animals than to their weight. Metabolic body weight that is $W^{0.75}$ (weight$^{0.75}$) is adopted as reference for surface area for ruminant's studies.

Fasting metabolism per unit of metabolic weight is higher in young animals than old and sex also affects the same.

Maintenance (ME):

M_m (Mj/ d) = (F + A) Km

F = Fasting metabolism

A = Activity allowance

F (Mj/ d) = 0.315 $W^{0.75}$ (AFRC 1993)

Km = efficiency of ME utilization

Additional energy required as allowance for different activities is:

Horizontal movement	=	3.5 J/ kg W/ meter
Vertical movement	=	28 J/ kg W/ meter
Standing for 24 hrs.	=	10 kJ/ kg W/ meter
Body position change	=	260 J/kg W
		Or 0.26 kJ/kg W

If a goat walks (moves) on lowland pasture @ 3000meters horizontally, move vertically 100 meters and stands for 12 hrs and make 12 positional changes daily then the activity requirement is:

$$A\ (Kj/\ d) = \frac{3000 \times 2.6}{1000} + \frac{28 \times 100}{1000} + \frac{10}{2} + \frac{260 \times 12}{1000}$$

$$= 7.8 + 2.8 + 5 + 3.1$$

$$= 18.7\ kj$$

$$\text{Or} = 019\ Mj/\ d$$

For hill and mountain goats, if they walk (move) 5 km, vertical distance 100 meters, 12 hours standing, 12 positional changes daily then

A (Kj/ d) = (13 + 2.8 + 5 + 3.1) W

= 23.9 W

Thus, the activity allowance Mj/ d for goats and hills grazing is 0.024 W.

Growth

AFRC (1998) give a function fitted to the total body energy content (Eg) of the castrate male kids in relation to their live weight:

$$\text{Eg (MJ)} = 4.972\ \text{W} + 0.1637\ \text{W}^2$$

Which can be differentiated to give the energy content of live weight gain.

$[\text{Evg}]\ \text{MJ/ kg} = 4.972 + 0.3274\ \text{W}$

The Eg requirement /d then given by:

Eg [MJ/D] = ? W x [Evg]

Milk:

The requirement of energy for milk production depends on the yield and energy value of the milk (Mj/kg). AFRC has proposed the following equation to estimate the energy value of goat's milk.

EVL (MJ/ kg) = 0.0406 fat + 1.509

= 0.376 Fat + 0.0209 protein + 0.948

When no value of milk composition is available than a value of 3.25 MJ/ kg may be used.

Pregnancy:

ARC (1980) equations for ewes are used in calculating the ME requirements for pregnant goats, using total weight of kids.

$\text{Log}_{10}(\text{Et}) = 3.222 - 4.979.\text{e}^{-0.00643t}$

Daily energy retention Ee as follows:

$\text{Ee MJ/ d} = 0.25\ \text{Wo (Et x } 0.07372\text{e}^{-0.000643t}$

Where 't' is number of days from conception, Wo' is the total weight of lambs at birth in kg.

The energy requirements for different physiological functions viz maintenance and growth, pregnancy and lactation have been given in Table 6, 7 and 8, respectively.

PROTEIN

Maintenance

Maintenance requirements (NP_m) of goats and cattle are the sum of their Basal Endogenous Nitrogen (BEN or NP_B) needs plus dermal losses as scurf and hair (NP_d).

$$NP_m \text{ (g/d)} = NP_B + NP_d$$
$$\text{(As } K_{nm} = 1.0\text{) or} = 2.30\ W^{0.75}$$
$$\text{BEN (g N/ d)} = 0.35W^{0.75}$$

$$MP \text{ (g/ d)} = \frac{6.25 \times 0.35\ W^{0.75}}{1.0} = 2.187\ W^{0.75}$$

$$MP_d \text{ (g/ d)} = \frac{6.25 \times 0.018\ W^{0.75}}{1.0} = 0.1125W^{0.75}$$

$$= 2.1875W^{0.75} + 0.1125W^{0.75}$$
$$= 2.2990W^{0.75}$$
$$= 2.30W^{0.75}$$

Growth

AFRC (1998) give a function fitted to the total net protein content (NP_f) of castrate male kids in relation to their live weight:

$$NP_f \text{ (g)} = 157.22\ W - 0.347\ W^2$$

Which can be differentiated to give the protein content of live weight gains

$$NP_f \text{ (g/ kg ? W)} = 157.22 - 0.694\ W$$

NP_f requirement per kg live weight gain is then 1/0.59 times thus,

$$= (157.22 - 0.694\ W) - 1.69$$
$$= 266 - 1.17\ W$$

Milk

Efficiency of utilization of absorbed amino acids for milk protein synthesis (Knl) has been specified as 0.68 so that,

$$MP_l \text{ (g/ kg milk)} = \frac{\text{True protein content of milk}}{0.68}$$

$$= 1.47 \times \text{TP of milk}$$

INRA (1988) quotes the true protein content of goat's milk as 29-g/ kg. For Barbari and Jamunapari the CP content is 3.76 and 3.74, respectively.

Pregnancy

ARC (1980) gives the daily tissue protein (Net protein) retention (NP_c) for kid pregnancy in sheep to produce a single (kg)

NP_c (g/ d) = TP_t x 0.06744 $e^{-0.00601t}$

Where 't' is the number of days from conception and TP_t in 'g' is given in ARC (1980)

$Log_{10} (TP_t) = 4.928 - 4.873\ e^{-0.00601t}$

Since the efficiency of utilization of MP for pregnancy is (K_{nc}) is 0.85 the requirement for pregnancy is given by

= 1.18 x

= 890.7g

= 24.38

The protein requirements for different physiological functions namely growth, pregnancy and lactation have been given in Table 9,10 and 11, respectively.

Minerals and Vitamins

To calculate the mineral requirement by factorial means it is essential to have the knowledge of mineral content of goat's body and its products. It was suggested that sheep data of ARC (1980) for mineral composition of foetus, adnexa and newborn lambs should be used for the goat. Ca and P are required mainly for the skeletal growth. Mineral requirement for fiber growth in goats are based on the fleece composition of sheep. Ca and P requirements of goat for maintenance are related to DMI and thus requirement of these two minerals vary with the dietary ME contents.

Ca and P requirements presented in Table (3) have been calculated based on several considerations like endogenous loss, composition of gain (g/Kg live weight), milk composition, true availability of the dietary Ca and P etc. Calcium and P requirements of growing goats have been given in Table (4)

Table: 3 Daily requirements of goats (g/d) for Ca and P

	AFRC (1991)		NRC (1981)	
	Ca	P	Ca	P
Maintenance (50 kg body weight) DMI2.5 %	1.3	1.6	4.0	2.8
Early lactation (60 kg body weight, 5 kg milk yield) DMI 5%	1.3	1.2	14.0	9.8
Late pregnancy (16 weeks) 60 Kg body weight, 2.0 kids of	5.1	4.0	6.0	4.2
4.0 kg body weight, DMI 2.5 % Young male- 20 kg, growth rate 200g/d, DMI 4.5%	6.0	4.2	4.5	2.8
Young female 30 kg body weight, growth 100g/d DMI 2.5%	3.4	2.5	3.0	2.1

Source: AFRC, 1998

Magnesium

Deficiency of Mg in goats causes tetany and hyperirritability, while excess Mg may reduce Ca absorption. Mg needs for maintenance are 3.5 mg/kg LW and 0.14 g Mg/Kg milk production. During the last 2 month of pregnancy 0.5 g Mg/d are recommended. The Mg requirement for live weight gain goats should be for 0.14 g/100 g live weight gain as recommended for cattle by ARC (1980).

Sodium, Potassium and Chloride

The metabolism of Na, K and Cl is closely interrelated in goats. Deficiencies symptoms include reduce appetite, depressed feed intake and emaciation. NRC (1981) recommended that diets of goats should have 5.0 g NaCl /Kg DM equivalent to 2 g Na/ Kg diet DM.

Trace minerals

Studies on the trace elements requirements except copper are very sparse. The requirements recommended for cattle (12 mg Cu/Kg DM) are adequate to meet the needs of goats. The requirements of trace elements are summarize in Table (5)

Table: 4 Calcium and Phosphorus requirements (g/d) for growing goats

Live weight	Energy contents of diets (qm)	Calcium requirement (g/d) Growth rate (g/d)					Phosphorus requirement (g/d) Growth rate (g/d)				
		0	50	100	150	200	0	50	100	150	200
10	0.7	-	2.2	3.7	5.3	6.9	-	0.8	1.5	2.2	2.8
20	0.7	-	2.2	3.5	4.9	6.3	-	0.9	1.5	2.2	2.8
30	0.6	1.1	2.4	3.7	5.0	6.4	0.9	1.6	2.4	3.3	4.1
40	0.5	1.5	2.8	4.1	5.5	-	1.5	2.3	3.3	4.3	-
60	0.5	1.9	3.2	4.5	-	-	2.1	3.0	4.1	-	-
80	0.5	2.2	3.6	-	-	-	2.6	3.7	-	-	-
100	0.5	2.6	-	-	-	-	3.1	-	-	-	-

Source: AFRC, 1998

Table: 5 Trace mineral requirements for goats

Element	**Requirement (mg/Kg DM of diet)**
Copper	10-20
Zinc	50-80
Cobalt	0.11-0.20
Iodine	0.15-2.0
Iron	30-40
Manganese	60-120
Selenium	0.05

Source: AFRC, 1998

Vitamin requirements

The information on the effects of vitamin on the production and milk goat composition appears to be limited. However a number of workers have recommended that in general rations should supply adequate amount of vitamins particularly the fat-soluble vitamins (Vitamin A, D and E). The requirements of vitamins for goats by NRC (1981) have been extrapolated from the cattle and sheep recommendations. The goats with full-developed rumen are capable to synthesize adequate amount of the B- complex, vitamin c and vitamin K like other ruminants. Young kids should be supplemented with B complex vitamins till the rumen is fully functionally developed. The requirements of vitamins given in this chapter have been taken from the NRC (1985).

Vitamin A Requirement of vitamin A for goats assumed to be 69 μg of â carotene/kg live weight/day or 47 IU of vitamin A/kg live weight/day. These values have been utilized as the basis to establish the vitamin requirement of all categories of goat except for late gestation and lactation where the requirements are 125 ug carotene /kg live weight/day or 85 IU of vitamin A/kg live weight/day. In the initial 6-8 weeks of lactation, the requirements of goats suckling twins are 147-μg â carotene/kg live weight/day or 100 IU of vitamin A/kg live weight/day.

Vitamin D and E

Besides early-weaned kids vitamin D requirements for all physiological stages of goat are 555 IU/100 kg live weight/day, while for weaned kids the requirements are 666IU/100 kg live weight/day.

Vitamin E requirement of growing and pregnant goats can be met by its level between 10.0 and 15.0-gm/kg dry matter of diet. Dietary Se level also effects vitamin E requirement since it reduces the vitamin E availability. Kids under 20 kg body weights should receive 20 IU/kg DM and kids above 20 kg live weight and pregnant goats should get 15 IU of vitamin E/kg DM of diet.

The ICAR (1998) requirements of DCP, TDN, ME and minerals for different body physiological functions i.e. maintenance, growth, pregnancy and lactation have been also given in Table 12, 13, 14 and 15, respectively.

Feeding of Goats

Lambs up to the age of three months need special attention. The suckling of colostrums during first few days of time is very important for the kids. Kids should be separated from their dams and kept away on starter ration at the age of one week, when weaning is practiced on established farm. However, under village conditions, where due care of the kids may not be done the weaning may increase the mortality rate.

Starter rations should be supplemented with antibiotics, as aureomycin and teramycin and the animals may be bottle fed 2 to 3 times a day. Kids may start nibbling green at 2 weeks of age, so free choice of legume hay, minerals in starter ration, and drinking water should be made available. The mineral mixture must contain, copper, iron, iodine, cobalt, and zinc along with other macro minerals, because milk may not contain the required quantities of these minerals and so the Anemia and other problems may occur if minerals are not added in suitable amounts.

The composition of starter ration

With grasses, or non-leguminous fodders or hay the concentrates (starter mixture) of higher digestible protein must be supplied because these grasses may not provide sufficient amount of protein. The expected rate of daily gains is 60 to 80 g in small breeds and about hundred grams in larger breeds.

At an age of 3 months, the kids may be put on a complete ration providing 9 – 10 % DCP and 62 to 65 % TDN. So good quality roughage alone will meet the requirements. After the age of 6 months the finisher ration may be given, according to the market demand feeding regime has to be regularized during finishing period when a fatty carcass is desired, the high energy rations containing lesser amount of roughages should be fed. For lean carcass production, high roughage diets rich in proteins should be fed.

If the animals are kept only on grazing and pasture conditions are not good as during late winter and summer in regions like Rajasthan the supplementation of concentrate should be done. The amount of concentrate may be adjusted as per the rate of gains in the animals.

Principles of Goat Feeding

A goat can consume substantially more feed than either cow or sheep, i.e. 3 to 4 percent of her body weight in dry matter compared to 2.5 to 3 percent in case of cattle and sheep. This means that goats can satisfy their maintenance and production requirement on good fodder and pasture. It is interesting to notice that each 45.4 kg body weight goat require 1.5 times as much feed per day for maintenance as a 45.4 kg cow. By feeding similar quantity of concentrate mixture (about 64 kg goat produced 9.1 litre more milk than cow.

Goats are known for their round the clock eating habits. Due to its smaller stomach capacity the daily feed allowance may be offered in 3 or 4 times, when the animals are stall-fed. It may also decrease the chances of bloat. As far as possible use of oil cakes should be avoided. Gram alone forms the best feed for goat. Mixture of crushed barley and peas form an ideal concentrate for goats. A little higher level of proteins seems to be beneficial for goat feeding. Urea may also be used in goats ration to the extent of 30 to 40 percent of the total protein requirement.

Fodders for goat feeding

Legumes are superior to ordinary grasses. Goats are exceptionally fond of leguminous crops. All green feeds should be put on higher platform. The feeding of wet grasses, fodders or leaves should be avoided. It is better to sundry these and then offer as wilted green. All left over greens should be removed from manger. Some of the important legumes relished by goats are Lucerne, berseem, cowpea, shem, guar, moong, urad, senji, kudzu, mooth, shaftal along with stylosanthes, siratro and desmenthus.

Fodder species viz. grasses; tree leaves, shrubs and others that are suited to the different habitats and can be fed to goat are listed below.

1. Humid Tropics

Grasses: *Chrysopogon fulvus, Chloris gayana, Dichanthium annulatum, Eremopogon foveolatus, Iseilema anthephoroides, I. laxum, Panicum antidotale, Setaria anceps, Sehima nervosum.*

Legumes: *Atylosia scarabaeoides, Clitoria tematea, Macroptelium atropurpureum, Stylosanthes guianensis, Desmodium sp., Neonotonia wightii etc.*

Trees & Shrubs: *Albizia chinensis, A. lebbeck, A. procera, Artocarpus heterophyllus, Azadirachta indica, Bauhinia purpurea, Gliricidia sepium, Moringa oleifera, Morus alba, Leucaena leucocephala, Sesbania grandiflora, S. sesban, Acacia mangium, A. auriculiformis, Pterocarpus indicus, Calliandra calothyrus.*

2. Semi-arid Tropics

Grasses: *Cenchrus ciliaris, C. setigerus, Dichanthium annulatum, Panicum antidotale.*

Legumes: *Desmodium triflorum, Macroptelium atropurpureum, Stylosanthes hamata.*

Trees & Shrubs: *Acacia nilotica, A. tortilis, Ailanthus excelsa, Albizia amara, A lebbeck, Azadirachta indica, Dichrostachys cinerea, Hardwickia binata, Leucaena leucocephala, Parkinsonia aculeate, Pithecolobium dulce, Prosopis juliflora, Sesbania sesban, Tamarix spp.*

3. Arid Tropics

Grasses: *Cenchrus ciliaris, Chrysopogon fulvus, Dichanthium annulatum, Heteropogon contortus, Lasiurus sindicus, Sehima nervosum.*

Legumes: *Atylosia scarabaeides, Desmodium ssp., Stylosanthes guianensis, S. hamata.*

Trees & Shrubs: *Acacia nilotica, A. nilotica, Ailanthus excelsa, Albizia amara, A. lebbeck, Bauhinia variegata Casuarina ,equisetifolia, Dichrostachys cinerea, Hardwickia binata, Zyzyphus mauritiana.*

4. Temperate and Sub-temperate Areas

Grasses: *Bormus inermis, Dactylis glomerata, Lolium multiflorum, L. perenne, Phelum pratense, Poa pratensis.*

Legumes: *Lespedeza juncea, Lotus corniculatus, Lupinus angustifolius, Trifolium incernatum, T. subterraneum.*

Trees & Shrubs: *Betula alnoides, Celtis australis, Morus serrata, Robinia pseudoacacia, Salix ssp*

Tropical grasses like Anjan/Buffel grass (*Cenchrus ciliaris*), *C. Setigerus,* marvel (*Dicanthium annulatum*), Sain (*Sehima nervosum*), *Panicum maximum*. have shown high yield potential for yield and quality. *Stylosanthes scabra, Stylosanthes hamata, Atylosia scarahaeoids, Desmenthus virgatus* legumes improve the pasture quality.

Table 6: Energy requirement (MJME/d) for maintenance[1] and growth[2] in castrate male

Body weight, W (kg)	Live-weight gain (kg/d)	Metabolisability (q_m) of diet		
		0.5	0.6	0.7
15	0	4.1	3.9	3.7
	0.1	6.3	5.8	5.3
	0.2	9.5	8.1	7.2
20	0	5.1	4.8	4.6
	0.1	7.6	7.0	6.5
	0.2	11.2	9.7	8.7
25	0	6.0	5.7	5.4
	0.1	8.9	8.2	7.6
	0.2	12.9	11.2	10.1
30	0	6.9	6.5	6.2
	0.1	10.2	9.3	8.7
	0.2	14.6	12.7	11.5

[1] Calculated from energy costs of fasting metabolism of 315 kJ/kg $W^{0.75}$ daily with the addition of 10% for activity, 5kJ/kg daily for 12h standing and 1.7 kJ/kg daily for 10 position changes.

[2] Calculated from energy retention model of ARC (1980) for cattle.

Source: AFRC (1998)

Table 7: Additional metabolisable energy requirements (MJME/d) above maintenance for dairy goats in the last 3 months of the pregnancy, assuming a requirement of 7.5 MJ ME/d energy deposited[1].

Month of pregnancy	Number of fetuses	Dairy goats			Fibre goats	
		1	2	3	1	2
3 (77 d)	Deposition[2] ME reqirement	0.11 0.83	0.18 1.35	0.25 1.88	0.07 0.53	0.10 0.75
4 (105 d)	Deposition ME requirement	0.26 1.95	0.43 3.23	0.59 4.43	0.15 1.13	0.23 1.73
5 (133 d)	Deposition ME requirement	0.49 3.68	0.82 6.15	1.09 8.18	0.29 2.18	0.44 3.30

[1] Based on efficiency utilization of ME for the conceptus 0.133 (ARC 1980).
Source: AFRC (1998)

Table 8: Requirements for energy (MJ ME/d) during the lactation of 65-kg goats neither losing nor gaining body energy reserves.

Milk yield (kg/d)	**Anglo-Nubian[2] Metabolisability of diet (q_m)**			**Saanan/Toggenburg[2] Metabolisability of diet (q_m)**		
	0.5	**0.6**	**0.7**	**0.5**	**0.6**	**0.7**
Maintenance[1]	12.5	11.9	11.3	12.5	11.9	11.3
1	18.3	17.3	16.5	17.4	16.5	15.7
2	24.1	22.9	21.7	22.3	21.2	20.1
3	30.1	28.5	27.1	27.3	25.9	24.6
4	36.2	34.2	32.5	32.4	30.7	29.1
5	42.3	40.0	38.0	37.5	35.5	33.7
6	48.6	45.9	43.6	42.8	40.5	38.4

[1] Energy cost of 315 kJ/kg $W^{0.75}$ daily plus 10% for activity, 5 kJ/kg for standing for 12 h/d and 3.12 kJ/kg for 12-position changed/d.

[2] Total ME requirement calculated using feeding level correction of 1.8% per multiple of maintenance, as for cows in ARC (1980).

Source: AFRC (1998)

Table 9: Requirement for truly absorbed amino acids (metabolisable protein, MP_f) for the growth of goat kids.

Body weight growth interval	**MP_f requirement (g/d)**		
	15-20 kg	**20-25 kg**	**25-30 kg**
Rate of growth			
100 g/d	24.6	24.0	23.4
200 g/d	49.2	48.0	46.8
300 g/d	73.8	72.0	70.2

Note: $NP_f = 157.22 - 0.694W$ and $k_{ng} = 0.59$
Source: AFRC (1998)

Table 10: Additional requirements of dairy and fibre goats for truly absorbed amino acids (metabolisable protein, MP_c) during the last 3 months of pregnancy.

Weeks before parturition	Weeks pregnant	Dairy (D) of Fibre (F)	MP_c requirements (g/d)		
			Single	Twins	Triplets
12-8	9-13	D F	4.4 2.8	7.2 4.2	10.0 -
8-4	13-17	D F	10.1 6.1	17.2 9.5	23.5 -
4-0	17 - term	D F	19.2 11.6	32.2 17.9	43.2 -

Note: MP_c values are (1/0.85) times protein depositions
Source: AFRC (1998)

Table 11: Metabolisable protein requirements (MP, g/d) of lactating, multiparous, 65-kg Anglo-Nubian and Saanen/ Toggenburg dairy goats.

Daily milk yield, kg/d	Anglo-Nubian			Saanen/ Toggenburg		
	Months 1[1]	Months 2-3	Months 4-9[2]	Months 1[1]	Months 2-3	Months 4-9[2]
1.	70	100	104	61	91	95
2.	118	148	152	99	129	133
3.	166	196	200	138	168	172
4.	213	243	247	176	206	210
5.	261	291	295	215	245	249
6.	309	339	343	253	283	287

[1] A deficit of 30 g MP/d is accepted in month 1 of lactation.

[2] Assuming live-weight gain 1.2 kg/month, equivalent to 4g MP/d. For primiparous goats, a further 9g/d should be added to allow for a growth rate of 2.2 kg/month.

Source: AFRC (1998)

Table 12: Daily nutrient requirements for maintenance of adult goats.

Body weight (wg)	DM (g)	DM AS% of body weight	Energy		DCP (g)	Ca (g)	P (g)
			TDN (g)	ME (Mcal)			
15	500	3.3	240	0.86	23	1.1	0.7
20	615	3.1	295	1.07	29	1.3	0.9
25	730	2.9	350	1.26	34	1.6	1.1
30	830	2.8	400	1.45	39	1.8	1.2
35	940	2.7	450	1.62	44	2.1	1.4
40	1040	2.6	500	179	48	2.3	1.5
45	1125	2.5	540	1.96	53	2.5	1.7
50	1230	2.4	590	2.12	57	2.2	1.8
55	1315	2.4	630	2.28	62	2.9	1.9
60	1410	2.3	675	2.43	66	3.1	2.1

Source: ICAR (1998)

Table 13: Daily nutrient requirements of growing kids

Body weight (wg)	Average daily gain	DM (g)	DM AS% of body weight	Energy		DCP (g)	Ca (g)	P (g)
				TDN (g)	ME (Mcal)			
10	50	380	3.8	265	0.96	27	2.0	1.4
	100	510	5.1	355	1.28	37	2.7	1.8
	150	635	6.3	445	1.60	47	3.4	2.3
15	50	510	3.4	330	1.19	33	2.7	1.8
	100	645	4.3	420	1.51	43	3.5	2.3
	150	785	5.2	510	1.84	53	4.2	2.8
20	50	640	3.2	385	1.39	39	3.3	2.2
	100	790	3.9	475	1.71	49	4.1	2.7
	150	985	4.9	590	2.13	59	5.1	3.4
25	50	760	3.0	440	1.59	44	3.8	2.5
	100	915	3.7	530	1.91	54	4.6	3.0
	150	1070	4.3	620	2.24	64	5.3	3.6

Source: ICAR (1998)

Table 14: Daily nutrient requirements of pregnant does.

Body weight (wg)	DM (g)	DM AS% of body weight	Energy		DCP (g)	Ca (g)	P (g)
			TDN (g)	ME (Mcal)			
15	700	4.7	385	1.38	42	2.1	1.4
20	865	4.3	475	1.72	52	2.6	1.7
25	1025	4.1	564	2.03	62	3.1	2.1
30	1170	3.9	645	2.33	71	3.5	2.3
35	1320	3.8	725	2.62	80	4.0	2.7
40	1460	3.6	802	2.89	88	4.4	2.9
45	1590	3.5	875	3.16	96	4.8	3.2
50	1725	3.4	984	3.42	104	5.2	3.5
55	1850	3.4	1018	3.67	112	5.5	3.7
60	1975	3.6	1086	3.92	120	5.9	3.9

Source: ICAR (1998)

Table 15: Daily nutrient requirements of lactating does.

Body weight (wg)	Milk yield	DM (g)	DM AS% of body weight	Energy		DCP (g)	Ca (g)	P (g)
				TDN (g)	ME (Mcal)			
20	0.5	865	4.3	468	1.69	51	4.3	2.9
	1.0	1185	5.9	640	2.31	74	5.9	3.9
25	0.5	868	3.9	523	1.89	56	4.8	3.2
	1.0	1290	5.2	695	2.51	79	6.4	4.3
30	0.5	1060	3.5	573	2.07	61	5.3	3.5
	1.0	1380	4.6	745	2.69	84	6.9	4.6
35	0.5	1155	3.3	623	2.25	66	5.8	3.9
	1.0	1470	4.2	795	2.87	89	7.3	4.9
40	0.5	1245	3.1	673	2.43	70	6.2	4.1
	1.0	1565	3.9	845	3.05	93	7.8	5.2
45	0.5	1320	2.9	713	2.57	75	6.6	4.4
	1.0	1640	3.6	885	3.19	98	8.2	5.3
50	0.5	1410	2.8	763	2.75	79	7.0	4.7
	1.0	1730	3.5	935	3.32	102	8.6	5.7
55	0.5	1490	2.7	803	2.90	84	7.4	4.9
	1.0	1805	3.3	975	3.52	107	9.0	6.0
60	0.5	1570	2.6	848	3.06	88	7.8	5.2
	1.0	1890	3.1	1020	3.68	111	9.4	6.3

Source: ICAR (1998)

Bibliography

ARC (1980). The nutrient requirements of ruminant livestock. C.A.B. International.

Brown, L.E. and Johnson, W.L. (1985). Intake and digestibility of wheat straw diets by goats and sheep. J. Anim. Sci., 60: 1318 – 1323.

Devendra, C. and Burns, M. (1983). Goat Production in the Tropics. CAB Slough, U. K.

Gall, C. (1981). Goat Production. Academic Press, London U K

ICAR (1998). Nutrient requirements of livestock and poultry. Indian Journal of Agricultural Research, New Delhi.

La, C.D. and Potchoiba, J.J. (1990). Feed intake and weight gain of growing goats fed diets of various energy and protein levels. J. Anim. Sci. 68: 1751-1759.

Monard Fehr, P. (1981). Nutrition and Feeding of Goats: Application to temperate climate conditions. In gloat Production, Gall, C. (Ed.). London, U. K. Academic Press

NRC (1981). Nutrient requirements of goats. Angora, dairy and meat goat in temperate and tropical countries. In: Nutrient requirements of domestic animals, No. 15, Washington, D. C., USA, National Academic Press.

NRC (1985b). Nutrient requirements for sheep: 6th Revise Edition, Washington, D. C., USA, National Academic Press.

Singh, S. N. and Sengar O. P. S. 1990. Final Technical Report. PL 480. Research Project on Goats. RBS college Agra, UP college Varanasi.

Steele, Mike. (1996). Goat. CTA Publication. Published by McMillan Education Ltd. London and Basingstoke

The Nutrition of Goats. 1981. AFRC Technical Committee on responses to nutrients. Report No. 10. CAB International Publication.

Quick, T.C. and Dehority, B.A. (1986). A comparative study of feeding behavior and digestive function in dairy goats, wool sheep and hair sheep. J. Anim. Sci. 65: 1516-1526.

IMPROVEMENT OF FIBROUS ROUGHAGES

Fibrous crop residues are an important component of the feed base for ruminants in India. Animal production growing rapidly but the pace of productivity has to match the needs of increasing human population. The present trend is shifting towards foods of animal origin and cereal requirement per se is decreasing in the cities. There is deficit of green, dry forage and concentrates in our country. Crop residue like wheat straw and paddy straw are low in nitrogen content and high in complex cell wall content and possesses low digestibility. A specific composition of carbohydrate polymers, cross linking with lignin and the physical arrangement of the cellular and vascular tissues reduce the utilization of this available major feed resource. The availability of major crop residues in the country has been given in Table 1.

The utility of the straw can be increased by the addition of nitrogen, minerals and vitamin, but to extract cell wall energy further it has to be exposed to physical, chemical or biological treatments. Due to inherent limitations, animals fed exclusively on straw can just survive for some time, but cannot even support the maintenance requirements. Feed intake of wheat straw is improved in cattle and buffalo, after soaking with water.

Ammoniation or use of sodium hydroxide exhibited promising results on straw treatment but these methodologies have not been adopted by farmers or feed industry. Chlorine dioxide is possibly the most efficient oxidant of lignin and has been reported to reduce 75% lignin from wheat straw. Most prominent oxidizing chemicals like sodium hypochlorite, sodium sulphite, bleaching powder have been used to delignify crop residues. However, systematic studies on the use of these chemicals on wheat straw are scanty.

The density of wheat straw as such is very low (55 to 83 kg/ M^3). Enhancement in rumen digestibility of straw has been reported when nitrogen and energy sources are added through concentrate ingredients. Feed intake as well as feed conversion efficiency tends to increase further when treated straw is fed in the form of complete feed. The large capacity of the rumen and the presence of active and large microbial population subsisting on plant materials are especially evolved for the digestion of cellulosic feed. This versatile symbiotic relationship has enabled the ruminants to play a major role in the economic development of agricultural regions of the world.

The ad libitum feeding system should be designed based on the complete diet principle. Complete diet provides constant environment in the rumen throughout the day. It is difficult to transport straw to a long distance from surplus to scarcity regions due to escalation of transport cost. However, the compounded complete feed having major component of treated straw need hardly 1/3rd space after compaction.

Complete feed method ensures the consumption of roughage and concentrate in a proper proportion and results in efficient rumen fermentation. The adoption of such feeding system not only reduces the cost of feeding but also enhances voluntary feed intake and productivity in ruminant.

Table 1: Availability of major crop residues (2000-2001)

Crop	Grain yield (million tonnes)	Factor	Crop residue / fibrous by products (million tonnes)
Wheat	69.68	1.3	90.58
Rice	93.20	1.3	121.16
Barley	1.43	1.3	1.86
Maize	13.16	3	39.48
Millets	11.23	4	44.95
Sorghum	7.55	4	30.22
Total	196.27		328.26

Methods

1. Ammonia treatment

The upgrading effects of ammonia on low quality roughages have been known for a long time. The chemical was not used in practice to a great extent till 1970. Since then a number of methods have been developed (Table 2).

Table 2: Methods for treatment of poor quality roughages with ammonia

Form of ammonia	Procedure for treatment	Conditions (optimum)
Anhydrous and aqueous ammonia	1. Stacks of rectangular or round bales covered with polyethylene and injected with ammonia.	3-3.5% NH_3 on DM basis, 15-20% moisture in the material, 1-8 weeks of treatment depending on temperature
	2. Round bales packed in polyethylene bags and treated individually with ammonia	As above

Form of ammonia	Procedure for treatment	Conditions (optimum)
Andhydrous	1. Treatment in ovens with added heat	Treated with heat (90°C) and ammoniated (3-5% NH_3) air for 17 h, standstill 5 h and fresh air circulation.
Urea	1. Loose material ensiled in earthen pits, trench silos, bamboo baskets or stacks	5% urea in a water solution, mixed with material at a 50:50% ratio ± at 20°C, one week or more.
	2. Chopped or ground material added urea before pelleting in an industrial plant	2-3% urea, at least 133°C, 15-20% moisture
	3. Wheat straw in stock of 100 kg applied with urea and kept air tight	3% urea, with 30 to 35 liters of water for 3 weeks or more.

Anhydrous ammonia

Anhydrous ammonia is dangerous to handle because of low boiling point (-33.4°C) at atmospheric pressure. It has to be kept pressurized. At concentration between 15- 28% in air, ammonia is explosive upon ignition.

Anhydrous ammonia is commonly used for treatment of straw in Norway, Denmark and some other European countries. The stacks of straw as such or in the form of bales are wrapped with sheets of polyethylene and sealed. Thereafter, a perforated injection pipe is forced through the plastic into the middle of the stack where the anhydrous ammonia is administered at about 30 kg per tonne. The liquid ammonia evaporates gradually and the ammonia gas penetrates the stack. In Northern Europe, a treatment time of up to eight weeks is often recommended. In warmer climates, which prevails in the tropics, the treatment time may be considerably shorter and only one to two weeks may be sufficient. Ammonia treatment can also be carried out in airtight boxes, rooms or houses.

Ammonia is an efficient preservative of high moisture straw, stover etc. In humid climates, a check on mould growth is one of the greatest advantages of this method. Storage with low moisture content (below 10%) may, on the other hand be a disadvantage. At low temperatures, the ammonia is effectively bound to the moisture of the material and difficult to aerate. Cold and high moisture straw will therefore liberate ammonia when brought in to a warm barn and cause environment pollution.

Aqueous ammonia

Aqueous ammonia is used commercially for straw treatment and it is easier to handle ammonia in aqueous form. If the straw is extremely dry, the added water may also have positive influence on the treatment effect. At normal pressure, one kg ammonia solution can hold 285g at 30°C and 342g at 20° C. Solution containing 25-35% ammonia are common. The solution, sufficient to give 3.0-3.5% ammonia in the straw, is normally injected in to stacks through a tube mounted at a metal pipe. The treatment time of about one week is sufficient.

2. Urea

Urea is more convenient chemical to handle than ammonia. It is also readily available in all parts of the country. Straw treatment with urea may be carried out on an industrial as well as on farm scale.

Industrial scale urea treatment has been developed in the Germany where urea is added to chopped or ground straw before entering the pellet press. The heat developed in the pressure causes the decomposition of urea into carbon dioxide and ammonia. At 133°C the decomposition of urea is almost complete. The liberated ammonia will then act upon the roughage.

Farm scale urea treatment has been tried in several places in the world but the most practical approach was developed in Bangladesh. Five per cent urea is dissolved in water and the solution is mixed with equal amount of rice straw (weight basis). A pit is dug in the ground to serve as a silo for the material during treatment. The bottom and walls of the pit may be lined with banana leaves or the like to prevent contamination with soil. Banana leaves, mats or a piece of plastic may be placed on top of the straw before it is covered with soil. Initially a treatment of three weeks is recommended, but more recent results indicated that treatment time may be reduced. Other type of containers like bamboo baskets or simply stacks of loose straw may be used.

3. Treatment with sodium hydroxide

An overview of some methods for NaOH treatment of low quality roughages is presented in Table 3.

Wet treatment

The modified Beckmann procedure for treatment of straw has been used by Norwegian farmers for more than 40 years. However, the problem related to the disposal of the rinsing water forced a majority of the user to stop this treatment for high yielding dairy cows as ammonia treated straw did not seem to be a good substitute for the Beckmann –treated straw. An alternate method "dip treated straw" of which 3-4 kg dry matter is normal in a daily ration for high yielding dairy cows.

TABLE 3: Methods for treatment of poor quality roughages with sodium hydroxide (Sundstol, 1988)

Moisture level	Procedure for treatment	Conditions (optimum)
Wet treatment	1. Immersion of material in a NaOH solution followed by rinsing (Backmann procedure)	1.5-2.5 NaOH solution, time of soaking about 12h, rinsing to neutrally.
	2. Immersion of material in a NaOH solution without rinsing (dip treatment method), but with a period of storage	1.5% NaOH solution, time of soaking 0.5h, time of 'ripening' 3-6 days
Semi wet treatment	1. Ensiling of the material with a NaOH solution in a container (silo) or stack.	40-70% moisture, 3-5% NaOH on DM basis, air tight, minimum one week of treatment
	2. Ensiling of straw with NaOH in a stack	40-50% niustyrem 1% NaOH and treatment period of 3 to 4 weeks
Dry treatment	1. Loose material sprinkled with a NaOH solution from a garden can or the like mixed.	Solution of 2% NaOH per 100 kg material, treatment time approximately 24 h.
	2. Treatment of material during collection or baling	
	3. Spraying of material with a NaOH solution in the bales.	0.8-1.0 kg of a 50% NaOH solution per bale of 10 kg, 1 week storage
	4. Chopped and sliced material mixed with a NaOH solution and blown into a heap where the temperature rises upto 80-100°C (Straw treatment machines)	150-180 kg of a 27% NaOH solution per ton DM, minimum 3 days of treatment
	5. Chopped or ground material is mixed with a NaOH solution and pressed into cubes or pellets (Industrial plants).	Conditions vary considerably. Solution containing from 27-47% NaOH is added to give 4-5% NaOH on DM basis. Pressed at more than 100 atmosphere temperature 70-90°C, then cooled.

For large scale treatment the solution may be pumped into another vessel to leave the straw in the first vessel for drainage before storage. The time needed for 'ripening' is dependent on the temperature, and a temperature above 10°C may be desirable during storage ('ripening').

The amount of sodium hydroxide needed for replenishment of the lye solution is normally 6-6.5 kg NaOH per 100 kg straw, but may vary with the type of material treated. It is therefore advisable to titrate the solution time to time to check that the strength is about 15g NaOH per liter.

Semi – wet treatment

Ensiling of straw and other low quality roughages with NaOH is used when the material has a moisture content intermediate to the wet treatment and the dry treated straw, i.e.40 – 70 % dry matter. This method may be of particular interest when weaker chemicals are applied and a long treatment time is needed, or the cost of NaOH is prohibitive.

Dry treatment

Industrial procedures for dry treatment of straw with NaOH have been developed in many countries. The general procedure is that the material is chopped or milled before a solution containing 200-300 g NaOH per litre is sprayed on the material. After a thorough mixing the material passes through a pellet press. The high temperature developed in the press (ring die) causes an immediate reaction between the material and the alkali. After leaving the press, 1 –5 % of the moisture evaporates. The material may be fed as it is or mix with concentrates either before or after pelleting to make 'complete feed'. The optimal concentration of NaOH seems to be 4 – 6% of straw dry matter.

'On the farm' procedures are available for dry treatment of straw and other low quality roughages with sodium hydroxide. The simplest form of treatment is that a sodium hydroxide solution is sprinkled over long form of straw, which is thereafter mixed thoroughly by hand using a fork. A lye solution may be sprayed over chopped straw in a mixer wagon ensuring a sufficient mixing of the material before feeding. Straw may also be treated during the process of baling.

Biological treatment

Enhancement in the digestibility of fermented straw has been observed with white rot fungi. Some fungi grow on straws but reduce the digestibility of fermented product. Agrocybe aegerita is unable to break down the lingo - cellulose complex thereby resulting in decreased substrate digestibility. Such organisms, which are generally brown rot and sweet rot fungi, utilize cellulose, hemicelluloses and other soluble carbohydrate fractions for their growth leaving behind residues of low digestibility. The effect of biological treatment by different fungal microorganisms on digestibility of poor quality roughages are presented in Table 4.

Eleven white rot fungi inoculated on barley straw, pea straw, sugarcane bagasse and sunflower hulls revealed that fungi needed to be selected specifically for particular by-product and conditions of incubation may need to be manipulated in order to achieve substantial increase in digestibility. Wheat straw digestibility increase from 16 to 34% at the expense of loses in lignin and cellulose by culturing Sporotrichum pulverulentum and even a higher digestibility of 40-50% could be obtained. With the removal of 52% lignin from Aspen wood through fermentation with Merulius tremellosus, cellulose digestibility increased from 18-53%.

Wheat straw incubated with edible mushroom, Pleurotus ostreatus increased digestibility by 6% units, through an incubation period of 28 days. However, 50% DM with doubled incubation period with Erwinia carotovora increase digestibility from 32.7 to 47.7%. Feeding of Pleurotus harvested wheat straw did not cause improvement in digestibility. In most of the studies, dry matter losses varied widely from 6 to 40% depending on the organism used, duration of fermentation, type of substrate and environmental conditions (Reade and McQueen, 1983). The losses should be minimum during fermentation process, because removal of digestible fractions reduces the nutritional quality and quantity of residual substrate.

Biodegraded straws are rich in cellulose, hemicellulose, and fungal mass (protein) which may be a suitable feed for ruminants because of their fermentative digestion in rumen. Whereas separated fungal mass would serve as protein feed for mono-gastric animals. Some of the fungal proteins are only slightly digestible, while others are highly digestible. Direct studies on effect of biodegraded straw on rumen and intestinal microbes are lacking. However, indirect observations suggested that some of the liberated phenolic acids may hamper growth of cellulose utilizing bacteria. But the quantum of liberated phenolic acids may not be enough and rumen anaerobic environment may not allow the growth of inoculated fungi, still direct studies on this aspect may reveal true picture. Thus, it is evident that lignin degradation, dry matter yield and protein enrichment are the factors responsible for increasing nutritive value of lignocellulosic materials.

Table 4: Digestibility of fungi colonized straws

Organism	Substrate	Digestibility control	Change in digestibility (% unit) due to fungal growth	Remark
Polyporus tulipiferal	Oat straw	38.9	+14.1	At 30° C
Polyporus spp	Oat straw	33.8	+ 9.0	30/35°C
Phanerochaete chryosporium	Oat straw	34.1	+ 9.3	35°C
Agrocybe ageerita	Wheat straw	-	- 20.2	60 day*
Lentinus edodes	Wheat straw	-	+ 27.5	60 day*
Lentinus edodes	Wheat straw	-	+ 19.4	160 day*
Pleurotus eryngii	Wheat straw	-	+22.3	60 day*
Pleurotus eryngii	Wheat straw	-	+19.7	90 day*
P. salmonso florida	Wheat straw	-	+11.0	60 day*
P. salmonso altranineus	Wheat straw	-	+9.6	60 day*
Stropharia	Wheat straw	-	+13.6	60 day*
Rugosoannulata	Wheat straw	-	+12.4	60 day*
Agaricus bisporus	Wheat straw	46.2	-12.2	Calculated from animal trail data
Sporotrichum pulveruletum	Wheat straw	37.8	+16.3	
S. pulveruletum (CBS 841.73)	Wheat straw	45.0	+5.0	20 day*
S. pulveruletum (CBS 841.73)	Wheat straw	23.0	+22.0	10 day*
-do-	Wheat straw	35.0	-5.0	60 day*
Diciomitus squalens	Wheat straw	60.0	+20.0	30 day*
Diciomitus squalens	Wheat straw	28.0	-12.0	10 day*

* Time span after culturing and estimation of digestibility.

Animal production studies

Growth in animals is governed by genetic and environmental factors. Among environmental factors, nutrition plays an important role. Straw treatment with caustic soda improves digestibility and intake by 24 and 40%, whereas straw treatment with urea (NH_3) enhance digestibility between 20-27%, respectively (Jackson, 1978). Urea treated straw would be able to support maintenance requirement and alkali treated wheat straw, supports an additional gain of about 100 – 200 g/ day. This indicates that to achieve more production in the form of live weight gain or milk production, the supplementation of nutrients is inevitable. Effect of different chemical treatments on intake, digestibility, nutritive value and body weight gain of animals is given in Tables 5 and 6.

Wheat straw and rice straw treated with 4% NaOH (dry method) when compared with lucerne ad lib. DMI (g/kg $w^{0.75}$) was 111, 118 and 110 g/day and daily gain 320, 320 and 410 g, respectively (Singh and Jackson, 1975). The diet in straw group had 15 % groundnut cake and 10% molasses.

Performance of calves on untreated and treated (3.3 kg NaOH/100 kg) straw was compared in calves receiving 800 g and 1000 g groundnut cake in untreated and treated group respectively. DM intake in treated group increased by 32% and gain were 0.25 and 0.42 kg/day in untreated and treated group, respectively. In more difficult situations

Table 5: Dry matter intake, digestibility and ME of untreated and treated wheat straw

Treatment/feed	Straw DMI/100 kg BW	Straw DMI g / $w^{0.75}$ kg BW	Straw DM Digestibility %	Straw TDN Kg
Wheat straw (6 Sahiwal x HF)	1.27	48.77	49.54	44.92
Wheat straw + UMMB *(0.445 kg/day) (6 Sahiwal x HF)	1.95	75.55	48.65	
Wheat straw (3 Karan Fries cattle)	1.12	44.0		40.12
Anhydrous ammonia (3%w/w) treated straw (3 Karan Fries cattle)	1.56	62.5		56.04
Wheat straw (4 Bos Indicus)	1.10	49.4		46.85
Urea treated wheat straw 1.60 (4% urea) (4 Bos Indicus)	72.20		54.77	
Wheat straw (4% urea treated) 1.8 (4 Haryana calves one year)	65.3	48.15	47.98	
Wheat straw (+ 0.4 kg DM/100 kg Body weight) (Buffalo heifers)		53		46
Wheat straw (2% urea)		65		48
Wheat straw (NaOH 5%)		51		55
Wheat straw (NaOH 5%+2%urea)		81		57

* 15% urea and 10% cotton seed cake resulting 9.41 % N (urea molasses mineral block)

where little or no supplementary forage or concentrate were fed, live weight gain of 180 g in treated straw group against 100 g/day in untreated group was reported. This experiment was conducted in Indian village conditions and straw treatment was done by Backmann method.

Buffalo heifer fed treated wheat straw prepared by dry method daily (5%NaOH) improved weight gain (70 g/day).Thus, in animals on treated straw supplemented with 2% urea, the weight gain was 150 g/ day against 130 g on wheat straw (Khurana and Naik, 1978). Each animal in this experiment was also given a green maize or sorghum forage supplement containing 0.5 kg DM/ 100kg body weight/ day.

Table 6: Dry matter intake, digestibility and daily gain (ADG) in animals fed urea treated straw rations.

	Digestibility			**DMI (g/kg $W^{0.75}$)**			**ADG (g)**
	UT	**T**	**Increase %**	**UT**	**T**	**Increase %**	**Average dally gain(g)**
(1)	**(2)**	**(3)**	**(5)**	**(6)**	**(7)**	**(9)**	**(10)**
1) Rice straw Buffalo calves. 4 week treated, 4%urea (40% concentrate in ration)	31	62	100	69	95	38	
2) Rice straw 5% urea	41	52	27			42	
3) Rice straw (3%) urea treated	41	51	24			33	
4) Rice straw (4%) urea treated		52[2]			40		
5) Rice straw (5%) urea treated	45[2]	56	24	46	61	33	
6) Rice straw (3%) urea treated	45	54	20	46	52	13	
7) Rice straw (5%) urea treated	48[2]	55	15				
8) Rice straw	42[1]		26	65		46	
9) Rice straw	53[1]		11			34	
10) Rice straw 5%urea (Buffaloes)	50[1]	52	4	2.11[3]	2.34	11	182 - 79
11) Rice straw 5%urea (Cattle)	44[1]	52	18	2.68[3]	2.70		34 – 7
12) Rice straw 3% urea (Buffaloes)	43[1]	53	23	2.03[3]	2.17	7	130 – 50
13) Rice straw 6% urea treated (Buffaloes)	43[1]	55	28	2.03[3]	2.52	24	130 – 210
14) Rice straw 4% urea treatment (Cross-bred cattle)	41[1]	48	17			44	
15) Rice straw treated with 6% urea + 1.5 kg concentrate						21	79 - 431
16) Rice straw 5% urea treated (6 Buffalo bull)				79.6	86.7	9	90 - 97
17) Wheat straw (Treated ammonia in commercial oven)	45[2]	56	24				
18) Wheat straw treated 3.2% urea (Buffaloes)	41.2[1]	49.2	19.4	1.31	1.69	29	

1-DM digestibility, 2- OM digestibility
3- DMI/100 kg body weight
UT=Un-treated; T= Treated

The report on Hereford cross steers (14 month age 360 kg weight) fed straw from 4 barley varieties and one wheat variety untreated but 2% urea supplemented and all these five substrates treated with 30g anhydrous ammonia/ kg indicated that gain in body weight were 110, 130, 400, 200, 270 g in untreated and 360, 330, 610, 600, 520 g/day in treated groups, respectively. In addition to straw diets, all the steers were given 1.5 kg concentrate consisting of 750 g sugar beet pulp, 200 g fishmeal and 50 g minerals. Crossbred calves offered wheat straw and one 1 kg concentrate (group I) and wheat straw treated with 3.2 % urea along with similar quantity of concentrate (group II) attain a daily gain was 196 and 256 g/day in group I and II, respectively. Supplementation of treated straws with higher quantities of concentrate further results in higher body weight gain of growing animals. Treatment of straw by chemical means also had beneficial effects on milk production and milk fat synthesis in lactating animals (Table 7) compared to untreated straw.

Table 7: Effect of chemical treatment on intake, milk yield and milk fat of cattle and buffalo

Treatment	Cattle			Buffalo		
	Intake	Milk yield	Fat %	Intake	Milk yield	Fat %
Untreated straw + 1.5 kg concentrate	5.2	2.42	4.6	10.0	6.71	2.17
Urea treated straw + 1.5 kg concentrate	8.6	3.41	4.91	10.0	7.54	2.97
Untreated straw + 6.9 kg concentrate	5.7	24.0	3.34	-	-	-
NaOH treated straw + 6.9 kg concentrate	7.4	24.6	3.64	-	-	-

Many studies have been carried out during last three decades on ways and means for upgrading of low quality roughages as feed for farm animals. Physical treatment, such as chopping has been used in practice for a long time and the main effect is the increased feed intake. Many types of chaff cutter including simplest, manually used, animal drawn and power operated are available. Paradoxically, in large areas of the country still those machines are not used. Steam treatment has shown the positive effect on the nutritive value of fibrous materials but the feasibility of this method is still questionable, its linking with sugar industry could be considered mainly for economical steam availability from the latter.

Soaking with water is common in many regions that lead to higher DM intake. Sodium hydroxide was the first chemical used in practical treatment of straw. Soaking of material in NaOH solution undoubtedly results in the improvement in the digestibility of straw but the corrosiveness of the chemical and the pollution due to final product are some of the disadvantages that make this method less attractive. Dry processes for NaOH treatment may be more convenient but the improvement is generally lower. The cost of NaOH is also prohibitive in India.

Ammonia or chemicals, that liberate ammonia, seems to be of great interest. In temperate and industrialized countries the stack method for treatment of straw with anhydrous or aqueous ammonia is predominant, whereas ammonia added to straw in the form of urea seems to be more appropriate for tropical and developing countries. Even, if ammonia treatment does not result in the improvement in the digestibility. It has a number of other advantages such as: (a) combined treatment and NPN supplementation, (b) combined storage and preservation (high moisture straw), (c) treatment is simple (d) final product is easy to handle (bales) (e) excess ammonia passed to urine, and may be used as manure.

Besides these two, there are a numbers of other chemicals, such as Ca $(OH)_2$, KOH, SO_2, H_2O_2, H_2SO_4, Na_2CO_3, O_3, ClO_2 have been applied with variable sources. The cost of chemicals makes the processes less adoptable. Microbiological treatment has not been used at commercial scale so far and an appreciable amount of research remains to be done before a practical method is designed. No doubt, this approach has a potential, and high priority may therefore, be given to research on this aspect. The process is likely become industrially variable in future.

The catalytic supplementation by green forage as leucaena leaf meals or green legume such as berseem remains of practical value to many of the resource poor farmers. The stack method has been standardized in the country and ammonia may be used in industrial process. The enriched straws thus prepared could be compressed in the form of blocks after using little molasses and concentrates. Market situation will ultimately lead to final application and selection of the technology.

Bibliography

Agosin, F. and Odier, E. 1985. Solid state fermentation, lignin degradation and resulting digestibility of wheat straw fermented by selected white rot fungi. Appl. Microbiol. Biotechnol., 21: 397-403.

Dolberg, F., Saadullah, M., Haque, M. and Ahmed, R, 1981. Storage of urea-treated straw using indigenous material. World Anim. Rev., 38: 37-41.

Greenhalgh, J.F.D., 1976. Improving the nutritive value of straw by alkali treatment. ARC Research Review, 2 (3): 67-70.

Jackson, M.G., 1977. The alkali treatment of straws. Anim. Feed ScL Technol., 2: 105-130.

Jackson, M.G., 1978. Treating straw for animal feeding - an assessment of its technical and economic feasibility. World Anim. Rev. 28: 38-43.

Kundu, S.S. 1989. Improving the nutritive value of wheat straw for buffaloes by urea treatment. Tropical Agriculture 66: 321-325.

Kundu, S.S. and Mudgal, V.D. 1985. Chemical changes and digestibility of chemically treated wheat straw. Indian J. Anim. Nutr. 2: 157-161.

Kundu, S.S., Kundu, S., Prasad, D.N. and Jakhmola, R.C. 1988. Lignin biodegradation to improve digestibility of straws. International J. Anim. Sci., 3: 1-16.

Kundu, S.S. and Sharma, D.D. 1995. Influence of bleaching powder and other chemical treatments of wheat straw on in sacco degradability and nutrient utilization in buffaloes. Buffalo J. 1: 1-13.

Kundu, S.S., Das, M.M. and Sharma, D.D. 1998. Densification for block making and evaluation of calcium hypochlorite and alkali treated wheat straw based feed blocks in comparison to urea treated straw feed. Buffalo J. 3: 291-302.

Kundu, S.S. and Sharma, D.D. 2003. Composition, in sacco degradability, feed intake and nutrient utilization in reverine buffaloes. Buffalo J. 19: 127-137.

Owen, E and Kategile, J.A. 1984. Straw etc. in the practical rations for sheep and goats. In: Sundstol, F and Owen, E. (editors). Straw and Other Fibrous By-products as Feed. Elsevier, Amsterdam, pp. 454-486.

Reade, A. E. and McQueen, R. E. 1983. Investigation of white rot fungi for conversion of poplar into a potential feed stuff for ruminants. Canadian J microbiology, 29: 457-463.

Reid, G.W., Orskov, E.R. and Kay, M. 1988. A note on the effect of variet, type of straw and ammonia treatment on digestibilityand on growth rate in steers. J. Anim. Prod. 47: 157-160.

Reid, I.D. 1985. Biological delignification of aspen wood by solid state fermentationwith the white rot fungi. Appl. Environ. Microbiol. 50: 133-139.

Rexen, F.P. and Bach-Knudsen, K.E., 1984. Industrial-scale dry treatment with sodium hydroxide. In: F. Sundstol and E. Owen (Editors), Straw and Other Fibrous By-Products as Feed. Elsevier, Amsterdam.

Saadullah, M., Haque, M. and Dolberg, F., 1981. Effectiveness of ammonification through urea in improving the feeding value of rice straw in ruminants. Trop. Anim. Prod., 6: 30-36.

Sihag, Z.S, Rathee, C.S. and Lohan, O.P. 1991. Factors affecting complete feed block formulation of straw based rations. Indian Journal of Animal Sciences 61: 1218-1221.

Singh, M and Jackson, M.G. 1975. Improved utilization of Agricultural waste material and industrial by productsas livestock feed. Research Progress Report, 1969-74, G.B. Pant Univ., Pantnagar.

Sundstol, F. 1988. Feed stuffs. 5. Straw and other fibrous by products. Livestock Prod. Sci., 19: 137-158.

Sundstol, F., Coxworth, E. and Mowat, D.N., 1978. Improving the nutritive value of straw and other low-quality roughages by treatment with ammonia. World Anim. Rev., 26: 13-21.

Sundstol, F., Said, A.N. and Arnason, J., 1979. Factors influencing the effect of chemical treatment on the nutritive value of straw. Acta Agric. Scand., 29: 179-190.

Sundstol, F. and Orskov, E. R. 1984. Straw and other Fibrous by-products as Feed. Elsevier publishing.

Wanapat, M., Sundstol, F. and Garmo, T.H., 1985. A comparison of alkali treatment methods to improve the nutritive value of straw.1. Digestibility and metabolizability. Anim. Feed. Sci. Technol., 12: 295-309.

Wanapat, M., Sundstol, F. and Hall, J.M.R, 1986. A comparison of alkali treatment methods to improve the nutritive value of straw. II. In sacco and in vitro degradation relative to in vivo digestibility. Anim. Feed. Sci. Technol., 14: 215-220.

Wilkinson, J.M., 1984. Farm-scale dry treatment with sodium hydroxide. In: F. Sundst0l and E. Owen (Editors), Straw and Other Fibrous By-Products as Feed. Elsevier, Amsterdam, pp. 162-180.

Yadav, K. K., Rathee, C.S. and Lohan, O.P. 1990. Effect of compaction of roughage based complete feed block on digestibility and rumen parameters. Indian J. Anim. Sci., 7: 27-30.

Zadrazil, F. and Brunnert, M. 1982. Solid state fermentation of lignocellulose containing plant residues with Sporotrichum pulverulentum and Merulius tremellosus. Eur. J. Appl. Biotechnol. 16: 45-51.

Zadrazil, F., 1977. The conversion of straw into feed by Basideomycetes. Eur. J. Appl. Biotechnol. 4: 291-294.

COMPLETE FEEDS AND DENSIFICATION

Ruminants are grazer and browser by their characteristic nature. They are well adapted to utilization of forages to meet their nutritional needs. This is possible because of the anatomy and function of complex four compartment stomach, where the microbial fermentation of feed and other vital process take place. They even require a substantial amount of fibre for good health and optimum performance. But with the advancement of agriculture and increase in animal population, the pasture/ grazing lands have decreased. The grasslands are also at their lowest ebb of production and there is hardly any managemental input. For efficient utilization of available herbages, cut and carry system has recently been introduced with the offering of forages to the animals either as such or in chaffed form. Some concentrates are also mixed with the available poor quality forages viz. mixed dry mature grasses, jungle grass hay, cereal straws and stovers etc. to receive higher production from animals. However, in certain areas these dry forages and concentrates are first soaked with water and then fed to animals. This is commonly known as 'sani' feeding which is also supplemented with kitchen wastes. While in complete feed system, the required nutrients are supplied through mixed diets to meet the requirements of maintenance, production and reproduction of animals. The objective of the system is to simplify the process of feeding to save labour, and to provide substrates of uniform composition. The complete feeds are also processed in different ways.

(a)	Chaffed mix	Forages are finely chaffed and evenly mixed with ground concentrate ingredients
(b)	Mash mix	Forages are ground and evenly mixed with ground concentrate ingredients
(c)	Pellets	Forages are ground, evenly mixed with ground concentrate ingredients and converted to pellets

Thus, changing the physical form of the diets also affects the performance characteristics of the animals and can be judged by an increased growth rate and increased milk, meat and wool production.

Conventional feeding

The common practice of feeding animals is to place forage/ roughage before the animal separately throughout the day and concentrate feeding is restricted to once a day or at the time of milking. However, in certain cases both forage and concentrate are given on a separate self-feeding system basis. But self-feeding of concentrate leads to inefficient use of concentrates, reduces fat content of the milk and create health problems. A feeding system that is fully *ad libitum* should, therefore, be based on the complete diet principle.

Complete diet feeding

The major factors limiting the extensive use of crop residues are high cell wall, lignin, low soluble carbohydrates, low nitrogen, low minerals and vitamins, and low intake as well as digestibility. The energy of these materials is poorly utilized since the presence of ligno-cellulose complexes which are either indigestible or act as a barrier between the potentially digestible fraction and the digesting enzymes. It is, therefore, appropriate to incorporate all the deficient nutrients in the ration along with crop residues so as to provide optimum environment for the rumen micro-flora to utilize locked energy in the residues. Such a system where all the nutrients are mixed together and fed at a time without offering any choice to the animal is called complete diet system. In this system even the flavour or unpalatable taste is masked by adding additives/ steaming or pelleting. Complete feeds ensure better consumption and avoid refusal of unpalatable portion of feedstuffs. More frequent feeding spreads the load on the rumen and maintains normal ratio of acetic to propionic acid, which favours normal butter fat content. A more even intake of feed into the rumen is also associated with less fluctuation in the release of ammonia, which ensures better utilization of protein in the feed. By adjusting the proportion of roughage and concentrate mixture (Table 1) in the ration, the requirement of various categories of animals can be met.

Complete diet feeding is widely practiced in the large dairy operations and feedlots of the United States, Israel and the United Kingdom. In the developing countries where there is a severe shortage of feedstuffs, the complete feed system can help in utilizing locally available crop residues, agro-industrial by product and wastes, more efficiently for raising livestock. In the developed countries, however, the good quality fodder, hays, silage, grains and other conventional ingredients are being utilized for the preparation of complete feeds. In some places, chemical treatment and mechanical processing of low quality crop residues have been successfully tried for making complete diets. The complete feeds containing straw are inferior to the rations containing either maize cobs or maize stover at an equal roughage level. Pelleting of complete rations have also proven better for animal production, showing greater nutritive value of the pelleted ration compared to ground ration at an equal DM intake.

Table 1.: Requirements for compounded feed mixture

S. No.	Characteristics	Requirement	
		Type 1	Type 2
I.)	Moisture, percent by mass, *Max*	11	11
II.)	Crude protein (N x 6.21), percent by mass, *Min*	22	20
III.)	Crude fat, percent by mass, *Min*	3.0	2.5
IV.)	Crude fiber, percent by mass, *Max*	7	12
V.)	Acid insoluble ash, percent by mass, *Max*	3.0	4.0

Pellets feeding

Mixing the feed ingredients into a complete diet seems to give rise to an increase in daily DM intake. Such an increase may arise from the extra processing of the diet that occurs during preparation and mixing of the diet or may be due to the fact that whole diet is constantly available. Again voluntary intake of ration in pellet form is more than ration in mash form. However, pelleting does not affect the digestibility of nutrients. But cubes or pellets of dried grass passes through the digestive tract more quickly and have a lower digestibility than the long material. This is caused in part by a more rapid rate of passage of the finely ground particles through the reticulo-rumen which affects the nature of microbial fermentation and reduces the time of exposure of feed to rumen microorganisms. Grinding and pelleting reduces refusal and waste of the more unpalatable rations of roughages. Feed wastage is 2 to 4% for ground hay as compared to 16.2% for hay fed in the long form.

Pelleted feeds also change the pattern of rumen fermentation considerably. There are thinner rumen walls, longer papillae, poorly attached often large and paddle shaped with eroded tips following the feeding of pelleted diet. Rumen contents are greater in volume and density, and appear foamy. There is lower fermentation of rumen content and lower bacteria and protozoa count on feeding of pelleted diets. Pelleting, especially of roughages, adversely affects rumen fermentation. Total VFA and butyric acid increases but acetic and propionic acids decreases with pelleting. Pellet fed animals are more liable to rumen acidosis. The molar proportion of acetic to propionic acid is 3.4: 1 and 1.5: 1 on traditional and pelleted diets, respectively. The nitrogen utilization is also improved on pelleted diets. Release of ammonia-N is low on pelleted substrate, but microbial synthesis, ^{35}S incorporation and total VFA production are more on unpelleted substrates. Again, in rumen liquor both ammonia-N and TCA-N increase on increasing the CP from 8 to 16% in pelleted rations. Percent incorporation of ^{35}S is also more on 12% CP ration.

Efficiency of feed utilization on complete feeds

The complete diet concept is based on group rather than individual feeding and associated with fewer changes in diet formulation according to milk yield. Experiments carried out in U.K. on flat rate feeding of concentrates to groups of dairy animals as compared with individual rationing suggest that yield and efficiency were not improved by individual rationing at any level of animal potential. When compared two systems of complete diet feeding, one with forage to grain DM ratio of 60: 40 and the other a variable ratio throughout the lactation period, there was no variation in milk yield and fat % in both the systems. The economic formulation of complete diets is usually based on the use of high quality forage and the optimum concentrate: forage DM ratio is not likely to exceed 60:40. For high yielding cow 50:50 ratio is usually adequate. It has been a long held belief by feed formulators that variety of sources of grains or cakes provide safeguard for nutrients availability in complete feeds.

The necessity for varying forage and concentrate ratio with stage of production in dairy animal has been reviewed in relation to complete diets and for the feeding of separate ingredients. The evidence suggests that where the basic forage is of good quality, there seems little justification for varying diet composition. A change to a complete diet with a lower energy concentration leads to an immediate reduction in intake and milk yield (Table 2), suggesting that intake is not regulated according to physiological requirements alone. It was observed that milk composition, feed efficiency and feed cost per kg FCM did not differ significantly among the complete feeds but feed cost per kg milk was lowest with the mash diet. On processed feed, CP digestibility increased while crude fibre digestion was low.

Daily DM intake was significantly higher on diet containing 78% DM in lactating animals fed on complete diets having variable moisture content. Milk yield, eating time, rumination time and changes in body weight did not differ due to DM of the feed. In another experiment silage and barley grains were either given together as a complete feed or with cereal available separately for two hours. DM intake as well as live weight gain were higher on rations containing silage and barley together. Complete feed with forest grasses gave 58 and 92 g of daily weight gains in lambs with mash or pelleted diets, whereas the weight gain observed was 55 and 86 g when sorghum straw was added.

Table 2: Performance of dairy cows fed on complete feeds/ total mixed rations with different levels of TDN and CP

Parameter	Group-I (TDN: CP, 55: 10)	Group-II (TDN: CP, 60: 12)	Group-III (TDN: CP, 65: 14)
Nutrient digestibility (%)			
Dry matter	74.99 ± 1.87	71.22 ± 1.38	72.39 ± 1.28
Crude protein	68.77 ± 1.19[a]	69.08 ± 1.06[a]	74.57 ± 1.13[b]
Crude fiber	74.12 ± 3.11	66.97 ± 1.65	72.11 ± 0.60
Feed consumption & milk production			
Dry matter intake (kg/ 100 kg BW)	3.88	3.49	2.61
Milk production (kg/day)	7.85 ± 0.56[a]	10.72 ± 0.37[b]	8.65 ± 0.68[b]
Fat (%)	4.38 ± 0.04	4.03 ± 0.24	4.26 ± 0.20
Protein (%)	3.67 ± 0.08	3.41± 0.08	3.69 ± 0.21

[abc] Figures carrying the different superscript in a row differ significantly
Source: Girdhar and Balaraman (2004)

Scope of complete diet system

In tropical and sub-tropical environment the roughage quality detoriorates faster and animals subsist mainly on low grade crop residues having lignocellulosic material as the main energy source. Such basic roughage could be used optimally if concentrate feed ingredients are blended to form complete feed. In labour scarce regions in the country, this system of feeding could perform economically since all the feed stuffs are to be fed together. This system could also work better when one or more unpalatable feed ingredients are to be included in the feed mixture. The complete feeds do not affect the composition of milk *per se* but improves feed intake and prevents development of acidosis thereby increasing milk fat. Waste materials like bagasse, cotton seed hulls and waste paper etc. have been successfully tried by various workers in the country.

Complete feed could be compacted to feed blocks and therefore, could be transported economically to cities or milk-shed areas where green fodders are not available abundantly. Complete feed could also be converted to briquettes and disposed in the market under different levels of productivity of the dairy animals.

Advantages

(i) Optimizes the functioning of rumen microflora

(ii) Maintains specific rumen environment throughout the day resulting better nitrogen utilization

(iii) Entertains large quantity of low-grade roughage in the diet. Ratio of concentrate to roughage could be controlled and kept within specific limits

(iv) Some low grade feed ingredients/industrial by-products could be used effectively in complete feeds

(v) Production of acidosis could be checked

(vi) Maintains a particular level of fat content in the milk and enhances milk production

(vii) Feed/nutrients intake can be closely monitored and recorded

(viii) Chemical treatment as well as mechanical processing of complete diets has proved more advantageous

(ix) High production potentials of animals can be sustained by incorporation of required components of undegradable protein

(x) Commercial production of complete feed blocks is cheaper than traditional concentrate thus reduces cost of animal production

(xi) Feeds according to the level/type of production can be easily formulated

(xii) Labour cost is reduced, in feeding

(xiii) Fodder production avoided, thus land saved for high priced crops

Disadvantages

(i) Requires skill to balance the nutrients in the complete feed

(ii) Require machinery for mixing or further processing to pelleting, briquetting or block making

(iii) Cost has to be worked for machinery and other process

DENSIFICATION

Densification is a process of compaction of biomass like grasses or crop residues along with feed ingredients so as to increase the density of the feedstuffs. The compacted feed may either be made available in the form of briquettes or cubes in case of grasses or leguminous fodder, pellets in case of compounded feed and blocks of different sizes in case of straws/crop residues alone or with feed ingredients. The complete feed is produced by the processing and mixing of all feed ingredients inclusive of roughage into a uniform blend, which ensures supply of the diet of the same composition to the animal. The complete feed is fed as a sole source of nutrients and controls the ratio of concentrate and roughage and thus reduces feeding cost.

Concept of densification

Crop residues are bulky and difficult to transport from surplus to scarcity regions of the country or even from the field to the storage place of the farmer. It is essential for ease of handling and storage to reduce bulk. One way to reduce transport cost is to compress or to densify the bulky feeds. Densities of crop residues can be increased from 65 to 75 kg/m^3 to 100 to 110 kg/m^3 by baling or even 300 to 500 kg/m^3 by block making or briquetting. The carrying capacity of a truck is doubled after baling and increased to manifold after compaction. Rate of compaction is governed by the nature of material, moisture content, pressure applied, type and level of binder, particle size of the material and dwell time. Relatively higher increase in bulk density by compaction was reported in paddy straw compared to wheat straw and in ungrounded straw compared to ground straw.

Feed blocks prepared with a lower level of wheat straw (67%) were more compact (density 0.53g/cm^3) and a pressure of 80 to 110 kg/cm^2 for 3 minutes appeared more effective for setting the feed mass. Moisture level higher than 18-20% resulted in a loose blocks from wheat straw, however, higher (40%) moisture was preferred in case of sugarcane bagasse. Alkali treated straw blocks were more compact than untreated straw. Higher durability and a lower post compression expansion in alkali treated straw feed blocks were reported than the untreated straw. The durability of blocks increased when 6-20% molasses was included. Urea treatment did not show significant effect on compaction of paddy straw. Treatment of straw combined with compaction had an added effect for efficient management of using low-density feedstuffs for animal production.

Feasibility of compaction technology

Commercial entrepreneurs or Farmers' co-operative organizations could use power driven hydraulic/mechanical compaction machines. Certainly these machines shall be economically viable. However, for using this technology by individual farmer, a modified bullock driven machine would be a better fit which may not give that high compaction of the material but easier to handle by the farmer. Sugar mills have adopted compaction technology to compact sugarcane bagasse. Rice husk compacted to briquettes is used for fuel purposes. Paper mills have been using high pressure baling machines for transporting straw, bagasse, waste paper etc.

Advantages of compaction

(i) Compaction increases the density of the crop residues and decreases the volume, which reduces the handling and transport cost as well as space for storage.

(ii) Prior to compaction, the particle size of the material is reduced, resulting increased digestibility and higher dry matter intake ensuring more intake of digestible material

(iii) By reducing particle size, the surface area of the crop residues is increased which projects higher efficiency in chemical and biological treatments

(iv) Pelleting of urea treated straw is reported to improve organic matter intake as well as its digestibility. This may be due to increase in temperature during pelleting, urea is decomposed and ammonia released reacts with the straw

(v) Densification results a blend of biomass and fractions of feed, which may be otherwise less palatable, when fed alone, can be consumed in enough quantities, thereby ensuring the use of less exploited materials

(vi) More efficient rumen fermentation is maintained in animals fed these feeds

(vii) Densification also makes commercialization of residue up-gradation technologies, convenient and feasible.

Disadvantages of compaction

(i) The process is expensive since both fuel and machinery are used and requires skill to operate.

(ii) With the reduction of particle size, the digestibility coefficients of proximate principles are reduced during pelleting.

(iii) Large thorough put of materials needed to make operation cost effective.

Feed formulae

A few typical feed mixtures that have been tried out at various national research institutes (given below) have given satisfactory results and may be used for compounding feed (concentrates) mixture using ingredients listed, in proportions indicated against each. In the formulae where the proportion of mineral mixture and common salt to be mixed has not been indicated, in such cases not less than 1 percent by mass of mineral mixture and common salt not exceeding 2 percent by mass should be added. However, the feed formulae should also conform to specific requirements.

S.No.	*Ingredients*	*Percentage*
1.	Wheat barn	50
	Rice polish	10
	Groundnut oil cake	20
	Sesamum (TIL) oil cake	20
2.	Tapioca	56
	Groundnut oilcake	44
3.	Cottonseed cake	15
	Groundnut oilcake	20
	Wheat barn	15
	Rice polish	15
	Tur husk	15
	Horse gram	20
4.	Cottonseed cake	80
	Wheat barn	10
	Oats	7
	Salt	2
	Mineral mixture	1
5.	Tapioca	40
	Gram	10
	Groundnut oilcake	37
	Wheat barn	13
6.	Wheat bran	27
	Groundnut oilcake	30
	Wheat cutting (unfit for human consumption)	10
	Gram chuni	20
	Gram husk	10
	Salt	2
	Mineral mixture	1
7.	Cottonseed oilcake	25
	Groundnut oilcake	25
	Rice polish	20
	Maize barn	30
8.	Cottonseed oilcake	17
	Guar meal	25
	Groundnut oilcake	20
	Wheat bran	15
	Tur chuni	5
	Urad chuni	18
9.	Cottonseed oilcake	20
	Guar meal	30
	Groundnut oilcake (solvent extract)	25
	Wheat bran	10
	Maize bran	15
10.	Guar meal	20
	Groundnut oilcake	30
	Rice polish	25
	Wheat bran	25
11.	Groundnut oilcake	35
	Tapioca spent pulp	35
	Wheat bran	30
12.	Tapioca (chips)	25
	Groundnut oilcake	30
	Gram chuni	25
	Rice bran	20
13.	*Jowar*	25
	Sesamum (*TIL*) oilcake	25
	Wheat bran	20
	Gram husk	10
	Horse gram	10
	Gram chuni	10
14.	Maize	30
	Groundnut oilcake	14
	Gram chuni	12
	Wheat bran	20
	Rice bran	10
	Urea	1
	Molasses	10
	Mineral mixture	2
	Salt	1
15.	Wheat barn	27
	Groundnut oilcake	30
	Gram chuni	20
	Gram husk	10
	Gram	10
	Salt	2
	Mineral mixture	1
16.	Wheat barn	25
	Rice barn (solvent extract)	15
	Gram	12
	Groundnut oilcake	25
	Groundnut oilcake	20

S.No.	*Ingredients*	*Percentage*
	(solvent extract)	
	Salt	2
	Mineral mixture	1
17.	Tapioca	25
	Gram	20
	Groundnut oilcake	25
	Wheat bran	20
	Gram husk	10
18.	Cottonseed oilcake	30
	Guar meal	20
	Groundnut oilcake	20
	Wheat bran	30
19.	Cottonseed oilcake	35
	Guar meal	25
	Groundnut oilcake	15
	Wheat bran	25
20.	Cottonseed oilcake	5
	Guar meal	30
	Groundnut oilcake	35
	Wheat barn	15
	Maize bran	15
21.	Barley	40
	Groundnut oilcake	17
	Wheat bran	30
	Urea	1
	Molasses	10
	Mineral mixture	1
	Salt	1
22.	Cottonseed oilcake	12
	Guar meal	30
	Wheat bran	25
	Maize gluten	8
	Damaged wheat	15
	Wheat sieving	10
23.	Cottonseed oilcake	30
	Guar meal	25
	Groundnut oilcake	10
	Wheat bran	30
	Urad chuni	5
24.	Cottonseed oilcake	25
	Groundnut oilcake	25
	Rice polish	20
	Maize bran	30
25.	Groundnut oilcake	40
	Rice bran	20
	Tapioca spent pulp	20
	Molasses	20
26.	Sesamum (*TIL*)	35
	oilcake	20
	Rice bran	15
	Gram husk	30
	Horse gram	
27.	Barley/Oats	44
	Groundnut oilcake	18
	Wheat bran	24
	Urea	1
	Molasses	10
	Mineral mixture	2
	Salt	1

Bibliography

BIS. (1979). Indian Standard: Specification for compounded feeds for cattle. Bureau of Indian Standards, New Delhi.

Girdhar, N. and Balaraman, N. (2004). Nutrient utilization and balances in lactating crossbred cows fed maize fodder based total mixed ration containing different levels of energy and protein. Indian J. Dairy Sci., 57 (4): 259-265.

Kundu, S.S., Singh, Sultan and Sharma, D.D. (2003). Densification to form complete feed blocks. Indian Farming 53 (4): 29-31.

Mudgal, V.D., Singhal, K.K. and Sharma, D.D. (2003). Advances in Dairy Animal Production. International Book Distributing Co., Lucknow, India.

Reddy, G.V.N. and Reddy, M.R. (1999). Nutrient utilization and rumen fermentation pattern of cotton seed hulls based complete diets in crossbred bulls. Indian J. Anim. Nutr., 16 (1): 6-11.

Reddy, M.R. (1986). Processing of crop residues for formulation of complete feeds. Proceeding Vth Animal Nutrition Research Workers' Conference, M.L. Sukhadia University, Udaipur.

Sharma, D. D. and Singhal, K.K. (1986). Efficiency of complete feeding system in ruminants. Proceeding Vth Animal Nutrition Research Workers' Conference, M.L. Sukhadia University, Udaipur.

FEEDING STANDARDS AND RATION COMPUTATION

Livestock are to play central role in agricultural development and to supply quality foods for all including city dwellers. Since the animals are generally raised by small farmers at a subsistence level of management and it is sometimes stated why feeding standards for these categories of animals. Who needs feeding standard? This question was also raised about a quarter century ago, and replied in next issue of animal feed science and technology journal. It suffices, and satisfies the authors to state that as we are lacking the feed stuffs or we do not have surplus to waste in this country, available resources must, therefore, be judiciously used for efficient animal production.

- Lack of one nutrient triggers the wastage of other nutrients. Carbohydrates, the energy source cannot be utilized without the optimum level of protein and reverse is also true. This holds true in case of almost all, about forty or more nutrients.
- Animals fed excess of requirements may lead to toxicity, as in case of inorganic elements.
- Extra feeding hampers reproduction in animals, say more energy results in fatty animals or higher protein increases ammonia and urea levels in blood, which ultimately affect reproduction in males as well as females.
- Undigested nutrients present in excess in feaces and urine especially nitrogen and phosphorus lead to the growth of undesirable flora in soil and water bodies, and modifies the soil characteristics. The green house gases are also produced and thereby add to pollution.
- Since the country has high fibrous feed materials and a vast population of livestock to feed upon, every kg of material should be in properly balanced to get efficient production.

Feeding standards are the systematically organized statements of the amounts, largely in tabular forms, of feeds and specific nutrients which should be provided to different species for various purposes such as growth, maintenance, reproduction, lactation and wool production. They serve as guides in feeding the animals and in estimating the adequacy of feed intakes and of feed supplies for animals. There are two terms which have been used in the feeding standards. One is the nutrient allowance and another is the requirements. The former gives an extra allowance of nutrient over the requirement which gives a margin of safety whereas latter term gives the requirement for optimum. Standards are essential for planning, budgeting not only for an individual animal but also for whole farm, district or the country as a whole.

History of feeding standards

The idea of expressing the nutrient requirement of livestock in terms of feeding standards has undergone slow but steady evolution, with the progress of our knowledge in the chemistry of food/ feed and the physiology of the digestion. German workers are credited for early development of feeding standards for farm animals. In 1810, Albrecht Thaer put forward a table of equivalents in which different feedstuffs in commonly use for livestock feeding were compared with the meadow hay as standard. This was termed as "Hay equivalent" and was obtained by comparing the amounts of feedstuff that were rendered soluble by treatments with warm water, dilute acid and dilute alkali, with the amount rendered soluble by equivalent treatment of a similar sample of average hay. Though this system had various demerits even then it spread very quickly throughout central Europe. Boussingault (1830–1845), Henneberg and Stohman (1860), Henry (1904), and Armsby (1917) all credited Thaer with this concept of originating hay equivalents. With the gradual advancement in the knowledge of chemistry, Grouven in 1859 proposed the first feeding standard based on total quantity of protein, fat and carbohydrate as measured by chemical analysis.

Digestible Nutrient System

Emil Von Wolff made considerable advances in 1864 by expressing the feeding value in terms of digestible protein, fat and carbohydrate required by various categories of farm animals. But during that period, production and maintenance requirements were not considered separately. Julius Kuhus in 1887 made distinction between maintenance and production parts of the ration. In 1897, Lehmann published a modified Wolff Lehmann feeding standard in which the views of Kuhus finding were supported. The next development took place in 1903 when Haecker published a digestible nutrient standard. In this system he made allowances for milk production according to its fat content and quality. The standard proposed by Haecker was then subsequently modified by Savage and Morrison in dairy industry of America.

Morrison Feeding Standard

Morrison combined all the standards studied by that time in one set and the proposed standard was expressed in terms of dry matter, digestible protein, total digestible nutrient, instead of giving separate recommendations for digestible carbohydrate and for digestible fat as in Wolff and Lehmann standard. These standards are known as "Morrison Feeding Standards".

National Research Council of USA

In 1945, NRC standards of USA came into existence. These standards are being constantly reviewed, and revised standards are published from time to time based on the work done on the nutrient requirements of several feeding trials. These nutrient requirements are based on the size of the animal and unit of production to be achieved. These standards are being followed for practical feeding of the livestock in a number of countries. The nutrient requirements of dairy animals based on these standards (NRC, 2001) are given in appendix. From time to time, the NRC revises these feeding standards keeping in view the available information and change in feeding practices.

Agricultural Research Council of U.K.

In U.K., the feeding standards are given in the Ministry of Agriculture's Bulletin "Ration for Livestock and Poultry Nutrition". Requirements are provided in three separate reports dealing with poultry, ruminants and pigs, each of these reports extensive summaries of the literature upon which the requirement are based. Since the first edition (Agricultural Research Council, 1965), the publication has been greatly revised. Last of these reports on nutrient requirements of goats were published in 1994.

Indian standards

On the basis of the work carried out in India, Sen and Ray published nutrient requirements for cattle and buffaloes in 1964 and revised by Ray and Ranjhan in 1978. Again on the basis of scientific information arising from experimental work carried in India over the past two decades, nutrient requirements of Indian livestock and poultry was ultimately published by ICAR in 1985 and no doubt form a strong guideline for feeding our livestock. Since then Ranjhan (1991) has compiled the nutrient requirements for cattle, buffaloes, pigs, sheep, goats and poultry. The ICAR has also given the nutrient requirement in the second revised edition (1998).

FEED ENERGY

Energy can be defined as the capacity to do work. This can involve physical activity, biochemical processes, nerve impulses, or transmission of substances across membrane barriers. Quantitatively, energy is the most important item in an animal's diet. All animals require energy. The amount varies according to their physiological functions and environmental conditions. There are complex interrelationships among the various energy fractions (Fig. 1) of a feed during their utilization by an animal. The carbohydrates, fats and proteins which the feed supplies to the animal body may all be used as energy for maintaining the body temperature as well as the vital functions like growth activity, production and reproduction. Based on the age and species of animal concerned, 70-85% of the total dry matter consumed is utilized for energy for these functions. The lack of adequate feed or poor quality feed which limits its consumption and/ or digestibility, also limits the energy supply and a serious problem to animal health and productivity in many countries.

Gross Energy (GE)

It is the total heat generated by oxidation of a feed sample in a bomb calorimeter. When a feed substance is completely burnt to its ultimate products like carbon dioxide, water and other gases, the heat generated is considered heat of combustion or gross energy. Although this measure is the starting point in determining the energy value of feeds for animal's use, there is no correlation between the amount of GE in a feed and its utilization by an animal. Gross energy does, however, provide a reference point from which GE digestibility can be calculated.

Digestible Energy (DE)

It describes the proportion of GE not recovered in the faeces. It is calculated by taking the GE of the feed consumed and subtracting the GE in the faeces. Because small amounts of the faecal energy come from endogenous sources (mucosa cells, micro flora residues etc.), this term is sometimes referred to as apparent digestible energy. In fact, this loss is a part of the maintenance requirement of the animal. While true digestible energy denotes the value arrived at by subtracting the faecal energy of feed origin only from the GE intake. The faecal energy forms a substantial part of the GE intake and varies from 40 to 50% in the case of roughages and 20 to 30% in the case of concentrates in cattle and sheep. Again this faecal energy loss per unit of feed intake increases with level of feeding.

Metabolizable Energy (ME)

It refers to the energy available to maintain the body functions of animal including minimum activity and the heat increment (HI). Approximately 75% of the ME requirement for maintenance is used to support fasting metabolism (FM). Fasting metabolism is calculated as a function of the body surface area of the animal. By using a conversion factor, body size can be converted from bodyweight to surface area or "metabolic body size" W $kg^{0.75}$. The FM per day = 70 kcal (W $kg^{0.75}$). Fasting metabolism requirements vary with respect to age and physiological function of the animal. For mature sheep and cattle the values are 58 and 80 kcal per W $kg^{0.75}$, respectively. Voluntary activity and other factors account for approximately all the remaining 25% of the maintenance requirement. Hence, the energy required for maintenance per day = 1.33 x 70 kcal (W $kg^{0.75}$). The ME remaining after satisfying the maintenance requirement is available for production. ME is determined by subtracting the urinary losses, losses from methane production and endogenous source from DE. The gaseous products of digestion result from fermentations which take place in the digestive tract. The gases comprise of largely methane with traces of hydrogen and hydrogen sulfide, which contain energy and result in an energy loss. This gaseous loss (6 to 10% of GE intake) is of great significance in farm animals like cattle, buffalo, sheep and goats. The urine also contains energy (4 to 5% of the GE intake in cattle) and leads to further loss and to be subtracted in arriving metabolizable energy.

Net Energy (NE)

It is determined by subtracting the losses due to digestion, fermentation (heat of fermentation, HF) and nutrient metabolism (heat of nutrient metabolism, HNM) from ME. In a living organism, chemical reactions occur constantly as an essential feature of life processes and some of the chemical energy involved is lost as heat energy. Hence, there is continual out-go of heat from the body and increases with the consumption of feed, termed as heat increment (HI). The total heat loss from the animal body may constitute 25 to 40% or even more of the GE intake. HI is also subjected to many variables following the nature of the ration, the kind of animal, the purpose for which it is fed, and others. Net energy is available for maintenance (NE_m) and production (NE_p). The NE_m is utilized to satisfy the needs of fasting metabolism, activity of maintenance, and temperature control (above or below critical temperature). The NE_p is used for either growth, fattening, or in the production of milk, eggs, wool and foetus or in work. Ruminants maintain a constant energy balance by altering food intake in response to change in physiological and environmental factors. Food intake is linked to energy requirements providing that the amount of digestible energy in the diet is above a critical low point.

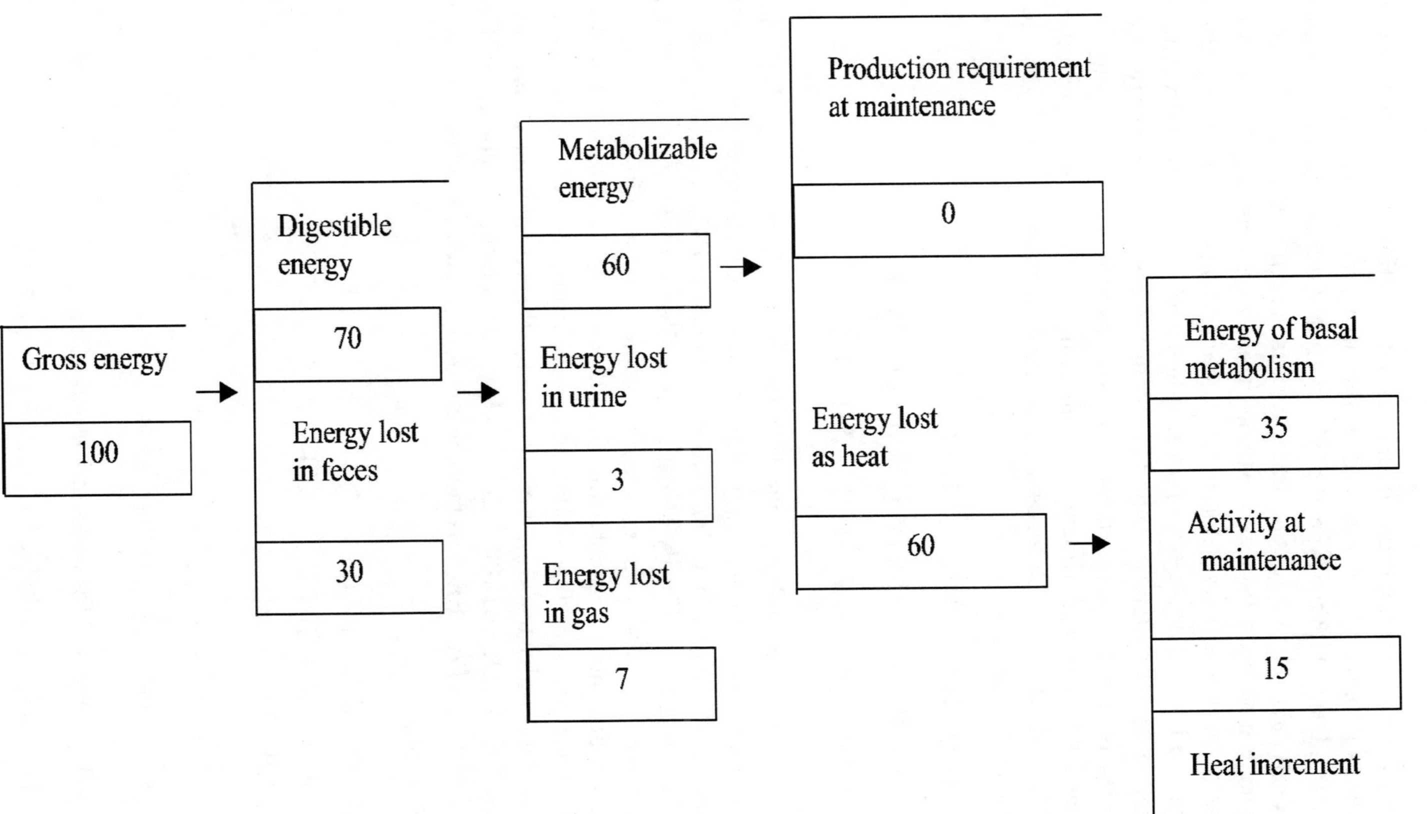

Figure 1: Partition of dietary energy at maintenance level of intake (Kearl, 1982)

Energy is the most important item in ruminant rations in terms of quantity. Great losses of energy are involved in the utilization of various feeds by ruminants and vary with the type of feed and kind of production to be obtained from the animals. Therefore, the recent approaches of expressing the energy value of feeds have been discussed. Although, other systems like total digestible nutrients (TDN), Kellners' Starch Equivalent, Armsby net energy value, Scandinavian Food Unit etc. are of great value. NRC (2001) for dairy animals has recently adopted different method to obtain and express feed energy values. Earlier feedstuffs were assigned total digestible nutrients (TDN) values that have been determined experimentally using similar feeds. The concentrations of DE, ME and NE_L for each feedstuff were then calculated from TDN values using different equations.

$$DE\ (M\ cal/\ kg) = 0.04409 \times TDN\ (\%)$$

$$ME\ (M\ cal/\ kg) = 1.01 \times DE\ (M\ cal/\ kg) - 0.45$$

$$NE_L\ (M\ cal/\ kg) = 0.0245 \times TDN\ (\%) - 0.12$$

Because of certain problems, the TDN values at maintenance level of intake (TDN_{1X}) are calculated from composition data rather than being experimentally determined. In this summative approach to derive the TDN_{1X}, the concentrations (percent of dry matter) of truly digestible non-fiber carbohydrate (NFC), CP, ether extract (EE), and NDF for each feed are estimated using different equations. Ether extract does not represent a nutritionally uniform fraction and therefore, does not have a constant digestibility across feedstuffs. Fatty acids (FA) are a uniform fraction with a true digestibility of 95 to 100 percent when diets contain 3% or less EE. A value of 100% digestibility was chosen, FA content of feed can be estimated as FA = EE -1. A more accurate approach would be to measure FA directly.

Truly digestible NFC (td NFC)

$$= 0.98\ (100 - [(NDF - NDICP) + CP + EE + Ash]) \times PAF$$

Truly digestible CP for forages (td CPf)

$$= CP \times \exp[-1.2 \times (ADICP/CP)]$$

Truly digestible CP for concentrates (td CPc)

$$= [1 - (0.4 \times (ADICP/CP))] \times CP$$

Truly digestible FA (td FA)

= FA Note: If EE<1, then FA = 0

Truly digestible NDF (td NDF)

$= 0.75 \text{ X } (NDFn - L) \times [1 - (L/NDFn)^{0.667}]$

In above equations, NDICP = neutral detergent insoluble N x 6.25, PAF = processing adjustment factor, ADICP = acid detergent insoluble N x 6.25, FA = fatty acids (i.e., EE – 1), L = acid detergent lignin, and NDFn = NDF – NDICP and all values are expressed as a percent of dry matter (DM).

Above equations are based on true digestibility, but TDN is based on apparent digestibility, therefore, metabolic faecal TDN must be subtracted from the sum of the digestible fractions. It has been determined that, on an average, metabolic faecal TDN equally 7. The TDN_{1x} is then calculated as follows-

TDN_{1x} (%) = td NFC + td CP + (td FA x 2.25) + td NDF - 7

Whereas, NE_L values are calculated based on actual intake and the digestibility of the entire diet. Since the digestibility of diets fed to animals is reduced with increasing feed intake. This reduces the energy value of any given diet as feed intake increases. This is particularly important for high producing animals where feed intake even exceeds 4 times of maintenance level of intake.

DE_{1X} (Mcal/ kg) =

(td NFC/ 100) x 4.2 + (td NDF/ 100) x 4.2 + (td CP/ 100) x 5.6 + (FA/ 100) x 9.4 – 0.3

- DE at productive levels of intake (DEp)

  DEp (Mcal/ kg) = DE_{1X} x Discount factor

Where, Discount factor =

$[(TDN_{1X} - [(0.18 \times TDN_{1X}) - 10.3]) \times \text{Intake})] / TDN_{1X}$

Intake = As incremental intake above maintenance

- ME at production level of intake (MEp)

  MEp (Mcal/ kg) = [1.01 x (DEp) – 0.45] + 0.0046 x (EE – 3)

  Where, DEp = Mcal/ kg, EE = % of DM

- NE_L at production level of intake (NE_{LP})

  NE_{LP} (Mcal/ kg) = [0.73 x MEp (Mcal/ kg) – 0.19]

Estimating DE of feeds

The gross energy of TDN is 4.409 M cal/ kg. But nutrients have different heats of combustion (like 4.2 M cal /kg for carbohydrates, 5.6 M cal/ kg for protein, 9.4 M cal/ kg for long chain fatty acids and 4.3 M cal/ kg for glycerol) and the gross energy value of TDN is not constant among feeds. The gross energy of TDN of a feed that has a high proportion of its TDN provided by protein will be greater than 4.409. On the contrary, the gross energy of TDN of a feed with a high proportion of its TDN provided by carbohydrate or fat will be less than 4.409. Hence, the calculation of DE as 0.04409 x TDN (percent) has been abandoned. Digestible energy is calculated by multiplying the estimated digestible nutrient concentrations by their heats of combustion. Since DE is based on apparent digestibility, while the calculations are based on true digestibility, a correction for metabolic faecal energy is done. The heat of combustion of metabolic faecal TDN is assumed to be 4.4 M cal/ kg; metabolic faecal DE = 7 x 0.044 = 0.3 M cal/ kg.

For most feeds

DE_{IX}(M cal/ kg) = (tdNFC/100) x 4.2 + (tdNDF/100) x 4.2 + (tdCp/100) x 5.6 + (FA/100) x 9.4 – 0.3

For animal protein meals

DE_{IX} (M cal/ kg) = (tdNFC/100) x 4.2 + (tdCP/100) x 5.6 + (FA/100) x 9.4 – 0.3 (2-8b)

For fat supplements with glycerol

DE_{IX} (M cal/ kg) = 9.4 x (FA digest x 0.9 x (EE100)) + (4.3 x 0.1 x (EE/ 100))

For fat supplements without glycerol

DE_{IX} (M cal/ kg) = 9.4 x FA digest x (EE/100)

Where, tdNFC, tdNDF, tdCP and FA are expressed as percent of DM

Again the method used to estimate the digestibility values of fatty acids are already accounted for the difference between apparent and true digestibility and thus 0.3 adjustments is not done.

Estimating DE at Actual Intake

The digestibility of diets fed to dairy animals is reduced with increasing feed intake. This reduces the energy value of any given diet as feed intake increases. This is particularly important for high producing dairy animals where it is not uncommon for feed intake to exceed 4 times the maintenance level of intake. The rate of decline in digestibility with level of feeding has been shown to be related to digestibility of the diet at maintenance. Diets with high digestibility at maintenance exhibit a higher rate of depression in digestibility with level of feeding than diet with low digestibility fed at maintenance. Accordingly, NRC (1989) has used a constant depression of 4 percent per multiple of maintenance to adjust maintenance energy values to 3X maintenance energy values. Using this method of discounting, the percentage unit decline in TDN for a diet containing 75 percent TDN_{IX} would be 3 percentage units per multiple of maintenance, while the depression for a diet containing 60 percent TDN_{IX} would be 2.4 units the differences in the rate depression in digestibility are generally negligible for diets having maintenance TDN values of 60 percent or less.

But studies on the relationship between digestibility at maintenance and the percentage unit decline in digestibility per multiple of maintenance feeding, it was found that the rate of decline in digestibility with level of feeding was a function of the maintenance digestibility of the diets fed: TDN percentage unit decline = 0.18 x TDN_{IX} – 10.3 ($r^2 = 0.85$). Because DE, not TDN, is used to calculate ME and NE_L, this equation was converted so that a percent discount, and not a TDN percentage unit discount, was calculated:

$$\text{Discount} = [(TDN_{IX} - [(0.18 \times TDN_{IX} - 10.3]) \times \text{intake})] / TDN_{IX}$$

Where TDN_{IX} is as a percent of dry matter and is for the entire diet, not the individual feed, and intake is expressed as incremental intake above maintenance (say for a cow consuming 3X maintenance, intake above maintenance = 2). For example, for a cow consuming a diet that contains 74 percent TDN_{IX} at 3X intake, digestibility would be expected to be 0.918 times the value obtained at maintenance.

For diets with 60 percent or less TDN_{IX} the discount is set to 1.0 (i.e., no discount was applied). Furthermore, a maximum discount is set so that discounted diet TDN could not be less than 60 percent. Data on effects on intake much greater than 4X maintenance are lacking. Again the effect of intake on digestibility is not linear, but rather the digestibility discount increases at a decreasing rate as feed intake increases. The possibility of a nonlinear response was one reason that the minimum discounted TDN was set at 60 percent. Data are needed on the effects of very high intake on digestibility. The adjustment is used continuously across all levels of feeding. The DE_{IX} for each feed is determined and then multiplied by the discount factor to calculate DE at productive levels of intake (DE_p).

Estimating ME at Actual Intake

ME at production levels of intake ME_P is calculated from DE_P. ME_P values of feeds with more than 3 percent EE are increased by 0.0046 per percentage unit increase in EE content above 3 percent. For feeds with less than 3 percent EE, there is no increase in ME_P content.

$$ME_P \text{ (M cal/ kg)} = [1.01 \times (DE_P) - 0.45] + 0.0046 \times (EE - 3)$$

Where DE_P is M cal/ kg and EE is percent of DM.

For fat supplements, ME_P (M cal/ kg) = DE_P (M cal/ kg)

Estimating NE_L at Actual Intake

An equation derived by Moe and Tyrrell (1972) to convert ME_P to NE_L at production levels of intake (NE_{Lp}) has been chosen to replace the previous TDN–based NE_L equation.

$$NE_{Lp} \text{ (M cal/ kg)} = [0.703 \times ME_P \text{ (M cal/ kg)}] - 0.19$$

A modification is made to adjust for improved metabolic efficiency of fat. The average efficiency of converting ME from fat to NE_L is 0.80.

$$NE_{Lp} \text{ (M cal/ kg)} = 0.703 \times ME_P - 0.19 + ([(0.097 \times ME_P + 0.19)/97] \times [EE - 3])$$

Where ME_P is M cal/ kg and EE is percent of DM

For fat supplements, NE_{Lp} (M cal/ kg) = 0.8 x ME_P (M cal /kg)

Estimating Net Energy of Feeds for Maintenance and Gain

The NE_M and NE_G content of feeds assumed dry matter intake at 3 times maintenance and are calculated by multiplying DE_{IX} (described above) by 0.82 to obtain ME. This ME value is then converted to NE_M and NE_G using the following relationship:

$$NE_M = 1.37\,ME - 0.138\,ME^2 + 0.0105\,ME^3 - 1.12$$

$$NE_G = 1.42\,ME - 0.174\,ME^2 + 0.0122\,ME^3 - 1.65$$

Where, ME, NE_M and NE_G are expressed in M cal/ kg.

The above equations are not appropriate for fat supplements. For those feeds, $ME_P = DE_P$, and the same efficiency (0.80) of converting ME to NE_L is used to convert ME to NE_M. The efficiency of converting ME to NE_G is set at 0.55 for fat supplements.

According to ARC/ AFRC, the energy values of feeds are expressed in terms of ME and the energy requirement of an animal as net energy (NE). Allowances of NE are given separately for each of the main function of the animal like maintenance, growth or milk production. A practical advantage of the system is that ME contents of feeds and forages can be determined easily by digestion trials, whereas the determination of the NE of feedstuffs requires time consuming and complicated experiments with respiration chambers or animal calorimeter.

Metabolizable Energy (ME)

Metabolizable energy intake is the gross energy (GE) of the feed minus that of the faeces (FE), urine (UE) and combustible gas (mostly methane, M_E). It is expressed as MJ/kg DM or MJ/d.

$$ME = GE - FE - UE - ME$$

Metabolizability of Feeds

The metabolizability of the GE of a feed at maintenance (q_m), is defined as the proportion of ME in the GE of the Feed.

$$q_m = ME/GE$$

Alternatively, the ME concentration of the feed dry matter (M/D) was proposed by ARC (1965), because of the paucity of GE data on ruminant feeds.

Fermentable metabolizable Energy (FME)

It is the ME content of the feed minus the ME present as total oils and fats (ME fat), and the ME contribution of fermentation acids (ME_{ferm})

$$FME\ (MJ/kg\ DM) = ME - ME_{fat} - ME_{ferm}$$

The average ME_{fat} = 35 MJ/kg DM, therefore, as little as 4% of oil reduces FME by 1.4 MJ/kg DM. The fermentation acids primarily lactic, acetic, propionic and butyric acids being the end product of microbial fermentation cannot yield any ATP under anaerobic fermentation in the rumen hence their values should be deducted from the feed ME value.

Efficiencies of ME utilization

Efficiency for maintenance	$K_m = 0.35q_m + 0.503$
Efficiency for lactation	$K_l = 0.35q_m + 0.420$
Efficiency for growth	
Growing ruminants	$K_f = 0.78q_m + 0.006$

Lactating ruminants	$K_g = 0.95k_l$
Efficiency for growth of the concepta	$K_c = 0.133$
Efficiency for utilization of mobilized body tissue for lactation	$K_t = 0.84$

Correction for feeding level

The high level of feeding increases the digesta outflow rates from rumen. Hence ME actually available to the animal reduced significantly. Therefore, it needed correction and the correction factor (C_L) is calculated as follows:

Lactating ruminants

$C_L = 1 + 0.018\ (L\text{-}1)$

Where, L is multiples of maintenances ME requirement.

Growing and fattening animals:

$R = B\ (1\text{-}e^{-kI}) - 1$

Where, R= retention of NE

I = ME intake

$B = K_m / (K_m - K_f)$

$K = K_m \times M\ (K_m / K_f)$

Calculation of ME requirements

Lactating ruminants

$$M_{mp}\ (MJ/d) = C_L\ (E_m/K_m + E_l/K_l + E_g/K_g + E_c/K_c)$$

Growing cattle and sheep:

$$M_{mp}\ (MJ/d) = (F/K) \times I_n\ \{B/(B\text{-}R - 1)\}$$

Principles of diet formulation

With diet formulation, the ME required is influenced by the calculated M/D (q_m) of the diet being formulated. If the diet specification gives the M/D required, (or both DM and ME intakes), then two component diets of forage and compound can be formulated using:

$$M/D = (ME\ compound) + (1 - x)\ (ME\ forage)$$

Or,

$$x = \frac{M/D - ME\ forage}{ME\ compound - ME\ forage}$$

Where, x is the decimal proportion of the compound in the diet.

NE system for diet formulation

The value for the joint efficiency of utilization of ME for maintenance and production (K_{mp}) is given by:

$$K_{mp} = E_{mp} / M_{mp}$$

The value for K_{mp} obtained is then used to calculate the Net Energy (E_{mp}) value for the feed when incorporated in the diet for the given level of animal production:

$$E_{mp} \text{ (MJ/d)} = M_{mp} \times K_{mp}$$

The calculated feed E_{mp} values can be used to formulate a diet supplying the required amount of E_{mp}, which should include the recommended 5% safety margin, and taking cognizance of DM intake limitations.

Protein utilization

NRC (2001) has adopted the model of Cornell Net Carbohydrate Protein System (CNCPS). In this model, feed CP is divided into five fractions (A, B_1, B_2, B_3, and C) which sum to unity. The five fractions have different rates of ruminal degradation. Fraction A (NPN) is the percentage of CP that is instantaneously solubilized at time zero, which is assumed to have a degradation rate (k_d) of infinity, it is determined chemically as that proportion of CP that is soluble in borate-phosphate buffer but not precipitated with the protein denaturant, trichloroacetic acetic (TCA). Fraction C is determined chemically as the percentage of total CP recovered with ADF (i.e., ADIN) and is considered to be undegradable. Fraction C contains proteins associated with lignin and tannins and heat-damaged proteins such as the Maillard reaction products. The remaining B fractions represent potentially degradable true protein. The amounts of each of these 3 fractions that are degraded in the rumen are determined by their fractional rates of degradation (k_d) and passage (k_p); a single k_p value is used for all fractions. Fraction B_1 is that percentage of total CP that is soluble in borate-phosphate buffer and precipitated with TCA. Fraction B_3 is calculated as the difference between the portions of total CP recovered with NDF (i.e., NDIN) and ADF (i.e., fraction C). Fraction B_2 is the remaining CP and is calculated as total CP minus the sum of fractions A, B_1, B_3, and C. reported ranges for the fractional rates of degradation for the three B fractions are: B_1 (120 – 400%/h), B_2 (3 – 16%/h), and B_3 (0.06 – 0.55%/h). The RDP and RUP values (percent of CP) for a feedstuff using this model are computed using the equations:

$$RDP = A + B_1 [k_dB_1 / (k_dB_1 + k_p)]$$
$$+ B_2 [k_dB_2 / (k_dB_2 + k_p)]$$
$$+ B_3 [k_dB_3 / (k_dB_3 + k_p)]$$

$$RUP = B_1 [k_p / (k_dB_1 + k_p)]$$
$$+ B_2 [k_p / (k_dB_2 + k_p)]$$
$$+ B_3 [k_p / (k_dB_3 + k_p)] + C$$

Although the most commonly used model to describe in situ ruminal protein degradation divides feed CP into three fractions (A, B, and C). Fraction A is the percentage of total CP that is NPN (i.e., assumed to be instantly degraded) and a small amount of true protein that rapidly escapes from the in situ bag because of high solubility or very small particle size. Fraction C is the percentage of CP that is completely undegradable; this fraction generally is determined as the feed CP remaining in the bag at a defined end-point of degradation. Fraction B is the rest of the CP and includes the proteins that are potentially degradable.

Similarly the present UK metabolisable protein system for the estimation of protein requirements of ruminants (AFRC, 1993) recognizes that protein, like the carbohydrate present in feed, can be considered as a number of fractions with different degradability characteristics. The immediately soluble protein, akin to WSC, is assumed to be readily utilized by the rumen microflora (QDN, quickly degraded nitrogen), while a second insoluble fraction is degraded more slowly (SDN, slowly degraded nitrogen) leaving a rumen undegradability fraction (UDN, undegradable nitrogen). Assessment of protein quality ideally requires a measure of these fractions coupled with the rate at which the SDN fraction is degraded. Various laboratory methods can be used to measure the immediate soluble fraction, and the amount of nitrogen remaining associated with ADF (ADIN, acid detergent in soluble nitrogen) used to indicate the size of the non-degradable fraction. Subtraction of these values from the total nitrogen/ protein will provide an estimate of the amount of SDN present. However, this is a crude approximation, and the various regression equations developed to predict the needs for growth, lactation or wool production require kinetic data. Typically, values for QDN, SDN, UDN and the necessary rate data are still obtained by the nylon bag method. As a result, ration formulation for ruminants may utilize actual CP values and, occasionally AIDN, but otherwise is based wholly on table values and is unable to take account of variations in raw ingredients.

Metabolizable Protein

After digestion and absorption of feed protein the total digestible true protein available to the animal for metabolism is called metabolizable protein (MP). It has two components:

i. Digestible Microbial True Protein (DMTP)

It is synthesized by the rumen microbes from fermentable energy, amino acids and NPN sources. About 0.25 part of Microbial Crude Protein (MCP) is present as nucleic acids which cannot be used by ruminants and remaining 0.75 parts as Microbial True Protein (MTP) which is digestible.

$$DMTP\ (g/d) = 0.75 \times 0.85 \times MCP$$
$$= 0.6375\ MCP\ (g/d)$$

ii. Digestible undegraded feed Protein (DUP)

It is that fraction of feed protein which is not degraded in the rumen (UDP) but digested and absorbed in lower intestines. The proportion of DUP in UDP varies from zero to 0.9. The digestibility of UDP can be predicted from the acid detergent insoluble nitrogen (ADIN) content of the feed.

Thus, MP (g/d) = 0.6375 MCP + DUP

Degradation of feed proteins in the rumen

In the MP system the key protein parameters, QDP, SDP and UDP are derived from measurements of the rates of feed protein degradation (dg) in the rumen. The dg can be calculated by following equation:

$$dg = a + b\ \{1 - e^{(-ct)}\}$$

where, a = water soluble N
b = potentially degradable
c = fractional rate of degradation of feed N per hour
t = time

Rumen retention time and outflow rate

Retention time in the rumen is highly correlated with the level of feeding of the animal (L). The fractional rumen outflow rates per hour (r) vary from 0.02 to 0.08 depending on the level of feeding as follows:

1) Maintenance level of feeding, r = 0.02/h
2) Higher but less than 2 x maintenance level of feeding, r = 0.05/h
3) Greater than 2 x maintenance level of feeding, r = 0.08/h

The equation used for calculation of this factor is:

$$r = - 0.024 + 0.179 \{1 - e^{(-0.278L)}\}$$

Influence of rumen outflow rate on degradability

The effective degradability (P), for any given rumen outflow rate (r), is:

$$P = a + (b \times c)/ (c + r)$$

Quickly degradable protein (QDP)

It is the fraction of total crude protein (CP) extracted by cold water, defined by the constant, a, consisting of mainly NPN and water soluble small protein molecules

$$QDP\ (g/kg\ DM) = a \times CP\ (g/kg\ DM)$$

Slowly degradable protein (SDP)

It is the function of level of feeding (L) and outflow rate and is calculated as follows:

$$SDP\ (g/kg\ DM) = \{(b \times c)/ (c + r)\}\ CP\ (g/kg\ DM)$$

Effective rumen degradable protein (ERDP)

The term ERDP has been defined as a measure of the total N supply that is actually captured and utilized by the rumen microbes for growth and synthetic purposes. ERDP content of a feed is defined as:

$$ERDP\ (g/kg\ DM) = 0.8 \times QDP + SDP$$

The total ERDP supply to the rumen microbes (g/d) is the sum of the weights of feed DM in kg/d (W_n) multiplied by the feed ($ERDP_n$) contents in g/kg DM calculated as follows:

$$ERDP\ (g/d) = W_1\ (ERDP_1) + W_2\ (ERDP_2) + W_3\ (ERDP_3)$$

Digestible undegradable protein (DUP)

$$UDP\ (g/d) = CP - RDP$$

Or,

$$UDP\ (g/d) = CP - (QDP + SDP)$$

Or,

$$UDP\ (g/kg\ DM) = 0.9\ \{(UDP) - 6.25\ (ADIN)\}$$

The level of feeding (L) also influences the ERDP and DUP values of feed.

Microbial protein synthesis in the rumen

In MP system different factors are taken into account for microbial protein syntheses are:

i) Energy supply to the microbes
ii) Nitrogen supply to the microbes
iii) Level of feeding to the animals
iv) Outflow rate, principally determined by level of feeding

i.) Energy supply to rumen microbes

Energy supply is normally the first limiting factor on microbial protein synthesis. For estimation of rumen microbial crude protein (MCP) yield (y), fermentable metabolizable energy (FME), MJ/d or MJ/kg DM should be used (AFRC, 1993).

The values of microbial crude protein synthesis (y), at different level of feeding (L) are as follows:

All animals (L = 1)	9g MCP/ MJ of FME
Growing sheep and cattle (L = 2)	10 g MCP/MJ of FME
Late pregnancy/ lactating ewes and cows (L = 3)	11g MCP/ MJ of FME

The empirical equation for estimation of microbial yield is

$$y\ (\text{g/MJ FME}) = 7.0 + 6.0\ \{1 - e^{(-0.35\ L)}\}$$

ii.) Nitrogen supply to microbes

Nitrogen supply to microbes may be limiting microbial protein synthesis. MP system increases the amounts of ERDP required to match the amount of FME supplied by the diet.

$$\text{MCP (g/d)} = \text{ERDP (g/d)}$$

Estimation of microbial crude protein supply

If ERDP is greater than or equal to the MCP supply, then

$$\text{MCP (g/d)} = \text{FME (MJ/d)} \times y\ (\text{g MCP} \times \text{MJ FME})$$

Estimation of digestible true protein supply

Having estimated microbial crude protein supply, corrections are required to estimate the amount of true protein that is absorbed in the lower digestive tract. This requires estimates of two fractions:

1) True protein content of MCP (MTP), is estimated to be 0.75 of MCP.
2) Digestibility of MTP (DMTP), is estimated to be a constant 0.85

Therefore,

DMTP (g/d) = 0.75 x 0.85 x MCP (g/d)

= 0.6375 MCP

Efficiencies of utilization of metabolizable protein

Efficiency of utilization of an ideal amino acid mixture (K_{aai}) is 1.0 for maintenance and 0.85 for all protein synthetic functions.

Relative value of absorbed amino acid mixtures (RV) vary according to the mixture of amino acids supplied to the tissues:

Growth	RV = 0.7
Pregnancy	RV = 1.0
Lactation	RV = 0.8
Wool	RV = 0.3

Working values for efficiency of MP utilization (K_n) estimated by combining values for K_{aai} and RV

Maintenance	$K_{nm} = 1.00$
Growth	$K_{ng} = 0.59$
Pregnancy	$K_{nc} = 0.85$
Lactation	$K_{nl} = 0.68$
Wool	$K_{nw} = 0.26$

RATION COMPUTATION

The bulk of feeds eaten by animals in our country are usually poor quality forages like cereal straws/ stovers and dry mixed grasses. When animals are fed such forages alone, they can not eat enough to obtain the energy, protein and minerals needed to produce good amount of milk. Thus it is necessary to include a concentrate source of energy, protein and minerals in the rations of animals. To supply all the needs for maintenance, growth, reproduction and lactation an animal must receive sufficient feed to furnish the necessary amount of above said nutrients and water.

Nutrients supplied by the ration (kg/ d) = Nutrients required by the animal (kg/ d)

When a ration is not balanced, the animal eats some nutrients in excess or in an insufficient amount, which leads to non-exploitation of animal's genetic potential. Although, all such imbalances are not nutritionally devastating, but economically none of them are acceptable because that results in a significant loss of production and loss of nutrients that might have been used more effectively otherwise.

Formulation

The formulation of a balanced ration consists of combining feeds that eaten in amounts needed to supply the daily requirements of the animals. A ration is balanced when all the required nutrients are present in the feed ingested by the animal during a period of 24 hours. A farmer desirous of producing concentrate mixture of his own should know the various types of feed ingredients required for making concentrate mixture ideal for livestock feeding. If the farmer is not well conversant with the quality of raw feed ingredients and unconventional feedstuffs available in the region, he may not be able to formulate an ideal concentrate mixture for his animals economically. The various feed ingredients in brief are as follows

(a) Primary source of energy (30-40%): Cereal grains like maize, barley, wheat, sorghum, oats etc

(b) Primary source of protein (25-30%): Oil seed cakes like groundnut cake, mustard cake, cotton seed cake, linseed cake, sesame cake and animal protein supplements like fish meal, meat meal, skim milk powder etc.

(c) Diluents (5-25%): Cereal by-products like wheat bran, rice bran etc. and pulse chunies like mung chuni, arhar chuni, massoor chuni, gram chuni etc.

(d) Common salt (1-2%)

(e) Mineral supplements (1-2%)

(f) Vitamin supplements (AD3 @ 20-30 g/ 100 kg concentrate mixture, containing 50, 000 I.U Vit. A and 5000 I.U Vit. D3 per g)

Steps for formulation

For the formulation of 100 kg balanced concentrate mixture following steps should be taken one by one-

- Crush or grind the selected feed ingredients like primary sources of protein (oil cakes) and energy (cereal grains) separately. Diluents (cereal and pulse by-products) do not require crushing or grinding.

- Since the primary sources of protein also provide the energy and the energy sources protein too. First weigh 25 to 30 kg of primary source of protein depending upon the availability and the quality of concentrate mixture to be prepared.
- 30 to 40 kg of primary source of energy (cereal grains) are then added as per the availability and also the quality of concentrate mixture to be prepared.
- Now add 1 kg of finely ground common salt and 2 kg commercial mineral mixture available in the market.
- Then subtract the total amount of primary sources of protein, energy and supplements (salt and mineral mixture) from 100 kg. The rest amount will be the diluents either from cereal by-products and/ or from pulse by-products.
- Finally mix all the above ingredients properly to make the concentrate mixture homogenous and put it in bags and store at a dry place.

Thus, during formulation of concentrate mixture, primary consideration is given to protein and energy content of the ration, which is satisfied by selecting suitable protein and grain supplements. While cereal by-products are palatable and laxative, furnishing good amount of minerals. Pulse chunies are also palatable and improve the nutritive values of concentrate mixture. The addition of salt (1-2%) to concentrate mixture increases the palatability and supplies sodium and chlorine to the animals. Minerals are vitally important when good quality greens are not available. When straw is used as a major source of roughage, both minerals and vitamin supplements are essential.

Characteristics of a good concentrate mixture

- A good concentrate mixture should contain 12-18% digestible crude protein (DCP) and 65-75% total digestible nutrients (TDN).
- Final concentrate mixture should contain 1 % common salt and 2% mineral and vitamin mixture
- Moisture content should not be more than 12-15%
- Crude fibre content should not be more than 18%
- Concentrate ingredients should be properly ground before mixing
- It should not be too dense or too bulky

Factors to be known about animals

In order to balance a ration and to feed a lactating/ producing animal properly, it is important to know the factors that affect the nutrient requirements of the animal. The important factors are-

- Body weight of animals
- Level of milk production
- Composition of the milk (% fat)
- Stage of lactation
- Number of lactations
- Average daily gain or change in body weight

Factors to be known about the feeds/ feed ingredients

- Moisture content and chemical composition on a dry matter basis (energy/ TDN, crude protein, minerals and vitamins)
- The specific attributes of the feed and the maximum percentage allowed in the ration
- If forage, then stage of maturity, the physical form of the fibre particles, the length of storage etc.
- If purchased feeds, then local availability, the cost per unit of nutrient, the quality of the feed and degree of processing (heating, pelleting, extrusion etc.)

Feeding schedule of concentrate mixture

As per the standard nutrient requirements for maintenance, milch animals (cows & buffaloes) require concentrate allowance @ 0.5 kg/ 100 kg body weight. Besides maintenance, they require extra allowance for other activities like lactation, reproduction, growth etc. When green forages like sorghum, bajra, oats or their silage are available, the amount of concentrate mixture can be reduced up to 25% and leguminous fodder like berseem, cowpea etc., if available in plenty, the amount of concentrate mixture can be reduced up to 50% of the maintenance requirement. Cows should be given 1 kg of concentrate mixture extra for every 3 kg of milk and buffaloes for every 2.5 kg of milk produced per day.

Few balanced and economical concentrate mixture with 12-16% DCP and 67-70% TDN are given herewith for the help of farmers.

	Ingredients	Quantity (kg)
(a)	Groundnut cake	20
	Barley	25
	Wheat bran	52
	Mineral mixture	02
	Common salt	01
(b)	Mustard cake	25
	Barley	20
	Rice bran	52
	Mineral mixture	02
	Common salt	01
(c)	Mustard cake	25
	Maize	20
	Wheat bran	52
	Mineral mixture	02
	Common salt	01
(d)	Cotton seed cake	25
	Oats	22
	Rice bran	30
	Mineral mixture	02
	Common salt	01

Computation of ration

A basic rule for feeding dairy animals is to encourage them to consume large quantities of quality forages. But they can not eat enough quantity to meet their requirements. In fact the main problem is that high producing animals simply can not digest and metabolize the necessary feed nutrients needed to sustain heavy milk yield from roughage alone. Concentrates must be fed to the good producers to fill the gap between the nutrients required by the animal and those furnished by the forage/ roughage. Hence, the kind and quality of forage that is fed will dictate the type of concentrate mixture needed. The genetic ability of the animal to produce milk and the amount of forage that is fed will determine the amount of concentrate that must be fed.

In computation of ration for animals, first the requirements of dry matter, protein and energy (TDN) in a day are ascertained and then these are met. The dry matter intake depends upon the body weight of the animals and kind of production. Dairy animals normally consume 2.0 to 3.0 kg dry matter for every 100 kg body weight (cows: 2.0-2.5% and buffaloes: 2.5-3.0% of body weight.

Hence, all its nutritional requirements like protein, energy, fats, mineral, vitamins etc should come from the total dry matter allotted to the animal. While calculating the requirements of CP/ DCP and TDN one has to consider the purpose for which the animal is to be fed. The requirement of nutrients for all such purposes may be obtained separately from appropriate standard requirement tables. One has to simply add up the additional/ production requirements to maintenance requirement following physiological conditions. While average CP/ DCP and TDN contents of the feedstuffs have been worked out by different research workers and given in appropriate tables. One has to select suitable concentrate ingredients and forages as available locally and is cheaper also.

For a dairy animal (cow or buffalo) producing different levels of milk, the rations are computed as follows-

A. Rabi season

(a) An animal (assumed body weight 425 kg) producing 6.0 kg of milk a day with 5% fat

Purpose/Ration	**Requirement**				**Availability**			
	CP (g)	**DCP (g)**	**TDN (kg)**	**ME (M cal)**	**CP (g)**	**DCP (g)**	**TDN (kg)**	**ME (Mcal)**
Maintenance	500	300	3.60	12.9				
Production	500	300	2.00	7.2				
30 kg berseem					1000	600	3.60	12.9
4.5 kg wheat straw					140	0	1.80	6.5
Total	**1000**	**600**	**5.60**	**20.1**	**1140**	**600**	**5.40**	**19.4**

N.B.: 1.0 kg CP = 0.60 kg DCP, 1.0 kg TDN = 3.60 M cal ME

(b) An animal (assumed body weight 425 kg) producing 10 kg of milk a day with 5% fat

Purpose/Ration	**Requirement**				**Availability**			
	CP (g)	**DCP (g)**	**TDN (kg)**	**ME (M cal)**	**CP (g)**	**DCP (g)**	**TDN (kg)**	**ME (Mcal)**
Maintenance	500	300	3.60	12.9				
Production	1000	600	4.00	14.4				
43 kg berseem					1333	800	4.80	17.3
4.5 kg wheat straw					140	0	1.80	6.5
Total	**1500**	**900**	**7.60**	**27.3**	**1473**	**800**	**6.60**	**23.8**

The above examples reveal that an animal producing upto 10 kg of milk a day can get its full nutritional requirements from fodder alone in rabi season. For additional milk production, concentrates should be fed.

B. Kharif season

Similarly, for Kharif season the ration may be computed as below

(a) An animal (assumed body weight 425 kg) producing 4.0 kg of milk a day with 5% fat

Purpose/Ration	**Requirement**				**Availability**			
	CP (g)	**DCP (g)**	**TDN (kg)**	**ME (M cal)**	**CP (g)**	**DCP (g)**	**TDN (kg)**	**ME (Mcal)**
Maintenance	500	300	3.60	12.9				
Production	400	240	1.60	5.7				
30 kg maize + cowpea						750	450	3.60
12.9								
10 kg bajra green					166	100	1.68	6.0
0.5 kg wheat straw					15	0	0.20	0.7
Total	**900**	**540**	**5.20**	**18.6**	**931**	**550**	**5.28**	**19.6**

Here also, for additional milk production, concentrates should be fed to the producing animals.

But when there is non-availability of green forages and animals are depending on poor quality forages like cereal straws and stovers, dry mature grasses etc, animals must be fed concentrate mixture. Since these roughages are poor sources of protein and energy besides being deficient in minerals and vitamins.

An animal (assumed body weight 400 kg) producing 10 kg of milk a day with 5% fat

Purpose/Ration	**Requirement**				**Availability**			
	CP (g)	**DCP (g)**	**TDN (kg)**	**ME (M cal)**	**CP (g)**	**DCP (g)**	**TDN (kg)**	**ME (Mcal)**
Maintenance	408	245	3.10	11.2				
Production	933	560	3.80	13.6				
5 kg concentrate mixture					1333	800	3.75	13.5
6 kg wheat straw					189	0	2.64	9.5
Total	**1341**	**805**	**6.90**	**24.8**	**1522**	**800**	**6.39**	**23.0**

N.B.: 1.0 kg CP = 0.60 kg DCP, 1.0 kg TDN = 3.60 M cal ME

With wheat straw although, concentrate mixture has been given @ 1.0 kg per 2.0 kg of milk, still the supply of TDN/ ME (energy) for animal producing 10 kg of milk is little short.

Hence, protein and energy requirements of the milch animals and their fulfillment, is of critical economic importance to our dairy farmers. There can be increase in profits by reducing feed costs through homemade concentrate mixture. Many readymade concentrate mixtures under different trade name available in the market are of substandard quality because manufacturers generally use poor quality ingredients to get more profit. These products are likely to be detrimental to animal health and production. Therefore, it is essential to formulate a good quality concentrate mixture from locally available cheaper sources at the users' level. The prerequisites to formulate a balanced concentrate mixture as well as the procedure are given herewith for the help.

Ration computation based on recent NRC approach

Say, Feeding of buffalo calves (body weight 250 kg) for expected gain of 680 g daily, to be fed sorghum (25% DM, 8.9% CP, 1.0 Mcal/ kg NEm, 0.56 NEg) along with common salt and a complete mineral mixture and concentrate supplements.

Requirements:

NEm = 4.84 Mcal/ day

NEg (680g ADG) = 2.296 Mcal

Protein (CP) requirement = 626 g

Feed (DM) intake = 6.24 kg

Assuming 6.0 kg DM of this will be from sorghum

Nutrient contributions from sorghum-

Protein (CP) = 6 x 0.089 g = 534 g

CP deficit = 626 – 534 = 92 g

Energy:

To provide 4.84 M cal NEm, requires = 4.48/ 1.00 =4.84 kg

Hence, 1.16 kg are used to provide NEg = 1.16 x 0.56 = 0.65 M cal

Total NEg needed = 2.296 M cal

NEg deficit = 2.296 – 0.65 = 1.646

Therefore, supplements should provide 92g CP and 1.646 M cal NEg per animal per day

Composition of supplements (assumed)

a) Barley	CP	= 13.5%
	NEm	= 2.06 M cal/ kg
	NEg	= 1.40 M cal/ kg
b) Cottonseed meal	CP	= 45.2%
	NEm	= 1.82 M cal/ kg
	NEg	= 1.19 M cal/ kg

To meet the nutrient deficiency:

CP = 92 x (100/ 41.4) = 224 g cottonseed meal

NEg deficit = 1.646 M cal

224 g cottonseed provides = 0.242 M cal

Barley must provide =1.646 – 0.242 = 1.404 M cal

Hence, 1.404 Mcal is provided by (1/ 1.23) x 1.404 = 1.14 kg barley

Therefore, supplement will contain-

Cottonseed meal	= 0.224 kg
Barley	= 1.140 kg
Total	= 1.364 kg

Hence, feed 1.36 kg of this supplement (barley: 83.5% & cottonseed meal: 16.5%) per animal per day. Thus, the new systems include various new findings and become more precise, but are more complex also.

Bibliography

AFRC (1993). Energy and Protein Requirements of Ruminants. An adversary manual prepared by the AFRC technical committee on responses to nutrients. CAB International, Walling ford, UK.

Banerjee, G.C. (1988). Feeds and Principles of Animal Nutrition. Rev. edn. Oxford & IBH Publishing Co. Pvt. Ltd., New Delhi.

Cheeke, P.R. (1991). Applied Animal Nutrition: Feeds and Feeding. Prentice Hall, Englewood Cliffs, New Jersey.

ICAR (1998). Nutrient Requirements of Livestock and Poultry. ICAR, New Delhi.

Kearl, L.C. (1982). Nutrient Requirements Ruminants in Developing Countries. International Feedstuff Institute. Utah State University, Logan, Utah, U.S.A.

Mudgal, V.D. (1992). Formulation of economic ration for dairy cattle under field conditions. Papers presented at the 'All India Dairy Husbandry Officers' Workshop', NDRI, Karnal, India.

NRC (2001). Nutrient Requirement of Dairy Cattle. National Research Council, Washington, D.C.

Pathak, N.N. and Kamra, D.N. (1989). A Text Book of Livestock Feeding in Tropics. Cosmo Publications, New Delhi.

HAY

Hay making seems to be ancient but most important way of conserving forages/fodder despite its dependence on suitable weather at harvest time and crop harvesting stage. Hay making is one of the traditional ways for conservation of forage and crop residues for lean and scarcity period utilization as well as to store in hay bank for the utilization during natural calamity.

Hay making is a process where moisture content of cut herbage is reduced from initial moisture of 70 - 90% to 15-20%. Hay is classed as roughage means a feed having more than 18% crude fibre and less than 20% crude protein in its dry matter. It is rarely a complete food and must be used as part of an overall feeding system. Crop residues, straws and stovers, mainly but not exclusively from cereals, are also important as lean-season feed for ruminants, and are often used in association with hay. Storing forage for use in the scarcity period is practiced since ancient time. Storing of straws and crop residues as animal feed is probably as old as cereal cultivation.

Drying of forages in sunlight and wind naturally till to day is the commonest way of conserving fodder, natural grass and crop residues, however auxiliary artificial drying is used, occasionally, in some of the more highly mechanized systems. Hay can be made by simple equipment, and, once correctly cured, is easy to transport and store, and can be fed with little or no wastage. Since hay is appropriate conserved forage for small-scale farmers and pastoralists with limited resources, it should be encouraged where climatic and economic conditions are suitable. Hay in baled form is often marketed as a cash crop. Alike other methods of conservation, drying reduces digestibility and intake, related to the time that the herbage is exposed to leaching and other losses in the field. These can be reduced by barn drying and high temperature drying of herbage.

Need of conserved fodder/feed

Grasslands, grazing lands and pasture constitute a vast part of the earth's surface but are rarely able to provide adequate (quantity and quality) year round feed to support higher livestock production. In such situation supplementation of forages/crop residues from cultivated land is inevitable to achieve the desired animal production. More over majority of the animals are maintained in mixed farming system and in such mixed situation crop residues have important role. Hay making from traditional grasslands is traditional, while hay also could be made from the cultivated fodder crops.

HAY MAKING: Hay making is a process wherein green perishable forage is converted in to a product that is easy to store and transport with out spoilage, where dry matter and nutrients losses are kept minimum. This includes moisture reduction from 70 - 90% to 20 - 25% or less. To make a good quality hay the crops should be harvested at a stage (preferably pre flowering) when maximum amount of nutrients are present in the plants. It should be prepared in such a way where leaves shattering are minimum and retention of crop green color is maximum. Hay should be free from odour and weeds and should not have more than 15% moisture to avoid incidences of fermentation and combustion.

Problems in Hay making vary according to the crop, climate and prevailing weather at harvest

- Under sub-humid and humid temperate conditions, the main problems are related to slowness of drying, so, to avoid loss by spoilage, the aim is to dry the crop as quickly as the conditions will allow.
- Under hot, dry conditions, in contrast, the problems are likely to be either shattering of the finer parts of the plant, through too rapid drying, or bleaching, with consequent loss of carotene and vitamins.

Hay Types Type of hay may vary as per the conditions, its use and level of technology. The main types of hay based on technology are:

- Long hay It is the traditional type of hay where crop is tied in small bundles, turns and stored in a stack form. This is the most common type of hay making.
- Chopped hay This type of hay is prepared where drying conditions are good and systems highly mechanized; it is less bulky and better for mechanical handling.
- Baled hay Initially prepare by hand (manually*)* and then by stationary machines. Big bales which can be individually handled by a tractor-mounted front-end loader are now the main kind in large-scale farming. Round bales are the simplest to make and most popular. Their shape sheds rain and resists water better than traditional bales.
- Hand-trussed hay is widespread in manual Hay making, often as a means of reducing shattering. *Wafered and pelleted hay* is dense and free-flowing, so it is easy to transport, handle and store. Field units are available, but expensive; they are used for high-quality legume hay in climates which allow rapid drying. Losses are lower than with baling.
- Dried grass: Here herbage is artificially dried at high temperatures. This type of hay is not economic.

- Barn-dried hay: The hay prepared by air drying the feed material with the help of fan. This type of hay is not widely prepared.

Principles of hay making

The drying process

Initially moisture loss is fast especially from the leaves, when the stomata remain open. As the herbage wilts, the stomata close and water removes through the waxy epidermis of the leaves and stems. The moisture inside the stems takes longer removal time. The rate of drying during hay making depends on several factors viz. weather: sunshine, wind and the moisture content of the air.

Drying should be done as quickly as possible to minimize losses. At mowing, the herbage will contain between 70 and 90% of moisture, and this will have to be reduced to between 12 and 20% before the hay can be stored safely. The safe moisture content will depend on a range of factors, including storage method; local climatic conditions where the hay is to be stored; size of stacks or bales; and the nature of the crop. As thumb rule 25% moisture is recommended as a maximum for long loose hay; 20% for chopped hay and 5% for pelleted, cured hay.

There are two types of limitation to water loss: In hay making (i) those which restrict movement of water from plant tissues to the air immediately surrounding the plant within the swath, and (ii) those which restrict the movement of water from the air within the swath to the ambient air. The most favorable conditions for drying are at the swath surface, where both radiation and air circulation are at a maximum.

THE MAIN OPERATIONS IN HAY MAKING

Hay making methods vary depending on crop and circumstances, but the main operations are more or less the same in all cases:

- mowing, which may be combined with conditioning;
- artificial conditioning for rapid drying;
- turning and tedding for even drying of the swath, help dissipate heat and reduce the danger of mould development and fermentation;
- Windrowing, i.e., putting the cut herbage into rows for further handling and collection, and sometimes also for protection at night. In hot arid conditions, windrowing protects the crop against shattering and bleaching;
- trussing or putting into cocks (small heaps) are intermediate stages of drying used in manual systems

- carting and storage, with or without baling. In traditional systems the cured long hay is carted and stored in stacks or barns. Baling before storage is much more common in modern mechanized systems.

LOSSES IN HAY MAKING

During normal hay making there are some inevitable losses. The hay prepared without bleaching and fermentation losses will be of good nutritive value. To minimize loss and spoilage in hay making it is essential to dry the forage quickly and evenly as possible with careful handling thereafter. Main losses occurred in hay making are.

- **Fermentation**

  As the crop is harvested soluble carbohydrates are started fermenting by plant enzymes resulting in carbon dioxide (CO_2) and water production. Carbohydrate fractions are the major nutrients that are altered in hay making however, protein is too affected. Proteins are hydrolyzed to amino acids that will be lost due to leaching by rain water. Enzymatic oxidation of the sap and the activity of bacteria and moulds on the crop surface cause losses, with generation of heat. If the crop is not aerated sufficiently to dissipate the heat, serious damage can occur. Turning and tedding must therefore be prompt and adequate.

- **Oxidation**

  Exposure of high moisture green fodder to sun for prolonged period without turning will lead to almost all loss of carotene. A green fodder has carotene content between 150-200 ppm (% dry matter) that will reduce to 5-10 ppm due to bleaching. Quick drying on tripod system ensures the maximum conservation of carotene. Sunlight has added advantage as it results in vitamin D2 formation on ergosterol irradiation.

- **Leaf loss**

  Leaf losses occur during tedding and field handling. Loss of leaves is more in leguminous (Lucerne, berseem, stylosanthes, cowpea etc) than cereals due to prolonged drying without turning. The leaves of unturned material (crops) become brittle and shatter. Since the leaves are rich in cell contents (protein, vitamin, minerals etc) than stem losses are more acute. Hay of leguminous fodder crops should be made near to the places it has to be stored so that losses of leaves are minimal. Transport of leguminous hay in early morning will ensure relatively lower leaves shattering due to overnight dew. There are about 5-9 % loss in dry matter during hay making.

➤ **Leaching losses**

These losses occur if rain falls on the crop during the hay making process. Re-wetting of partially dried hay is much more deleterious than rain on newly-cut herbage, and can cause both leaching and increased mould damage. Due to leaching there is loss of protein, soluble carbohydrates (NFE) and some soluble nutrients (minerals). In leaching there is maximum loss of carotene (95%) followed by protein (28%), energy (25%) with 12-25 % of dry matter.

➤ **Further mechanical loss**

These losses occur during collection, transport and baling. Their severity depends partly on the skill of the person who performs these operations.

➤ **Spoilage in the stack or bale**

The loss in stack or bale may result in loss of total material that is stored at high moisture level in over large units and poorly designed stacks that allow rain penetration or do not allow some ventilation. The ideal moisture content for stacking or baling depends on the crop and the site. When the herbage feels crisp in the hand and does not show moisture when twisted, it is probably at 25- 30% moisture and ready to bale. Fine-leaved, thin-stemmed herbage dries most quickly; large coarse grasses with thick stems and nodes (e.g., Sudan grass) can still have a lot of juice in the stems after the leaves are quite dry. In extreme cases, the fermentation may raise temperatures to levels where spontaneous combustion occurs. Poorly cured or mouldy hay, apart from losses, will lead to poor intake or refusal by stock, and may contain mycotoxins. Mouldy hay gives rise to the human allergic disease known as "farmers' lung."

➤ **Wastage during utilization includes losses from opened stacks and poor feeding techniques.**

The losses that occur in hay making also vary with the drying method. Nutrients losses due to different drying methods have been shown in Table 1.

Table1: Nutrients losses in hay process

Conservation Process	Dry matter		Starch equivalent		Digestible protein	
	No. of samples	Loss (%)	No. of samples	Loss (%)	No. of samples	Loss (%)
Dried on the ground	125	21.1	126	42.0	119	33.8
Dried on appliances	101	17.6	101	38.6	101	29.2
Barn drying	5	15.0	5	35.0	5	25.0

Hay preservatives

Preservatives are the substances (mainly chemicals) used to store hay at high moisture levels and to inhibit the mould growth. Propionic acid is one of the substances used as hay preservative. Propionic acid applied at a rate of 10 g/kg water (i.e. 3 kg/t of forage with a moisture content of 300g/kg), will prevent moulding and preserve the hay. Anhydrous ammonia or urea treatment, which was originally developed for improving the digestibility of straw, has given good results in hay by improving its stability under anaerobic conditions, as well as improving its nutritive value. The ammonia, in addition to excluding air, has a fungicidal effect, and also increases the crude protein content of the hay.

Timing of harvest and stage of growth

Two main factors determine the timing of Hay making one is the weather of the season and another stage of crop maturity. Pasture natural or sown will be managed by grazing or taking silage cuts as necessary, so that the herbage reaches a suitable stage for curing when good weather is expected. Where sown forage is to be cured, the crop will be chosen and its cultivation organized to match good Hay making weather. If a farm intends to make a lot of hay, consideration must be given to growing a range of species and cultivars to extend the season during which the various crops will be in peak condition for Hay making, and thus make best use of the available equipment and labour.

All steps in Hay making plays an important role to produce a high quality hay but the stage of maturity of the crop is the major factor determining the quality of the final product. Herbage quality drops off rapidly with increasing maturity, especially after flowering, but very young, leafy grass is less easy to handle than more mature herbage. The protein and energy contents of fodder are high at early stage of crop growth but yield is less. Thus a compromise between yield and quality has to be made, and, usually, cutting at early to mid-flowering will give a reasonable product. As the crop matures beyond full stage protein and other nutrients reduced. This results in lowered nutrients digestibility. The optimum time for harvesting a crop for hay making is 1/3 to ½ bloom stage, in cereals grain milk stage, while in legumes tender pod formation stage. Further Kharif crops may be harvested in early September (termination of rainy season), while Rabi crops may be harvested in February/March when temperature starts rising. Where tripod drying or barn-drying facilities are available, a younger product can be dried. In "good" Hay making climates, younger material can be cured more successfully than in areas with moist climates, or where there is danger of tropical storms. Cereals for hay are cut at a more mature stage than are pasture grasses, often when the grains are fully formed. Hay from natural pasture is more influenced by the natural growth cycle than are sown crops, and harvesting seasons are often traditional. The aim is always to have both high-quality herbage and a high probability of suitable Hay making weather coincide. Where there is an overall scarcity of forage, farmers may make hay even when the quality of the crop is poor. For example, in the monsoonal tropics and subtropics, a lot of hay is made from very over-mature herbage because the weather during the season when the grass is at its best quality is unsuitable for drying. The product is poor, but even hay which is no better than straw is highly valued.

Crops for hay making

Crops with thin, soft and pliable stem are more suitable and convenient for hay making unlike the hard thick stemmed crops. Oat is the best suited crop for hay making because of its thin and hollow stem and narrow leaf to stem ratio. Cultivated fodder crops like berseem, lucerne, cowpea, sunhemp along with range legumes like stylosanthes, siratro are other legumes suitable for hay making. Care should be taken to ensure minimum leaf losses (shattering) while preparing hay from leguminous crops. Hay from leguminous crops is more nutritive and usually used to supplement the straws and stovers and other fibrous feeds fed to animals. Annual and perennial crops (mainly grasses) also form god hay provided they are harvested at optimum stage of their growth. In Indian conditions most of the hay is prepared from the crops like sorghum, maize and millets normally harvested after grain harvesting.

Methods of hay making

There are three common methods of hay making depending up on the amount, condition and facilities available. These are namely field curing, barn drying and artificial drying.

Field curing

It is the most simple and common method followed by farmers to prepare hay. Here the crop after harvesting is left for sun drying in the field. The crop is turned in the field till the desired moisture level is attained. The crop after attaining desirable moisture level (15%) is collected in small bundles and their after stored. Here chances of leaf losses are more if not cured and handled properly. Care should also be taken to protect the crop from rodents and stray animals.

Barn Drying

In this method partially dried hay is cured inside the barns in mows. Heated or unheated air is blown on to the moves till 75-80% dry matter is achieved. During field curing the moisture is reduced up to 35-40 % and then taken to barns for further drying. With unheated air the hay is fully cured between 7-14 days, while less time is required with heated air. The hay prepared by this method is having more green leaves and also more number of leaves as the incidences of leaf losses are less in this method.

Artificial drying (Dehydration)

In this method the crop is initially chopped (fresh or wilted) and then dried in artificial driers. This method is followed for commercial hay production particularly when alfalfa leaf meal is used as vitamin supplement to poultry and swine. The hay prepared by this method is of high quality as there is minimum loss of plant nutrients.

Practice and Equipment

The following steps should be followed to ensure hay quality and to avoid spoilage.

- Heavy applications of nitrogenous fertilizers should be limited as it increases density of sward, which hinders drying and can cause lodging, with subsequent harvesting difficulties and losses.
- Damage by wheel traffic can lead to severe reductions in stand production and life, either through soil compaction or, especially in lucerne, by damaging crowns and re-growth when field operations are delayed.

Cocks, tripods and rack-drying

Where there is danger of re-wetting, which is very damaging, there is an intermediate stage where the partially-dried herbage is put into small heaps "cocks" - to dry further with less danger of rain affecting the whole mass.

Under sub-optimal conditions, the cocks may be built on wooden tripods to improve ventilation, or the hay may be put on drying racks. They are also useful in hot dry areas where the best quality hay is made by putting it up in cocks, which reduces both shattering and weathering and produces a green leafy hay high in carotene. Racks and tripods, however, require a cheap supply of poles, not always readily found. Fully mechanized systems attain a similar effect through careful swath handling and windrowing.

Swath handling

Tedding immediately after mowing boosts the drying rate. Conditioning or crushing should be done immediately after mowing. Timing of swath handling is critical and, at a time of peak demand for traction and labour, it is desirable to have a machine with a high work rate and a low power demand.

In cool, temperate conditions the swath has to be raised for aeration. Under hot dry conditions, it may be beneficial to spread it over the greatest possible area to benefit from the sun. In most temperate situations it is necessary, however, to tease out, mix and turn the swath as drying proceeds. Unless the weather is warm and sunny, the swath should be set up to catch any air movement across the field. In very hot, arid conditions, too-rapid drying of the leaves leads to shattering and bleaching, but windrowing will palliate this. Drying can be speeded up by collecting the half-dried hay into narrow, compact rows overnight, and spreading it out over all the ground once the soil between the rows has dried in the morning. The swath must always be collected into windrows for carting or baling.

STORAGE

Once the hay is made it can be kept for long periods if properly made and correctly stored; in contrast, it can deteriorate rapidly and even be lost by careless storage. The aim in storing hay is to keep it dry and to protect it from wastage due to rats, pests, stray livestock, fire or wind.

Storage foundation should be dry round the year not at the season/period of storage. They should be accessible all year round to the type of transport used, and protected from stray livestock and any fire risk. Where possible, it is better to store close to the point of use rather than in the field. Barn storage is ideal, of course, but is not always feasible.

Stacking

Stacks of loose material are the traditional method of storing large quantities of hay and, apart from barn storage, were the principal system on large as well as small farms until the introduction of the pick-up baler revolutionized commercial Hay making.

Once the hay has been field-dried to a suitable stage, it is carted to the stack-yard and built into stacks. The hay must be carefully built on to carts in the field and taken off in the same order as it was loaded. Under no circumstances should it be tipped off the cart if it has to be hand-forked to the stack, as random hay is very difficult to pull from the heap. Stacks may be cylindrical or rectangular in shape, depending on the quantities involved and the condition of the hay. When building the stack, the centre must be tightly packed, otherwise the stack will sag as it settles and rain may penetrate. The sides should slope outwards slightly to the eaves and the head thereafter rise gently to a point or ridge. Ropes or nets are used to protect the stacks against wind. Thatching with coarse grass or straw is common in areas where rain is expected during storage. This should not be done immediately, because it is necessary to allow dissipation of heat and moisture following stacking.

Storage of hay and straw in the forks of trees is widespread, especially for straw, in order to keep it out of reach of wandering livestock. House-top (roof) storage is very common on smallholdings in semi-arid areas.

In parts of India and Nepal, *feral* wooden shelters are commonly constructed near the homestead for milch stock, usually buffaloes. *Terai* are flatroofed and serve as storage platforms for straw and hay, as well as protecting the stock from the sun.

Fire Risks

Hay is easily set on fire, as are dry crop residues, so care must be taken in siting stack-yards and storage so as to minimize the risks. Stored hay may be accidentally ignited by household fires, by smokers' carelessness, by sparks from bush fires, and so on. Care in siting, cleanliness - including tidying up of loose hay which could carry fire - and avoidance of smoking or other sources of fire near or in the stack-yard will help prevent accidents. Where dry fodder is stored on or close to houses, extra care is needed. Spontaneous combustion can occur when hay is stored at too high a moisture content.

Bibliography

Mcdonald,, P., Edwards, R. A., Greenhalgh, J. D. F. and Morgan, C. A. 2002. Animal Nutrition. 6th edition, Pearson Education, Singapore Pte. Ltd., Indian Branch, 482 F. I. E. Patparganj, Delhi, India

Nash, M. J. 1985. Crop conservation and Storage. Oxford, Pergmon Press

Sullivan, J. T. 1973. Drying and Storing herbage as hay. In Butler, G. W. and Bailey R. W. (eds). Chemistry and Biochemistry of Herbage, vol. 3, London Academic Press

Suttle, J. M. 2000. Hay and straw conservation. For small scale farming and pastoral conditions. FAO, Publication: Plant production and protection Series No. 29.

SILAGE

Climatib factors play an important role in acute scarcity of green fodders at one end and surplus green fodder availability on other. The surplus green fodder is available during July–October and December–April followed by an acute scarcity period between mid October to mid December and mid April to mid July. Thus it becomes imperative to conserve the surplus green fodder at appropriate time with a view to tide over the lean periods. Unless such measures are taken it is not feasible to sustain and economize livestock production during this lean period. Silage making is one of the ways to provide quality feed to animals. Every effort should be made to preserve the forages as silage at the appropriate stage of the growth.

Silage: Silage is a high quality succulent feed resulted from the fermentation of green fodder stored and preserved anaerobically. Silage making preserve the fodder crop more or less in its original condition with minimum deterioration and loss of nutrients.

Ensiling: The process of conserving forages as silage is called ensiling.

Silo: A silo is an air tight to semi–air tight structure to store and preserve high moisture feeds as silage.

Haylage: This is a form of preserved forage with characteristics between hay and silage (40 and 50 % moisture content). The haylage is a material obtained after ensiling of grass and cultivated forages through controlled microbial fermentation under strict anaerobic conditions. In haylage making there is limited bacterial growth and fermentation.

Silage making: In the classical process of silage making, soluble carbohydrates of fodders get converted to lactic acid and other organic acids by the action of lactic acid bacteria. The resultant product is acidic, ideally have a pH value around 4.0. At this pH level growth of butyric acid producing bacteria is inhibited and the degradation of protein to NH_3 and amines is prevented.

Silage Characteristics (Quality)

There are several factors associated with the growth stage of fodder and fermentation process that governs the quality of silage.

1. **Growth stage of fodder:** Good quality silage can be obtained if the crop is harvested at its optimum growth stage. The optimum growth stage for silage making of some fodders is given below.

Crop	Stage
Natural grasses	Boot to early heading
Oat	Flowering to dough
Sorghum and Bajra	Flowering to dough
Maize	Milk to dent
Teonsinte	Boot
Napier bajra hybrid and Guinea grass	1 meter tall

Harvest a crop for silage making when nutrients accumulation is highest with enough dry matter. A crop with 30–35% dry matter conserves into high quality silage.

2. **pH. :** The pH in high moisture silage should be less than 4.2, wilted 4.5 and in high dry matter silage no criteria has been recommended so far.
3. **Lactic acid:** The concentration of lactic acid in high moisture silage should be between 5 – 9%. In high dry matter silages small amount of this acid will result in good silage.
4. No fungal colony should be visible.
5. Aroma should be good, free from NH_3, butyric acid smell.
6. No caramelized or tobacco type smell in high dry matter silages.
7. Greenish in colour.

Advantages of Silage

1. More total digestible nutrients per unit land.
2. No losses of shattering, leaching/ bleaching.
3. Silage can be made in almost any kind of weather.
4. Less fire hazard with silage compared to hay and other dry crops.
5. No waste in feeding, as cobs, shucks, and coarser stems are eaten.
6. More palatable and digestible.
7. Maintains feed succulence with minimal losses of nutrients.
8. During silage making weeds are also utilized along with main fodder crop.
9. The organic acids produced during silage making process are identical to those produced in digestive tract of ruminant animals and thus are used in the same fashion.

10. Silage can be made from thick stem crops those are not suitable for hay making.
11. Nutritive value of ensiled forage could be increased by adding suitable additive.

Silage Making Methods

There are several methods of silage making but all these methods follow the same principle. The following steps are involved in silage making:

1. The fodder crops to be used for silage making should be harvested or wilted in field to attain a moisture level of 65–75% (25–35% DM).
2. Harvested fodder can be ensiled as such or chopped. But the silage crops are usually chopped into fairly small pieces (2.5–5 cm). Chopping facilitates the good packaging of fodder and handling of silage thereafter.
3. Fill the silo/ trench with chopped fodder.
4. Ensure even distribution of chopped fodder in the pit/ trench. Press (trample) the chopped material either by feet or with a tractor or bullocks as per the size of pit to exclude air as much as possible. Pack the fodder up to 1 meter (3 – 4 feets) at the top of silo or above ground level of pit.
5. Every ½ meter thick longer of chaffed fodder be regularly pressed (tramp led).
6. Cover the fodder with 14 – 15 cm thick layer of karbi or wheat busha/ paddy straw from all sides. Put mud on it and finally plaster it to seal the material preventing the entry of air and water. See that the silo is completely air tight.
7. Alternatively, a plastic sheet may also be used to cover the packed fodder. Edge of the sheet may be sealed by mud plastering.
8. Keep occasional watch and if there is any crack or hole plug it immediately. Silage would be ready within 45 days.
9. Ground shelled corn, cane molasses, limestone, urea and or various commercial preparations are sometimes added to reduce fermentation losses. Salt and urea are added @ 0.5 and 1%, respectively to cereals and grasses to improve their palatability and nitrogen content. In grass silages molasses are added at 3–3.5% to improve the sugar content.
10. Open the silo pit from one end only and take out the required quantity, while the remaining silage is kept covered.

Silage Feeding:

Initially for few days animals may not relish the taste of silage. Help them to develop the taste by mixing 5 – 10 kg of silage in their green fodder ration for the first one week. Once the taste is developed 20 – 34 kg of silage along with other fodders may be fed per head/ day.

Crops for Silage Making

Practically any crop can be used for silage making. However, the quality of silage depends on the kind of crop and its stage of harvest.

Crops rich in soluble sugars/ carbohydrates (more than 10%) are most suitable for silage making. These are oat, maize, bajra, sorghum, hybrid napier and guinea grass. These crops are rich in sugar and carbohydrates and low in protein. Good quality silage can be prepared from a mixture of grasses/ cereals and legumes (berseem, Lucerne, stylosanthes, etc.) in a ratio of 3:1. Cultivated and natural grasses can be conserved as good silage with the addition of molasses at 3–3.5%. Silage preparation from leguminous crops is some what difficult as they are low in soluble carbohydrate, high protein, high moisture content and have high buffering capacity that resist pH change. Good quality silage can be prepared from the leguminous crops by specific treatment. The treatment used to improve the silage quality of leguminous crops is mainly based on three principles.

a. Increasing of soluble carbohydrate contents
b. Inhibition of undesirable fermentation by the addition of specific additives
c. Reducing moisture content either by wilting or by addition of dry roughages having higher water absorption capacity.

At the harvesting stage, crop should have about 35% dry matter. In most of the forages this dry matter level is achieved when they are harvested between flowerings to milk stage. Good quality silage can be prepared from fresh berseem and chapped paddy straw in different proportions. The fodder combinations like turnip sorghum (stover), berseem wilted to 35 % dry matter, various species of stylosanthes and subabbol have produced good quality silage.

Products of Silage Fermentation

Organic acids – primarily lactic, acetic, and butyric acid.
Ethanol in varying amounts.
Fermentation gases – primarily CO_2, CH_4, CO, NO and NO_2
Water
Heat

Nutrients Losses in Silage Making: (During Ensiling)

The main losses that occur during ensiling processes are as follows:

1. Field losses

Fodder harvested and used for silage making on same day results in negligible losses of nutrients. Up to 24 hour of wilting, DM loss are not more than 1–2%. Wilting more than 48 hours causes considerable losses of nutrients depending upon weather conditions. The main nutrients that are affected due to wilting are water soluble carbohydrate and proteins that are converted to amino acids. Dry matter losses to the tune of 6 and 10% have been reported after 5 days and 10 days of wilting, respectively in the field situations.

2. Oxidative losses

These losses are caused by the combined action of plant and microbial enzymes where substances like sugars in the presence of oxygen (O_2) lead to the formation of CO_2 and H_2O. In a rapidly filled and sealed silo oxygen trapped within the plant tissue is of less significance as it causes not more than 1% of dry matter loss. Long exposure of ensiled material to O_2 from the sides and upper surface causes formation of inedible composted material.

3. Fermentation losses

During fermentation, biochemical changes occur primarily in soluble carbohydrate and protein that are converted to different organic acids and amino acids due to lactic acid bacteria. However, the overall dry matter and energy losses are low. In lactic acid bacteria fermentation losses of dry matter and energy losses are less than clostridial and enterobacterial fermentation as the action of these later microorganisms leads to more production of gases like CO_2, hydrogen and NH_3 and thus losses of nutrients are more.

4. Effluent losses

Since the silage is prepared with a high moisture content fodder. If the moisture content of a crop at ensiling time is high, the production of liquid and effluent is more. Secondly, if the silo is not fully covered the rain water will enter resulting in higher effluent production. This effluent carries some of the water soluble nutrients (mainly sugars soluble nitrogen compounds and minerals) with it. The extent of effluent losses may vary from 0–10% depending upon the moisture level of ensiled fodder and weather condition. The nutrient losses in silage making have been given in Table 1. The systematic record of various losses during silage preparation are given in Table 2.

Table1: Nutrients losses in silage process

Conservation Process	Dry matter		Starch equivalent		Digestible protein	
	No. of samples	Loss (%)	No. of samples	Loss (%)	No. of samples	Loss (%)
Ordinary	486	16.7	64	32.0	71	37.9
Stimulated	246	16.7	31	21.8	31	17.8
Acidified	307	12.9	45	24.5	71	13.8
Inhibited	66	16.9	4	29.6	4	49.0

Table 2: Losses in silage during different biological processes

Biological process	Judgment	Approximate loss (%)
Respiration	Unavoidable	1-2
Fermentation	Unavoidable	1-4
Effluent	Mutual	5-7
Prewilting	Unavoidable	2-5
Secondary fermentation	Avoidable	0-5
Aerobic transformation	Avoidable	0-10
Aerobic transformation after unloading	Avoidable	0-15
Total conservation losses	Avoidable	7-35

Source: Zimmer, 1991.

Fermentation Reactions in Ensiling

On fresh herbage (fodder) aerobic bacteria and fungi are main micro-organisms. As the anaerobic conditions develop in the silo these aerobic micro organisms are replaced by anaerobic microbes particularly the lactic acid bacteria, clostridia and enterobacteria. The fermentation end products produced due to action of these microorganisms during ensiling are as follows:

1. Lactic acid bacteria

Homofermentative

1 Glucose + 2ADP => 2 lactic acid + 2 ATP

1 Fructose + 2 ADP => 2 lactic acid + 2 ATP

1 Pentose + 2 ADP => 1 lactic acid + 1 acetic acid + 2 ATP

Heterofermentative

1 Glucose + ADP => 1 lactic acid + 1 ethanol + 1 CO_2 + ATP

3 Fructose + 2 ADP => 1 lactic acid + 2 mannitol + 1 acetic acid + 1 CO_2 + 1H_2O + 2 ATP

1 Pentose + 2 ADP => 1 lactic acid + 1 acetic acid + 2 ATP

2. Clostridia

Saccharolytic

2 lactic acid => Butyric acid + 2 CO_2 + 2H_2

Proteolytic

Glutamic acid => Acetic acid + pyruvic acid + NH_3

Lysine => Acetic acid + butyric acid + 2 NH_3

Decarboxylation

Arginine => Putrescine + CO_2

Glutamic acid γ – amino butyric acid + CO_2

1 Alanine + 2 glycine => 3 acetic acid + 3 NH_3 + 1 CO_2

3 Alanine => 2 propionic acid + 1 acetic acid + 3 NH_3 + 1 CO_2

Histidine => Histamine + CO_2

1 Leucine => 1 isovaleric acid + NH_3 + 1 CO_2

Lysine => Cadavarine + CO_2

3. Enterobacteria

Glucose => Acetic acid + ethanol + 2 CO_2 + 2H_2

Since the lactic acid bacteria are facultative anaerobes, they grow rapidly as the fodder is chopped and packed for silage. These bacteria ferment the water soluble carbohydrates to organic acids (mainly lactic acid) that reduce the pH of ensiled material. The homofermentative lactic acid bacteria are more efficient in lactic acid production from hexose sugars than heterofermentative microbes. The sacchrolytic clostridia ferment already formed lactic acid and residual water soluble carbohydrates to butyric acid that raises pH of ensiled material and make the way for proliferation of putrefactive organisms. On the other hand proteolytic clostridia ferment the products (amino acids) of protein hydrolysis to amines, ammonia, CO_2, butyric and fatty acids. Clostridia for their active growth require wet conditions. The most important way to reduce the clostridia is to reduce the moisture to less than 70 % and other is to increases the acidity of the silage. Clostridia that develop under water logged conditions cause considerable destruction of proteins.

As the enterobacteria are facultative anaerobes alike the lactic acid producing bacteria they compete with lactic acid bacteria for soluble carbohydrates and convert these to acetic acid, methanol, hydrogen and NH_3.

Silage Additives

Silage quality is dependent on several factors like level of moisture content, concentration of soluble carbohydrates, proteins, kind of fodder and its harvest stage. Besides these pre requisites (requirement of ensiling process), silage making seems to be the major factor affecting silage quality. Except in perfect situations, either of the conditions/ requirements is lacking. In order to fill up this gap, the need of silage additives becomes inevitable. Silage additives are the substances that either enhance or inhibit the fermentation process of ensiling and also sometimes add in the nutritive value of the final product of ensiled fodder. Based on their action silage additives are of two types:

Fermentative Stimulants – such as sugar rich material (molasses), inoculants (freeze dried culture of lactic acid bacteria) and certain enzymes (cellulases and hemicellulases) that boost the proliferation of lactic acid bacteria.

Fermentative inhibitors – are usually the chemical substances that partially or completely inhibit the growth of undesirable micro-organisms. These are mainly acids like HCl, H_2SO_4 (Sulphuric acid). But nowadays formic acid is preferred over mineral acids as it is less corrosive. Use of formalin (40% formaldehyde solution) has also been demonstrated as a fermentative inhibitor.

Addition of sodium metabisulphite in powder form to unwilted forage crops has produced silage of superior quality i. e. in colour, palatability and carotene contents.

Types of silos

Generally following types of silos are used for silage making:

1. Pits 2. Bunkers 3. Trenches 4. Towers 5. Trench–cum bunker 6. Buns

The shape and size of silo varies with the quantity of the forage to be ensiled and the economic feasibility. In village conditions, normally pits and bunkers are used as silo. However, at commercial dairy farms and government livestock farms where large quantities of silage are to be made towers or buns are common silos. The silos may be kacha or pacca and may be of rectangular and cylindrical shape.

Characteristics of a Silo

The silo may be constructed below or above the ground but it should meet the following specification:

1. Silo should have air tight walls, impermeable with no cracks to prevent the percolation of water from outside. The walls may be made with polythene lining.
2. The silo should be of sufficient depth, should not be shallow. However, the depth of silo is determined by the water table of locality. It should always be above the water level. It should be about 2–3 meters above ground as most of the filling in Indian condition is done manually.
3. It should be located on an elevated ground.
4. Calculate the size of silo based on the quantity of fodder to be ensiled.
5. Silo should not be constructed in low lying, water logged areas, near to water ponds where chances of water entry to silos are more.

Silage Quality

Based on the criteria of odour, colour, pH level and concentration of fermentation products during ensiling, Shepherd *et al.*, (1948) distinguished the silage with following three classes.

Criteria	Very Good Silage	Good Silage	Fair Silage
Taste	Acidic taste	Acidic taste	Butyric acid taste
Butyric acid	Absent	Traces	More
pH	3.5 – 4.2	4.2 – 4.5	4.8
Lactic acid	1 – 2%	-	-
NH_3 – N% of total-N	< 10%	10 – 15%	20%
Colour	Greenish-brown	Dark – brown	Black

A.I.V. Method of Silage Making

This method is after the name of Professor A.I. Vertanen who proposed the silage using acid additives in 1925 in Finland. The method makes the use of equal ratio of sulphuric acid and hydrochloric acid. The mixture of both these acids is added to green fodder to bring the pH below 4.0. Later addition of carbohydrate rich materials (germinated maize and molasses) was realized to maintain the silage quality for longer period.

Chemical Changes in Ensiled Fodder

During the ensiling processes a number of chemical changes occur in the green crops. These changes occur due to the action of micro-organisms (particularly the bacteria and enzymes both of plant origin and produced in fermentation). Once the chopped fodder is compacted and sealed in the silo, the plant continues to respire for a shorter period and rapidly utilizes the oxygen entrapped in ensiled material. During respiration some of the sugars are also used resulting in CO_2 production. *Escherichia coli* and *Aerobacter aerogenes* are the main bacteria involved at this stage and produce acetic acid from sugars. The production of organic acids prevents the growth of undesirable bacteria (cause putrification). As the temperature rises due to CO_2 production, the bacteria present on the surface of ensiled mass gets activated and utilizes carbohydrates to produce CO_2 and other organic acids like lactic acid, acetic and butyric with small amounts of alcohol.

During fermentation proteins are hydrolyzed to amino acids, however, in poorly preserved silage the amino acid are further broken down to produce amines (Histamine, tryptamine, phenylethylamine)

Betaine, adenine and pentamethylene diamine are the primarily products of putrefaction. During ensiling few minerals namely Ca, Na, Mg and K present in green material may form salts with lactic acid and volatile acids. Brown colour of silage is due to the action of organic acids on the chlorophyll that convert chlorophyll in to the brown Mg free pigment-phaeophytin. The aroma of silage is due to the combination of acids with alcohol formed during ensiling.

Problems on Silage Making

It has been observed that, even when all precautions are taken, the production of good quality silage cannot be assured under tropical conditions. It is well known that silage production at low temperature (internal temperature about 100^{0}F) is superior to that at high temperature. The predominance of acetic acid (in contrast with lactic acid) is one important characteristics of silage made in the tropical countries. In this connection it may be mentioned that at least some lactic acid producing bacteria (e.g. *Lactobacillus plantarum*) do not survive at 109^{0}F. This suggests that efforts should be made to develop special strains of lactic acid producing bacteria, which may with stand temperatures around 120^{0}F. Inoculation of chaffed herbages with such bacteria is expected to result in the rapid production of lactic acid, provided that the forage contains sufficient amount of soluble carbohydrates. It has been emphasized that if the rate of pH fall is slow, Clostridial action might cause damage to the silage in the very beginning of the ensiling process. Amines produced under such conditions might affect the health of animals. The keeping quality of silage under hot environment needs early investigation. It has been reported that silage stability is not related to its quality. Appreciable loss of nitrogenous materials and reduction in the digestibility of crude protein take place during the ensiling process. This important problem so far received scant attention.

Bibliography

Stark, B.A. and Wilkinson, J.M. (eds) 1987. Development in silage. Marlow, Chalcombe Publications.

McDonald, P; Henderson, A.R. and Heron, S.J.E. 1991. The biochemistry of silage. 2nd edn. Marlow, Chalcombe Publication.

Bernett, A. J. G. 1954. Silage Fermentation. Academic Press, New York.

Singh, A.P. (1980). Preservation and evaluation of ensiled forages, Ph.D, thesis, Banaras Hindu University, Varanasi, India

Singh, A.P. and Rekib, A. (1986b). Feeding value of grass silage prepared with different levels of urea. Indian J. Animal Science, 56: 572 – 575.

Verma, A; Yadav, B.P.S. and Sampath, K.T. (1987). An ensilage technology for the tribal farmers of North-Eastern hill region. Indian J. Animal Science, 57: 1306-1309.

Zimmer, E. 1971. Factors effecting silage fermentation in soils. International Silage Research conference, held at Washington, D.C. from Dec. 6 to 8.

CODEX ALIMENTARIUS AND LIVESTOCK

With the establishment of the World Trade Organization (WTO) on 1 January 1995, after the protracted Uruguay Round negotiations spanning 1986-94, the reach of multilateral rules and disciplines related to trade in goods were substantially extended to trade in services, knowledge (Intellectual Property Rights) and trade related investments. The Uruguay Round negotiations also resulted in a separate agreement on agricultural products. The rules prescribed by General Agreement on Tariff and Trade (GATT) and its associate agreements were to be subscribed to by all the member countries of WTO as a 'Single Undertaking'. The Government of India ratified the WTO agreements after taking into account the perceived balance of rights and obligations conferred by these agreements. The WTO provides a rule based, transparent and predictable multilateral trading system which protects the member countries from the pressures of stronger trading partners. WTO rules envisage non-discrimination in the form of National Treatment and Most Favoured Nation (MFN) treatment to Member countries. The WTO rules also ensures that India's products once imported into the territory of other WTO members would not be discriminated vis-à-vis the domestic products in those countries. The MFN treatment ensures that members do not discriminate among various WTO members not only in their tariff regimes but also in respect of various other rules, regulations, incentives etc. The three main functions of WTO are (i) make international trade as free as possible by ensuring that the ground rules are made predictable and transparent, (ii) act as a forum for trade negotiations among countries, and (iii) work as a neutral agency for settlement of trade disputes.

After the establishment of WTO, its highest decision making body-the Ministerial Conference meets every two years starting from 1996. The latest in the series was the Cancun meet (Sept.10-14, 2003) which was preceded by Doha (Nov.9-14, 2001). After the Cancun Ministerial Conference ended in deadlock, the WTO Members agreed to keep the Doha Round trade negotiations alive. The Doha round of Ministerial Declaration on agricultural sector mainly aimed at: (i) substantial improvement in market access, (ii) reduction of, with a view to phasing out , all forms of export subsidies, (iii) substantial reduction in trade-distorting domestic support and (iv) more effective special and differential treatment for developing countries by taking into account their developmental needs including food security and rural development.

The long term objective of the Agreement on Agriculture was to establish a fair and market oriented trading system through a programme of fundamental reform encompassing strengthened rules and specific commitments on domestic support, export subsidies and protection in order to correct and prevent restrictions and distortions in world agricultural markets. The agriculture sector in many developed countries continued to be characterized by high level of support and protection. Simultaneously, the market access to developing countries is being resisted by the developed countries and barriers on export of agricultural commodities including livestock products is being erected in the disguise of Sanitary and phytosanitary Measures (SPS) and Technical Barriers to Trade (TBT), both of which are in essence, non-tariff barriers (NBT).

The application of Sanitary and Phytosanitary Measures (SPS) which came into force with the establishment of WTO, confers the right of WTO Member countries to apply measures to protect human, animal and plant life and health. Member countries were asked to apply only those measures for protection that are based on scientific principles, only to the extent necessary and not in a manner which may constitute a disguised restriction on international trade.

A Technical barrier to trade (TBT) is a specific type of NTB that inhibits or prevents trade through the use of regulations or standards. These measures generally focus on ensuring that imported products satisfy domestic tastes, preferences and requirements with respect to quality or appropriate consideration of environmental concerns during manufacture, packaging and/or shipment of the products. A technical regulation can be defined as any document stating product characteristics or their related processes and production methods with which an imported product must conform. Safety regulations and environmental protection regulations are just a few examples. As compliance is mandatory, any import that does not meet the stated requirements of a technical regulation is not permitted entry into the country. The TBT is generally perceived to encompass issues related to product weight, size of packaging material, ingredient or identity standards, mandatory nutritional labeling and shelf life restrictions. However, in future the TBT may also cover the aspects related to ethical issues in animal production, which are broadly covered under organic animal production.

The increase in volume of international trade in products of animal origin and the prospects of India making its presence felt may be seriously jeopardize if our products do not conform to the SPS measures and TBT. One way to beat these NTB is to get concessions in the multilateral negotiations held under the banner of WTO, but in the long run we need to have standards in place which conforms to international practice. The Codex standards serve as a benchmark for comparison of national sanitary and phytosanitary measures. While it is not compulsory for member states to apply Codex standards, it is in their best interests to harmonize their national food standards with those elaborated by Codex.

The Codex Alimentarius Commission (CAC) is an intergovernmental body under FAO/ WHO that coordinates food standards at the international level. Its main objective are to protect the health of consumers and ensure fair practices in food trade. The CAC has proved to be most successful in achieving international harmonization in food quality and safety requirements. It has formulated international standards for a wide range of food products and specific requirements covering pesticide residues, food additives, veterinary drug residues, hygiene, food contaminants, labeling etc. Codex work has created worldwide awareness of food safety, quality and consumer protection issues and has achieved international consensus on how to deal with them scientifically through a risk based approach. As a result there has been a continuous appraisal of the principles of food safety and quality at the international level. The Codex Committees, which are important in food trade, are:

(i) Codex Committee on Food Hygiene: Public health concern related to higher pathogencity of resistant strain of salmonella and camplyobactor in food and related profile and risk assessment policy should be defined. Diseases like, Bovine Spongiform Encephalopathy (BSE), Chronic Wasting Syndrome (CWS) similar to Crautzfeldt Jacob Disease (CJD)-a rare brain disease is appearing in man and game animals need to be checked for their entry into food chain.

(ii) Codex Committee on Residue of Veterinary Drugs in Foods: Residues of veterinary drugs appear in food because of their use in animals. Use of antibiotics in animal feed to contain spread of diseases and growth promoter has been banned in many countries. The European Union has sought an outright ban of antibiotics use.

(iii) Codex Committee on Food Additives and Contaminants: This committee has requested the Netherlands to revise discussion paper on Dioxins and its permissible level in feed. Dioxin compounds include polychlorinated biphenols (PCB), phenoxy herbicide and the most toxic in this group is 2,3,7,8-tetrachloradibenzo-p-dioxin (TCDD). It is carcinogenic, neurotoxic and cause birth defects. WHO has recently recommended its level should be less than 1.4 picogram for adult man.

(iv) Codex Committee on Pesticide Residue

The Codex Alimentarius covering subjects relevant to livestock and livestock products are covered under following volume:

Vol. 1 : General requirements

Vol 1B: General requirements (food hygiene)

Vol. 2A : Pesticide residues in food (general texts)

Vol. 2B : Pesticide residues in food (maximum reside limits)

Vol. 3 : Residues of veterinary drugs in foods

Vol. 10 : Meat and meat products; soups and broths

Vol. 12 : Milk and milk products

Vol. 13: Methods of analysis and sampling

Importance of Sanitary and Phytosanitary Measures (SPS) and Technical barriers to Trade (TBT) for Livestock sector in India

The Indian Livestock industry is poised to take off in a big way in coming decades. The livestock revolution unfolding on Indian horizon has set the alarm bells ringing among some of the developed countries. The country which once was importing huge quantity of milk and milk products has now become the leading producer of milk in the world. Very soon the country may have a substantial quantity of milk and milk products ready for export. But the countries which already have surplus milk and milk products may resist it by raising the non-tariff barriers by bringing into the issues related to quality aspects of the products. They have also in the past tried to beat up the non issues like berating the milk from India as being produced by strange black animals (buffalo). Therefore, we cannot let ourselves be caught unawareness on the emerging issues where we can hold a significant share in the world market in products of animal origin. Similarly, the export potential of buffalo meat in middle east countries is yet to be realized where again the quality aspects will come to haunt us if we continue to ignore the hygiene and quality control aspects.

Therefore, the implementation of Codex Alimentarius needs attention, not only for production of food for human beings but also for animals. In the endeavour to produce livestock products for International trade, the attention has to be paid at the beginning of food chain which concerns with way the animals are fed and raised. The adoption of Codex Alimentarius for 'Good Animal Feeding Practices' (Annexure-I) will determine the quality of livestock product, which will allow us to catch a significant share in the world market. Even otherwise, the domestic consumers stand to get an assured high quality of products of animal origin. Moreover, the health of our livestock will see a marked improvement with fewer veterinary problems, simultaneously breaking the chain of several microorganisms and contaminants in the food chain.

The movement for sustainable agriculture has added a new dimension through 'organic farming' which is based on specific and precise standards of production which aim at achieving optimal agro-ecosystems which are socially, ecologically and economically sustainable. "Organic" is a labeling term that denotes products that have been produced in accordance with organic production standards and certified by a duly constituted certification body or authority. Organic agriculture is based on minimizing the use of external inputs, avoiding the use of synthetic fertilizers and pesticides. To achieve the goals of organic farming, animals have to be an important component in organic farming systems from the outset, both conceptually and in practice, and high animal health and welfare status of the animals have to be maintained. Animals should be an integral component of the organic farm, part of a system in which all parts interact to their mutual benefit, and where a harmony is created between land, animals and the people. The farm strives to be a closed system, producing feed for its own animals and incorporating their manure into crop production. Organic farming is driven by its own positive goals and ideals, and is not only a challenge to what its supporters find objectionable (such use of chemical fertilizers and pesticides) in mainstream agriculture. It represents a philosophy for production, distribution and consumption of food in a responsible way that offers a sustainable framework for the environment and all living organisms, on both local and global level. Its core values are interrelated and form a broad ecological notion of sustainability (Vaarsat et al., 2004).

With the recent advances in the animal production system and the consumer preference for organically produced livestock products, for which they are ready to pay a premium, the Codex Alimentarius has also formulated a guidelines. The ethical issues, which the organic production system takes care of, may also play dominant role in trade of livestock products. We should also need to be vigilant and prepare against this TBT coming in way of our products getting a foothold in international market. The Codex Alimentarius on 'Practices for Organic Animal Production' (Annexure-II) should be followed by the specialized commercial livestock enterprises, with an eye to capture the market for Organically produced animal food within and outside the country should follow the principles as given in the Annexure. This can also find place in interior villages, which are still using little or no synthetic materials.

HACCP in feed industry

The Hazard Analysis and Critical Control Point (HACCP) which forms an important component in the implementation of 'Good Animal Feeding Practices' and has a wider ramifications for the feed industry, has been described in detail. Application of HACCP in the Feed industry will ensure quality livestock feed available to animals to give animal product of International standards. Simultaneously, it will also help explore the opportunities for the export of their product. The products used in country's market will affect the health and well being of the people positively.

HACCP is a system that identifies, evaluates and controls hazards, which make a food unsafe for human consumption. HACCP is a systematic approach for the control of food safety throughout the commodity system of food chain. Before HACCP, most of the food systems were based on end product testing which did not assure 100 per cent safety. It requires understanding of the relationship between cause and effect in order to be more proactive.

The Development of HACCP system was started in 1959 when Pittsburgh Company was asked to develop in cooperation with National Aeronautics Space Agency (NASA), US Army and US Air Force, a food that could be used in the space capsules. The demand was for such an advanced food, that would not be contaminated with bacterial/viral pathogen, toxins, chemicals or physical hazards, that could cause an illness or injury. Such a hazard might result in catastrophic space mission.

The existing technique of quality control at that time could not assure that there would not be a problem. While looking for a new system, Bauman and his colleagues developed a new preventive system that enabled the management control over raw materials, process, environment, personnel, storage and distribution as early in the system as they could. If a system could be established together with record keeping, one should be able to give a product with high degree of assurance that it would be safe. For practical purpose, if this system were implemented correctly, there would be no testing requireds of finished products other than for monitoring purposes (Pierson and Corlett, 1996).

The basic principle of HACCP signaled a shift in emphasis from end product testing to preventive control of critical aspects of food production. The HACCP concept has been refined in and International Life Science Institute (ILSI) Europe gave the definition:

HACCP is a concept which permits a systematic approach to the identification and assessment of hazards and risks associated with the manufacture, distributions and use of food product, as well as the definition of preventive measures for their control.

The application of HACCP in food production has been well known in developed countries. However, application of HACCP in feed production is not that known. Although the hazard for animals may not be as critical as for human, the growing concern on the animal product contamination in food chain-which may be caused by contamination of feed, make the concept of HACCP quite relevant for the livestock feed industry. Nutritional quality of feed should be guaranteed in order to achieve maximum and consistent animal production. This has relevance concerning not only the health of the population consuming the products of animal origin but also the International trade will be increasingly governed by safety of food products as defined by the Codex guidelines.

The HACCP system, which is science based and systematic, identifies specific hazards and measures for their control to ensure the safety of food. HACCP is a tool to assess hazards and establish control systems that focus on prevention rather than relying mainly on end product testing. Any HACCP system is capable of accommodating change, such as advances in equipment design, processing procedures or technological developments. The HACCP system consist of the following seven element:

1. Conduct hazard analysis
2. Determine the Critical Control Points (CCPs)
3. Establish critical limits
4. Establish system to monitor control of Critical Control Points
5. Executions of corrective actions when critical limits are not met
6. Establish procedures for verification that HACCP system is working
7. Record keeping

Conducting hazard analyses:

It involves a systematic study of the ingredients, the final ration or products, condition of processing, handling, packaging, storage, distribution and consumer use. The study will cover identification of the hazard, assessment of the severity and the risk. From this information, it will be to determine the critical control points in the system that has to be monitored. In conducting the hazard analysis, wherever possible the following should be included:

- the likely occurrence of hazards and severity of their adverse health effects;
- the qualitative and/or quantitative evaluation of the presence of hazards;
- survival or multiplication of microorganisms of concern;
- production or persistence in foods of toxins, chemicals or physical agents; and,
- conditions leading to the above.

Biological, chemical, physical and nutritional hazard

In the HACCP for feed, the term of hazard refers to any aspect of the feed production chain that is unacceptable because it is a potential cause of feed safety problem. National Advisory Committee on Microbiological Criteria for Food defines: "Hazard as any biological, chemical and physical properties that may cause an unacceptable consumer risk"

Similar principle may be applied to animal feed that any hazard can be classified to biological, chemical and physical however additional hazard is also important to animal feed, which is nutritional hazard.

Biological hazard

Animal feed can also be a source of biological hazard causing food borne illness to human beings. The source of biological hazard in animal feed could be fungi, bacteria and insects. Salmonella, which cause illness in human beings through infected egg or broiler meat, affects an estimated 2-4 million people annually, however, only 40,000 cases are reported each year. Growth of fungi or mould is a major problem in tropics because of high temperature and humidity. Poor harvesting of cereals encourages growth of fungi/ mould on raw materials like corn, rice bran, oil cakes etc., which together with poor pelleting conditions may encourage fungi to grow in final product. Fungi not only reduces the nutritive value of feed but also cause a severe problem when mycotoxin is produced. High mortalities in turkeys have been reported in the beginning of aflatoxin cases. The toxin, however, is classified as chemical hazard. Aspergillus sp. causes aspergillosis to the bird and several cases has been reported to occur in broiler.

Another biological hazard is infestation of insect due to poor storage condition or sanitation. Infestation of Tribollium or Sitophyllus are common problem during storage of feed. Many farmers may reject feed that contain insect. Although insects are not dangerous to animals, they cause a reduction in the nutritive value of feed.

Table 1: Source of biological hazard and their control measures

Pathogen	Control measures
Bacillus cereus	Proper holding and cooling temperatures of foods; thermal processing of shelf-stable canned food.
Campylobacter jejuni	Proper pasteurization or cooking; avoiding cross contamination of utensils, equipment; freezing; atmospheric packaging.
Clostridium botulinum	Thermal processing of shelf-stable canned food; addition of nitrate and salt to cured processed meats; refrigeration

Pathogen	Control measures
	of perishable vacuum packaged meats; acidification below pH 4.6; reduction of moisture below water activity of 0.93.
Clostridium perfringens	Proper holding and cooling temperatures of foods; proper cooking times and temperatures; adequate cooking and avoidance of cross- contamination by unsanitary equipment.
Listeria monocytogenes	Proper heat treatments; rigid environmental sanitation programme; separation of raw and ready – to – eat production areas and product.
Salmonella spp.	Proper heat treatments; separation of raw and cooked product; proper employee hygiene; fermentation controls; decreased water activity; withdrawing feed from animals before slaughter; avoiding exterior of hide from contacting carcass during skinning; antimicrobial rinses; scalding procedures; disinfecting knives.
Staphylococcus aureus	Employee hygiene; proper fermentation and pH control; proper heat treatment and post-process product handling practices; reduced water activity.
Yersinia enterocolitica	Proper refrigeration; heat treatments; control of salt and acidity; prevention of cross- contamination.
Mycobacterium tuberculosis	Heat treatment and proper employee hygiene
Eschericia coli	Proper heat treatment and prevention of contamination

Chemical hazard

The chemical hazard in animal feed can occur due to i) Naturally occurring chemical and/ or ii) added chemicals. Naturally occurring chemicals in feed include toxicants are found in feed ingredients or secondary by-product, produced during metabolism of feed.

The growth of microorganisms produces several toxins, which include different mycotoxins. Biochemical reactions due to enzyme or oxidation may results in biogenic amine in fish product and peroxide in fat containing materials. Chemicals in feed may be added intentionally or contaminated during handling of raw materials. Several feed additives have been used to improve the quality of feed such as minerals, vitamins, antibiotic, feed preservative, growth promoters, processing modifier, enzyme etc. Most of these chemicals may improve animal production when used at recommended level. However, excessive or improper use may cause hazard not only to animal but also to human. Contaminated chemicals that can be detected in feed include: insecticide, pesticide, toxic element etc. They may be detected in fumigated feedstuff or the minerals that contain toxic chemicals (Vanadium or Pb in Rock phosphate etc).

Control of chemical hazard includes :

- control before receiving
- control before use
- control storage and handling condition
- making inventory all chemicals in facility

Physical hazard

Physical hazard is caused by extraneous matter not normally found in feed. They may cause illness or injury to animals or health risk. Materials which pose physical hazard include sand, wood, stone, plastic, metal, insect, rubber, paper. Methods involved in controlling physical hazard include raw material specification and inspection along with vendor certification and guarantee. Various preventive measures are available to find and remove physical hazard including metal removal, feed cleaner screen etc. Effective pest control and foreign object removal from plant environment are also essential. Proper receiving, distribution, storage procedures as well as packaging materials, handling practice may be evaluated for potential hazard. Employee education and practice should involve knowledge and prevention of physical hazard.

Determining Critical Control Points

There may be more than one CCP at which control should be applied to address the same hazard. The determination of a CCP in the HACCP system can be facilitated by the application of a decision tree, which indicates a logic reasoning approach.

Establish critical limits for each CCP

Critical limits must be specified and validated if possible for each Critical Control Point. In some cases more than one critical limit will be elaborated at a particular step. Criteria often used include measurements of temperature, time, moisture level, pH, pressure etc.

Establish a monitoring system for each CCP

The monitoring procedures should be able to detect loss of control at the CCP. Where possible, process adjustments should be made when monitoring results indicate a trend towards loss of control at a CCP. The adjustments should be taken before a deviation occurs. The monitoring procedures for CCPs need to be done rapidly because they relate to on-line processes and there will not be time for lengthy analytical testing. Physical and chemical measurements are often preferred to microbiological testing because they may be done rapidly and can often indicate the microbiological control of the product.

Establish corrective actions

Specific corrective actions must be developed for each CCP in the HACCP system in order to deal with deviations when they occur. The actions must ensure that the CCP has been brought under control. Actions taken must also include proper disposition of the affected product.

Establish verification procedures

Verification and auditing methods, procedures and tests, including random sampling and analysis, can be used to determine if the HACCP system is working correctly. The frequency of verification should be sufficient to confirm that the HACCP system is working effectively. Examples of verification activities include:

Review of the HACCP system and its records;

Review of deviations and product dispositions;

Confirmation that CCPs are kept under control.

Establish Documentation and Record Keeping

Efficient and accurate record keeping is essential to the application of a HACCP system. HACCP procedures should be documented. Documentation and record keeping should be appropriate to the nature and size of the operation.

Documentation examples are:

Hazard analysis;
CCP determination;
Critical limit determination.
Record examples are:
CCP monitoring activities;
Deviations and associated corrective actions;
Modifications to the HACCP system.

Harmonizing Indian laws along Codex Alimentarius

The Government of India is seized upon the need for harmonizing laws with the International practice so that our livestock products stand the stringent test in the International market. To regulate and control the standards in food a 'Food Safety & Standard Authority of India' (FSSAI) is proposed to be established involving the Ministries of Agriculture, Commerce, Consumer's Affairs, Food Processing, Health and Law& Justice. The scientific committee and scientific panels to be set up under FSSAI include panels on Food Additives, Plant Health, GMOs, Dietetic Products, Biological Hazards, Contaminants in Food Chain and a Panel on Labeling.

Adoption of Codex Alimentarius by the livestock sector in India by adopting 'Practices on Good Animal Feeding, and 'Practices for Organic Animal Production' will address to all the issues related to the food and feed safety of consumers. It will open up the opportunities for the livestock products from India to enter into the world market creating more employment opportunities and income to people. The setting up of 'Food Safety & Standard Authority of India' (FSSAI) by the Government will be a positive step in this direction to regulate this emerging sector.

Bibliography

Boring, S.A. (1995). Soybean processing quality control. In Practical handbook of soybean processing and utilization (Ericson, D.R., ed). AOCS press and USB St. Louis, MO.

Codex Alimentarius (2005). Practices on Good Animal Feeding. www.codexalimentarius.net. Accessed January 2005.

FAO/ WHO (2001). Codex Alimentarius –Food Hygiene: Basic Texts. Joint FAO/WHO Food standards Programme, Codex Alimentarius Commission, Rome.

FAO /WHO (2001). Codex Alimentarius – Organically Produced Foods. Joint FAO/WHO Food standards Programme, Codex Alimentarius Commission, Rome.

Farnum, N.R. (1975). Modern statistical quality control and improvement. Outbury Press, Belmount Calfornia.

ILSI (1997). A simple guide to understanding and applying the hazard analysis critical control point concept. ILSI Europe Sci. Com. And Microbiology.

Leslie, A. (1997). Quality control in feedmill. Procedures for an effective programm. Proc. 5th Regional ASA Feed Tech. & Nutr. Workshop 25-29 May 1997, Chiangmai, Thailand.

Pierson, M.D. and Corlett, D.A. (1996). HACCP Principles and Application. Chapman & Hall, N.V.

Tangendjaja, B. (2003) The application of Hazard Analysis and Critical Control Point Principle in Feed Manufacturing. American Soybean Association, New Delhi.

Vaarst, M., Roderick, S., Lund, V and Lockeretz, W. (2004). Animal Health and Welfare in Organic Agriculture. CABI Publishing, UK.

ANNEXURE-I

Codex Alimentarius on Good Animal Feeding Practice

The code on Good Animal Feeding tries to establish a feed safety system for food producing animals which covers the whole food chain, taking into account relevant aspects of animal health and the environment in order to minimize risks to consumers' health. This Code applies in addition to the principles of food hygiene already established by the Codex Alimentarius Commission taking into account the special aspects of animal feeding. The objective of this Code is to help ensure the safety of food for human consumption through adherence to good animal feeding practice at the farm level and good manufacturing practices (GMPs) during the procurement, handling, storage, processing and distribution of animal feed and feed ingredients for food producing animals. This Code of Practice applies to the production and use of all materials destined for animal feed and feed ingredients at all levels whether produced industrially or on farm. It also includes grazing or free-range feeding, forage crop production and aquaculture.

Those issues of animal welfare other than food safety related animal health are not covered. Environmental contaminants should be considered where the level of such substances in the feed and feed ingredients could present a risk to consumers' health from the consumption of foods of animal origin. While recognizing that, in its totality, a feed safety system would address animal health and environmental issues, in addition to consumers' health, this Code of Practice, in fulfilling the Codex mandate of consumer protection, only addresses food safety. Notwithstanding this, best efforts have been made to ensure that the recommendations and practices in this Code of Practice will not be detrimental to the more general animal health and environmental aspects of animal feeding.

General Principles and Requirements

Feed and feed ingredients should be obtained and maintained in a stable condition so as to protect feed and feed ingredients from contamination by pests, or by chemical, physical or microbiological contaminants or other objectionable substances during production, handling, storage and transport. Feed should be in good condition and meet generally accepted quality standards. Where appropriate, good agricultural practices, good manufacturing practices (GMPs) and, where applicable, Hazard Analysis and Critical Control Point (HACCP) principles' should be followed to control hazards that may occur in food. Potential sources of contamination from the environment should be considered.

Parties that produce feed or feed ingredients, those that rear animals for use as food and those that produce such animal products need to collaborate to identify potential hazards and their levels of risk to consumers' health. Such collaboration will enable the development and maintenance of appropriate risk management options and safe feeding practices.

Feed Ingredients

Feed ingredients should be obtained from safe sources and be subject to a risk analysis where the ingredients are derived from processes or technologies not hithelio evaluated from a food safety point of view. The procedure used should be consistent with the Working Principles for Risk Analysis for Application in the Framework of the Codex Alimentarius. Manufacturers of feed additives in particular should provide cleat' information to the user to permit correct and safe use. Monitoring of feed ingredients should include inspection and sampling and analysis for undesirable substances using risk-based protocols. Feed ingredients should meet acceptable and. if applicable, statutory standards for levels of pathogens, mycotoxins, pesticides and undesirable substances that may give rise to consumers' health hazards.

Labelling

Labelling should be clear and informative as to how the user should handle, store and use feed and feed ingredients. Labelling should be consistent with any statutory requirements and should describe the feed and provide instructions for use. Labelling or the accompanying documents should contain, where appropriate:

- information about the species or category of animals for which the feed is intended; the purpose for which the feed is intended;
- a list of feed ingredients, including appropriate reference to additives. in descending order 01 proportion:
- contact information of manufacturer or registrant;
- registration number if available; directions and precautions for use;
- lot identification:
- manufacturing date; and
- use before or expiry date

Traceability/Product Tracing and Record Keeping of Feed and Feed Ingredients

Traceability/product tracing of feed and feed ingredients, including additives, should be enabled by proper record keeping for timely and effective withdrawal or recall of products if known or probable adverse effects on consumers' health are identified. Records should be maintained and readily available regarding the production. distribution and use of feed and feed ingredients to facilitate the prompt trace-back of feed and feed ingredients to the immediate previous source and trace-forward to the next subsequent recipients if known or probable adverse effects on consumers' health are identified.

Special Conditions Applicable to Emergency Situations

Operators should, as soon as reasonable, inform the competent authorities in the country if they consider that a feed or feed ingredient does not satisfy the feed safety requirements established in this Code. The information should be as detailed as possible and should at least contain a description of the nature of the problem. a description of the feed or feed ingredients, the species for which it is intended. the lot identifier. the name of the manufacturer and the place of origin. The competent authorities and operators should immediately take effective measures to ensure that those feed or feed ingredients do not pose any danger to consumers' health.

As soon as it becomes likely that a particular feed or feed ingredient is to be traded internationally and may pose a danger to consumers' health. the competent authorities of the exp0l1ing countries should notify, at least, the competent authorities of the relevant importing countries. The notification should be as detailed as possible and should at least contain the par1iculars indicated in the previous paragraph.

Inspection and Control Procedures

Feed and feed ingredients manufacturers and other relevant parts of industry should practice self-regulation/auto-control to secure compliance with required standards for production, storage and transpol1. It will also be necessary for risk-based official regulatory programmes to be established to check that feed and feed ingredients are produced, distributed and used in such a way that foods of animal origin for human consumption are both safe and suitable. Inspection and control procedures should be used to verify that feed and feed ingredients meet requirements in order to protect consumers against food-borne hazards. Inspection systems should be designed and operated on the basis of objective risk assessment appropriate to the circumstances. Preferably the risk assessment methodology employed should be consistent with internationally accepted approaches. Risk assessment should be based on current available scientific evidence.

Inspection And Control Procedures

* Monitoring of feed and feed ingredients, whether by industry or official inspection bodies, should include inspection and sampling and analysis to detect unacceptable levels of undesirable substances.

Health Hazards Associated With Animal Feed

* All feed and feed ingredients should meet minimum safety standards. It is essential that levels of undesirable substances are sufficiently low in feed and feed ingredients that their concentration in food for human consumption is consistently below the level of concern. Codex Maximum Residue Limits and Extraneous Maximum Residue Levels set for feed should be applied. Maximum residue limits set for food, such as those established by the Codex Alimentarius Commission, may be useful in determining minimum safety standards for feed.

Feed Additives And Veterinary Drugs Used In Medicated Feed

* Feed additives and veterinary drugs used in medicated feed should be assessed for safety and used under stated conditions of use as pre-approved by the competent authorities.
* Veterinary drugs used in medicated feed should comply with the provisions of the Codex Recommended International Code of Practice for the Control of the Use of Veterinary Drugs 10 .
* Borderlines between feed additives and veterinary drugs used in medicated feed may be set to avoid misuse.
* Feed additives should be received, handled and stored to maintain their integrity and to minimise misuse or unsafe contamination. Feed containing them should be used in strict accordance with clearly defined instructions for use.
* Antibiotics should not be used in feed for growth promoting purposes in the absence of a public health safety assessment.11

Feed And Feed Ingredients

* Feed and feed ingredients should only be produced, marketed, stored and used if they are safe and suitable, and, when used as intended, should not represent in any way an unacceptable risk to consumers' health. In particular, feed and feed ingredients contaminated with unacceptable levels of undesirable substances should be clearly identified as unsuitable for animal feed and not be marketed or used.

* Feed and feed ingredients should not be presented or marketed in a manner liable to mislead the user.

Undesirable Substances

* The presence in feed and feed ingredients of undesirable substances such as industrial and environmental contaminants, pesticides, radionuclides, persistent organic pollutants, pathogenic agents and toxins such as mycotoxins should be identified, controlled and minimised. Animal products that could be a source of the Bovine Spongiform Encephalopathy (BSE) agent 12 should not be used for feeding directly to, or for feed manufacturing for, ruminants. Control measures applied to reduce unacceptable level of undesirable substances should be assessed in terms of their impact on food safety.
* The risks of each undesirable substance to consumers' health should be assessed and such assessment may lead to the setting of maximum limits for feed and feed ingredients or the prohibition of certain materials from animal feeding.

Production, Processing, Storage, Transport and Distribution of Feed and Feed Ingredients

The production, processing, storage, transport and distribution of safe and suitable feed and feed ingredients is the responsibility of all participants in the feed chain, including farmers, feed ingredient manufacturers, feed compounders, truckers, etc. Each participant in the feed chain is responsible for all activities that are under their direct control, including compliance with any applicable statutory requirements.

Feed and feed ingredients should not be produced, processed, stored, transported or distributed in facilities or using equipment where incompatible operations may affect their safety and lead to adverse effects on consumers' health. Due to the unique characteristics of aquaculture, the application of these general principles must consider the differences between aquaculture and terrestrial-based production.

Where appropriate, operators should follow GMPs and, where applicable, HACCP principles to control hazards that may affect food safety. The aim is to ensure feed safety and in particular to prevent contamination of animal feed and food of animal origin as far as this is reasonably achievable, recognising that total elimination of hazards is often not possible.

The effective implementation of GMPs and, where applicable, HACCP-based approaches should ensure, in particular, that the following areas are addressed.

Premises

* Buildings and equipment used to process feed and feed ingredients should be constructed in a manner that permits ease of operation, maintenance and cleaning and minimises feed contamination. Process flow within the manufacturing facility should also be designed to minimise feed contamination.
* Water used in feed manufacture should meet hygienic standards and be of suitable quality for animals. Tanks, pipes and other equipment used to store and convey water should be of appropriate materials which do not produce unsafe levels of contamination.
* Sewage, waste and rain water should be disposed of in a manner which avoids contamination of equipment, feed and feed ingredients.

Receiving, Storage And Transportation

* Chemical fertilizers, pesticides and other materials not intended for use in feed and feed ingredients should be stored separately from feed and feed ingredients to avoid the potential for manufacturing errors and contamination of feed and feed ingredients.
* Processed feed and feed ingredients should be stored separately from unprocessed feed ingredients and appropriate packaging materials should be used. Feed and feed ingredients should be received, stored and transported in such a way so as to minimize the potential for any cross-contamination to occur at a level likely to have a negative impact on food safety.
* The presence of undesirable substances in feed and feed ingredients should be monitored and controlled.
* Feed and feed ingredients should be delivered and used as soon as possible. All feed and feed ingredients should be stored and transported in a manner which minimizes deterioration and contamination and enables the correct feed to be sent to the right animal group.
* Care should be taken to minimize deterioration and spoilage at all stages of handling, storage and transport of feed and feed ingredients. Special precautions should be taken to limit fungal and bacterial growth in moist and semi-moist feed. Condensation should be minimized in feed and feed ingredient manufacturing and processing facilities. Dry feed and feed ingredients should be kept dry in order to limit fungal and bacterial growth.
* Waste feed and feed ingredients and other material containing unsafe levels of undesirable substances or any other hazards should not be used

as feed, but, should be disposed of in an appropriate manner including compliance with any applicable statutory requirements.

Personnel Training

* All personnel involved in the manufacture, storage and handling of feed and feed ingredients should be adequately trained and aware of their role and responsibility in protecting food safety.

Sanitation And Pest Control

* Feed and feed ingredients, processing plants, storage facilities and their immediate surroundings should be kept clean and effective pest control programmes should be implemented.
* Containers and equipment used for manufacturing, processing, transport, storage, conveying, handling and weighing should be kept clean. Cleaning programmes should be effective and minimise residues of detergents and disinfectants.
* Machinery coming into contact with dry feed or feed ingredients should be dried following any wet cleaning process.
* Special precautions should be taken when cleaning machinery used for moist and semi-moist feed and feed ingredients to avoid fungal and bacterial growth.

Equipment Performance And Maintenance

* All scales and metering devices used in the manufacture of feed and feed ingredients should be appropriate for the range of weights and volumes to be measured, and be tested regularly for accuracy.
* All mixers used in the manufacture of feed and feed ingredients should be appropriate for the range of weights or volumes being mixed and be capable of manufacturing suitable homogeneous mixtures and homogeneous dilutions, and be tested regularly to verify their performance.
* All other equipment used in the manufacture of feed and feed ingredients should be appropriate for the range of weights or volumes being processed, and be monitored regularly.

Manufacturing Controls

* Manufacturing procedures should be used to avoid cross-contamination (for example flushing, sequencing and physical clean-out) between

batches of feed and feed ingredients containing restricted or otherwise potentially harmful materials (such as certain animal by-product meals, veterinary drugs). These procedures should also be used to minimise cross-contamination between medicated and non-medicated feed and other incompatible feed. In cases where the food safety risk associated with cross-contamination is high and the use of proper flushing and cleaning methods is deemed insufficient, consideration should be given to the use of completely separate production lines, transfer, storage and delivery equipment.

* Pathogen control procedures, such as heat treatment or the addition of authorised chemicals, should be used where appropriate, and monitored at the applicable steps in the manufacturing process.

Recalls

* Records and other information should be maintained to include the identity and distribution of feed and feed ingredients so that any feed or feed ingredient considered to pose a threat to consumers' health can be rapidly removed from the market and that animals exposed to the relevant feed can be identified.

On–Farm Production and Use of Feed and Feed Ingredients

* This section provides guidance on the cultivation, manufacture, management and use of feed and feed ingredients on farms and in aquaculture.
* This section should be used in conjunction with the applicable requirements of Sections 4 and 5 of this Code.
* To help ensure the safety of food used for human consumption, good agricultural practices 13 should be applied during all stages of on-farm production of pastures, cereal grain and forage crops used as feed or feed ingredients for food producing animals. For aquaculture the same principles should apply, where applicable.

  Three types of contamination represent hazards at most stages of on-farm production of feed and feed ingredients, namely :

- Biological, such as bacteria, fungi and other microbial pathogens;
- Chemical, such as residues of medication, pesticides, fertilizer or other agricultural substances; and
- Physical, such as broken needles, machinery and other foreign material.

Agricultural Production of Feed

* Adherence to good agricultural practices is encouraged in the production of natural, improved and cultivated pastures and in the production of forage and cereal grain crops used as feed or feed ingredients for food producing animals. Following good agricultural practice standards will minimize the risk of biological, chemical and physical contaminants entering the food chain. If crop residuals and stubbles are grazed after harvest, or otherwise enter the food chain, they should also be considered as livestock feed. Most livestock will consume a portion of their bedding. Crops that produce bedding material or bedding materials such as straw or wood shavings should also be managed in the same manner as animal feed ingredients. Good pasture management practices, such as rotational grazing and dispersion of manure droppings, should be used to reduce cross-contamination between groups of animals.

Site Selection

* Land used for production of animal feed and feed ingredients should not be located in close proximity to industrial operations where industrial pollutants from air, ground water or runoff from adjacent land would be expected to result in the production of foods of animal origin that may present a food safety risk. Contaminants present in runoff from adjacent land and irrigation water should be below levels that present a food safety risk.

Fertilizers

* Where manure fertilization of crops or pastures is practised, an appropriate handling and storage system should be in place and maintained to minimize environmental contamination, which could negatively impact on the safety of foods of animal origin. There should be adequate time between applying the manure and grazing or forage harvesting (silage and hay making) to allow the manure to decompose and to minimize contamination.

* Manure, compost and other plant nutrients should be properly used and applied to minimize biological, chemical and physical contamination of foods of animal origin which could adversely affect food safety.

* Chemical fertilizers should be handled, stored and applied in a manner such that they do not have a negative impact on the safety of foods of animal origin.

Pesticides and other Agricultural Chemicals

* Pesticides and other agricultural chemicals should be obtained from safe sources. Where a regulatory system is in place, any chemical used must comply with the requirements of that system.
* Pesticides should be stored according to the manufacturer's instructions and used in accordance with Good Agricultural Practice in the Use of Pesticides (GAP)14 . It is important that farmers carefully follow the manufacturer's instructions for use for all agricultural chemicals.
* Pesticides and other agricultural chemicals should be disposed of responsibly in a manner that will not lead to contamination of any body of water, soil, feed or feed ingredients that may lead to the contamination of foods of animal origin which could adversely affect food safety.

Manufacturing of Feed on-Farm

Feed ingredients

* On-farm feed manufacturers should follow the applicable guidelines established in sub-section 4.1 of this Code when sourcing feed ingredients off the farm.
* Feed ingredients produced on the farm should meet the requirements established for feed ingredients sourced off the farm. For example, seed treated for planting should not be fed.

Mixing

* On-farm feed manufacturers should follow the applicable guidelines established earlier.
* In particular, feed should be mixed in a manner that will minimize the potential for cross-contamination between feed or feed ingredients that may have an effect on the safety or withholding period for the feed or feed ingredients.

Monitoring Records

* Appropriate records of feed manufacturing procedures followed by on-farm feed manufacturers should be maintained to assist in the investigations of possible feed-related contamination or disease events.
* Records should be kept of incoming feed ingredients, date of receipt and batches of feed produced in addition to other applicable records set out in sub-section 4.3 of the Code.

Good Animal Feeding Practice

* Good animal feeding practices include those practices that help to ensure the proper use of feed and feed ingredients on-farm while minimising biological, chemical and physical risks to consumers of foods of animal origin.
* Water
* Water for drinking or for aquaculture should be of appropriate quality for the animals being produced. Where there is reason to be concerned about contamination of animals from the water, measures should be taken to evaluate and minimise the hazards.

Pasture grazing

* The grazing of pastures and crop lands should be managed in a way that minimises the avoidable contamination of foods of animal origin by biological, chemical and physical food safety hazards.
* Where appropriate, an adequate period should be observed before allowing livestock to graze on pasture, crops and crop residuals and between grazing rotations to minimise biological cross-contamination from manure.
* Where agricultural chemicals are used, operators should ensure that the required withholding periods are observed.

Feeding

* It is important that the correct feed is fed to the right animal group and that the directions for use are followed. Contamination should be minimised during feeding. Information should be available of what is fed to animals and when, to ensure that food safety risks are managed.
* Animals receiving medicated feed should be identified and managed appropriately until the correct withholding period (if any) has been reached and records of these procedures must be maintained. Procedures to ensure that medicated feed are transported to the correct location and are fed to animals that require the medication should be followed. Feed transport vehicles and feeding equipment used to deliver and distribute medicated feed should be cleaned after use, if a different medicated feed or non-medicated feed or feed ingredient is to be transported next.

Stable Feeding and Lot/Intensive Feeding Units

* The animal production unit should be located in an area that does not result in the production of food of animal origin that poses a risk to food safety. Care should be taken to avoid animal access to contaminated land, and to facilities with potential sources of toxicity.

Hygiene

* The animal production unit should be designed so that it can be adequately cleaned. The animal production unit and feeding equipment should be thoroughly cleaned regularly to prevent potential hazards to food safety. Chemicals used should be appropriate for cleaning and sanitising feed manufacturing equipment and should be used according to instructions. These products should be properly labelled and stored away from feed manufacturing, feed storage and feeding areas.
* A pest control system should be put in place to control the access of pests to the animal production unit to minimise potential hazards to food safety.
* Operators and employees working in the animal production unit should observe appropriate hygiene requirements to minimise potential hazards to food safety from feed.
* To ensure food safety, necessary precautions should be taken regarding culturing methods, culturing sites, technologies, materials and feed used to minimize contamination in order to reduce food hazards.

Methods of Sampling and Analysis

Sampling

* Sampling protocols should meet scientifically recognized principles and procedures.

Analysis

* Laboratory methods developed and validated using scientifically recognized principles and procedures should be used.16 When selecting methods, consideration should also be given to practicability, with preference given to those methods which are reliable and applicable for routine use. Laboratories conducting routine analyses of feed and feed ingredients should ensure their analytical competency with each method used and maintain appropriate documentation.

ANNEXURE-II

Codex Alimentarius on 'Principles and practices for Organic Animal Production'

General Principles
Livestock Sources/Origin
Conversion
Nutrition
Health Care
Livestock Husbandry, Transport and Slaughter
Housing and Free-Range Conditions
Manure Management
Record Keeping and Identification
Species Specific Requirements

General Principles

1. Where livestock for organic production are maintained, they should be an integral part of the organic farm unit and should be raised and held according to these guidelines.
2. Livestock can make an important contribution to an organic farming system by:

a) improving and maintaining the fertility of the soil;

b) managing the flora through grazing;

c) enhancing biodiversity and facilitating complementary interactions on the farm; and increasing the diversity of the farming system.

3. Livestock production is a land related activity. Herbivores must have access to pasture and all other animals must have access to open-air runs; the competent authority may allow exceptions when the animals' physiological state, inclement weather conditions, and state of the land so permit, or the structure of certain ′traditional′ farming systems restrict access to pasture, providing the welfare of the animals can be guaranteed.
4. Stocking rates for livestock should be appropriate for the region in question taking into consideration feed production capacity, stock health, nutrient balance, and environmental impact.

5. Organic livestock management should aim to utilize natural breeding methods, minimize stress, prevent disease, progressively eliminate the use of chemical allopathic veterinary drugs (including antibiotics), reduce the feeding of animals with products of animal origin (e.g. meat meal), and maintain animal health and welfare.

Livestock Sources/Origin

6. The choice of breeds, strains and breeding methods shall be consistent with the principles of organic farming, taking into account in particular:

a) their adaptation to the local conditions;

b) their vitality and resistance to disease;

c) the absence of specific diseases or health problems associated with some breeds and strains (porcine stress syndrome, spontaneous abortion etc).

7. Livestock used for products satisfying Section 1.1 (a) of these guidelines must come, from birth or hatching, from production units complying with these guidelines, or have been the offspring of parents raised under the conditions set down in these guidelines. They must be raised under this system throughout their life.

- Livestock may not be transferred between organic and non-organic units. The competent authority can establish detailed rules for the purchase of livestock from other units complying with these Guidelines.

- Livestock existing on the livestock production unit, but not complying with these Guidelines, may be converted.

8. When an operator can demonstrate to the satisfaction of the official or officially recognized inspection/certification body that livestock satisfying the requirements indicated in the previous paragraph are not available, the official or officially recognized inspection/ certification body may allow livestock not raised according these guidelines under circumstances such as:

a) for considerable expansion of the farm, when a breed is changed or when new livestock specialization is developed;

b) for the renewal of a herd, e.g., high mortality of animals caused by catastrophic circumstances;

c) males for breeding.

The competent authority may set the specific conditions under which livestock from non-organic sources may be allowed or not allowed, taking into account that animals be brought in as young as possible as soon as they are weaned.

9. These livestock qualified by the derogations indicated in the previous paragraph must comply with the conditions set out in paragraph 12. These conversion periods must be observed if the products are to be sold as organic according to Section 3 of these guidelines.

Conversion

10. The conversion of the land intended for feeding crops or pasture must comply with the rules set out in Part A paragraphs 1, 2, and 3 of this Annex.

11. The competent authority may reduce the conversion periods or conditions established in paragraph 10 (for the land) and/or paragraph 12 (for livestock and livestock products) in the following cases:

a) pasture, open-air runs and exercise areas used by non-herbivore species;

b) for bovine, equine, ovine and caprine coming from extensive husbandry during an implementation period established by the competent authority or dairy herds converted for the first time;

c) if there is simultaneous conversion of livestock and land used only for feeding within the same unit, the conversion period for both livestock, pasture and/or land used for animal feed, may be reduced to two years only in the case where the existing livestock and their offspring are fed mainly with products from the unit.

12. Once the land has reached organic status and livestock from a non-organic source is introduced, and if the products are to be sold as organic, such livestock must be reared according to these Guidelines for at least the following compliance periods:

Bovine and equine

Meat products: 12 months and at least ¾ of their life span in the organic management system;

Calves for meat production: 6 months when brought in as soon as they are weaned and less than 6 months old;

Milk products: 90 days during the implementation period established by the competent authority, after that, six months.

Ovine and caprine Meat products: six months;

Milk products: 90 days during the implementation period established by the competent authority, after that, six months.

Porcine:

Meat products: Six months.

Poultry/laying hens

Meat products: whole of life span as determined by the competent authority;

Eggs: six weeks.

Nutrition

13. All livestock systems should provide the optimum level of 100% of the diet from feedstuffs (including 'in conversion' feedstuffs) produced to the requirements of these guidelines.

14. For an implementation period to be set by the competent authority, livestock products will maintain their organic status providing feed, consisting of at least 85% for ruminants and 80% for non-ruminants and calculated on a dry matter basis, is from organic sources produced in compliance with these Guidelines.

15. Not withstanding the above, where an operator can demonstrate to the satisfaction of the official or officially recognized inspection/certification body that feedstuffs satisfying the requirement outlined in paragraph 13 above are not available, as a result of, for example, unforeseen severe natural or manmade events or extreme climatic weather conditions, the inspection/certification body may allow a restricted percentage of feedstuffs not produced according to these guidelines to be fed for a limited time, providing it does not contain genetically engineered/modified organisms or products thereof. The competent authority shall set both the maximum percentage of non-organic feed allowed and any conditions relating to this derogation.

16. Specific livestock rations should take into account:

- the need of young mammals for natural, preferably maternal, milk;
- that a substantial proportion of dry matter in the daily rations of herbivores needs to consist of roughage, fresh or dried fodder, or silage;

- that polygastric animals should be not fed silage exclusively;
- the need for cereals in the fattening phase of poultry; - the need for roughage, fresh or dried fodder or silage in the daily ration for pigs and poultry.

17. All livestock must have ample access to fresh water to maintain the full health and vigour of the livestock.
18. If substances are used as feedstuffs, nutritional elements, feed additives or processing aids in the preparation of feedstuffs, the competent authority shall establish a positive list/s of substances in compliance with the following criteria:

General Criteria:

a) substances are permitted according to national legislation on animal feeding;

b) substances are necessary/essential to maintain animal health, animal welfare and vitality; and

c) such substances:

- contribute to an appropriate diet fulfilling the physiological and behavioural needs of the species concerned; and
- do not contain genetically engineered/modified organisms and products thereof; and
- are primarily of plant, mineral or animal origin.

Specific Criteria for Feedstuffs and Nutritional Elements:

a) feedstuffs of plant origin from non-organic sources can only be used, under the conditions of paragraphs 14 and 15, if they are produced or prepared without the use of chemical solvents or chemical treatment;

b) feedstuffs of mineral origin, trace elements, vitamins, or provitamins can only be used if they are of natural origin. In case of shortage of these substances, or in exceptional circumstances, chemically well-defined analogic substances may be used;

c) feedstuffs of animal origin, with the exception of milk and milk products, fish, other marine animals and products derived therefrom should generally not be used or, as provided by national legislation. In any case, the feeding of mammalian material to ruminants is not permitted with the exception of milk and milk products;

d) synthetic nitrogen or non-protein nitrogen compounds shall not be used.

Specific Criteria for Additives and Processing Aids:

a) binders, anti-caking agents, emulsifiers, stabilizers, thickeners, surfactants, coagulants: only natural sources are allowed;

b) antioxidants: only natural sources are allowed;

c) preservatives: only natural acids are allowed;

d) colouring agents (including pigments), flavours and appetite stimulants: only natural sources are allowed;

e) probiotics, enzymes and microorganisms are allowed;

f) antibiotics, coccidiostatics, medicinal substances, growth promoters or any other substance intended to stimulate growth or production shall not be used in animal feeding.

19. Silage additives and processing aids may not be derived from genetically engineered/modified organisms or products thereof, and may be comprised of only:

- sea salt;
- coarse rock salt;
- yeasts;
- enzymes;
- whey;
- sugar; or sugar products such as molasses;
- honey;
- lactic, acetic, formic and propionic bacteria, or their natural acid product when the weather conditions do not allow for adequate fermentation, and with approval of the competent authority.

Health Care

20. Disease prevention in organic livestock production shall be based on the following principles:

a) the choice of appropriate breeds or strains of animals as detailed in paragraph 6 above;

b) the application of animal husbandry practices appropriate to the requirements of each species, encouraging strong resistance to disease and the prevention of infections;

c) the use of good quality organic feed, together with regular exercise and access to pasture and/or open-air runs, having the effect of encouraging the natural immunological defence of the animal;

d) ensuring an appropriate density of livestock, thus avoiding overstocking and any resulting animal health problems.

21. If, despite the above preventative measures, an animal becomes sick or injured it must be treated immediately, if necessary in isolation and in suitable housing. Producers should not withhold medication where it will result in unnecessary suffering of the livestock, even if the use of such medication will cause the animal to lose its organic status.

22. The use of veterinary medicinal products in organic farming shall comply with the following principles:

a) where specific disease or health problems occur, or may occur, and no alternative permitted treatment or management practice exists, or, in cases required by law, vaccination of livestock, the use of parasiticides, or therapeutic use of veterinary drugs are permitted;

b) phytotherapeutic (excluding antibiotics), homeopathic or ayurvedic products and trace elements shall be used in preference to chemical allopathic veterinary drugs or antibiotics, provided that their therapeutic effect is effective for the species of animal and the condition for which the treatment is intended;

c) if the use of the above products is unlikely to be effective in combating illness or injury, chemical allopathic veterinary drugs or antibiotics may be used under the responsibility of a veterinarian; withholding periods should be the double of that required by legislation with, in any case, a minimum of 48 hours;

d) the use of chemical allopathic veterinary drugs or antibiotics for preventative treatments is prohibited.

23. Hormonal treatment may only be used for therapeutic reasons and under veterinary supervision.

24. Growth stimulants or substances used for the purpose of stimulating growth or production is not permitted.

Livestock Husbandry, Transport and Slaughter

25. Maintenance of livestock should be guided by an attitude of care, responsibility and respect for living creatures.

26. Breeding methods should be in compliance with the principles of organic farming taking into account:
   a) the breeds and strains suitable for raising under local conditions and under an organic system;
   b) the preference for reproduction through natural methods, although artificial insemination may be used;
   c) that embryo transfer techniques and the use of hormonal reproductive treatment shall not be used;
   d) that breeding techniques employing genetic engineering must not be used.
27. Operations such as attaching elastic bands to the tails of sheep, tail-docking, cutting of teeth, trimming of beaks and dehorning are generally not allowed in the organic management system. Some of these operations may, however, be authorized in exceptional circumstances by the competent authority or its delegate, for reasons of safety (e.g. dehorning in young animals) or if they are intended to improve the health and welfare of the livestock. Such operations must be carried out at the most appropriate age and any suffering to the animals must be reduced to a minimum. Anaesthetic should be used where appropriate. Physical castration is allowed in order to maintain the quality of products and traditional production practices (meat-type pigs, bullocks, capons, etc) but only under these conditions.
28. The living conditions and the management of the environment should take into account the specific behavioural needs of the livestock and provide for:
   - sufficient free movement and opportunity to express normal patterns of behaviour;
   - company of other animals, particularly of like kind;
   - the prevention of abnormal behaviour, injury and disease;
   - arrangements to cover emergencies such as the outbreaks of fire, the breakdown of essential mechanical services and the disruption of supplies.
29. The transport of living stock should be managed in a calm and gentle way and in a manner which avoids stress, injury and suffering: the competent authority should establish specific conditions in order to meet these objectives and may establish maximum transport periods. In transporting livestock, the use of electric stimulation or allopathic tranquilizers is not permitted.

30. The slaughter of livestock should be undertaken in a manner which minimizes stress and suffering, and in accordance with national rules.

Housing and Free-Range Conditions

31. Housing for livestock will not be mandatory in areas with appropriate climatic conditions to enable animals to live outdoors.

32. Housing conditions should meet the biological and behavioural needs of the livestock by providing:
   - easy access to feeding and watering;
   - insulation, heating, cooling and ventilation of the building to ensure that air circulation, dust level, temperature, relative air humidity and gas concentration are kept within limits which are not harmful to the livestock;
   - plentiful natural ventilation and light to enter; 33. Livestock may be temporarily confined during periods of inclement weather, when their health, safety or well being could be jeopardized, or to protect plant, soil and water quality.

34. The stocking density in buildings should:
   - provide for the comfort and well being of the livestock having regard for the species, the breed and the age of the livestock;
   - take into account the behavioural needs of the livestock with respect to the size of the group and the sex of the livestock;
   - provide them with sufficient space to stand naturally, lie down easily, turn round, groom themselves, and assume all natural postures and movements such as stretching and wing flapping.

35. Housing, pens, equipment and utensils should be properly cleaned and disinfected to prevent cross infection and the build-up of disease carrying organisms.

36. Free-range, open-air exercise areas, or open-air runs should, if necessary, provide sufficient protection against rain, wind, sun and extreme temperatures, depending on the local weather conditions and the breed concerned.

37. The outdoor stocking density of livestock kept on pasture, grassland, or other natural or semi-natural habitats, must be low enough to prevent degradation of the soil and over-grazing of vegetation.

Mammals

38. All mammals must have access to pasture or an open-air exercise area or run which may be partially covered, and they must be able to use those areas whenever the physiological condition of the animal, the weather conditions and the state of the ground permit.

39. The competent authority may grant exceptions for :
    - the access of bulls to pasture or, in case of cows to an open-air exercise area or run during the winter period;
    - the final fattening phase.

40. Livestock housing must have smooth, but not slippery floors. The floor must not be entirely of slatted or grid construction.

41. The housing must be provided with a comfortable, clean and dry laying/ rest area of sufficient size, consisting of a solid construction. Ample dry bedding strewn with litter material must be provided in the rest area.

42. The housing of calves in individual boxes and the tethering of livestock are not permitted without the approval of the competent authority.

43. Sows must be kept in groups, except in the last stages of pregnancy and during the suckling period. Piglets may not be kept on flat decks or in piglet cages. Exercise areas must permit dunging and rooting by the animals.

44. The keeping of rabbits in cages is not permitted.

Poultry

45. Poultry must be reared in open-range conditions and have free access to open-air run whenever the weather conditions permit. The keeping of poultry in cages is not permitted.

46. Water fowl must have access to a stream, pond or lake whenever the weather conditions permit.

47. Housing for all poultry should provide an area of solid construction covered with litter material such as straw, wood shavings, sand or turf. A sufficiently large part of the floor area must be available to laying hens for the collection of droppings, Perches/higher sleeping areas of a size and number commensurate with the species and size of the group and of the birds and exit/entry holes of an adequate size must be provided.

48. In the case of laying hens, when natural day length is prolonged by artificial light, the competent authority shall prescribe maximum hours respective to species, geographical considerations and general health of the animals.

49. For health reasons, between each batch of poultry reared buildings should be emptied, and runs left empty to allow the vegetation to grow back.

Manure Management

50. Manure management practices used to maintain any area in which livestock are housed, penned or pastured should be implemented in a manner that:

a) minimizes soil and water degradation;

b) does not significantly contribute to contamination of water by nitrates and pathogenic bacteria;

c) optimizes recycling of nutrients; and

d) does not include burning or any practice inconsistent with organic practices.

51. All manure storage and handling facilities, including composting facilities should be designed, constructed and operated to prevent contamination of ground and/or surface water.

52. Manure application rates should be at levels that do not contribute to ground and/or surface water contamination. The competent authority may establish maximum application rates for manure or stocking densities. The timing of application and application methods should not increase the potential for run-off into ponds, rivers and streams.

Record Keeping and Identification

53. The operator should maintain detailed and up-to-date records as given below:

(i) All livestock should be identified individually or, in the case of small mammals or poultry, by herd or flock or in the case of bees by hive. Written and/or documentary accounts should be kept to enable tracking of livestock and bee colonies within the system at all times and to provide adequate traceback for audit purpose. The operator should maintain detailed and up-to-date records of:

7a) breeding and/or origins of livestock;

b) registration of any purchases;

c) the health plan to be used in the prevention and management of disease, injury and reproductive problems;

d) all treatments and medicines administered for any purpose, including quarantine periods and identification of treated animals or hives;

e) feed provided and the source of the feedstuffs;

f) stock movements within the unit and hive movements within designated forage areas as identified on maps;

g) transportation, slaughter and/or sales.

h) extraction, processing and storing of all bsee products.

(ii) Storage, on the unit, of input substances, other than those whose use is with paragraph 4.1(b) of these guidelines is prohibited.

(iii) The official or officially recognized certification body or authority should ensure that a full physical inspection is undertaken, at least once a year, of the unit. Samples for testing of products not listed in these guidelines may be taken where their use is suspected. An inspection report should be drawn up after each visit. Additional occasional unannounced visits should also be undertaken according to need or at random.

(iv) The operator should give the certification body or authority, for inspection purposes, access to the storage and production premises and to the parcels of land, as well as to the accounts and relevant supporting documents. The operator should also provide the inspection body with any information deemed necessary for the purposes of the inspection.

(v) Products referred to in Section 1 of these guidelines which are not in their packaging for the end consumer should be transported in a manner which should prevent contamination or substitution of the content with substances or product not compatible with these guidelines and the following information, without prejudice to any other indications required by law:

- the name and address of the person responsible for the production or preparation of the product;
- the name of the product; and
- that the product is of organic status.

(vi) Where an operator runs several production units in the same area (parallel cropping), units in the area producing crop, crop products not covered by

Section 1 should also be subject to the inspection arrangements as regards the dash points of paragraph 4 and paragraphs 6 and 8 above. Plants of indistinguishable varieties as those produced at the unit referred to in paragraph 3 above should not be produced at these units.

If derogations are allowed by the competent authority, the authority must specify the types of production and circumstances for which derogations are granted and the supplementary inspection requirements, such as unannounced site visits; extra inspections during harvest; additional documentary requirements; assessment of an operation's ability to prevent co-mingling, etc., which are to be implemented.

Pending further review of these guidelines in accordance with Section 8, member countries can accept parallel cropping of the same variety, even if it is not distinguishable, subject to adequate inspection measures being applied.

(vii) In organic livestock production, all livestock on one and the same production unit must be reared in accordance with the rules laid down in these Guidelines. However, livestock not reared in accordance with these Guidelines may be present on the organic holding provided that they are separated clearly from livestock produced in accordance with these Guidelines. The competent authority can prescribe more restrictive measures, such as different species.

(viii) The competent authority may accept that animals reared in accordance with the provisions of these Guidelines may be grazed on common land, provided that:

a) this land has not been treated with products other than those allowed in accordance with Section 4.1 (a) and (b) of these Guidelines, for at least three years;

b) a clear segregation between the animals reared in accordance with the provisions of these Guidelines, and the other animals can be organized.

(ix) For livestock production, the competent authority should ensure, without prejudice to the other provisions in this Annex, that the inspections related to all stages of production and preparation up to the sale to the consumer ensure, as far as technically possible, the traceability of livestock and livestock products from the livestock production unit through processing and any other preparation until final packaging and/or labelling.

GLOSSARY

Abomasum : The fourth compartment of ruminant stomach, also known as true stomach.

Absorption : The passage of materials across a biological membrane.

Acetonemia : A condition charcterised by an abnormally elevated concentration of ketone bodies in the body tissues and fluids.

Achromotricia : Loss of pigment in hair.

Actin : Muscle fiber fraction that complexes with myosin to bring about muscle contraction.

Ad libitum : At pleasure or feed choice by the animal.

Adaptation : The adjustment of an organism to a new or changing environmental condition.

Additive Premix : (As defined by FDA) An article that be diluted for safe use in a feed additive concentrate, a feed additive supplement or a complete feed.

Additive : An ingredient or combination of ingredients in micro-quantities added to the basic feed mix or parts.

ADF : Fiber extracted with acidic detergent in a technique employed to appraise the quality of forage.

Adiabatic : No gain or loss of heat.

Adipose : Of a fatty nature.

Adrenal Gland : One of the endocrine glands of the body located near the kidney.

Aerial Part : The above-ground part of a plant.

Aerobic : A term usually applied to microorganisms that require oxygen to live and reproduce.

After Birth : The membranes expelled from the uterus following delivery of a foetus.

Air Assist : (Process) The use of air in a grinding system to increase grinding efficiency, control practice size, pressure buildup, heating, dusting and moisture condensation.

Alfalfa Leaf Meal : The powdered from the dried alfalfa (lucerne) leaves. The meal on an average contains about 20 per cent crude protein and less than 18 per cent crude fibre.

Alopecia : Loss of hair.

Ambient Temperature : Temperature of fluid (usually air) that surrounds object on all sides.

Amino Acid : Any one of class of organism compounds which contain both the amino (NH_3) group and the carboxyl (COOH) group.

Ammoniated : Combined or impregnated with ammonia or an ammonium compound.

Amylase : Any one of several enzymes which aid in the hydrolysis of starch to maltose, for example, pancreatic amylase (amylopsin) and salivary amylase (Ptyalin).

Anabolism : Synthesis of compounds.

Anaerobic : Oxygen not present.

Anemic : Lacking in size and or number of red blood cells.

Aneurysim : Dilatation of the wall of artery.

Anorexia : Lack or loss of appetite.

Antibiotics : A class of drugs, usually synthesized by a living microorganism and in proper concentration, inhibits the growth of other microorganisms.

Antibody : Substance produced in the body that acts against disease.

Antigen : A high-molecular-weight substance (usually protein) that, when foreign to the bloodstream of an animal, stimulates the formation of a specific antibody and reacts specifically in vivo or in vitro with its homologous antibody.

Antioxidant : Substances having the property of protecting other substances from oxidation such as vitamin E protects unsaturated fatty acids by oxidizing themselves.

Artificially Dried (Process) : Moisture removed by other than natural means.

Aspirated, Aspirating : (Process) having removed chaff, dust, or other light materials by use of air.

Atrophy : A wasting away of a part of the body.

Balanced : A term that is applied to diet, ration, or feed having all known required nutrients in proper amount and proportion based upon recommendations of recognized authorities in the field of animal nutrition.

Basal Metabolism : Quantity of chemical energy expanded for body maintenance measured under specific conditions, e.g. standard metabolism, tissue activity of physiochemical changes of a resting animal.

Biochemical Oxygen Demand (BOD) : The quantity of oxygen used by bacteria in the oxidation of organic matter in a specific time, at a specified temperature, and under specified conditions.

Biological value : The efficiency with which a protein furnishes the proper proportions and amounts of the essential amino acids. A protein which has a high biological value is said to be good quality.

Blending (Process) : To mingle or combine two or more ingredients or feeds.

Bloat : A disorder of ruminants caused by excess gas in the rumen.

Blocked, Blocking (Process) : Having agglomerated individual ingredients or mixture in to a large mass.

Blocks : (Process Form) Agglomerated feed compressed or chemically hardened into a solid mass cohesive enough to hold its form.

Blood meal : Ground and dried meal of blood having CP ranges from 79-85%, used as protein-rich ingredient in animal feed.

Boiler Efficiency : The heat of condensation divided by the heat of combustion of the fuel used to produces the steam.

Bomb Calorimeter : An instrument used to determine the gross energy content of a material.

Bone meal row : Dried and ground bone meal obtained from un-decomposed bones which have first been boiled in hot water.

Bone Meal Steamed : The ground product sterilized by cooking un-decomposed bones with steams under pressure.

Bran : The pericarp or seed coat

Brand Name : Any word, name, symbol or device, or any combination thereof, identifying the commercial feed of a distributor and distinguishing it form that of others.

Bricks : Agglomerated feed compressed into a solid mass cohesive enough to hold its form and weighing less than two pounds.

British Thermal Unit (BTU) : An engineering unit of heat energy equal to the quantity of heat necessary to raise the temperature of one pound of water one degree Fahrenheit.

Buffer : Any substance that can counteract changes in free acid or alkali concentration.

By-Product : (Part) Secondary products produced in addition to the principle product.

Cake (Physical Form) : The mass resulting from the pressing of seeds, meat, or fish in order to remove oils, fats, or other liquids.

Calorimeter : An instrument for measuring energy.

Canned (Process) : A term applied to a feed which has been processed, packaged, sealed and sterilized for preservation in cans or containers.

Canula : A tube for insertion into a body cavity.

Carcass : The body of dead animal less the viscera and usually the head, skin and lower legs.

Carnivore : Animals that eats primarily animal material.

Carotene : A yellow organic compound that is a precursor of vitamin A.

Carriers : A edible material to which ingredients are added to facilitate uniform incorporation of the latter into feeds.

Cartilage : The gristle or connective tissue attached to the end of bone.

Catabolism : The conversion of complex substances into more simple compounds by living cells.

Catalyst : A substance that speed up the rate of a chemical reaction but is not itself used up in the reaction.

Chaff : (Part) Glumes, husks, or other seeds covering together with other plants part separated from seed in threshing or processing.

Chemical Oxygen demand (COD) : The amount of oxygen required for the oxidation of organic and inorganic matter by a strong oxidizing agent under acidic conditions.

Chipped, Chipping : (Process) Cut or broken into fragments; also meaning prepared into small thin slices.

Chopped, Chopping : (Process) Reduced in particle size by cutting with knives or other edged instruments.

Chuni : The compound consists primarily of the broken pieces of endosperm including germ and a portion of husks obtained as by-product during the processing of pulse grains for human consumption.

Cleaned, Cleaning : (Process) Removal of material by such methods as scalping, aspirating, magnetic separation, or by any other method.

Clipped, Clipping : (Process) Removal of the ends of whole grain.

Coefficient of digestibility : The percentage value of a food nutrient that has been absorbed.

Coenzyme : A partner required by some enzymes to produce enzymatic activity.

Collagen : Fraction for the fibers of connective tissue, cartilage of bone, forms gelatin when boiled.

Combustion : The combination of substances with oxygen accompanied by the liberation of heat.

Comfort zone : The temperature range at which no demand is made on temperature regulating mechanism.

Commercial Feed : The term "commercial feed" means all materials which are distributed for use as feed or for mixing in feed.

Compensatory growth : Faster than normal weight gains following a period of under- feeding.

Complete Feed : A nutritionally adequate feed for animals by specific formula is compounded to be fed as the sole ration.

Composite Sample : A combination of individual samples taken at selected intervals to minimize the effect of the variability of an individual sample.

Concentrate : A feed used with another to improve the nutritive balance of the total and intended to be further diluted and mixed to produce a supplement or a complete feed.

Condensed, Condensing (Process) : Reduced to denser form by removal of moisture.

Conditioned, Conditioning (Process) : Having achieved predetermined moisture characteristics and/or temperature of ingredients or a mixture of ingredients prior to further processing,

Congenital : Existing at birth.

Cooked, Cooking (Process) : Heated in the presence of moisture to alter chemical and/or physical characteristics or to sterilize.

Cooled, Cooling : (Process) Temperature reduced by air movement, usually accompanied by a simultaneous drying action.

Coprophagy : The ingestion of own feaces.

Cost Center : A segment of a mill for which the costs of one or more related activities are accumulated.

Cracked, Cracking (Process) : Particle size reduced by a combined breaking and crushing action.

Creatinine : A nitrogenous compound arising from protein metabolism and secreted in the urine.

Crimped, Crimping : (Process) Rolled by use of corrugated rollers. It may entail tempering or conditioning and cooling.

Critical temperature (lower) : That temperature below which heat production increases in response to a fell in environmental temperature.

Crude fibre : The more fibrous, less digestible fraction of a feed and consists primarily of cellulose, hemicellulose and lignin.

Crude protein : Total ammonia nitrogen x 6.25, based on the fact that feed protein on the average contains 16.0 per cent nitrogen.

Crumbled, Crumbling : (Process) Pellets reduced to granular form.

Crumbles : (Physical form) Pelleted feed reduced to granular form.

Crushed, crushing : (Process) See rolled, rolling.

Cubes : (Physical form) See pellets.

Cut, Cutting (Process) : See chopped, chopping.

Decortication : Removal of the bark, hull, husk, or shell from a plant seed .

Degree Day : 65°F minus mean temperature of the day (F).

Dehulled, Dehulling (Process) : Having removed the outer covering from grains or other seeds.

Dehydrated, Dehydrating (Process) : Having been freed of moisture by thermal means.

Dermatitis : Inflammation of the skin.

Desiccate : To dry completely.

Dextrin : An intermediate polysaccharide product obtained during starch hydrolysis.

Diarrhoea : Abnormal frequency and liquidity of faecal discharge.

Diet : Feed ingredient or mixture of ingredients including water, which is consumed by animals.

Digestible energy : The part of the gross energy of a feed which does not appear in the feaces.

Digestion : The processes involved in the conversion of feed into absorbable forms.

Diluent (Physical form) : An edible substance used to mix with and reduce the concentration of nutrients and/or additives in make them more acceptable to animals, safer to use, and more capable of being mixed uniformly in a feed.

Diverticulitis : Inflammation of a pouch of the intestine.

Dried, Drying : (Process) Materials from which water or other liquid has been removed.

Drugs : (As defined by FDA) A substance (a) intended for use in the diagnosis, cure, mitigation, treatment or prevention of disease in human or other animals or (b) a substance other than food intended to affect the structure or any function of the body of human or other animals.

Dry Matter : The part of feed which is not water, sometimes referred to as total solids. This is the sum of the crude protein, crude fat, crude fibre, nitrogen free extract and ash.

Dust : (Part) Fine, dry, pulverized particles of matter usually resulting from the cleaning or grinding of grain,

Dystrophy : Degeneration.

Edema : Swelling of a part of or of the entire body due to an accumulation of an excess of water.

Elastin : The base of elastic tissue.

Emaciated : An excessively thin condition of body.

Emulsifier : (Part) A material capable of causing fat or oils to remain in liquid suspension.

Emulsify : To disperse small drops of one liquid into another liquid.

Endemic : Occurring in low incidence but more or less constantly in a given population.

Endocrine : Pertaining to internal secretions.

Endogenous : Generated from within. For example, endogenous feacal calcium denotes the calcium in the feaces that is from the body and not from the food.

Endometrium : The mucous membrane that lines the uterus.

Energy : The capacity to perform work.

Ensilage : The same as silage.

Enteritis : Inflammation of the intestine.

Enzyme : One of a class of organic compounds, formed by living cells, capable of producing or accelerating specific organic reactions. An organic catalyst.

Epidemic : When many people in a given region are attacked by some disease at the same time.

Essential amino acid : Any one of several amino acids that are needed by animal cannot be synthesized by them in the amount need and so must be present in the protein of the feed as such.

Essential fatty acid : A fatty acid that cannot be synthesized in the body or that cannot be made in sufficient quantities for the body's need.

Astivation (summer hibernation) : A dormant state in which some desert animals pass the hottest time of summer.

Estrogens : Estrous-producing hormones secreted by the ovaries.

Etiology : The causes of a disease or disorder.

Evaporated, Evaporating (Process) : Reduced to a denser form, concentrated as by evaporation or distillation,

Excreta : The products of excretion-primarily feaces and urine.

Exogenous : Originating from outside of the organism.

Expanded, Expanding (Process) : Subjected to moisture, pressure, and temperature to gelatinize the starch portion. When extruded, its volume is increased, due to abrupt reduction in pressure,

Expeller process : A process for the mechanical extraction of oil from seeds, involving the use of a screw press.

Extraction, Mechanical : (Process) Removal of fat or oil from materials by heat and mechanical pressure.

Extraction, Solvent : (Process) Removal of fat or oil from materials by organic solvents.

Extruded (Process) : A process by which feed has been pressed, pushed, or protruded through constrictions under pressure.

Fauna : The animal life present. Frequently refer to the overall protozoa population present.

Feed additive : A material added to animal rations, but which may not supply essential nutrients.

Fistula : An opening, often surgically prepared for nutritional studies. For example : rumen fistula or esophageal fistula.

Flakes (Physical form) : An ingredient rolled or cut into flat with prior steam conditioning.

Flora : The plant life present. In nutrition generally refers to the bacteria present in the digestive tract.

Flour : (Part) Soft, finely ground and bolted meal obtained from the milling of cereal grains, other seeds, or products. It consist ally of the starch and gluten of the endosperm.

Formula Feed : Two or more ingredients proportioned, mixed, and processed according to specifications.

Free Choice : A feeding system where animals are given unlimited access to the separate components or groups of components constituting the diet.

Gelatinizing (Process) : A process where starch is ruptured by a combination of moisture, pressure, and in some instances by mechanical shear.

Gluconeogenesis : Formation of glucose from protein or fat.

Glycogen : A polysaccharide with the formula $(C_6H_{10}O_5)_n$ which is formed in the liver and muscle and depolymerised to glucose to serve as a ready source of energy when needed by the animal. Also called as animal starch.

Glycogenesis : Conversion of glucose into glycogen.

Glycogenolysis : Conversion of glycogen into glucose.

Glycolysis : The sequence of reactions from glucose to pyruvic acid that is common to carbohydrate catabolism under both anaerobic and aerobic conditions.

Goiter : An enlargement of the thyroid gland located in the neck and caused by an iodine deficiency.

Goitrogens : Substances which interfere with thyroxin production in the thyroid gland.

Grab Sample : A single sample taken at neither a set time nor a specific flow.

Grain : (Part) Seed from cereal plants.

GRAS : Acronym for the phrase "Generally recognized as safe." A substance, which is generally recognized as safe by experts qualified to evaluate the safety of the substance for its intended

Grits (part) : Coarsely ground grain, from which the bran and germ have been removed, usually screened to uniform particle size.

Groats (Part) : Grain from which the hulls have been removed.

Ground, Grinding (Process) : Reduced in particle size by impact, shearing, or attrition.

Groundwater : That portion of the water beneath the surface of earth that can be collected with wells or similar facilities.

Heat increment : The heat which is unavoidably produced by an animal incidental with nutrient digestion and utilization was originally called work of digestion.

Heat-Processed, Heat-Processing - (Process) Subjected to a method of preparation involving the use of elevated temperatures with or without pressure.

Herbivore : An animal that eats primarily plant material.

Hibernation : A dormant state in which some animals pass the winter, in which body temperature drops to slightly above freezing and metabolic activity falls to a very low level.

Homeostasis : Maintenance of uniformity and stability.

Homeotherm : Warm-blooded animals in which the body temperature relatively constant.

Homogenized, Homogenizing : (Process) Particles broken down into evenly distributed globules small enough to remain emulsified for long periods of time.

Hood : A shaped inlet designed to capture contaminated air and conduct it into the exhaust duct system.

Hormone : A chemical substance secreted by an endocrine gland which has a specific effect on the activities of other organs.

Horsepower (hp) : A unit of power equal to 746 watts.

Hulls : (part) Outer covering of grain or other seed.

Humidity, Absolute : The weight of water vapour per unit volume, pounds per cubic foot or grams per cubic centimeter.

Humidity, Relative : The ratio of the actual partial pressure of the water vapour in a space to the saturation pressure of pure water at the same temperature.

Hydrolyzed, Hydrolyzing : (Process) Complex molecules having been split to simpler units by chemical reaction with water, usually by catalysis.

Hydrophobic : Water hating.

Hydrophylic : Water loving.

Hypocalcemia : Decreased level of calcium in the blood.

Hypokalemia : Decreased level of potassium in the blood.

Hyponatremia : Decreased level of sodium in the blood.

Immiscible : Not mixing.

In situ : Within the body.

Ingredient : The term "feed ingredient" means each of the constituent materials making up a commercial feed.

Irradiation : The act of treating with ultraviolet light.

Jejunum : The middle protein of the small intestine which extends from the duodenum to the ileum.

Keratin : A sulphur- containing protein which is the primary component of epidermis, hair, wool, hoof, horn, and the organic matrix of the teeth.

Keratosis : Any horny growth, such as a wart, causing the cornification, or hardening, of the epithelial skin layers.

Kernel : A de-hulled seed.

Ketosis : A condition characterized by an abnormally elevated of ketone (acetone) bodies in body fluids and tissues-acetonemia.

Kibbled, Kibbling : (Process) Cracked or crushed baked dough, or extruded feed that has been cooked prior to or during the extrusion process.

Land Application : The controlled application of waste water onto the land surface to achieve a designated degree of treatment through natural plant/soil/water matrix.

Leachate : The liquid that percolates through a landfill.

Legume : Refers to those crops that can absorb nitrogen directly from the atmosphere through bacteria that live in roots.

Lignin : An indigestible compound which along with cellulose is a major component of the cell wall of certain plant materials such as wood, hulls, straws, and overripe hays.

Limited feeding : Feeding animals to maintain weight and growth but not enough to fatten or increase production. Feeding animals less than they would like to eat.

Limiting amino acid : The essential amino acid of protein that shows the greatest percentage deficit in comparison with the amino acids contained in the same quantity of another protein selected as standard.

Lipids : A broad term for fats and fat-like substances.

Mash : (Physical form) A mixture of ingredients in meal form. Similar term : Mash feed.

Meal : A feed ingredient having a particle size somewhat larger than flour.

Medicated Feed : Any feed which contains drug ingredients intended or represented for the cure, mitigation, treatment, or prevention of diseases of animals other than man or which contain drug ingredients intended to effect the structure or any function of the body of animals other than man.

Metabolic body size : An expression relating energy metabolism to body weight which has a relationship to body surface. It is usually expressed as a power of body weight such as $W^{0.75}$ or some other exponent.

Metabolism : The sum total of the chemical changes in the body, including the building up (anabolic, assimilation) and the breaking down process. Transformation by which energy is made available to body use.

Metabolite : Any substance produced by metabolism.

Metabolizable Energy : Digestible energy minus the energy of the urine and fermentation gases.

Micro : A unit of length, e.g., particle diameter, equal to 0.0001millimeter.

Micro-Ingredients : Vitamins, minerals, antibiotics, drugs, and other materials normally required and measured in small units.

Micron : Unit of length in metric system; a thousandth part of one millimeter, approximately 1/ 25,000 of an inch (.00004 inch.)

Milieu : Mixture.

Mill By-Product (Part) : A secondary product obtained in addition to the principal product in milling practice.

Mill Dust : (Part) Fine feed particles of undetermined origin resulting from handling and processing feed and feed ingredients.

Mill Run (Part) : The state in which a material comes from the mill, un-graded and usually un-inspected.

Miscible : Mixable.

Mixing : (Process) To combine by agitation two or more materials to a specific degree of dispersion.

Moiety : Component, part.

Mono : One.

Mucosa : Mucous membrane, mucosal surface of the intestine on the absorptive side.

Mycotoxin : A fungus or bacterial toxin some times present in feed material.

Myoglobin : Muscle hemoglobin.

Myosin : Muscle fiber fraction that complexes with actine to bring about muscle contraction.

Net Energy : This is that part of metabolizable energy over the use of which the animal has complete control. It is metabolizable energy minus heat increment.

Net Protein Ratio : The difference between the average final body weights of a test group of animals fed a protein diet and that of a control group receiving a protein free diet divided by the amount of protein taken by the test group.

Nitrogen Free Extract : The part of feed dry matter, which is not crude protein, crude fat, crude fiber or ash. It consists mainly of sugar and starches. Sometimes referred as NFE.

Non-polar : Immiscible with water.

Non-protein : Any one of a group of ammonia nitrogen containing compounds, which are not true protein. Urea is a common example.

Oligo : A few.

Omnivore : Literally means eats everything. Animal that eats both plant and animal origin materials.

Opisthotonus : A condition in which the head tilts back, which may be caused by a deficiency of thiamine. Also called star-gazing.

Ossification : The process of bone formation.

Osteoblasts : Bone forming cells.

Osteoclasts : Bone resorbing cells.

Osteomalacia : A weakening of the bones due to calcium, phosphorous, or vitamin D deficiency.

Osteoporosis : An abnormal porousness of the bones the result of calcium, phosphorous, or vitamin D deficiency.

Parakeratosis : Any abnormality of the outermost horny layer of the skin.

Pearled, Pearling : (Process) De-hulled grains reduced by machine brushing, or abrasion, into small smooth particles.

Pelleted, Pelleting (Process) : Having agglomerated feed by compacting and forcing through die openings.

Pellets, Soft : (Physical form) Pellets containing a large percentage of liquids requiring immediate dusting and cooling.

Pellets : (Physical form) Agglomerated feed formed by compacting and forcing through die openings by a mechanical process. Similar terms : Pelleted feed, hard pellet.

Penultimate : Just before the last one.

pH : A term that expresses the intensity of the acid or alkaline condition of a liquid; a measure of the negative logarithm of the hydrogen ion concentration.

Phagocytes : Any cell that can ingest particles or cells that are foreign or harmful to the body.

Plasma : Fluid portion of the blood, similar to the serum except it contains fibrogen and other clotting factors.

Polar : Miscible with water.

Poly : Many.

Polymers : Multiple units of single compounds.

Polyneuritis : Multiple nutrients (degeneration of nerves).

Pophyrin : Porphin-ring derivative from complexes such as hematin and cytochrome.

Popped, Popping : (Process) The expansion of whole or cracked grain by heat. It may entail pressure as well.

Porphyrinuria : Porphyrin in the urine.

Postabsorptive state : No undigested or unabsorbed feed remaining in the digestive tract.

Premix : A uniform mixture of one or more micro-ingredients with diluent and/or carrier. Premixes are used to facilitate uniform dispersion of the micro-ingredients in a larger mix.

Pressed, Pressing : (Process) Compacted or molded by pressure; also meaning having fat, oil or juices extracted under pressure.

Product : (Part) A substance produced from one or more other substances as a result of chemical or physical change.

Prolactin : The lactation-stimulating hormone of the anterior lobe of the pituitary gland.

Protein (Part) : Any of a large class of naturally-occurringcomplex combinations of amino acids.

Prothrombin : Plasma fraction that is converted to thrombin during the clotting the blood.

Puffed, Puffing : (Process) The expansion of whole, cracked, or processed grain by pressure and heating.

Pulverized, Pulverizing : (Process) See ground, grinding. Range Cake - (Physical form) See cake.

Purified Diet : A mixture of the known essential dietary nutrients in a pure form that is fed to experimental animals in nutrition studies.

Rachitic : Adjective used to denote the condition of rickets.

Radiant heat : Heat transmitted by radiation (as from the sun) in contrast to that transmitted by conduction or convection.

Rancid : A term which is used to describe the fat which have undergone partial decomposition.

Range Cubes (Physical form) : Large pellets designed to be fed on the ground. Similar term : Range wafer.

Ration : The amount of the total feed, which is provided to one animal over a 24 hour period.

Reduction : Chemically, the removal of oxygen from, or the addition of hydrogen to, a substance.

Rennin : Enzyme that coagulates milk.

Retarded growth : Slower than normal rate of growth.

Rolled, Rolling : (Process) Having changed the shape and/or size of particles by compressing between rollers. It may entail tempering or conditioning.

Rumen culture : A liquid or dried preparation consisting of microbes from the rumen.

Runoff : That portion of storm water or land applied wastewater that flows over the land surface and reaches a sewer or receiving stream.

Scalped, Scalping (Process) : Having removed larger material by screening.

Scoured, Scouring : (Process) Removal of the ends of wholegrain.

Scratch : (Physical Form) Cleaned whole, cracked, or cut grain, usually in a mixture. Similar terms : Scratch grain, scratch feed.

Screened, Screening : (Process) Having separated various sized particles by passing over and/or through screens.

Self Fed : A feeding system where animals have continuous access to some or all component(s) of a ration, either individually or as mixtures.

Separating (Process) : Classification of particle by size, shape and density.

Serosa : Membrane lining the pericardial, pleural, and peritoneal cavities.

Serum : Fluid portion of blood similar to plasma except fibrinogen and other clotting factors have been removed.

Sifted (Process) : Material that has been passed through Wire sieves to separate particles in different sizes. The separation of finer material then would be done by screening.

Sizing : (Process) See screening.

Solubles : Liquid containing dissolved substances obtained from processing animal or plant materials. It may contain some fine suspended solids.

Solvent Extracted : (Process) A product from which oil has been removed by solvents.

Specific dynamic effect, SDE, SDA : The increase in heat production from consumption of feed usually referred to as heat increment.

Specific Gravity : Weight of a liquid compared to water, which is assigned a value of 1.0.

Spray Dehydrated : (Process) Material which has been dried by spraying on the surface of a heated drum. It is, recovered by scraping from the drum.

Standard Atmosphere : Condition when air is at 14.7 psi pressure at 68°F temperature.

Standard Conditions : Used as a basis for air conditioning calculations : Temperature at 68°F, pressure of 29.92 inches of Hg, and relative humidity of 30 percent.

Steamed, Steaming (Process) : Having treated ingredients with steam to alter physical and/or chemical properties.

Steatites : Inflammation of the adipose tissue.

Substrate : A substance upon which an enzyme acts. Same as zymolyte.

Supplement : A feed used with another to improve the nutritive balance or performance of the total and intended to be : (I) Fed undiluted as a supplement to other feeds; or (2) Offered free choice with other parts of the ration separately available; or (3) Further diluted and mixed to produce a complete feed.

Syndrome : A medical term meaning a set of symptoms that occur together.

Temperature, Dew-Point : The temperature at which the condensation of water vapor in a space begins for a given state of humidity and pressure as the temperature of the vapor IS reduced. The temperature corresponding to saturation (100 percent relative humidity) for a given absolute humidity at constant pressure.

Temperature, Dry-Bulb : The temperature of a gas or mixture of gases indicated by an accurate thermometer after corrections for radiation.

Temperature, Wet-Bulb : Thermodynamic wet-bulb temperature is the temperature at which liquid or solid water, by evaporating into air, can bring the air to saturation adiabatically at the same temperature. Wet-bulb temperature (without qualification) is the temperature indicated by a wet-bulb psychrometer constructed and used according to specifications.

Tempered, Tempering (Process) : See conditioned, conditioning.

Tetany : Spasmodic muscle contractions, convulsions.

Thermobin : Plasma fraction that converts fibernogen to fibrin in the clotting the blood.

Thermogenesis : Chemical production of heat in the body.

Thermoneutrality : The temperature range of comfort zone in which the body's temperature regulating mechanisms are not required to function to either cool or warm the body.

Thrombosis : The obstruction of a blood vessel by the formation of blood clot.

Toxic : Of a poisonous nature.

Trace Minerals : Mineral nutrients required by animals in micro amounts only.

Tracer Element : A radioactive element used in biological and other research to trace the fate of a substance.

Ultimorbranchial : Thyroid primordial derivatives.

Vacuum : Reduction in pressure below atmospheric pressure.

VFA : Volatile fatty acid commonly used in reference to acetic, propionic and butyric acid produced in rumen.

Villi : Small thread like projections attached to the interior side of the wall of the small intestine.

Vitamins : Organic compounds that function as parts of enzyme systems essential, for the transformation of energy and the regulation of metabolisms of the body.

Wafered, Wafering (Process) : Having agglomerated a feed of a fibrous nature by compressing into a form usually having a diameter or cross section measurement greater than its length.

Wet-Milled : (Process) Steeped in water with or without sulfur dioxide to soften the kernel in order to facilitate the separation of the various component parts.

Wet-Rendered, Wet-Rendering : (Process) Cooked with steam under pressure in closed tanks.

Xanthine : An intermediate in the metabolism of purines; related to uric acid.

Xeropthalmia : Dry infected eye condition caused by lack of vitamin A.

Zein : A protein of low biological value present in maize deficient in lysine and tryptophan.

Zymogen : The inactive form of an enzyme.

Chemical Composition of Feedstuffs

Table:1 Proximate and cell wall fractions of some important feeds, forages, tree leaves and shrubs

Name	CP	CF	EE	NFE	Ash	NDF	ADF	Cellulose	Lignin
Grasses									
H contortus	3.7		1.1		7.5	48.9	43.9	38.3	4.6
S nervosum	3.5		1.4		12.8	71.5	43.8	36.3	5.1
C fulvus	3.5		1.3		8.4	72.9	40.6	33.5	6.6
C ciliaris	4.1		1.5		10.9	69.4	40.9	34.1	5.7
D. annulatum	4.3		1.5		10.6	71.1	42.4	34.8	4.9
Guinea grass	8.2	36.8	1.4	37.6	11.4	73.2	40.0	30.7	4.0
Setaria grass	15.0	35.0	1.8	34.8	15.6	73.4	50.1	33.7	12.0
Straws & Stovers									
Wheat straw	13.2	42.9	1.0	36.6	10.6	83.7	53.6	39.8	10.5
Paddy straw	6.0	32.3	2.7	47.3	10.0	66.1	60.3	36.4	8.5
Pearl millet stover	8.0	35.7	1.2	44.0	7.6	76.2	44.1	33.2	5.4
Maize stover	7.3	31.7	1.6	47.6	6.5	73.2	45.7	36.1	8.1
Sorghum stover	6.0	36.1	1.5	38.1	7.5	76.2	44.9	36.2	4.7
Tree leaves									
A. lebbek	20.8		4.7		9.5	47.7	29.1	17.9	11.0
L. leucocephala	21.8		5.2		11.8	27.1	16.7	10.6	6.4
H. binata	15.9		2.5		8.2	35.9	26.9	18.2	8.5
G. optiva	16.7		4.2		13.5	34.4	21.0	17.1	3.9
A. pendulla	12.3		4.1		9.0	37.6	22.2	15.5	6.9
Azdirachta indica	17.6		2.4		9.9	45.9	35.4	23.6	11.8
Albizia anara	27.6		5.6		11.1	45.2	26.2	12.3	14.1
Dalbergia sissoo	18.7		4.6		13.7	48.2	26.5	19.7	14.1
Ficus religiosa	19.8		3.8		12.7	48.2	33.8	22.9	14.6
Butea monosperma	14.0		3.8		10.5	65.4	48.3	28.1	20
Holoptelea integrifolia	25.9		2.1		2.1	34.7	20.2	14.4	6.5
Shrubs									
A. catechu	11.8		4.1		9.2	31.5	19.5	9.4	6.8
S. virosa	11.3		1.9		9.6	25.8	16.5	13.1	3.4
Z. xylophyrus	11.8		2.4		8.1	26.3	13.0	8.6	5.0
D. cinerea	17.2		1.5		11.6	40.3	28.9	13.0	13.4
H. isora	13.4		3.1		11.5	39.7	25.1	17.9	6.0
Legume									
Berseem	16.2		4.5		13.0	36.3	26.3	19.9	5.6
Cowpea	10.1		4.5		14.3	63.1	43.6	27.2	9.6
Stylosanthes	13.1		1.2		8.7	54.2	38.1	31.3	6.3
Khair	13.7		1.5		8.7	48.1	25.4	15.5	9.1
Lucerne	26.4	17.5	1.7	35.5	17.4	45.0	21.3	18.2	7.4
Desmenthus	24.2	32.5	2.1	30.4	9.5	51.0	36.3	22.0	17.5
Dolichose	17.4	26.9	2.2	42.8	10.2	10.8	35.1	31.1	8.0

Name	CP	CF	EE	NFE	Ash	NDF	ADF	Cellulose	Lignin
Non-legume									
Wheat straw	10.7		1.5		12.3	63.2	37.9	30.0	5.7
Guinea	7.0		10		12.1	76.2	47.4	38.5	5.3
Maize	9.4		1.4		7.1	62.1	32.0	27.4	3.1
Sorghum	6.4		1.6		7.1	71.7	32.4	29.7	5.0
Barley	14.7	41.1	1.5	20.5	9.0	68.0	35.7	33.2	2.7
Oat	9.5	38.7	1.8	40.8	6.0	72.0	46.6	41.5	7.0
Energy Sources									
Deoiled rice bran	6.7		0.5		16.6	67.7	51.8	26.6	14.1
Maize grain	11.9		5.8		0.6	22.5	4.3	3.1	1.3
Barley	15.4		2.6		2.9	22.7	6.9	5.7	4.9
Wheat bran	9.0		4.7		4.7	36.1	10.1	6.3	2.6
Protein Sources									
Linseed cake	25.9		2.5		9.8	44.8	21.0	11.8	6.7
Mustard seed cake	35.0		8.2		7.9	24.5	17.1	10.3	5.3
Cotton seed cake	30.3		2.4		5.8	40.3	29.5	21.9	8.1
Soybean cake	47.6		6.4		6.4	44.1	9.3	8.0	0.4
Groundnut cake	42.9		7.7		5.5	16.7	9.7	7.6	2.6

IGFRI, Jhansi

Table: 2 protein and carbohydrate fraction of some important feeds, forages, tree leaves and shrubs

	Protein fraction				Carbohydrate fraction		
Name	**NDIP%**	**ADIP%**	**SP%**	**NPN%**	**NDF %**	**Starch%**	**NSP %**
Grasses							
H contortus	4.6	54.9	38.2	31.3	48.9	78.7	
S nervosum	5.1	52.2	47.8	25.5	71.5	53.9	
C fulvus	6.6	69.0	65.7	24.9	72.9	45.6	
C ciliaris	5.7	55.8	47.62	31.4	69.4	26.7	
D. annulatum	4.9	40.7	30.64	38.7	71.1	17.9	
Guinea grass	4.0	35.7	21.4	25.5	73.2	47.6	54.3
Setaria grass	12.0	37.3	6.9	40.6	73.4	51.0	43.0
Straws & Stovers							
Wheat straw	10.5	53.3	14.8	50.4	83.7	34.3	72.3
Paddy straw	8.5	55.3	42.1	16.2	66.1	63.1	93.33
Pearl millet stover	5.4	43.7	25.6	27.6	76.2	50.1	90.8
Maize stover	8.1	36.7	16.7	30.1	73.2	27.3	90.1
Sorghum stover	4.7	54.0	30.1	21.5	76.2	72.0	22.4
Tree leaves							
A. lebbek	11.0	34.2	18.2	35.1	47.7	58.8	
L. leucocephala	6.4	27.7	19.3	35.1	27.1	81.5	

		Protein fraction				Carbohydrate fraction	
Name	**NDIP%**	**ADIP%**	**SP%**	**NPN%**	**NDF %**	**Starch%**	**NSP %**
H. binata	8.5	39.1	18.7	50.4	35.9	76.3	
G. optiva	3.9	41.5	18.5	39.7	34.4	66.8	
A. pendulla	6.9	30.5	20.9	18.8	37.6	59.5	
Azdirachta indica	11.8				45.9		
Albizia anara	14.1				45.2		
Dalbergia sissoo	14.1				48.2		
Ficus religiosa	14.6				48.2		
Butea monosperma	20				65.4		
Holoptelea integrifolia	6.5				34.7		
Shrubs							
A. catechu	6.8	42.8	25.0	22.8	31.5	54.1	
S. virosa	3.4	27.8	21.3	16.9	25.8	45.3	
Z. xylophyrus	5.0	54.0	35.4	18.2	26.3	36.1	
D. cinerea	13.4	33.2	31.1	35.1	40.3	16.5	
H. isora	6.0	62.0	21.4	31.6	39.7	71.6	
Legume							
Berseem	5.6	25.9	7.7	48.6	36.3	67.2	
Cowpea	9.6	55.0	20.8	5.6	63.1	41.7	
Stylosanthes	6.3	28.8	12.4	18.3	54.2	37.8	
Khair	9.1	39.2	19.8	11.5	48.1	77.2	
Lucerne	7.4	19.5	8.7	25.6	45.0	74.7	49.3
Desmenthus	17.5	27.6	10.8	26.1	51.0	20.8	85.6
Dolichose	8.0	19.1	18.6	22.4	10.8	65.6	74.5
Non-legume							
Wheat straw	5.7	30.4	7.2	49.1	63.2	40.7	
Guinea	5.3	50.5	24.9	25.1	76.2	92.9	
Maize	3.1	47.1	11.2	21.2	62.1	24.6	
Sorghum	5.0	51.4	16.2	22.2	71.7	80.7	
Barley	2.7	38.1	7.2	15.3	68.0	63.5	91.9
Oat	7.0	45.1	15.6	20.5	72.0	47.3	61.3
Energy Sources							
Deoiled rice bran	14.1	47.2	19.2	39.0	67.7	34.3	
Maize grain	1.3	14.1	4.4	14.3	22.5	35.8	
Barley	4.9	13.0	2.6	36.3	22.7	34.2	
Wheat bran	2.6	22.1	3.0	18.4	36.1	23.6	
Protein Sources							
Linseed cake	6.7	25.9	6.4	23.5	44.8	25.0	
Mustard seed cake	5.3	8.8	5.1	39.0	24.5	35.2	
Cotton seed cake	8.1	8.4	5.0	21.2	40.3	38.7	
Soybean cake	0.4	20.9	1.7	9.0	44.1	11.8	
Groundnut cake	2.6	6.9	2.1	60.5	16.7	14.8	

IGFRI, Jhansi

Table : 3 Fractionations of Carbohydrate and protein in feedstuffs

Feed Name			Carbohydrate fractions				Protein fractions					
	Dry Matter % AF	NDF % DM	Lignin % NDF	Starch % NSC	CP % DM	Sol. CP % CP	NPN % SOLP	NDFIP % CP	ADFIP % CP	UIPIx % CP	Fat & DM	Ash % DM
1. Alfalfa *(Medicago sativa)* Hay, Brazil	89.5	66.9	18.10	10.0	19.0	30.0	93.0	1.0	0.5	10.0	1.9	10.4
2. Alfalfa *(Medicago sativa)* Silage, Brazil	49.7	43.5	18.30	10.0	18.6	45.0	100.0	19.8	10.4	19.9	1.7	9.6
3. Bermudagrass *(C. dactylon)* Fresh, Brazil	24.7	79.8	8.40	6.0	9.4	25.9	25.4	4.5	1.7	11.5	1.4	7.2
4. Bermudagrass *(C. dactylon)* Hay, Brazil	88.2	75.5	7.50	6.0	10.6	28.9	96.0	35.9	5.4	38.8	1.5	6.8
5. Bermudagrass *(C. dactylon)* Hay, Florida	89.4	81.0	9.60	6.0	9.8	15.6	96.0	47.3	23.1	51.1	1.5	7.0
6. Black oats *(Avena strigosa)* - Hay, Brazil	88.6	75.8	7.50	90.0	13.7	30.0	93.0	30.0	10.0	32.8	3.6	6.4
7. Brewers - Dry residue, Brazil	90.2	52.3	9.20	100.0	17.6	4.0	75.0	40.0	3.5	48.6	3.3	5.4
8. Buffelgrass *(Cenchrus ciliaris)* Hay, Brazil	92.0	75.7	11.20	0.0	3.7	24.3	5.0	40.5	21.6	43.9	3.0	4.7
9. Cassava *(Manihot utilissima)*- Residue, Brazil	18.8	20.2	12.00	13.6	2.2	25.0	45.0	30.0	5.0	34.7	0.2	1.4
10. Congograss *(B. ruziziensis)* Hay, Brazil	91.2	75.5	10.60	6.0	3.4	38.2	2.4	38.2	17.6	41.1	2.1	6.3
11. Corn *(Zea mays)* Cracked, Brazil	87.9	10.9	7.30	90.0	9.2	17.8	30.0	6.8	2.3	40.4	4.0	1.9
12. Corn *(Zea mays)* grain, Brazil	88.0	13.4	2.80	90.0	9.5	11.0	73.0	10.7	4.0	24.4	4.4	1.5
13. Corn *(Zea mays)* Silage, Brazil	31.4	53.2	8.10	100.0	7.1	41.4	98.3	14.6	10.8	37.0	2.5	4.2
14. Cottonseed *(Gossypium spp)*- Meal, Brazil	90.1	46.9	15.10	90.0	37.9	20.0	40.0	9.3	1.5	21.0	1.5	5.8
15. Gliricidia *(Gliricidia sepium)*- Hay, Honduras	88.5	37.3	24.40	10.0	24.4	31.6	96.0	25.4	11.9	26.8	3.2	8.6
16. Guineagrass *(P. maximum)*-Fresh, Fert., Mex.	18.4	66.9	6.80	38.1	9.6	28.7	39.2	34.5	8.1	17.7	3.0	12.7
17. Guineagrass *(P. maximum)*- Hay, Brazil	89.3	67.7	6.90	6.0	8.4	3.6	96.0	37.0	11.6	41.9	1.5	7.9
18. Guineagrass *(P. maximum)*- Hay, Honduras	88.8	66.6	6.80	6.0	14.9	35.5	96.0	39.0	4.4	40.2	1.5	13.4
19. Guineagrass *(P. maximum)*-Pasture, Honduras	91.3	66.7	6.10	8.0	10.3	32.4	3.4	43.7	9.2	29.1	3.2	12.4
20. Guineagrass *(P. maximum)*- Silage, Honduras	87.5	66.7	9.60	100.0	10.5	39.0	90.0	33.3	14.3	28.9	2.8	17.1
21. Koroniviagrass *(B. humidicola)*- Fresh, Mex.	22.9	75.7	8.10	21.2	6.0	47.0	50.5	21.2	7.3	17.8	1.6	6.6
22. Koroniviagrass *(B. humidicola)*-Fresh, Fert., Mex.	14.0	70.3	7.20	35.1	9.8	39.9	53.6	17.1	4.3	15.5	2.0	10.6
23. Koroniviagrass *(B. humidicola)*- Hay, Brazil	93.6	79.0	7.50	0.0	4.4	25.0	5.0	36.2	16.0	45.2	0.9	5.3

Feed Name	Dry Matter % AF	NDF % DM	Carbohydrate fractions Lignin % NDF	Starch % NSC	CP % DM	Sol. CP % CP	Protein fractions NPN % SOLP	NDFIP % CP	ADFIP % CP	UIPIx % CP	Fat & DM	Ash % DM
24. Leucaena *(L. leucocephala)* Hay, Brazil	94.7	64.3	21.40	38.0	11.7	25.0	5.0	33.9	14.5	44.6	0.7	4.8
25. Molasses Sugarcane, Mex.	85.8	0.0	0.00	0.0	4.2	98.0	100.0	0.0	0.0	0.4	2.2	11.6
26. Molassesgrass *(M. minutiflora)*- Fresh, Brazil	30.1	78.3	6.80	0.0	7.1	25.0	5.0	3.4	1.7	37.6	2.2	8.3
27. Molassesgrass *(M. minutiflora)*-Fresh, Fert., Mex.	17.6	70.2	4.70	0.0	8.6	27.9	50.1	33.0	4.9	16.4	2.5	12.8
28. Molassesgrass *(M. minutiflora)*- Hay, Brazil	90.8	81.5	10.30	0.0	3.9	25.0	5.0	2.0	1.0	21.1	2.9	6.1
29. Napiergrass *(P. purpureum)*- Fresh, Brazil	23.7	74.0	9.60	8.0	6.6	46.0	2.2	2.2	0.9	30.1	2.3	9.4
30. Napiergrass *(P. purpureum)*- Hay, Brazil	91.3	65.8	7.30	10.0	10.4	32.4	95.0	39.7	10.0	41.0	1.7	11.8
31. Napiergrass *(P. purpureum)*- Silage, Brazil	23.8	78.4	12.20	100.0	5.4	50.0	90.0	52.6	26.1	40.2	2.3	7.7
32. Palisadegrass *(B. brizantha)*- Fresh, Brazil	25.1	76.4	8.30	5.0	7.1	41.0	2.4	4.6	2.1	38.6	1.6	7.6
33. Palisadegrass *(B. brizantha)*-Fresh, Mex.	25.1	67.8	5.30	32.7	7.1	45.9	70.7	15.9	3.8	12.3	2.0	8.0
34. Palisadegrass *(B. brizantha)*-Fresh, Fert., Mex.	19.8	63.7	5.90	26.5	11.7	37.5	68.3	10.1	2.9	12.8	2.8	11.1
35. Palisadegrass *(B. brizantha)*- Hay, Brazil	88.8	75.2	5.90	0.0	5.9	38.2	5.0	33.7	10.5	40.0	1.4	7.2
36. Pangolagrass *(D. decumbens)*- Fresh, Mex. 1	26.8	69.5	7.50	28.2	8.9	41.9	36.3	32.5	5.4	16.5	2.4	8.6
37. Pangolagrass *(D. decumbens)*- Fresh, Mex. 2	26.8	70.0	7.30	31.5	7.0	36.7	38.3	31.3	6.1	19.2	1.8	8.0
38. Pangolagrass *(D. decumbens)*- Hay, Brazil	88.9	72.3	9.60	6.0	7.3	30.9	96.0	38.2	16.4	40.1	2.3	6.9
39. Pangolagrass *(D. decumbens)*- Hay, Honduras	89.9	63.9	7.20	6.0	14.2	29.6	96.0	42.9	8.1	43.9	1.6	13.8
40. Paspalum *(P. fasciculatum)* Fresh, Fert., Mex.	18.5	63.5	5.70	55.8	11.9	15.1	87.1	41.9	6.5	21.0	1.2	14.2
41. Paspalum *(P. fasciculatum)* Hay, Florida	89.4	80.8	7.30	6.0	9.9	12.7	25.4	75.0	29.7	73.2	1.6	5.5
42. Pearlmillet *(P. americanum)* Grain, Brazil	88.0	27.3	7.10	90.0	15.1	53.0	19.0	32.8	18.9	33.5	3.8	2.0
43. Signalgrass *(B. Decumbens)*- Fresh, Brazil	28.9	75.8	7.50	5.0	7.2	42.0	4.8	2.1	0.8	36.2	1.2	8.2
44. Signalgrass *(B. Decumbens)*- Hay, Brazil	88.8	84.2	10.00	0.0	5.5	38.5	5.0	13.3	6.1	30.6	2.3	7.0
45. Siratro *(M. atropupureum)* Hay, Brazil	92.0	47.5	22.70	10.0	18.8	28.7	96.0	16.0	8.0	20.9	3.0	8.2
46. Sorghum *(S. vulgare)*-Grain cracked, Brazil	87.2	24.2	9.20	100.0	9.9	12.0	33.0	1.9	0.1	15.4	8.7	0.9

Feed Name	Dry Matter % AF	NDF % DM	Carbohydrate fractions Lignin % NDF	Starch % NSC	CP % DM	Sol. CP % CP	Protein fractions NPN % SOLP	NDFIP % CP	ADFIP % CP	UIPIx % CP	Fat & DM	Ash % DM
47. Sorghum *(S. vulgare)*-Grain, Brazil	88.2	11.2	2.22	90.0	8.9	13.0	41.9	15.0	5.0	37.4	5.1	2.0
48. Sorghum *(S. vulgare)*-Residue, Honduras	88.2	66.2	10.00	6.0	8.8	51.3	96.0	20.9	12.8	23.5	2.3	11.1
49. Sorghum *(S. vulgare)*-Silage, Brazil	28.8	61.6	9.40	100.0	6.2	45.0	100.0	19.6	16.8	25.0	3.5	5.3
50. Soybean *(Glycine max)* Hulls, Brazil	89.8	62.7	3.20	90.0	13.4	18.0	72.0	21.1	5.8	30.6	2.0	5.5
51. Soybean *(Glycine max)*- Meal, Brazil	87.8	14.1	19.00	90.0	47.0	35.9	11.1	2.8	1.6	16.5	5.5	6.2
52. Soybean *(Glycine max)*- Straw, Brazil	87.4	56.8	1.90	100.0	11.9	70.0	100.0	33.3	27.0	31.4	2.8	10.5
53. Stargrass *(C. plectostachyus)*- Fresh, Mex.	30.2	76.8	7.60	28.0	6.7	35.5	31.4	42.7	10.8	22.4	1.1	7.7
54. Stargrass *(C. plectostachyus)*- Fresh, Fert., Mex.	21.5	71.6	7.30	22.6	10.0	35.7	40.1	34.0	8.0	19.4	1.6	11.0
55. Sugarcane *(S. officinarum)*- Bagasse, Brazil	15.6	75.6	11.30	100.0	2.6	20.0	95.0	75.0	65.0	74.9	1.8	1.9
56. Sugarcane *(S. officinarum)*- Fresh, Brazil	29.7	57.1	11.0	0.0	2.5	55.0	100.0	16.0	9.0	19.2	1.4	2.9
57. Wheat *(Triticum spp)* Hay, Brazil	89.0	55.0	3.50	100.0	4.1	20.0	95.0	75.0	65.0	75.0	0.9	6.1
58. Wheat *(Triticum spp)* Middling, Brazil	88.3	43.0	11.20	90.0	17.6	53.0	19.0	3.0	0.5	18.4	5.3	5.6

Source: NRC (2001)

Nutrient Requirement of Dairy Cattle (NRC, 2001)

Table 1: Daily Nutrient requirement of small breed cows (live weight = 454 kg) in early lactation (intake estimated at 11 days in milk). Values are appropriate for the diet below with 78% TDN[a]

Milk (kg)	Fat (%)	True protein (%)	DMI (kg)	LW change (kg)	NE_L (Mcal)	RDP (g)	RUP (g)	RDP (%)	RUP (%)	CP (%)
15	4.0	3.0	9.4	-0.3	19.0	1060	500	11.3	5.3	16.6
15	4.0	3.5	9.4	-0.3	19.4	1060	630	11.3	6.7	18.0
15	4.0	4.0	9.4	-0.4	19.8	1060	760	11.3	8.1	19.4
15	4.5	3.0	9.7	-0.3	19.7	1090	490	11.2	5.1	16.3
15	4.5	3.5	9.7	-0.4	20.1	1090	620	11.2	6.4	17.6
15	4.5	4.0	9.7	-0.5	20.5	1090	750	11.2	7.7	18.9
15	5.0	3.0	9.9	-0.4	20.4	1110	480	11.2	4.8	16.0
15	5.0	3.5	9.9	-0.5	20.8	1110	610	11.2	6.2	17.4
15	5.0	4.0	9.9	-0.5	21.2	1110	740	11.2	7.5	18.7
30	4.0	3.0	12.9	-1.4	30.1	1410	1170	10.9	9.1	20.0
30	4.0	3.5	12.9	-1.6	30.9	1410	1430	10.9	11.1	22.0
30	4.0	4.0	12.9	-1.7	31.8	1410	1690	10.9	13.1	24.0
30	4.5	3.0	13.5	-1.5	31.5	1460	1150	10.8	8.5	19.3
30	4.5	3.5	13.5	-1.7	32.3	1460	1410	10.8	10.4	21.2
30	4.5	4.0	13.5	-1.9	33.2	1460	1670	10.8	12.4	23.2
30	5.0	3.0	14.0	-1.6	32.8	1510	1140	10.8	8.1	18.9
30	5.0	3.5	14.0	-1.8	33.7	1510	1400	10.8	10.0	20.8
30	5.0	4.0	14.0	-2.0	34.6	1510	1660	10.8	11.9	22.7

[a] Diet used for this table consisted of 15% immature legume silage, 33% normal corn silage, 34% ground high moisture shelled corn, 12% soybean meal (48% crude protein), 2.5% tallow, 1.5% menhaden fish meal, and 2% mineral and vitamin mix. Requirements are dependent upon the diet fed. Requirements shown do not include nutrients needed for live weight change. Live weight change is based on assumed NE_L intake minus requirements. Requirements for RUP do not include protein provided by loss in body reserves or required for gain in body reserves. Requirement for total CP assumes RDP and RUP are met. Requirement for total CP will increase if RDP requirement is not met.

Table 2: Daily Nutrient requirement of small breed cows (live weight = 454 kg) in midlactation (intake estimated at 90 days in milk). Values are appropriate for the diet below with 68% TDN[a]

Milk (kg)	Fat (%)	True protein (%)	DMI (kg)	LW change (kg)	NE_L (Mcal)	RDP (g)	RUP (g)	RDP (%)	RUP (%)	CP (%)
10	4.0	3.0	12.4	0.9	15.3	1240	230	10.0	1.9	11.9
10	4.0	3.5	12.4	0.8	15.6	1240	320	10.0	2.6	12.6
10	4.0	4.0	12.4	0.8	15.9	1240	420	10.0	3.4	13.4
10	4.5	3.0	12.7	0.9	15.7	1270	230	10.0	1.8	11.8
10	4.5	3.5	12.7	0.8	16.0	1270	320	10.0	2.5	12.5
10	4.5	4.0	12.7	0.8	16.3	1270	410	10.0	3.2	13.2
10	5.0	3.0	12.9	0.9	16.2	1290	220	10.0	1.7	11.7
10	5.0	3.5	12.9	0.8	16.5	1290	310	10.0	2.4	12.4
10	5.0	4.0	12.9	0.8	16.8	1290	400	10.0	3.1	13.1
20	4.0	3.0	16.0	0.4	22.7	1560	680	9.8	4.3	14.1
20	4.0	3.5	16.0	0.3	23.2	1560	860	9.8	5.4	15.2
20	4.0	4.0	16.0	0.2	23.8	1560	1040	9.8	6.5	16.3
20	4.5	3.0	16.5	0.4	23.6	1610	660	9.8	4.0	13.8
20	4.5	3.5	16.5	0.3	24.2	1610	840	9.8	5.1	14.9
20	4.5	4.0	16.5	0.2	24.8	1610	1030	9.8	6.2	16.0
20	5.0	3.0	17.0	0.4	24.5	1660	650	9.8	3.8	13.6
20	5.0	3.5	17.0	0.2	25.1	1660	830	9.8	4.9	14.7
20	5.0	4.0	17.0	0.1	25.7	1660	1010	9.8	5.9	15.7
30	4.0	3.0	19.5	-0.1	30.1	1870	1130	9.6	5.8	15.4
30	4.0	3.5	19.5	-0.3	30.9	1870	1400	9.6	7.2	16.8
30	4.0	4.0	19.5	-0.4	31.8	1870	1670	9.6	8.6	18.2
30	4.5	3.0	20.3	-0.2	31.5	1940	1110	9.6	5.5	15.1
30	4.5	3.5	20.3	-0.3	32.3	1940	1380	9.6	6.8	16.4
30	4.5	4.0	20.3	-0.5	33.2	1940	1650	9.6	8.1	17.7
30	5.0	3.0	21.1	-0.2	32.8	2000	1090	9.5	5.2	14.7
30	5.0	3.5	21.1	-0.4	33.7	2000	1360	9.5	6.4	15.9
30	5.0	4.0	21.1	-0.6	34.6	2000	1630	9.5	7.7	17.2

Table 3: Daily Nutrient requirement of small breed cows (live weight = 680 kg) in midlactation (intake estimated at 90 days in milk). Values are appropriate for the diet below with 68% TDN[a]

Milk (kg)	Fat (%)	True protein (%)	DMI (kg)	LW change (kg)	NE_L (Mcal)	RDP (g)	RUP (g)	RDP (%)	RUP (%)	CP (%)
25	3.0	2.5	19.6	1.0	26.0	1940	620	9.9	3.2	13.1
25	3.0	3.0	19.6	0.8	26.8	1940	840	9.9	4.3	14.2
25	3.0	3.5	19.6	0.7	27.5	1940	1070	9.9	5.5	15.4
25	3.5	2.5	20.3	0.9	27.2	2000	600	9.9	3.0	12.9
25	3.5	3.0	20.3	0.8	27.9	2000	820	9.9	4.0	13.9
25	3.5	3.5	20.3	0.6	28.7	2000	1050	9.9	5.2	15.1
25	4.0	2.5	21.0	0.9	28.4	2060	580	9.8	2.8	12.6
25	4.0	3.0	21.0	0.7	29.1	2060	810	9.8	3.9	13.7
25	4.0	3.5	21.0	0.6	29.8	2060	1030	9.8	4.9	14.7
35	3.0	2.5	22.7	0.6	32.2	2210	990	9.7	4.4	14.1
35	3.0	3.0	22.7	0.4	33.2	2210	1300	9.7	5.7	15.4
35	3.0	3.5	22.7	0.2	34.2	2210	1620	9.7	7.1	16.8
35	3.5	2.5	23.6	0.5	33.8	2290	960	9.7	4.1	13.8
35	3.5	3.0	236.	0.3	34.8	2290	1280	9.7	5.4	15.1
35	3.5	3.5	23.6	0.1	35.9	2290	1600	9.7	6.7	16.4
35	4.0	2.5	24.5	0.4	35.4	2370	940	9.7	3.8	13.5
35	4.0	3.0	24.5	0.2	36.5	2370	1260	9.7	5.1	14.8
35	4.0	3.5	24.5	0	37.5	2370	1570	9.7	6.4	16.1
45	3.0	2.5	25.7	0.1	28.3	2470	1370	9.6	5.3	14.9
45	3.0	3.0	25.7	-0.1	39.7	2470	1780	9.6	6.9	16.5
45	3.0	3.5	25.7	-0.4	41.0	2470	2180	9.6	8.5	18.1
45	3.5	2.5	26.9	0	40.4	2570	1340	9.6	5.0	14.6
45	3.5	3.0	26.9	-0.2	41.8	2570	1750	9.6	6.5	16.1
45	3.5	3.5	26.9	-0.5	43.1	2570	2160	9.6	8.0	17.6
45	4.0	2.5	28.1	-0.1	42.5	2670	1310	9.5	4.7	14.2
45	4.0	3.0	28.1	-0.3	43.8	2670	1720	9.5	6.1	15.6
45	4.0	3.5	28.1	-0.6	45.2	2670	2130	9.5	7.6	17.1

Table 4: Nutrient requirements of lactating dairy cows as determined using sample diets

Days in milk **Dry matter intake input**	**Jersey= 454 kg Bwt, mature Bwt = 454 kg, BCS = 3.0, 65 mos age milk, fat = 4.2%, milk true protein = 3.6%, lactose = 4.8%** **Default environmental conditions**					
	90 Model predicted	**90 Model predicted[h]**	**90 Model predicted**	**50 Model predicted[h]**	**120 Model predicted[h]**	**90 Model predicted +50%**
Milk production (kg)	25	35	40	35	35	35
Milk production (lbs)	55	77	88	77	77	77
Dry matter intake (kg)	18	21.7	23.5	19.8	22.2	22.7
Dry matter intake (lbs)	39.8	47.7	51.7	43.6	48.8	49.9
Daily wt change (kg)	0	-0.2	-0.5	-0.7	-0.1	0
Days to gain one condition score	3777	-	-	-	-	4247
Days to lose one condition score	-	241	110	80	532	-
Energy[a]						
NE_L (Mcal/ day)	27.7	35.6	39.5	35.6	35.6	35.6
NE_L (Mcal/ kg)	1.54	1.64	1.68	1.8a	1.6	1.57
NE_L (Mcal/ lbs)	0.7	0.74	0.76	0.82	0.73	0.71
Protein						
Metabolizable protein (g/d)	1991	2639	2965	2579	2656	2672
Diet % MP	11.1	12.2	12.6	13	12	11.8
Rumen degradable protein (g/d)	1747	2125	2288	1971	2167	2206
Diet % RDP	9.7	9.8	9.7	10	9.8	9.7
Rumen undegradable protein (g/d)	1151	1632	1865	1670	1619	1611
Diet % RUP	6.4	7.5	7.9	8.4	7.3	7.1
% RDP + % RUP (crude protein)[b]	16.1	17.3	17.6	18.4	17.1	16.8
Fiber and carbohydrate[c]						
NDF, min %	25-33	25-33	25-33	25-33	25-33	25-33
ADF, min %	17-21	17-21	17-21	17-21	17-21	17-21
NFC, max %	36-44	36-44	36-44	36-44	36-44	36-44
Minerals						
Absorbable calcium (g/day)	50.7	65.2	72.4	65.2	65.2	65.2
Dietary Ca %	0.57	0.57	0.63	0.66	0.54	0.53
Absorbable phosphorous (g/ day)	41.4	54.1	60.4	52.2	54.6	55.1
Dietary P%	0.33	0.37	0.36	0.44	0.35	0.34
Mg[d]%	0.18	0.19	0.2	0.21	0.19	0.19
Cl%	0.24	0.26	0.27	0.28	0.25	0.25
K[e]%	1.02	1.03	1.04	1.07	1.03	1.02
Na %	0.2	0.2	0.2	0.22	0.2	0.19
S %	0.2	0.2	0.2	0.2	0.2	0.2
Co mg/ kg	0.11	0.11	0.11	0.11	0.11	0.11
Cu mg/ kg[f]	10	10	10	11	10	9
I mg/ kg[g]	0.44	0.4	0.34	0.4	0.36	0.35
Fe mg/ kg	14	16	17	18	16	15
Mn mg/ kg	12	12	12	12	12	12
Se mg/ kg	0.3	0.3	0.3	0.3	0.3	0.3
Zn mg/ kg	45	49	51	54	48	47
Vitamin A (IU/ day)	49500	49500	49500	49500	49500	49500
Vitamin D (IU/ day)	13500	13500	13500	13500	13500	13500
Vitamin E (IU/ day)	360	360	360	360	360	360
Vitamin A (IU/ kg)	2772	2300	2123	2520	2247	2198
Vitamin D (IU/ kg)	755	627	579	687	613	600

Days in milk Dry matter intake input	Jersey= 454 kg Bwt, mature Bwt = 454 kg, BCS = 3.0, 65 mos age milk, fat = 4.2%, milk true protein = 3.6%, lactose = 4.8% Default environmental conditions					
	90 Model predicted	90 Model predicted[h]	90 Model predicted	50 Model predicted[h]	120 Model predicted[h]	90 Model predicted +50%
Vitamin E (IU/ kg)	20	17	16	18	16	16
Some diets used in model to generate tables. Ingredients listed as kg/ day DM						
Corn silage, normal	8.96	7.77	7.39	7.1	7.96	8.15
Soybean meal, solv. 48% CP	2.16	2.78	1.67	2.54	2.85	2.91
Legume forage silage, mid-maturity	2.67	3.1	-	2.83	3.18	3.25
Corn grain, steam flaked	2.6	4.91	5.88	4.48	5.03	515
Calcium carbonate	0.06	0.04	0.03	0.04	0.04	0.04
Monosodium phosphate (1H2O)	0.01	0.01	0.03	0.01	0.01	0.01
Soybean meal, expellers	-	-	1.16	-	-	-
Legume forage hay, immature	-	-	4.59	-	-	-
Sodium chloride	0.1	0.1	0.12	0.09	0.1	0.1
Grass hay, c-3, mid-mat	0.85	0.95	0.97	-	-	-
Vitamin and mineral premix	0.5	0.5	0.5	0.45	0.51	0.52
Bermudagrass hay, coastal	-	-	-	-	-	-
Cottonseed, whole with lint	-	1.02	1.64	0.94	1.05	1.07
Tallow	-	0.24	0.21	0.22	0.24	0.25
Calcium soaps of fatty acids	-	0.18	0.21	0.17	0.19	0.19
Blood meal, ring dried	-	0.11	0.1	0.1	0.11	0.12
Sorghum, sudan type, silage	-					
Sample diet evaluation	-					
NE_L (Mcal/ kg)	1.54	1.59	1.57	1.62	1.58	1.57
Undiscounted TDN %	69	73	75	73	73	73

[a] Recommended energy content of early lactation rations must be limited to prevent rumen acidosis. Cow must therefore, be expected to utilize body reserves to meet energy needs at highest levels of milk production. See fiber and NFC restrictions.

[b] Equivalent to crude protein requirement only if RDP and RUP are perfectly balanced.

[c] These are the minimum fiber (or maximum NFC) concentrations needed to maintain rumen health and milk fat test.

[d] Assumes that active transport of magnesium across the rumen wall is intact. High dietary potassium and excess non-protein nitrogen often interfere with Mg absorption. Under these conditions dietary Mg should be increased to 0.3%-0.35%.

[e] Heat stress may increase the need for potassium.

[f] High dietary molybdenum, sulphur, and iron can interfere with copper absorption increasing the requirement.

[g] Diets high in goitrogenic substances increase the iodine requirement.

[h] Diet composition is the same in all four cases of the Jersey cow producing 35 kg milk. Amount of dry matter consumed varies with days in milk and the use of predicted vs. actual dry matter intake in the model

Table 5: Nutrient requirements and required diet nutrient concentrations for fresh cows fed and example fresh-cow ration

Days in milk **Dry matter intake input**	**Jersey= 454 kg Bwt, mature Bwt = 454 kg, BCS= 3.0, 65 mos age milk, fat = 4.2%, milk true protein = 3.6%, lactose = 4.8% Default environmental conditions**		
	11 Model predicted +20%	**11 Model predicted**	**11 Model predicted +20%**
Milk production (kg)	35	25	25
Milk production (lbs)	77	55	55
Dry matter intake (kg)	18.8	11.9	14.3
Dry matter intake (lbs)	41.4	26.2	31.5
Daily wt change (kg)	-0.6	-1.4	-0.7
Days to gain one condition score			
Days to lose one condition score	143	41	83
Energy[a]			
NE_L (Mcal/ day)	34.8	27.7	27.7
NE_L (Mcal/ kg)	1.85	2.33	1.93
NE_L (Mcal/ lbs)	0.84	1.06	0.88
Protein[b]			
Metabolizable protein (g/d)	2254	1801	1875
Diet % MP	12.0	15.1	13.1
Rumen degradable protein (g/d)	1931	1244	1469
Diet % RDP	10.3	10.5	10.3
Rumen undegradable protein (g/d)	1045	1265	1202
Diet % RUP	5.6	10.6	8.4
% RDP + % RUP (crude protein)[c]	15.9	21.1	18.7
Fiber and carbohydrate[c]			
NDF, min %	25-33	25-33	25-33
ADF, min %	17-21	17-21	17-21
NFC, max %	36-44	36-44	36-44
Minerals			
Absorbable calcium (g/day)	64	51	51
Dietary Ca %	0.68	0.80	0.70
Absorbable phosphorous (g/ day)	52.0	35.0	37.7
Dietary P%	0.37	0.40	0.36
Mg[e]%	0.24	0.27	0.22
Cl%	0.33	0.36	0.30
K[f]%	1.14	1.19	1.10
Na %	0.28	0.31	0.26
S %	0.2	0.2	0.2
Co mg/ kg	0.11	0.11	0.11
Cu mg/ kg[g]	13	15	12
I mg/ kg[h]	0.64	0.67	0.56
Fe mg/ kg	19	21	17
Mn mg/ kg	17	19	15

Days in milk Dry matter intake input	Jersey= 454 kg Bwt, mature Bwt = 454 kg, BCS= 3.0, 65 mos age milk, fat = 4.2%, milk true protein = 3.6%, lactose = 4.8% Default environmental conditions		
	11 Model predicted +20%	**11 Model predicted**	**11 Model predicted +20%**
Se mg/ kg	0.3	0.3	0.3
Zn mg/ kg	60	67	56
Vitamin A (IU/ day)	75000	49900	49900
Vitamin D (IU/ day)	21000	13600	13600
Vitamin E (IU/ day)	545	363	363
Vitamin A (IU/ kg)	3978	4193	3490
Vitamin D (IU/ kg)	1085	1143	951
Vitamin E (IU/ kg)	29	31	25
Sample diet used in model to generate tables. Ingredients listed as % DM			
Corn silage, normal			
Corn grain, steam flaked			
Soybean meal, expellers			
Soybean meal, solv. 48% CP			
Legume forage hay, immature			
Cottonseed, whole with lint			
Calcium soaps of fatty acids			
Blood meal, ring dried			
Calcium carbonate			
Monosodium phosphate (1 H2O)			
Sodium chloride			
Vitamin and mineral premix			
Sample "fresh cow" diet evaluation			
NDF%			
Forage NDF%			
ADF%			
NFC%			
Undiscounted TDN			
Diet NE_L (Mcal/ kg), dependent on DMI	1.70	1.72	1.69
Crude protein %			

[a] Recommended energy content of early lactation rations must be limited to prevent rumen acidosis. Cow must therefore, be expected to utilize body reserves to meet energy needs at highest levels of milk production. See fiber and NFC restrictions.
[b] It will be nearly impossible to meet the metabolizable protein needs of the high producing fresh cow due to low dry matter intake and the difficulty formulating rations with such high RUP.
[c] Equivalent to crude protein requirement only if RDP and RUP are perfectly balanced.
[d] These are the minimum fiber (or maximum NFC) concentrations needed to maintain rumen health and milk fat test.
[e] Assumes that active transport of magnesium across the rumen wall is intact. High dietary potassium and excess non-protein nitrogen often interfere with Mg absorption. Under these conditions dietary Mg should be increased to 0.3%-0.35%.
[f] Heat stress may increase the need for potassium.
[g] High dietary molybdenum, sulphur, and iron can interfere with copper absorption increasing the requirement.
[h] Diets high in goitrogenic substances increase the iodine requirement.

Table 6: Example diet incorporating dietary guidelines suggested in chapter 9 for transitioning a Heifer (entering first lactation) during the later dry period to acclimate her to the lactating ration and to reduce.

This animal must be entered into the model as a replacement heifer to accommodate growth requirements. Tissue requirements for a heifer (270 d pregnant, weighing 625 kg with conceptus, mature body weight of 680 kg, consuming 10.6 kg dry matter/ day, gaining 300 g body weight plus 660g conceptus weight each day, and current body condition score of 3.3) and nutrient densities of an example ration that follows the recommendation guidelines.

	Guidelines for close-up heifer diet Recommendations that differ from requirement	Nutrient required by tissues (as assessed with the example diet)	Example close-up heifer ration Nutrients supplied
Energy			
NE_L (Mcal/ day)		14.8	16.9
NE_L (Mcal/ kg)	1.54 – 1.62[a]	1.40	1.59
Protein			
Metabolizable protein (g/d)		888	1027
Diet % MP		8.4	9.7
Rumen degradable protein (g/d)		1052	1067
Diet % RDP		9.9	10.2
Rumen undegradable protein (g/d)		336	511
Diet % RUP		3.2	4.9
% RDP + % RUP (crude protein)	13.5-15.0[c]	13.1	15.0
Fiber and carbohydrate[d]			
NDF, min %	33[c]	25-33	39.1
ADF, min %	21[c]	17-21	23.4
NFC, max %	43[c]	36-43	39.4
Minerals			
Absorbable calcium (g/day)		22.4	24.5
Dietary Ca %		0.40	0.44
Absorbable phosphorous (g/ day)		16.6	25.4
Dietary P%	0.3-0.4[e]	0.23	0.37
Mg%	0.35-0.40[f]	0.14	0.40
Cl%		0.16	0.44
K %	0.55[g]	0.55	1.54
Na %		0.12	0.13
S %		0.20	0.19
Co mg/ kg		0.11	0.11
Cu mg/ kg		16	16
I mg/ kg		0.4	0.4
Fe mg/ kg		26	26
Mn mg/ kg		22	22
Se mg/ kg		0.3	0.3
Zn mg/ kg		30	30

	Guidelines for close-up heifer diet Recommendations that differ from requirement	**Nutrient required by tissues (as assessed with the example diet)**	**Example close-up heifer ration Nutrients supplied**
Vitamin A (IU/ day)	75000[h]	68750	75000
Vitamin D (IU/ day)	20000[h]	18750	20000
Vitamin E (IU/ day)	1200[h]	848	1200
Vitamin A (IU/ kg)		6486	7075
Vitamin D (IU/ kg)		1769	1887
Vitamin E (IU/ kg)		80	113
Dietary cation-anion difference (Na + K) – (Cl + S), meq/kg		24	203
Sample diet used in model to generate tables	Ingredients (kg DM/day)		Example Close-up heifer diet
	Ingredients (kg DM/day)		
	Corn silage, normal		4.58
	Soybean meal, solv. 48% CP		1.07
	Grass silage, C-3, mid-mat		2.44
	Corn grain, ground high moist		1.04
	Beet sugar pulp, dried		0.95
	Sodium chloride		0.02
	Vitamin and mineral premix		0.39
	Magnesium oxide		0.04
	Calcium phosphate (Di-)		0.02

Table 7: Daily nutrient requirements (DM basis) of small breed (mature weight = 450 kg) non-bred heifers

BW kg	ADG kg/ d	DMI kg/d	TDN %	NEm Mcal/d	NEc Mcal/d	ME Mcal	RDP g/d	RUP g/d	RDP %	RUP %	CP[a] %	Ca g/d	P g/d
100	0.3	3.0	56.5	2.64	0.47	6.0	255	110	8.6	3.7	12.4	14	7
	0.4	3.0	58.6	2.64	0.64	6.4	270	143	9.0	4.7	13.7	18	8
	0.5	3.1	60.7	2.64	0.82	6.7	284	175	9.3	5.7	15.0	21	10
	0.6	3.1	62.9	2.64	1.00	7.0	298	207	9.6	6.7	16.3	25	11
	0.7	3.1	65.2	2.64	1.19	7.3	210	239	10.0	7.7	17.7	28	12
	0.8	3.1	67.7	2.64	1.37	7.6	323	270	10.4	8.7	19.0	31	13
150	0.3	4.0	56.5	3.57	0.63	8.2	346	95	8.6	2.4	11.0	15	8
	0.4	4.1	58.6	3.57	0.87	8.7	366	124	9.0	3.0	12.0	19	10
	0.5	4.1	60.7	3.57	1.11	9.1	385	152	9.3	3.7	12.9	22	11
	0.6	4.2	62.9	3.57	1.36	9.5	403	180	9.6	4.3	13.9	25	12
	0.7	4.2	65.3	3.57	1.61	9.9	421	207	10.0	4.9	14.9	28	13
	0.8	4.2	67.7	3.57	1.86	10.3	437	234	10.4	5.5	15.9	31	14
200	0.3	5.0	56.5	4.44	0.79	10.2	429	81	8.6	1.6	10.3	17	10
	0.4	5.1	58.6	4.44	1.08	10.7	454	106	9.0	2.1	11.1	20	11
	0.5	5.1	60.7	4.44	1.38	11.3	478	131	9.3	2.6	11.8	23	12
	0.6	5.2	62.9	4.44	1.68	11.8	500	156	9.6	3.0	12.6	26	13
	0.7	5.2	65.3	4.44	1.99	12.3	522	179	10.0	3.4	13.4	29	14
	0.8	5.2	67.7	4.44	2.31	12.8	543	202	10.4	3.9	14.2	32	15
250	0.3	5.9	56.5	5.24	0.93	12.0	508	69	8.6	1.2	9.8	19	11
	0.4	6.0	58.6	5.24	1.28	12.7	537	91	9.0	1.5	10.5	21	12
	0.5	6.1	60.7	5.24	1.63	13.4	565	113	9.3	1.9	11.1	24	13
	0.6	6.1	62.9	5.24	1.99	14.0	592	135	9.6	2.2	11.8	27	14
	0.7	6.2	65.3	5.24	2.36	14.6	617	155	10.0	2.5	12.5	30	15
	0.8	6.2	67.7	5.24	2.73	15.2	642	175	10.4	2.8	13.2	32	16
300	0.3	6.7	56.5	6.01	1.07	13.8	582	58	8.6	0.9	9.5	20	12
	0.4	6.9	58.6	6.01	1.46	14.6	616	79	9.0	1.1	10.1	23	13
	0.5	7.0	60.7	6.01	1.87	15.3	648	98	9.3	1.4	10.7	26	14
	0.6	7.0	62.9	6.01	2.28	16.0	678	117	9.6	1.7	11.3	28	15
	0.7	7.1	65.3	6.01	2.70	16.7	707	135	10.0	1.9	11.9	31	16
	0.8	7.1	67.7	6.01	3.13	17.4	736	151	10.4	2.1	12.5	34	17

[a] Crude protein required only if ration is perfectly balanced for RDP and RUP.

Table 8: Daily nutrient requirements (DM basis) of small breed (mature weight = 450 kg) non-bred heifers

BW kg	ADG kg/d	DMI kg/d	TDN %	NEm Mcal /d	NEc Mcal /d	ME Mcal	RDP g/d	RUP g/d	RDP %	RUP %	CP[a] %	Ca g/d	P g/d
240 days pregnant (Conceptus weight of 39 kg and ADG of 0.4 kg/day)													
300	0.3 (0.7)[b]	7.7	56.5	5.42	0.96	15.7	663	291	8.6	3.8	12.4	36	19
	0.4 (0.8)	7.7	58.6	5.42	1.32	16.4	693	310	9.0	4.0	13.0	39	20
	0.5 (0.9)	7.7	60.8	5.42	1.68	17.0	721	329	9.3	4.2	13.5	41	21
	0.6 (1.0)	7.7	63.1	5.42	2.06	17.7	748	346	9.7	4.5	14.1	44	22
	0.7 (1.1)	7.7	65.5	5.42	2.44	18.3	774	364	10.0	4.7	14.7	47	23
	0.8 (1.2)	7.7	68.1	5.42	2.82	18.9	798	380	10.4	5.0	15.4	49	24
	0.9 (1.3)	7.6	70.9	5.42	3.21	19.4	822	395	10.8	5.2	16.1	52	24
350	0.3 (0.7)	8.6	56.2	6.18	1.10	17.5	739	282	8.6	3.3	11.9	38	20
	0.4 (0.8)	8.7	58.3	6.18	1.50	18.3	773	299	8.9	3.4	12.4	40	21
	0.5 (0.9)	8.7	60.5	6.18	1.92	19.0	805	315	9.3	3.6	12.9	43	22
	0.6 (1.0)	8.7	62.8	6.18	2.35	19.8	836	330	9.6	3.8	13.4	46	23
	0.7 (1.1)	8.7	65.3	6.18	2.78	20.4	865	345	10.0	4.0	14.0	48	24
	0.8 (1.2)	8.6	67.8	6.18	3.22	21.1	893	358	10.4	4.2	14.5	51	25
	0.9 (1.3)	8.5	70.6	6.18	3.66	21.8	921	371	10.8	4.3	15.1	53	25
400	0.3 (0.7)	9.5	56.0	6.91	1.23	19.2	813	275	8.6	2.9	11.5	40	21
	0.4 (0.8)	9.6	58.1	6.91	1.68	20.1	851	291	8.9	3.0	11.9	42	22
	0.5 (0.9)	9.6	60.3	6.91	2.15	21.0	887	305	9.2	3.2	12.4	45	23
	0.6 (1.0)	9.6	62.6	6.91	2.62	21.8	921	319	9.6	3.3	12.9	47	24
	0.7 (1.1)	9.6	65.0	6.91	3.11	22.5	953	331	9.9	3.5	13.4	50	25
	0.8 (1.2)	9.5	67.6	6.91	3.60	23.3	985	342	10.3	3.6	13.9	52	26
	0.9 (1.3)	9.4	70.3	6.91	4.09	24.0	1015	352	10.8	3.7	14.5	55	26
450	0.3 (0.7)	10.4	55.8	7.62	1.35	20.9	884	273	8.5	2.6	11.2	41	22
	0.4 (0.8)	10.5	57.9	7.62	1.85	21.9	926	288	8.9	2.8	11.6	44	23
	0.5 (0.9)	10.5	60.1	7.62	2.37	22.8	965	301	9.2	2.9	12.1	46	24
	0.6 (1.0)	10.5	62.4	7.62	2.89	23.7	1003	313	9.5	3.0	12.5	49	25
	0.7 (1.1)	10.5	64.8	7.62	3.42	24.5	1038	324	9.9	3.1	13.0	51	26
	0.8 (1.2)	10.4	67.4	7.62	3.96	25.4	1073	333	10.3	3.2	13.5	54	27
	0.9 (1.3)	10.3	70.1	7.62	4.51	26.1	1106	341	10.7	3.3	14.0	56	28

[a] Crude protein required only if ration is perfectly balanced for RDP and RUP
[b] Values in parantheses are ADG (lbs/d)

Table 9: Daily energy and protein requirements of young replacement calves fed only milk or milk replacer

Live weight (kg)	Gain (g)	DMI[a] (kg)	Energy NE_M[b] (Mcal)	Protein NE_G[c] (Mcal)	ME[d] (Mcal)	DE[e] (Mcal)	Vitamin ADP[f] (g)	CP[g]	A[h] (IU)
25	0	0.24	0.96	0	1.12	1.17	18	20	2750
	200	0.32	0.96	0.26	1.50	1.56	65	70	2750
	400	0.42	0.96	0.60	2.00	2.08	113	121	2750
30	0	0.27	1.10	0	1.28	1.34	21	23	3300
	200	0.36	1.10	0.28	1.69	1.76	68	73	3300
	400	0.47	1.10	0.65	2.22	2.31	115	124	3300
40	0	0.34	1.37	0	1.59	1.66	26	28	4400
	200	0.43	1.37	0.31	2.04	2.13	73	79	4400
	400	0.55	1.37	0.72	2.63	2.74	120	129	4400
	600	0.69	1.37	1.16	3.28	3.41	168	180	4400
45	0	0.37	1.49	0	1.74	1.81	28	30	4950
	200	0.46	1.49	0.32	2.21	2.30	76	81	4950
	400	0.59	1.49	0.75	2.82	2.94	123	132	4950
	600	0.74	1.49	1.21	3.50	3.64	170	183	4950
50	0	0.40	1.62	0	1.88	1.96	31	33	5500
	200	0.45	1.62	0.34	2.37	2.47	78	84	5500
	400	0.63	1.62	0.77	3.00	3.13	125	135	5500
	600	0.78	1.62	1.26	3.70	3.86	173	185	5500

[a] Dry matter intake necessary to meet ME requirements for calves fed milk replacer composed primarily of milk proteins and containing ME at 4.75 Mcal/ kg of dry matter.

[b] NE_M (Mcal) = 0.086 $LW^{0.75}$, where LW is live weight in kilograms

[c] NE_G (Mcal) = (0.84 $LW^{0.355}$ x LWG^{12}) x 0.69, where LW and LWG (live weight gain) are in kilograms

[d] ME (Mcal) = 0.1 $LW^{0.75}$ + (0.84 $LW^{0.355}$ x LWG^{12}), where LW and LWG are in kilograms

[e] DE (Mcal) = ME/ 0.96

[f] ADP (apparent digestible protein, g/d) = 6.25 [1/BV (E + G + M x D) – M x D]. BV (biological value) is assumed to be 0.8. E (endogenous urinary nitrogen) is 0.2 $LW^{0.75}$/ d, where LW is in kilograms. M (metabolic fecal nitrogen) is 1.9 g/kg of dry matter intake (D). G (nitrogen in live weight gain) is 30 g/kg of LWG.

[g] CP (crude protein) = ADP/ 0.93. The digestibility of undenatured milk proteins is assumed to be 93 per cent.

[h] Vitamin A (IU) = 110 IU/ kg of LW.

Table 10: Daily energy and protein requirements of calves fed milk and starter or milk replacer and starter.

Live weight (kg)	Gain (g)	DMI[a] (kg)	Energy NE_M[b] (Mcal)	NE_G[c] (Mcal)	ME[d] (Mcal)	DE[e] (Mcal)	Protein ADP[f] (g)	CP[g]	Vitamin A[h] (IU)
30	0	0.32	1.10	0	1.34	1.43	23	26	3300
	200	0.42	1.10	0.28	1.77	1.89	72	84	3300
	400	0.56	1.10	0.65	2.33	2.49	122	141	3300
35	0	0.36	1.24	0	1.50	1.61	25	29	3850
	200	0.47	1.24	0.30	1.96	2.09	75	87	3850
	400	0.61	1.24	0.68	2.55	2.73	125	145	3850
40	0	0.40	1.37	0	1.66	1.78	25	33	4400
	200	0.51	1.37	0.31	2.14	2.29	78	90	4400
	400	0.66	1.37	0.72	2.76	2.95	128	148	4400
	600	0.83	1.37	1.16	3.44	3.68	178	205	4400
45	0	0.44	1.49	0	1.81	1.94	31	36	4950
	200	0.56	1.49	0.32	2.31	2.47	80	93	4950
	400	0.71	1.49	0.75	2.96	3.16	130	151	4950
	600	0.88	1.49	1.21	3.67	3.93	180	209	4950
50	0	0.47	1.62	0	1.96	2.10	33	38	5500
	200	0.60	1.62	0.34	2.48	2.65	83	96	5500
	400	0.76	1.62	0.77	3.15	3.37	133	154	5500
	600	0.94	1.62	1.26	3.89	4.17	183	212	5500
	800	1.13	1.62	1.78	4.69	5.02	233	270	5500
55	0	0.51	1.74	0	2.11	2.25	36	41	6050
	200	0.63	1.74	0.35	2.64	2.83	85	99	6050
	400	0.80	1.74	0.80	3.33	3.57	135	157	6050
	600	0.99	1.74	1.30	4.10	4.39	185	215	6050
	800	1.18	1.74	1.84	4.93	5.27	236	273	6050
60	0	0.54	1.85	0	2.25	2.41	38	44	6600
	200	0.67	1.85	0.36	2.80	3.00	88	102	6600
	400	0.84	1.85	0.83	3.51	3.76	138	159	6600
	600	1.04	1.85	1.34	4.31	4.61	188	217	6600
	800	1.24	1.85	1.90	5.16	5.52	238	275	6600

[a] These data apply to calves fed milk replacer (MR) plus starter. MR contains ME at 4.75 Mcal/kg of DM and starter ME at 3.28 Mcal/kg. It is assumed that MR provided 60 percent and starter 40 percent of dry matter intake; thus, dry matter consumed contained ME at 4.16 Mcal/kg. The DMI here is the total necessary to meet ME requirements and is not intended to predict voluntary intake.

[b] NE_M (Mcal) = 0.086 $LW^{0.75}$, where LW is live weight in kilograms

[c] NE_G (Mcal) = (0.84 $LW^{0.355}$ x LWG^{12}) x 0.69, where LW and LWG (live weight gain) are in kilograms

[d] ME (Mcal) was computed as follows:

ME (maintenance) = NE_M / 0.825. Efficiency of use of ME for maintenance (0.825) was computed as average of efficiencies of 0.86 for MR and 0.75 for starter, weighted according to proportions of ME supplied by each feed.

ME (gain) = NE_G Efficiency of use of ME for gain (0.652) was computed as weighted average of efficiencies of 0.69 and 0.57 for MR and starter, respectively.

[e] DE (Mcal) = ME/ 0.934. Efficiency of conversion of DE to ME is assumed to be 0.96 for MR and 0.88 for starter.

[f] ADP (apparent digestible protein, g/d) = 6.25 [1/BV(E + G+ M x D) – M x D]. BV (biological value) = 0.764 (weighted average of MR = 0.8 and starter = 0.70); E (endogenous urinary nitrogen, g) = $0.2LW^{0.75}$; G (nitrogen content of gain, g) = 30 g/kg gain; M (Metabolic fecal nitrogen, g/d) = 2.46 x dry matter intake, D kg). Metabolic fecal nitrogen for MR assumed to be 1.9 g/kg of DMI and for starter 3.3 g/kg of DMI.

[g] CP (crude protein, g) = ADP/ 0.8645. Digestibility of protein was assumed to be weighted average of 93 percent for MR and 75 percent for starter; MR was assumed to contain 21 percent CP and starter 18 percent CP.

[h] Vitamin A (IU) = 110 IU/kg of LW.

Table 11: Daily energy and protein requirements of veal calves fed only milk or milk replacer.

Live weight (kg)	Gain (g)	DMIa (kg)	NE_M[b] (Mcal)	Energy NE_G[c] (Mcal)	ME[d] (Mcal)	DE[e] (Mcal)	Protein ADP[f] (g)	CP[g]	Vitamin A[h](IU)
40	0	0.34	1.37	0	1.59	1.66	26	28	4400
	300	0.49	1.37	0.51	2.32	2.42	97	104	4400
	600	0.69	1.37	1.16	3.28	3.41	168	180	4400
50	0	0.40	1.62	0	1.88	1.96	31	33	5500
	300	0.56	1.62	0.55	2.67	2.79	102	109	5500
	600	0.78	1.62	1.26	3.71	3.86	172	185	5500
	900	1.02	1.62	2.05	4.85	5.05	244	262	5500
60	0	0.45	1.85	0	2.16	2.25	35	38	6600
	300	0.63	1.85	0.58	3.00	3.13	106	114	6600
	600	0.86	1.85	1.34	4.10	4.27	177	190	6600
	900	1.12	1.85	2.18	5.32	5.54	248	267	6600
70	0	0.51	2.08	0	2.42	2.52	39	42	7700
	300	0.70	2.08	0.62	3.32	3.45	110	119	7700
	600	1.94	2.08	1.42	4.48	4.66	181	195	7700
	900	1.21	2.08	2.31	5.76	6.01	253	272	7700
	1200	1.50	2.08	3.26	7.14	7.44	324	348	7700
80	0	0.56	2.30	0	2.68	2.79	44	47	8800
	300	0.76	2.30	0.65	3.61	3.76	115	123	8800
	600	1.02	2.30	1.49	4.83	5.03	186	200	8800
	900	1.30	2.30	2.42	6.18	6.44	257	276	8800
	1200	1.61	2.30	3.42	7.63	7.95	328	353	8800
90	0	0.62	2.51	0	2.92	3.04	48	51	9900
	300	0.82	2.51	0.68	3.90	4.06	119	128	9900
	600	1.09	2.51	1.55	5.17	5.39	190	204	9900
	900	1.38	2.51	2.55	6.62	6.85	263	283	9900
	1200	1.70	2.51	3.56	8.09	8.42	332	357	9900
100	0	0.67	2.72	0	3.16	3.29	52	55	11000
	300	0.88	2.72	0.70	4.18	4.35	122	132	11000
	600	1.16	2.72	1.61	5.50	5.72	194	208	11000
	900	1.46	2.72	2.62	6.96	7.25	265	285	11000
	1200	1.80	2.72	3.70	8.52	8.88	336	362	11000
	1500	2.14	2.72	4.84	10.17	10.59	408	438	11000
110	0	0.72	2.92	0	3.40	3.54	55	60	12100
	300	0.94	2.92	0.72	4.45	4.63	126	136	12100
	600	1.22	2.92	1.66	5.81	6.05	198	212	12100
	900	1.54	2.92	2.71	7.32	7.63	269	289	12100
	1200	1.88	2.92	3.83	8.94	9.32	340	366	12100
	1500	2.24	2.92	5.00	10.65	11.09	412	443	12100

Live weight (kg)	Gain (g)	DMIa (kg)	Energy NE_M[b] (Mcal)	NE_G[c] (Mcal)	ME[d] (Mcal)	DE[e] (Mcal)	Protein ADP[f] (g)	CP[g]	Vitamin A[h](IU)
120	0	0.76	3.12	0	3.63	3.78	59	64	13200
	300	0.99	3.12	0.75	4.71	4.91	130	140	13200
	600	1.29	3.12	1.72	6.12	6.39	201	217	13200
	900	1.62	3.12	2.80	7.68	8.00	273	293	13200
	1200	1.97	3.12	3.69	9.34	9.74	329	353	13200
	1500	2.34	3.12	5.16	11.10	11.56	416	447	13200
130	0	0.81	3.31	0	3.85	4.01	63	67	14300
	300	1.05	3.31	0.77	4.97	5.17	134	144	14300
	600	1.35	3.31	1.77	6.41	6.68	205	220	14300
	900	1.69	3.31	2.88	8.02	8.35	276	297	14300
	1200	2.05	3.31	4.06	9.74	10.14	348	374	14300
	1500	2.43	3.31	5.31	11.54	12.02	420	451	14300
140	0	0.86	3.50	0	4.07	4.24	66	71	15400
	300	1.10	3.50	0.79	5.22	5.43	137	148	15400
	600	1.41	3.50	1.82	6.70	6.98	209	224	15400
	900	1.76	3.50	2.95	8.35	8.70	280	301	15400
	1200	2.13	3.50	4.17	10.11	10.53	352	378	15400
	1500	2.52	3.50	5.45	11.97	12.45	423	455	15400
150	0	0.90	3.69	0	4.29	4.46	70	75	16500
	300	1.15	3.69	0.81	5.46	5.69	141	152	16500
	600	1.47	3.69	1.86	6.98	7.27	212	228	16500
	900	1.82	3.69	3.02	8.67	9.03	284	305	16500
	1200	2.21	3.69	4.27	10.48	10.91	355	382	16500
	1500	2.61	3.69	5.58	12.38	12.90	427	459	16500

[a] The DMI necessary to meet ME requirements when veal calves are fed a milk replacer containing ME at 4.75 Mcal/kg of DM.

[b] NE_M (Macl) = 0.086 $LW^{0.75}$, where LW is live weight in kilograms.

[c] NE_G (Mcal) = (0.84 $LW^{0.355}$ x $LWG^{1.2}$) x 0.69, where LW and LW gain (LWG) are in kilograms.

[d] ME (Mcal) = 0.1 $LW^{0.75}$ + (0.84 $LW^{0.355}$ x $LWG^{1.2}$) x 0.69, where LW and LWG are in kilograms.

[e] DE (Mcal) = ME/0.93.

[f] ADP (apparent digestible protein, g/d) = 6.25 [1/BV(E + G + M x D) – M x D]. BV (biological value) is assumed to be 0.8. E (endogenous urinary nitrogen) is 0.2 $LW^{0.75}$/d, where LW is in kilograms. M (metabolic fecal nitrogen) is 1.9 g/kg of dry matter intake (D). G (nitrogen in live weight gain) is 30 g/kg LWG.

[g] CP (crude protein) = ADP/0.93. The digestibility of unsaturated milk protein is assumed to be 93 percent.

[h] Vitamin A (IU) = 110 IU/kg of LW.

Table 12: Guidelines for Total Soluble Salts (TSS) in Water for Cattle.

TSS (mg/L)	Comments
<1000	Safe and should pose no health problems
1000 – 2999	Generally safe but may cause a mild temporary diarrhea in animals not accustomed to the water.
3000 – 4999	Water may be refused when first offered to animals or cause temporary diarrhea. Animal performance may be less than optimum because water intake is not maximized.
5000 – 6999	Avoid these waters for pregnant or lactating animals, may be offered with reasonable safety to animals where maximum performance is not required.
7000	These waters should not be fed to cattle. Health problems and/or poor production will result.

Source: National Research Council (1974).

Table 13: Water Hardness Guidelines.

Category	Hardness (mg/L)
Soft	0 – 60
Moderately hard	61 – 120
Hard	121 – 180
Very hard	> 180
[a] grain/gal = 17.1 mg/L.	

Source: National Research Council (1980).

Table 14: Nitrate in Water

Nitrate (NO_3) (mg/L)	Nitrate Nitrogen (NO_3-N) (mg/L)	Guidelines
0 – 44	0 – 10	Safe for consumption by ruminants
45 – 132	10 – 20	Generally safe in balanced diets with low nitrate feeds
133 – 220	20 – 40	Could be harmful if consumed over long period
221 – 660	40 – 100	Cattle at risk; and possible death
661	100	Unsafe – possible death; should not be used as a source of water.

Source: National Research Council (1974).

Table 15: Generally Considered Safe Concentrations of Some Potentially Toxic Nutrients and Contaminants in Water for Cattle.

Item	Upper-limit guidelines
Aluminum	0.5
Arsenic	0.05
Boron	5.0
Cadmium	0.005
Chromium	0.1
Cobalt	1.0
Copper	1.0
Fluorine	2.0
Lead	0.015
Manganese	0.05
Mercury	0.01
Nickel	0.25
Selenium	0.05
Vanadium	0.1
Zinc	5.0

Source: National Research Council (1974). Environmental Protection Agency (1997).

Table 16: Mineral and vitamin concentrations recommended for diets of young calves, compared with average for fresh whole milk (DM basis)

Nutrient[a]	Milk Replacer[b]	Starter Feed	Grower Feed	Whole Milk
Minerals				
Ca (%)	1.00	0.70	060	0.95
P (%)	0.70	0.45	0.40	0.76
Mg (%)	0.07	0.10	0.10	0.10
Na (%)	0.40	0.15	0.14	0.38
K (%)	0.65	0.65	0.65	1.12
Cl (%)	0.25	0.20	0.20	0.92
S (%)	0.29	0.20	0.20	0.32
Fe (mg/kg)	100[c]	50	50	3.0
Mn (mg/kg)	40	40	40	0.2-0.4
Zn (mg/kg)	40	40	40	15-38
Cu (mg/kg)	10	10	10	0.1-1.1
I (mg/kg)	0.50	0.25	0.25	0.1-0.2
Co (mg/kg)	0.11	0.10	0.10	0.004-0.008
Se (mg/kg)	0.30	0.30	0.30	0.02-0.15
Vitamins				
A (IU/kg of DM)	9000	4000	4000	11500
D (IU/kg of DM)	600	600	600	307
E (IU/kg of DM)	50	25	25	8

[a] B-complex vitamins are necessarily only in milk-replacer diets. Required concentrations (mg/ kg of DM): thiamin, 6.5; riboflavin, 6.5; pyrodoxin, 6.5; pantothenic acid, 13.0; niacin, 10.0; biotin, 0.1; folic acid, 0.5; B_{12}, 0.07; Choline, 1000.

[b] Required concentrations specified for milk replacer fed at 0.53 kg of DM per day to 45-kg calf. Assuming ME content of 4.75 Mcal/kg, this amount of milk replacer would provide energy – allowable growth of 0.3 kg/d. Concentrations of minerals and vitamins specified will provide adequate daily amounts of minerals and vitamins. User is cautioned that feeding larger or smaller amounts of milk replacer, or same amount of milk replacer to larger or smaller calf, changes expected growth and, consequently, requirements for many vitamins and minerals.

[c] For veal calves, decrease to less than 50 mg/ kg of DM.

Table 17: Energy, protein, calcium and phosphorous concentrations in feedstuffs commonly used in formulation of milk replacers for young calves[a]

Feed	International feed number		DM (%)	GE (Mcal/kg of DM)		DE (Mcal/kg of DM)		ME (Mcal/kg of DM)	
ME/GE (q)	NE_M (Mcal/kg of DM)	NE_G (Mcal/kg of DM)		CP	EE	Ca	P	Ash	
									% of DM
Whole milk	5-01-168	12.5	5.76	5.59	5.37	0.93	4.62	3.70	25.4
30.8	1.00	0.75	6.3						
Skim milk, fresh	5-01-170	10	4.31	4.19	4.02	0.93	3.46	2.77	35.5
0.3	1.35	1.02	6.9						
Skim milk, powder	5-01-175	94	4.38	4.25	4.08	0.93	3.51	2.82	37.4
1.0	1.29	1.08	6.9						
Whey-powder	4-01-182	93	3.92	3.80	3.65	0.93	3.14	2.52	13.5
1.0	0.76	0.68	8.1						
Whey protein concentrate	-	93	4.48	4.35	4.17	0.93	3.59	2.88	37.1
2.2	0.54	0.60	6.0						
Whey, fresh	4-08-134	7	3.89	3.78	3.62	0.93	3.12	2.50	14.2
0.7	0.73	0.65	8.7						
Whey, delactosed	4-01-186	93	3.65	3.54	3.40	0.93	2.92	2.34	17.9
0.7	1.71	1.12	16.5						
Whey permeate	-	98	3.66	3.55	3.41	0.93	2.93	2.35	3.7
0	1.77	0.97	9.0						
Casein	5-01-162	91	5.45	5.29	5.08	0.93	4.37	3.50	92.7
0.7	0.40	0.35	4.0						

[a] Data from NRC (1989); Toullee (1989); Tomkins and Jaster (1991).

Table 18: Energy, protein, fiber and mineral composition of three milk replacers (MR), a starter feed and a grower feed for young calves.

Feed	GE[a] (Mcal/kg DM)	DE[a] (Mcal/kg of DM)	ME[a] (Mcal/kg of DM)	NE_M (Mcal/kg of DM)	NE_G (Mcal/kg of DM)	CP (%)	EE (%)	ADF (%)	NDF (%)	Ca (%)	P (%)
MR-1	4.61	4.47	4.29	3.69[b]	2.96[c]	22	10	-	-	1.0	0.70
MR-2	5.10	4.95	4.75	4.09[b]	3.28[c]	20	20	-	-	1.0	0.70
MR-3	5.07	4.91	4.72	4.06[b]	3.26[c]	18	20	-	-	1.0	0.70
Starter	4.49	3.69	3.28	2.46[d]	1.78[e]	18	3	11.6	12.8	0.7	0.45
Grower	4.36	3.65	3.24	2.43[d]	1.61[e]	16	3	8.0	18.0	0.6	0.40

[a] Energy values calculated as follows:
Gross energy (GE) is calculated from composition and heat of combustion. For milk replacers, GE (kcal/kg) = 0.057 CP + 0.092 fat + 0.0395 lactose. For starter and grower, GE (kcal/kg) – 0.057 CP + 0.092 EE + 0.0415 carbohydrate.
For MR, digestible energy (DE) = 0.97 GE. For starter and grower, DE is calculated as sum of digestible protein, fat and carbohydrates, each multiplied by heat of combustion.
For MR, metabolizable energy (ME) calculated as 0.93 GE (ME/GE of whole milk has been measured at 0.93; Roy, 1980). For starter and grower feeds, ME = (1.01 x DE – 0.45) + (0.0046EE – 3).
[b] NE_M = 0.86 ME.
[c] NE_G = (0.38q + 0.337) x ME. Based on q of 0.93 yielding an efficiency of 0.69 for ME use
[d] NE_M = ME x 0.75
[e] NE_G = ME x 0.57

Index

H

I

J

K

L

M

ABOUT THE AUTHORS

Dr. S.S Kundu is currently Principal Scientist and Head, Plant Animal Relationship Division at IGFRI, Jhansi (U.P.). He has over 28 years of experience in the field of ruminant nutrition research. He was instrumental in initiating and shaping the work on feed technology, developing the feed block technology and feed microscopy for quality assurance. Dr. Kundu has undergone advanced training in the field of ruminant nutrition at the Netherlands and U.K. He is pioneer worker in the field of carbohydrate, protein fractionations of animal feeds in the country. In addition to books on "Buffalo Production Under Different Climatic Regions, Animal Feed Technology, Complete Feed Block (Bulletin)", he has more than 150 research publications in the Journal of National and International repute. He has guided many post graduate and Ph. D. students at NDRI, Karnal. He is Vice President (Animal Nutrition Society of India), and also life member of eight scientific societies.

Dr Sultan Singh is a Scientist (Animal Nutrition) in PAR at IGFRI, Jhansi (UP). His main areas of research in last 12 years are rumen metabolism and microbiology and evaluation of feeds and fodders. Transinoculation a mean of rumen manipulation and standardization of technique for in vitro dry matter digestion are accredited to his accomplishments. Recipient of two awards viz. ICAR Team Award (1994-96), for outstanding research in Agriculture and allied sciences and ICAR Young Scientist Award (1997-98) for his outstanding contribution in field of Animal Nutrition. He has over 50 research papers published in National and International journals and 3 Technical bulletins to his credit.

Dr. S.K. Mahanta, Scientist (Animal Nutrition), PAR Division, IGFRI, Jhansi has 10 years of research, extension and training experience in farm animal nutrition and feeding. He specializes in forage evaluation and development of forage based feeding system for different categories of ruminants. He has over 40 research papers of national and international repute. He is also in editorial board of two national journals and an active member of many scientific societies.

Dr. G.H. Pailan is currently working as scientist (Animal Nutrition) in the PAR Division at IGFRI, Jhansi (U.P.). He did B. V. Sc & A.H. from B.C.K.V., West Bengal, M. Sc. (A.N.) and Ph. D in Animal Nutrition from NDRI Karnal. He has about 10 years experience in the field of ruminant nutrition with specialization in protein nutrition and nutrition of grazing livestock. He is recipient of Dr. S.K. Ranjhan Award for best Ph. D thesis in Animal Nutrition and Allied Subjects for the biennial 1995-1997. He has over 35 research publications in national and international journals.